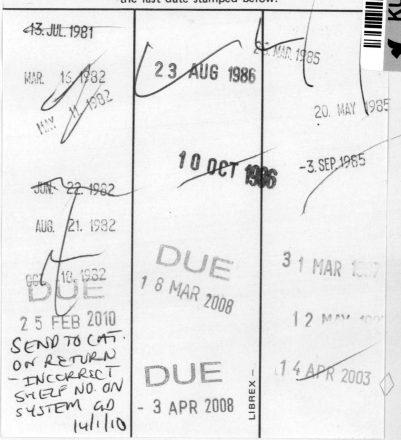

This book is to be returned on or before
the last date stamped below.

13. JUL. 1981

MAR. 13 1982

MAY 11 1982

23 AUG 1986

MAR. 1985

20. MAY 1985

JUN. 22 1982

AUG. 21. 1982

10 OCT 1986

-3. SEP. 1985

OCT 18 1982

DUE
25 FEB 2010

DUE
1 8 MAR 2008

3 1 MAR 1987

12 MAY

SEND TO CAT.
ON RETURN
— INCORRECT
SHELF NO. ON
SYSTEM GD
14/1/10

DUE
- 3 APR 2008

LIBREX —

14 APR 2003

Mining Geostatistics

Mining Geostatistics

A. G. JOURNEL
and
CH. J. HUIJBREGTS

Centre de Geostatistique
Fontainebleau, France

1978

ACADEMIC PRESS

London   New York   San Francisco

A Subsidiary of Harcourt Brace Jovanovich, Publishers

ACADEMIC PRESS INC. (LONDON) LTD.
24/28 Oval Road
London NW1

United States edition published by
ACADEMIC PRESS INC.
111 Fifth Avenue
New York, New York 10003

Library of Congress Catalog Card Number: 77-92823
ISBN: 0-12-391050-1

Printed in Great Britain by J. W. Arrowsmith Ltd, Winterstoke Road, Bristol BS2 2NT

Foreword

by G. MATHERON

The distribution of ore grades within a deposit is of mixed character, being partly structured and partly random. On one hand, the mineralizing process has an overall structure and follows certain laws, either geological or metallogenic: in particular, zones of rich and poor grades always exist, and this is possible only if the variability of grades possesses a certain degree of continuity. Depending upon the type of ore deposit, this degree of continuity will be more or less marked, but it will always exist; mining engineers can indeed be thankful for this fact because, otherwise, no local estimation and, consequently, no selection would be possible. However, even though mineralization is never so chaotic as to preclude all forms of forecasting, it is never regular enough to allow the use of a deterministic forecasting technique. This is why a scientific (at least, simply realistic) estimation must necessarily take into account both features – structure and randomness – inherent in any deposit. Since geologists stress the first of these two aspects, and statisticians stress the second, I proposed, over 15 years ago, the name *geostatistics* to designate the field which synthetizes these two features and opens the way to the solution of problems of evaluation of mining deposits.

Practicians are well aware that such problems are numerous and often very complex. They sometimes infer from this that the corresponding mathematical theory is necessarily very complicated. Actually, they should set their minds at ease, for this is luckily not the case. The mathematical formalism necessary for geostatistics is simple and, in some cases, even trite. The user of A. Journel and Ch. Huijbregts' book will come across nothing more than variances and covariances, vectors and matrices; in short, only linear calculations comparable to those which any computer user manipulates every day. After the preliminary effort of familiarizing themselves with the notations, they will note, possibly to their surprise, that they can follow all of the authors' steps without difficulty.

I might address a symmetric warning to statisticians. Those unfamiliar with the complexity and specificity of mining problems might find this treatise banal and of little interest. I feel that they would be mistaken. At this point, a comparison with geophysics could be instructive (besides, the very term "geostatistics" was formed after the model of "geophysics" or of "geochemistry").

If a theoretical physicist were to read a treatise on geophysics, he would risk getting the disappointing impression that it is but a collection of particular cases of the theories of potential and of wave propagation: a bit of gravitation and magnetism, quite a lot of electromagnetism, acoustics, etc., nothing very original, really. And yet, without special training, that very physicist, be he a Nobel prize winner, would admit to be unable to interpret the results of a seismic prospection campaign.

In the same way, a specialist in mathematical statistics, having paged through the present book, would not find much more than variances, covariances, optimal linear estimators, etc., and would find no justification in this for creating a separate field. However, let us submit to this specialist the following problem, which is quite a standard one for the mining engineer (in geostatistical terminology, it is the problem of the estimation of recoverable reserves).

"We have, at a certain depth, an interesting copper deposit. The geologists who have studied the deposit have arrived at certain conclusions about its genesis and structure, which you will find in this report. The available numerical data consist of a square grid of drill-holes 200 m apart. The core sections were analysed metre by metre, and, thus, we have a total of some 5000 chemical analyses. We are considering open-pit selective mining. This means that, during the mining of our approximately 200 Mm^3 deposit, which has been mentally divided into 50 000 blocks 20 m $\times$ 20 m $\times$ 10 m, we have the choice of sending each newly excavated block either to waste, or to the processing plant. Taking account of an economic pre-feasibility study, we have decided that the blocks with a mean ore grade greater than $x = 0.4\%$ Cu should be sent to the plant, and that the rest should be set aside with the waste. This cut-off grade x is subject to revision, since the market price of copper as well as the mining and processing costs can vary. x should be considered as a parameter, and the evaluations should be made under various hypotheses, i.e., with $x = 0.3\%$, 0.4%, 0.5%, for example.

"Moreover, we must consider the fact that our block by block decision is based, not on the unknown true block grade but on an estimator, this estimator being based on information which will be much richer than that now available (core drillings on a 200 m $\times$ 200 m grid): at the time of selection, we shall have the results of the chemical analyses of blast holes on a 5 m $\times$ 5 m grid. Based upon this final information, we shall build an estimator z_i^* for each of our blocks, and we shall send this block to the plant only if its estimated grade z_i^* is greater or equal to the cut-off grade x.

"We would like to know, as a function of x, the tonnage $T(x)$ of the finally selected blocks, and their mean grade $m(x)$; this time, of course, we are referring to the mean value of the true mean grades of the selected blocks, and not of their estimated mean grades z_i^*. On top of this, the evaluation, based upon the rather poor information we have now (drilling on a 200-m spacing), will inevitably be erroneous. So we shall also want to know the orders of magnitude of the errors committed in the evaluation of $T(x)$ and $m(x)$, for example through their estimation variances."

In the very same way our Nobel prize physicist above admits being unable to interpret the results of a seismic prospection campaign, I think that a statistician not versed in mining practice would honestly have to admit his inability to give a realistic solution to the above problem. Actually, this is a rather simple and favourable case: only one metal, a well-explored deposit with a regular sampling grid, no missing data points nor preferential or biased data, etc. One should also note that the mere statement of the problem itself takes up about a page. A statistician who is not familiar with mining may well be discouraged before he can even get a good idea of the problem at hand.

In fact, the difficulties are several. First of all, there is a violent contrast in masses: even though the experimental data are numerous (5000 samples), they represent but a minute part of the total mineralized mass (about 50 tonnes out of hundreds of millions). Then there is a volume (or support) effect: the 20 m × 20 m × 10 m blocks are not at all distributed (in the spatial sense as well as in the statistical sense) in the same way as the kilogram-size core sections. Finally, there is the fact that the criteria to be used for the future selection of blocks will not be based on the true grades of the blocks, nor upon the data now at hand, but on an intermediate level of information (the blast-holes). One can, thus, understand why the need to solve such problems would instigate, over the years, the creation of a field as specialized as geophysics, with its own terminology and special techniques – a field which carries the name of geostatistics.

André Journel and Charles Huijbregts, who have worked at the Centre de Géostatistique at Fontainebleau since its creation in 1968, are surely among the persons who have contributed the most to elevating this field to its present level of practical and operational efficiency. They have personally achieved studies and estimations upon more than 30 ore deposits of the most diverse types. The vast experience which they thus accumulated, and the close and permanent contact they maintained with the everyday realities of mining, account for the primarily operational and concrete nature of this book. Here is a practical treatise, written by practising mining engineers, for the benefit of other practicians. The authors have deliberately – and justly so – simplified to the utmost the purely mathematical part of their *exposé*, while remaining as rigourous as possible. They have concentrated their efforts upon two main points: first of all, clearly define the problems at hand, and we have seen that this is no mean task; second, show how one should go about solving them in practice. Practicians will appreciate this double approach: clarifying the objectives and efficient use of the techniques. Moreover, the mathematical formalism is laced with concrete comments which make intuitive sense to the practician. Each time a new notion is introduced, its definition is immediately followed by examples that illustrate its meaning.

This book is structured according to the most typical problems which occur in practice, progressing from the simplest to the most complicated: structural analysis, estimation of *in situ* resources, estimation of recoverable reserves, simulation of deposits. In each instance, several examples of applications to actual deposits of the most diverse types are given: the practician will be led by the hand through the labyrinth and forewarned of traps and accidents which lurk at each corner.

In Chapter VII, devoted to conditional simulations, the strictly "linear" code of the first six chapters is dropped. The object of such a simulation is to build a numerical model which has the same structure as the real deposit. More precisely, the numerical model must have the same histogram, variogram and cross-correlations as the real deposit, and must coincide with the experimental data values at the locations where they were taken. The authors describe the novel technique developed by the Centre de Géostatistique. These conditional simulations will be used more and more often to solve difficult non-linear problems which defy the use of any other technique (e.g., forecasting the fluctuations of ore characteristics at the mill feed, homogenization obtained after judicious ore blending and/or stockpiling, etc.). Here, geostatistics no longer confines itself to predicting once and for all the recoverable reserves, but helps in optimal planning and control of the mining operations, at the daily, monthly or yearly scale. The last chapter gives general indications about "non-linear" geostatistics, which is still under elaboration.

In summary, A. Journel and Ch. Huijbregts' book constitutes a systematic state-of-the-art account of geostatistics and traces its progress over the last ten years (progress to which the authors have largely contributed). This treatise, intended for practicians, is surely destined to become the reference text for the field during the 1980's.

Contents

ix

Introduction

This book is an attempt to synthetize the practical experience gained in mining geostatistics by the researchers from the Centre de Morphologie Mathématique in Fontainebleau, France† and by mining engineers and geologists around the world who were kind enough to let us know of their experience. This book is designed for students and engineers who wish to apply geostatistics to practical problems, and, for this reason, it has been built around typical problems that arise in mining applications. Several FORTRAN programs are given, and the techniques developed are illustrated by a large number of case studies. A modest knowledge of integral calculus, linear algebra, statistics, notions of stochastic processes, variance, covariance and distribution functions is required, as well as a geological and mining background. With this background, the elements of mining geostatistics can be understood and the approximations necessary for an experimental approach can be accepted.

Definition and basic interpretation of geostatistics

Etymologically, the term "geostatistics" designates the statistical study of natural phenomena. G. Matheron (1962) was the first to use this term extensively, and his definition will be retained: "Geostatistics is the application of the formalism of random functions to the reconnaissance and estimation of natural phenomena".

A natural phenomena can often be characterized by the distribution in space of one or more variables, called "regionalized variables". The distribution of grades in the three-dimensional space, for example,

---

† Namely and particularly J. Serra, A. Maréchal, J. Deraisme, D. Guibal. Special thanks to P. A. Dowd who contributed extensively to the English version from the original French draft.

1

characterizes at least some of the aspects of a mineralization. The distribution of altitude in a horizontal space characterizes a topographical surface.

Let $z(x)$ be the value of the variable z at the point x. The problem is to represent the variability of the function $z(x)$ in space (when x varies). This representation will then be used to solve such problems as the estimation of the value $z(x_0)$ at a point x_0 at which no data are available, or to estimate the proportion of values $z(x)$ within a given field that are greater than a given limit (e.g., cut-off grade).

The geostatistical solution consists of interpreting each value $z(x_i)$ (z in lower case) as a particular realization of a random variable $Z(x_i)$ (Z in upper case) at the point x_i. The set of these auto-correlated random variables $\{Z(x)$, when x varies throughout the deposit or domain $D\}$ constitutes a random function. The problem of characterizing the spatial variability of $z(x)$ then reduces to that of characterizing the correlations between the various random variables $Z(x_i)$, $Z(x_j)$ which constitute the random function $\{Z(x), x \in D\}$. This *fundamental interpretation* is justified *a posteriori* if it results in coherent and acceptable solutions to the various problems encountered in practice.

The adaptability of geostatistics

Geostatistics, and the probabilistic approach in general, is particularly suited to the study of natural phenomena. Among the many techniques and mathematical models available to the practitioner of geomathematics, a particular technique or approach will be chosen if, from experience, it is known to be best suited to the problem concerned. A Fourier analysis will suit one problem, while, in other cases, geostatistical tools are best suited; in some cases, a mathematical formulation may even be rejected for simple qualitative experience.

In the mining field, geostatistics provides a coherent set of probabilistic techniques, which are available to each person (geologist, mining engineer, metallurgist) involved in the mining project; the proper use of these techniques depends on these people. Just as the efficiency of the axe and the saw depends on the arm of the lumberman, so the efficiency of any geostatistical study depends on the liaison between the various people and departments involved. The proper conclusions of an analysis of spatial variability will be reached only if there is a close contact with the geologist or, better still, if the analysis is carried out by the geologist himself. The evaluation of recoverable reserves has no meaning unless technically feasible processes of selection and mining are clearly defined; in other words, the mining engineer must be involved. Outside the mining field, geostatistics applied to forestry requires the ability to interpret spatial correlation between the densities of growth of different species of trees; geostatistics applied to bathymetry must take into account all the problems related to the localization of the hydrographic survey ship at sea. Thus, a book on geostatistics cannot be dissociated from its practical context, in this case mining.

This instrumental nature of geostatistics both opposes and completes the explanatory aspects of the standard, deterministic, geological approach. For example, even if a metallogenic study is sometimes successful in explaining the overall genesis of a given deposit, such a deterministic approach will be clearly insufficient on a smaller scale, such as that of metal concentrations. In such a case, it is said that the grades are "randomly" distributed within the defined genetic outline.

But nothing is more precise than this "random" distribution. It will differ from a gold placer deposit to a porphyry-copper deposit, and the random distribution of the grades within the deposit will condition its survey and its selective mining, which represent considerable investments. On the smaller scale of mining, the probabilistic approach can relay the deterministic and genetic approaches by which the deposit was delineated on a larger scale. Our lumberman is not a botanist; similarly, the mining geostatistician accepts the reality imposed by nature, and is content to study the consequences of this reality on the mining project he must prepare. A probabilistic approach can, of course, be guided by genetic considerations, and the best mining geostatistician is the one who is able to combine his geological knowledge and his technical mining skill with a good use of the probabilistic language.

Plan of the book

The acceptance of geostatistics, and mining geostatistics in particular, rests on the coherence and effectiveness of the solutions it provides to the various problems encountered in practice. Thus, the primary aim of this book is to convince the various people involved in a mining project of the practical interest of geostatistics. An engineer will not be convinced in the same manner as a mathematician or a researcher; he will first want to know how the techniques have fared in practice. For this reason, before beginning any theoretical development, a practical presentation of geostatistics, by way of its mining applications, is given in Chapter I. This chapter can be read separately from the rest of the book, and is intended as a *résumé* for readers who are pressed for time, and anxious to make a quick decision on the usefulness of the proposed techniques. In the remainder of the book, each new concept is introduced by a practical example drawn from the study of a real deposit.

The theoretical structure of the book begins with Chapter II, which presents the theory of regionalized variables and the main tool of geostatistics – the variogram – and the two variances of estimation and dispersion.

Any geostatistical study begins with a structural analysis, which consists of modelling the *in situ* spatial variabilities of the parameters studied, e.g., grades, thicknesses, accumulations. Chapters III and IV show how a structural analysis is carried out in practice from the construction of variograms to their interpretation and fitting by models. Both chapters insist on the "skilled" nature of the structural analysis. Even though many computer programs have been written for this purpose, a structural analysis must never be carried out blindly; it must integrate the qualitative experience of the geologist and carry it through to the subsequent resource-reserves estimation and the eventual planning of the mining project.

It is obvious that a massive porphyry-copper deposit cannot be estimated in the same way as a sedimentary phosphate deposit. The estimation procedure must take into account both the structure of the spatial variability peculiar to each deposit, and the particular manner in which the deposit was surveyed. It is only by taking these peculiarities into account that a confidence interval can be ascribed to each estimated value. Kriging takes these factors into account, and is presented in Chapter V, along with the geostatistical approach to the estimation of the *in situ* resources.

The next step is to evaluate the proportion of these *in situ* resources that can be recovered within a given technological and economical framework. This estimation of recoverable reserves must take the nature of the proposed selection and mining methods into account. Chapter VI presents the geostatistical formalization of these problems of selection and estimation of recoverable reserves.

Since geostatistics interprets the regionalized variable $z(x)$ as a particular realization of a random function $Z(x)$, the possibility arises of drawing other realizations of this same random function, in much the same way as casting a die several times. This new, simulated realization represents a numerical model of the real spatial variability $z(x)$. Within certain limits, the consequences of various planned mining methods can then be examined by simulating these methods on the numerical model. The techniques for simulating deposits and the practical uses of these simulations are given in Chapter VII.

Each chapter begins with a summary of its contents, so as to allow the reader to sort out items of particular interest. For this purpose, the reader can also make profitable use of the indexes of geostatistical concepts, quoted deposits and FORTRAN programs. At a first reading, the reader is advised to omit the various theoretical demonstrations. These demonstrations are necessary for the logical coherence of the book, but not for a preliminary overall understanding of geostatistics.

First notations

For the uninitiated, one of the main difficulties of geostatistics is the notation. These notations are precisely defined as they enter into the work, and are summarized in Index D. However, it may prove helpful at this stage to introduce the following basic notations.

A deposit is always a regionalized phenomenon in the three-dimensional space. To simplify notation, the coordinates (x_u, x_v, x_w) of a point are usually denoted x. Thus, the integral (extended over the three-dimensional space) of the function $f(x_u, x_v, x_w)$ is written as

$$\int_{-\infty}^{+\infty} f(x)\,dx = \int_{-\infty}^{+\infty} dx_u \int_{-\infty}^{+\infty} dx_v \int_{-\infty}^{+\infty} f(x_u, x_v, x_w)\,dx_w.$$

The vector with three-dimensional coordinates (h_u, h_v, h_w) is denoted h, with a modulus

$$r = |h| = \sqrt{(h_u^2 + h_v^2 + h_w^2)}$$

and direction α. Note that a direction in a three-dimensional space is specified by two angles, its longitude and its latitude.

Lower-case letters, z or y, are reserved for real values as, for example, the grade $z(x)$ at point x or the density $y(x)$ at the same point. The corresponding upper-case letters, Z or Y, denote the interpretation of this real value as a random variable or function, as, for example, the random function $Z(x)$ with expectation $E\{Z(x)\}$. The real value $z(x)$ is said to be a particular realization of the random variable or function $Z(x)$.

An estimated value is distinguished from a real value, known or unknown, by an asterisk $*$. Thus, $z^*(x)$ represents the estimated value of $z(x)$, and $Z^*(x)$ represents the interpretation of this estimator as a random variable or function.

I

Geostatistics and Mining Applications

SUMMARY

This chapter is intended to be both an introduction to, and an overview of, geostatistics as applied to the mining industry. Each successive problem encountered from the exploration phase to the routine operating phase is presented along with its geostatistical approach. The problems of characterizing variability structures of variables of interest within the deposit are solved with the "variogram". Optimal local estimates with corresponding confidence intervals are found by "kriging". The choice of various working techniques (machinery, need for stock-piling or blending) can be facilitated by simulating them on a geostatistically determined numerical model of the deposit.

The main steps in proving a deposit are outlined schematically in section I.A along with some typical evaluation problems arising from such steps.

Section I.B introduces the geostatistical terminology from a practical and intuitive point of view. Such key notions as regionalized variables, variogram, estimation and dispersion variances, are presented.

The geostatistical approach to several of the problems encountered in section I.A is given in section I.C, which also outlines the sequence of steps in a geostatistical study.

I.A. THE STEPS INVOLVED IN PROVING A DEPOSIT

In this section, we shall consider the various steps involved in proving a deposit from prospection to production. The sequence of these steps and the critical nature of certain problems will obviously depend on the type of deposit.

Geological surveys

A preliminary survey of a province or favourable metallogenic environment, by such various means as geological mapping, geochemistry or
geophysics, will usually result in the location of a certain number of
prospects. If one of these prospects locates a mineralized zone, the next
step is to determine its value and extent by a more detailed survey involving, for example, drilling, trenches or even, sometimes, drives. The
locations of these samples are generally not regular, but are based on
geological considerations such as the need to verify the existence of the
mineralization and to provide a qualitative description of it (e.g.,
mineralogy, tectonics, metallogenic type). The results of such sampling
may be a proper basis on which to plan further reconnaissance, but they are
not entirely suited for quantitative estimations of tonnage and yardage.
The first estimations will tend to be more qualitative than quantitative, as
will be the geologist's image of the deposit.

Systematic survey on a large grid

The second stage of the survey usually consists of drilling on a large and
more or less regular grid. In general, this grid will cover what is considered
to be the mineralized zone and should be sufficient to evaluate the overall
in situ resources (tonnages and mean grades). If promising results are
obtained from this second stage, the survey may be extended to some in-fill
drilling or by drilling in adjacent zones. At this stage, the geologist and
project engineer are faced with two fundamental quantitative problems.

The first one is how to deduce the required global estimations from the
available information, which consists of qualitative geological hypotheses
and quantitative data from core samples, trenches, pits, etc. In general,
estimations of a geometric nature such as mineralized surface and yardage
will make more use of qualitative information than estimations of a
"concentration" nature such as mean grades and densities.

The second problem is to express the confidence that can be given to
these estimations. In practice, it is common to classify the *in situ* resources
in different categories such as, for example, those recommended by the US
Bureau of Mines: measured ore, indicated ore and inferred ore. The
boundaries between these categories do not always appear clearly. The
geologist knows from experience that the precision of an estimation, such
as mean grade, does not only depend on the number of the available assays
but also on the *in situ* variability of the grade in the deposit and the manner
in which this variability has been sampled. He also knows that the concepts
of continuity and zone of influence, which are used in the evaluation of

resources, depend on the particular deposit and the type of ore considered. And, finally, he knows that for a given ore the variability of assays of core samples, cuttings or channel samples can be markedly different.

In this chapter, it will be shown how geostatistics takes these various concepts into account by means of the variogram and structural analysis.

Sampling on a small grid

Once the deposit has been deemed economically mineable, the next step is to define the technological and economical framework for its exploitation.

The total *in situ* resources of a deposit are seldom entirely mineable, both for technological reasons (depth, accessibility) and for economic reasons (mining cost, cut-off grade). To define recoverable reserves within a given technological and economic context, or, conversely, to define the type of mining which will result in maximum profit from the deposit it is necessary to have a detailed knowledge of the characteristics of this deposit. These characteristics include the spatial distribution of rich- and poor-grade zones, the variability of thicknesses and overburden, the various grade correlations (economic metals and impurities).

To determine these characteristics, it is usual to sample on a small grid, at least over the first production zone. The results of this sampling compaign are then used to construct block estimations within this zone. These local estimations set the usual problem of determining estimators and evaluating the resulting estimation error. There will also be problems associated with the use of these local estimations: the evaluation of recoverable reserves depends on whether the proposed mining method is block-caving with selection units of the order of $100 \text{ m} \times 100 \text{ m} \times 100 \text{ m}$, or open-pit mining with selection units of the order of $10 \text{ m} \times 10 \text{ m} \times 10 \text{ m}$.

From experience, the mining engineer knows that, after mining, the actual product will differ from the estimations of the geologist. In general, he knows that tonnage is underestimated and, more importantly, quality is overestimated. In practice, this usually results in a more or less empirical "smudge factor" being applied to the estimation of recoverable reserves, which may reduce the estimated quantity of recovered metal from 5 to 20%. This is not a very satisfactory procedure, because it does not explain the reasons for these systematic biases between the geologist's estimations and actual production; in addition, it is far too imprecise when evaluating a marginally profitable new deposit.

Given the size of present day mining projects and the cost of their investments, the risks of such empirical methods are no longer acceptable, and it is imperative to set and rigorously solve such problems as those given below.

(i) The definition of the most precise local estimator and its confidence interval, taking account of the peculiarities of the deposit and the ore concerned. It is obvious that a gold nugget deposit cannot be evaluated in the same way as a sedimentary phosphate deposit.

(ii) The distinction between *in situ* resources and mineable reserves. It should be possible to model rigorously the effect of the mining method on the recovery of the *in situ* resources.

(iii) The definition of the amount of data (drill cores, exploration drives) necessary for a reliable estimation of these recoverable reserves.

Mine planning

Once the estimations of resources and recoverable reserves have been made, the next step is to plan the mining project itself. This involves choosing mining appliances, type of haulage, requirements of stockpiling and blending, mill planning, etc. Some of these choices are conditional, in all or part, upon the structures of the variabilities of the ore characteristics, either *in situ* or at any of the various stages of mining, haulage, stockpiling and mill feeding. It will be shown later how geostatistics can be used to provide representations of these variabilities by means of numerical models of the deposit. The effects of various extraction and blending processes can then be tested by simulating them on these numerical models.

I.B. THE GEOSTATISTICAL LANGUAGE

I.B.1 Regionalized variable and random function

A mineralized phenomenon can be characterized by the spatial distribution of a certain number of measurable quantities called "regionalized variables". Examples are the distribution of grades in the three-dimensional space, the distribution of vertical thicknesses of a sedimentary bed in the horizontal space, and the distribution of market price of a metal in time.

Geostatistical theory is based on the observation that the variabilities of all regionalized variables have a particular structure. The grades $z(x)$ and $z(x+h)$ of a given metal at points x and $x+h$ are auto-correlated; this auto-correlation depends on both the vector h (in modulus and direction) which separates the two points, and on the particular mineralization considered. The variability of gold grades in a placer deposit will differ from that of gold grades in a massive deposit. The independence of the two grades $z(x)$ and $z(x+h)$ beyond a certain distance h is simply a particular case of auto-correlation and is treated in the same way.

The direct study of the mathematical function $z(x)$ is excluded because its spatial variability is usually extremely erratic, with all kinds of discontinuities and anisotropies. It is equally unacceptable to interpret all the numerical values $z(x)$, $z(x')$ as independent realizations of the same random variable Z, because this interpretation takes no account of the spatial auto-correlation between two neighbouring values $z(x)$ and $z(x + h)$. Both the random and structured aspects of the regionalized variable are expressed by the probabilistic language of "random functions".

A random function $Z(x)$ can be seen as a set of random variables $Z(x_i)$ defined at each point x_l of the deposit D: $Z(x) = \{Z(x_i), \forall x_i \in D\}$. The random variables $Z(x_i)$ are correlated and this correlation depends on both the vector h (modulus and direction) separating two points x_i and $x_i + h$ and the nature of the variable considered. At any point x_i, the true grade $z(x_i)$, analysed over a core sample, for example, is interpreted as one particular realization of the random variable $Z(x_i)$. Similarly, the set of true grades $\{z(x_i), \forall x_i \in D\}$ defining the deposit or the zone D is interpreted as one particular realization of the random function $\{Z(x_l), \forall x_l \in D\}$.

The next section will deal with the operational aspect of this formalization by describing how the spatial variability of the regionalized variable $z(x)$ can be characterized.

I.B.2. The variogram

Consider two numerical values $z(x)$ and $z(x + h)$, at two points x and $x + h$ separated by the vector h. The variability between these two quantities is characterized by the variogram function $2\gamma(x, h)$, which is defined as the expectation of the random variable $[Z(x) - Z(x + h)]^2$, i.e.,

$$2\gamma(x, h) = E\{[Z(x) - Z(x + h)]^2\}. \tag{I.1}$$

In all generality, this variogram $2\gamma(x, h)$ is a function of both the point x and the vector h. Thus, the estimation of this variogram requires several realizations, $[z_k(x), z_k(x + h)]$, $[z_{k'}(x), z_{k'}(x + h)]$, $\ldots$, $[z_{k''}(x), z_{k''}(x + h)]$, of the pair of random variables $[Z(x), Z(x + h)]$. Now, in practice, at least in mining applications, only one such realization $[z(x), z(x + h)]$ is available and this is the actual measured couple of values at points x and $x + h$. To overcome this problem, the *intrinsic hypothesis* is introduced. This hypothesis is that the variogram function $2\gamma(x, h)$ depends only on the separation vector h (modulus and direction) and not on the location x. It is then possible to estimate the variogram $2\gamma(h)$ from the available data: an estimator $2\gamma^*(h)$ is the arithmetic mean of the squared differences between two experimental measures $[z(x_i), z(x_i+h)]$ at any two points separated by the vector h; i.e.,

$$2\gamma^*(h) = \frac{1}{N(h)} \sum_{i=1}^{N(h)} [z(x_i) - z(x_i+h)]^2, \qquad (\text{I.2})$$

where $N(h)$ is the number of experimental pairs $[z(x_i), z(x_i+h)]$ of data separated by the vector h.

Note that the intrinsic hypothesis is simply the hypothesis of second-order stationarity of the differences $[Z(x) - Z(x+h)]$. In physical terms, this means that, within the zone D, the structure of the variability between two grades $z(x)$ and $z(x+h)$ is constant and, thus, independent of x; this would be true, for instance, if the mineralization within D were homogeneous.

The intrinsic hypothesis is not as strong as the hypothesis of stationarity of the random function $Z(x)$ itself. In practice, the intrinsic hypothesis can be reduced, for example, by limiting it to a given locality; in such a case, the function $\gamma(x, h)$ can be expressed as two terms: $\gamma(x, h) = f(x) \cdot \gamma_0(h)$, where $\gamma_0(h)$ is an intrinsic variability constant over the zone D and $f(x)$ characterizes an intensity of variability which depends on the locality x, cf. the concept of proportional effect, treated in section III.B.6.

The variogram as a tool for structural analysis

"Structural analysis" is the name given to the procedure of characterizing the structures of the spatial distribution of the variables considered (e.g., grades, thicknesses, accumulations). It is the first and indispensable step of any geostatistical study. The variogram model acts as a quantified summary of all the available structural information, which is then channelled into the various procedures of resource and reserve evaluation. Thus $\gamma(h)$ can be regarded as injecting extensive geological experience into the sequence of studies involved in a mining project.

In the proving of any mining project, it is imperative to have good communication between all the departments involved (geology, mining techniques and economy, processing). It is a waste, for instance, to have a geologist carry out an excellent study of geological structures, only to be followed by a project engineer evaluating resources by such arbitrary procedures as polygons of influence or inverse-squared distance weighting. Such evaluations are independent of the deposit and its geology, and make no distinction, for example, between a gold placer deposit and a porphyry–copper deposit. The variogram model $\gamma(h)$ can help in this essential communication between the geologist and the project engineer.

Continuity

In the definition of the variogram $2\gamma(h)$, h represents a vector of modulus $|h|$ and direction α. Consider a particular direction α. Beginning at the origin, $\gamma(0) = 0$, the variogram increases in general with the modulus $|h|$. This is simply an expression of the fact that, on average, the difference between two grades taken at two different points increases as the distance $|h|$ between them increases. The manner in which this variogram increases for small values of $|h|$ characterizes the degree of spatial continuity of the variable studied. Figure II.10 shows profiles of four piezometer readings, giving the height of a water table, measured at four different points (denoted by 3, 4, 33 and 18), as a function of time. The four corresponding semi-variograms have markedly different shapes, from which a geostatistician would be able to deduce the degree of continuity of the four piezometer profiles.

Anisotropies and zone of influence

In a given direction α, the variogram may become stable beyond some distance $|h| = a$ called the range, cf. Fig. I.1. Beyond this distance a, the mean square deviation between two quantities $z(x)$ and $z(x + h)$ no longer depends on the distance $|h|$ between them and the two quantities are no longer correlated. The range a gives a precise meaning to the intuitive concept of the zone of influence of a sample $z(x)$. However, there is no reason for the range to be the same in all directions α of the space. In Fig. I.1, for instance, the vertical range characterizing the mean vertical dimension of the mineralized lenses differs from the horizontal ranges. For a given distance $|h|$, the horizontal variogram presents a weaker variability than the vertical variogram: this expresses the horizontal sedimentary character of the phenomenon considered.

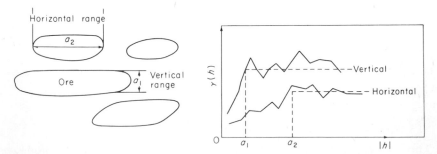

FIG. I.1. Structural anisotropy expressed by the variogram.

I.B.3. The estimation variance

The variogram defined in (I.1) can be viewed as an estimation variance, the variance of the error committed when the grade at point x is estimated by the grade at point $x + h$. From this elementary estimation variance, $2\gamma(h) = E\{[Z(x) - Z(x + h)]^2\}$, geostatistical techniques can be used to deduce the variance of estimation of a mean grade Z_V by another mean grade Z_v. This estimation variance is expressed as

$$E\{[Z_V - Z_v]^2\} = 2\bar{\gamma}(V, v) - \bar{\gamma}(V, V) - \bar{\gamma}(v, v). \tag{I.3}$$

The mean grades Z_v and Z_V can be defined on any supports, e.g., V may represent a mining block centred on the point y, and v the set of N drill cores centred on points $\{x_i, i = 1 \text{ to } N\}$. The mean grades are then defined by

$$Z_V = \frac{1}{V(y)} \int_{V(y)} Z(x)\, dx \quad \text{and} \quad Z_v = \frac{1}{N} \sum_{i=1}^{N} Z(x_i).$$

The notation $\bar{\gamma}(V, v)$, for example, represents the mean value of the elementary semi-variogram function $\gamma(h)$ when the end-point x of the vector $h = (x - x')$ describes the support V and the other end-point x' independently describes the support v, i.e.,

$$\bar{\gamma}(V, v) = \frac{1}{NV} \sum_{i=1}^{N} \int_{V(y)} \gamma(x_i - x)\, dx.$$

The general formula (I.3), shows that the estimation variance, i.e., the quality of the estimation, depends on all four of the following.

  (i) The relative distances between the block V to be estimated and the information v used to estimate it. This is embodied in the term $\bar{\gamma}(V, v)$.
  (ii) The size and geometry of the block V to be estimated. This is embodied in the term $\bar{\gamma}(V, V)$.
  (iii) The quantity and spatial arrangement of the information v, which is embodied in the term $\bar{\gamma}(v, v)$.
  (iv) The degree of continuity of the phenomenon under study, which is conveyed by its characteristic semi-variogram $\gamma(h)$.

Thus, the theoretical formula (I.3) expresses the various concepts that we intuitively know should determine the quality of an estimation.

Confidence interval

The estimation variance itself is not enough to establish a confidence interval for the proposed estimate, the distribution of the errors must also be known. In mining applications however, the standard 95% Gaussian confidence interval $[\pm 2\sigma_E]$ is used, where σ_E^2 is the estimation variance.

Experimental observation of a large number of experimental histograms of estimation errors has shown that the Gaussian distribution tends to underestimate the proportion of low errors (within $\pm\sigma_E/2$) and of high errors (outside the range $\pm 3\sigma_E$). However, the standard Gaussian confidence interval $[\pm 2\sigma_E]$ in general correctly estimates the 95% experimental confidence interval. Figure II.17 shows the histogram of estimation errors of 397 mean block grades from the Chuquicamata deposit in Chile. The arithmetic mean of the errors is $\pm 0.01\%$ Cu for a mean grade of 2% Cu. The experimental variance of the 397 errors is equal to 0.273 $(\%\ Cu)^2$, which is close to the geostatistically predicted value of 0.261. The predicted 95% confidence interval is, according to the Gaussian standard, $[\pm 2\sqrt{(0.261)}] = [\pm 1.02\%\ Cu]$. In fact, 96% of the observed errors fall within this interval.

Change of support (volume)

The elementary variogram $2\gamma(h)$, estimated from experimental data, is defined on the support of these data, e.g., core samples of given cross-section and length. By means of formula (I.3), the variogram $2\gamma_V(h)$ defined on any other support V can be deduced from the variogram $2\gamma(h)$, e.g., V may be the support of the mining block. In fact,

$$2\gamma_V(h) = E\{[Z_V(x) - Z_V(x+h)]^2\}$$

appears as the variance of estimation of the mean grade of the block V centred on point x by the mean grade of the block V centred on the point $x + h$. This formalism for the change of support has great practical applications. Since future mining will be carried out on the basis of blocks and not drill cores, it is of great benefit to be able to evaluate the spatial variability of the mean grades of these blocks by means of the variogram $2\gamma_V(h)$ defined on the block support.

I.B.4. The dispersion variance

There are two dispersion phenomena well known to the mining engineer. The first is that the dispersion around their mean value of a set of data collected within a domain V increases with the dimension of V. This is a logical consequence of the existence of spatial correlations: the smaller V, the closer the data and, thus, the closer their values. The second is that the dispersion within a fixed domain V decreases as the support v on which each datum is defined increases: the mean grades of mining blocks are less dispersed than the mean grades of core samples.

These two phenomena are expressed in the geostatistical concept of dispersion variance. Let V be a domain consisting of N units with the same support v. If the N grades of these units are known, their variance can be calculated. The dispersion variance of the grades of units v within V, written $D^2(v/V)$, is simply the probable value of this experimental variance and is calculated by means of the elementary variogram $2\gamma(h)$ through the formula

$$D^2(v/V) = \bar{\gamma}(V, V) - \bar{\gamma}(v, v). \qquad (I.4)$$

Taking the generally increasing character of the variogram into account, it can be seen that $D^2(v/V)$ increases with the dimension of V and decreases with the dimensions of v.

This formula can be used, for example, to calculate the dispersion variance of the mean grades of production units when the size of the units (v) varies or when the interval of time considered varies (i.e., V varies).

I.B.5. Coregionalizations

The concepts of variogram and various variances, which so far have been defined for the regionalization of one variable, can be generalized to spatial coregionalizations of several variables. In a lead–zinc deposit, for example, the regionalizations of the grades in lead, $Z_1(x)$, and in zinc $Z_2(x)$, are characterized by their respective variograms $2\gamma_1(h)$ and $2\gamma_2(h)$. However, these two variabilities are not independent of each other and we can define a cross-variogram for lead and zinc:

$$2\gamma_{12}(h) = E\{[Z_1(x) - Z_1(x+h)][Z_2(x) - Z_2(x+h)]\}. \qquad (I.5)$$

Once the cross-variogram is calculated, we can form the matrix

$$\begin{bmatrix} \gamma_1 & \gamma_{12} \\ \gamma_{12} & \gamma_2 \end{bmatrix},$$

called the coregionalization matrix. This coregionalization matrix can then be used to estimate the unknown lead grade $Z_1(x)$ from neighbouring lead and zinc data.

Sometimes, two different types of data relating to the same metal can be used in block estimation. One set may be the precise data taken from core samples, while the other may be imprecise data taken from cuttings and blast-holes. A study of coregionalization of these two types of data will determine the proper weights to be assigned to them during evaluation.

I.B.6. Simulations of deposits

The regionalized variable $z(x)$, which is the subject of the geostatistical structural analysis, is interpreted as one possible realization of the random function $Z(x)$. It is possible to generate other realizations $z_s(x)$ of this random function $Z(x)$ by the process of simulation. This simulated realization $z_s(x)$, will differ from point to point from the real realization $z(x)$, but, statistically, will show the same structure of variability (same histograms and variograms); moreover, at the experimental data locations x_i, the simulated values will be equal to the true values, i.e., $z_s(x_i) = z(x_i)$, $\forall x_i \in$ data set. The simulated and actual realization will, thus, have the same concentrations of rich and poor values at the same locations. The two realizations can be viewed as two possible variants of the same genetic phenomenon characterized by the random function $Z(x)$. The simulated realization, however, has the advantage of being known at all points and not only at the data points x_i.

Various methods of extraction, stoping, blending, stockpiling, etc., can then be simulated on this numerical model $\{z_s(x), x \in D\}$ of the deposit or zone D. Using feedback, the effect of these simulated processes on recovery, fluctuations of ore characteristics, and on the running of the mine and

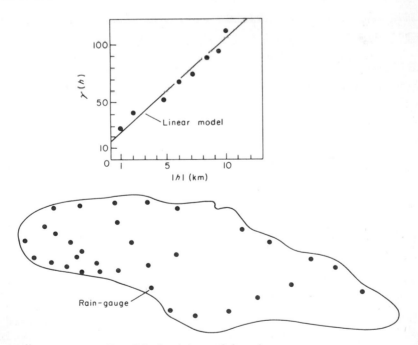

FIG. I.2. Locations of the rain gauges.

mill in general can be used to modify certain parameters of the mining project.

The techniques for the simulation of a regionalization $z(x)$ can be extended to simulations of coregionalizations, i.e., to the simulations of the simultaneous spatial distributions of the various characteristic variables (thicknesses, grades, density) of a mineralization.

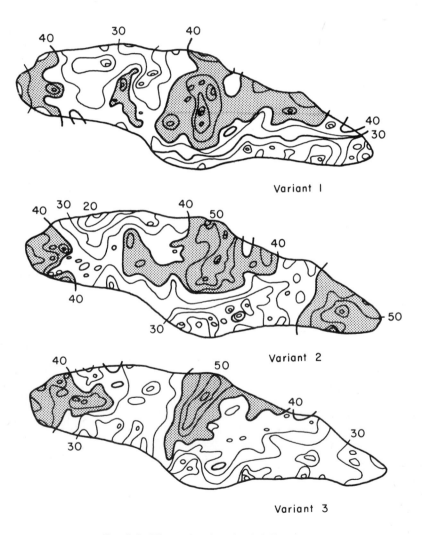

FIG I.3. Three simulated rainfalls.

Case study, after J. P. Delhomme and P. Delfiner (1973)

This was a study of the regionalization of water heights (measured in rain gauges) after rain showers, but it could equally well apply to the regionalization of the vertical thickness of a sedimentary bed.

Figure I.2 shows the location of rain gauges in the Kadjemeur basin in Tchad (Central Africa). After one particular rainfall, the height of water in each rain gauge was measured. These data were then used to construct an experimental semi-variogram to which a linear model was fitted. Figure I.3 shows three simulated rainfalls, each of which has the same linear variogram as the real rainfall and the same water heights at the actual rain gauge locations. At points other than these rain gauge locations, the three simulations can differ markedly from each other. This is noticeably evident in the extreme right-hand corner of the basin, where there are no rain gauges.

I.C. SOME TYPICAL PROBLEMS AND THEIR GEOSTATISTICAL APPROACH

As previously mentioned, the initial geological surveys of a deposit result in essentially qualitative information, and any quantitative information will generally be on a non-systematic grid. In general, it is not possible to construct sufficiently representative histograms and variograms on the basis of such information. However, even if it is of a purely qualitative nature (e.g., structural geology) this information can be used during the geostatistical structural analysis, which will later be carried out on the data obtained from the first systematic sampling campaign. The geology is used as a guide in the construction and interpretation of the various experimental variograms. These variograms then channel this geological knowledge into the various studies involved in a mining project.

I.C.1. Global estimation

Once the first systematic campaign is completed, it is usual to proceed to estimations of global *in situ* resources, such as tonnages of ore and overburden, quantities of metals and mean grades. These estimations must also include confidence intervals which will determine the point at which the *in situ* resources are sufficiently well estimated to proceed to the next stage of evaluation, i.e., that of the recoverable reserves.

At this stage of global evaluation of resources, there are no specific geostatistical methods for determining estimators. For example, a mineralized surface may be estimated by interpolation between positive and negative drill-holes. Mineralized thickness or mean grade may be estimated by the arithmetic mean of the positive data on a pseudo-regular grid, or by any of the traditional methods of extension of profiles or cross-sections. However, through the estimation variance, geostatistics can quantify the reliability of each of these traditional methods and so provide a criterion for choice between them. The estimation variance can also be used to objectively classify resources and to predict the number, location and distribution of additional information required to improve the quality of estimations, e.g., to move a given tonnage from the "possible" to the "proved" category, cf. A. G. Royle (1977).

I.C.2. Local estimation

Once the deposit has been deemed globally mineable, the next phase is local, block by block estimation. This local estimation gives an idea of the spatial distribution of the *in situ* resources, which is necessary for the evaluation of the recoverable reserves. This local estimation is usually carried out on the basis of samples obtained from a smaller sampling grid, e.g., through in-fill drilling. The estimation variance of a given panel, cf. formula (I.3), can be calculated as a function of the grid size and, thus, indicates the optimal sampling density to be used for the local estimation.

Once this second, denser sampling campaign is completed, the next problem is to determine the best possible estimate of each block. This amounts to determining the appropriate weight to be assigned to each datum value, whether inside or outside the panel. The determination of these weights must take into account the nature (core samples, channel samples) and spatial location of each datum with respect to the block and the other data. It must also take into account the degree of spatial continuity of the variable concerned, expressed in the various features of the variogram, $2\gamma(|h|, \alpha)$.

For example, suppose that the mean grade Z_V of the block V is to be estimated from a given configuration of N data Z_i, as shown on Fig. I.4. The geostatistical procedure of "*kriging*" considers as estimator Z_V^* of this mean grade, a linear combination of the N data values:

$$Z_V^* = \sum_{i=1}^{N} \lambda_i Z_i. \tag{I.6}$$

The kriging system then determines the N weights $\{\lambda_i, i = 1 \text{ to } N\}$ such that the following hold.

(i) The estimation is unbiased. This means that, on average, the error will be zero for a large number of such estimations, i.e.,

$$E\{Z_V - Z_V^*\} = 0.$$

(ii) The variance of estimation $\sigma_E^2 = E\{[Z_V - Z_V^*]^2\}$ is minimal. This minimization of the estimation variance amounts to choosing the (linear) estimator with the smallest standard Gaussian 95% confidence interval ($\pm 2\sigma_E$).

Consider the example shown on Fig. I.4, in which structural isotropy is assumed, i.e., the variability is the same in all directions. Provided that all the $N = 8$ data are of the same nature, and all defined on the same support, kriging gives the following more or less intuitive results:

(i) The symmetry relations, $\lambda_2 = \lambda_3$, $\lambda_5 = \lambda_6$;
(ii) The inequalities, $\lambda_1 \geqslant \lambda_i$, $\forall i \neq 1$; $\lambda_7 \leqslant \lambda_4$, i.e., the datum Z_4 screens the influence of Z_7; $\lambda_8 \geqslant \lambda_4$, i.e., part of the influence of Z_4 is transferred to the neighbouring data Z_5, Z_6 and Z_7.

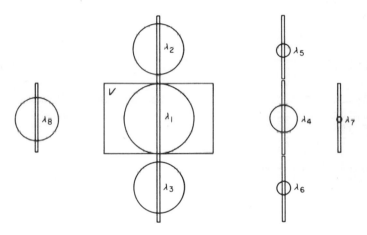

FIG. I.4. Kriging configuration of block V.

I.C.3. Selection and recoverable reserves

The terms "selection" and "recoverable reserves" have little meaning unless they are used within the context of precisely defined technological conditions for recovery. From experience, the mining engineer knows that the quantity and quality of recovery improves as:

(i) The mining selection unit becomes smaller (although this intuition is true only for highly selective mining);
(ii) The information available at the time of selection improves, to allow better discrimination between rich and poor units.

These two essential concepts are formalized by geostatistics in the following way.

1. Recovery depends on the support of the selection unit

Consider the two histograms shown on Fig. I.5(a). The first one represents the distribution of the mean grades of 5 m × 5 m × 5 m units v in the deposit D. These units may represent, for example, selective underground mining by small blocks. The second histogram represents the distribution of the mean grades of 50 m × 50 m × 50 m units V in the same deposit D, which may represent, for example, mining by block-caving. It is obvious that the first histogram (v) will be much more dispersed than the second one (V),

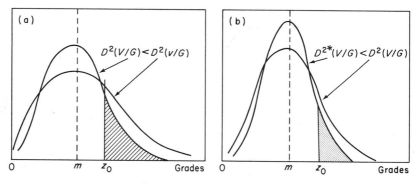

FIG I.5. Recoveries estimated from cut-off grades applied to histograms. (a) Influence of the support; (b) influence of the standard of information.

and that the same cut-off grade z_0 applied to both histograms will result in very different recoveries. These different dispersions are characterized by the two dispersion variances $D^2(v/D)$ and $D^2(V/D)$ with $D^2(v/D) > D^2(V/D)$. By assuming some hypotheses related to the conservation of the type of distribution law, it is possible to deduce these two histograms from the experimental histogram defined on the support c of core sample lengths. For any given selection unit size, the estimated histogram can then be used to determine the recoveries corresponding to different cut-off grades z_0. These recoveries are then used to build the so-called "grade-tonnage" curves.

2. Recovery depends on the standard of available information

The preceding histograms concerned the real grades of units v and V. In practice, however, it seldom happens that the true grades are available at the time of selection. Selection is made on the basis of the estimated grades of these units, which will inevitably result in a recovery different from the optimum recovery based on the true grades.

Figure I.5(b) shows the histogram of the true grades of the units V superimposed on the histogram of their estimated values. As actual selection will be made on the basis of these estimated values, the proportion of recoverable units V must be read from the histogram of the estimated grades (dotted area on Fig. I.5(b)) and not from the histogram of the true grades (hatched area on Fig. I.5(a)).

If the estimated grades are obtained by kriging, then the dispersion variance $D^{2*}(V/D)$ of the histogram of the estimated grades can be calculated from a relation known as the "smoothing effect" due to kriging:

$$D^{2*}(V/D) \simeq D^2(V/D) + \sigma_E^2, \qquad (I.7)$$

where $D^2(V/D)$ is the dispersion variance of the true grades given by formula (I.4), and σ_E^2 is the estimation variance given by the kriging system.

Knowing the two dispersion variances, $D^{2*}(V/D)$ and $D^2(V/D)$, the estimates of the two histograms on Fig. I.5(b) can be constructed and the effect on recovery of selection, based on estimated grades rather than real ones, can be evaluated. A balance can then be made between the profit in improving the recovery and the cost of increased and better information for selection.

The preceding recoveries, estimated from the histograms, take no account of the spatial locations of the recovered units. It is well known that a rich unit (with a grade greater than the cut-off z_0) in a poor zone will not be mined, and a number of poor units in a rich zone will be mined. Thus, a precise evaluation of recoverable reserves should take into account all the technological parameters of the envisaged mining method, e.g., pit contours, conditions and cost of accessibility, haulage, etc. The geostatistical solution to this problem is to work on a numerical model of the deposit. This model will have the same spatial distribution of grades, and, thus, can be considered as a deposit similar to the real one. The simulation of the foreseen mining process on this model will, thus, provide the required evaluation of recoveries and "smudge" factor.

I.C.4. Working out the mining project

Once the general framework of the mining method has been defined and the corresponding recoverable reserves have been estimated, the next problem is a more precise determination of the parameters of the mining project. Some of these parameters depend on the variability of the ore characteristics. Examples are as follows:

(i) The number and size of simultaneous stopes, and whether or not blending of the production from these stopes is required. These choices will of course affect the trucking equipment.

(ii) The need for stockpiling, for control or blending, and the characteristics of the stockpiles.

(iii) The flexibility of the mill, to allow absorption of the fluctuations in quantity and quality of the recovered ore.

The choice of such parameters involves a study of the fluctuations over time of the mined ore after each of the mining phases, trucking and stockpiling. The structure of the *in situ* variabilities will be partially destroyed after mining and haulage, and the modified structures must be known to determine the blending procedures necessary to meet the mill feed constraints. It may be possible to make such choices from experience gained on similar already mined deposits, but the analogy can sometimes be limited.

When there is no such "similar" deposit, the geostatistical approach consists in building a numerical model of the deposit on which the various processes of mining, trucking and blending can be simulated: their consequences on the production and the mill feed can then be observed, and, by feedback, the parameters adopted for the various mining processes can be corrected. These simulations are, of course, only as reliable as the numerical model of the deposit. This model must faithfully reproduce all the features of the spatial distribution of the characteristic variables (grades, thicknesses, accumulations, but also morphology and tectonics) of the real deposit. The better the characteristics of variability and morphology of the real deposit are known, the better the numerical model will be. Modelling does not create information, it can only reproduce the available structural information in a simple numerical manner, easy to handle.

I.C.5. Mine planning in the routine phase

Once routine mining of the deposit has begun, the information about the deposit becomes increasingly more detailed, both qualitatively (e.g., the working faces can be seen) and quantitatively (data from blast-holes, mill feed grades, etc.). Using these data, the estimations can be constantly updated to improve overall operation. These estimations may be the basis for short- or long-term mine planning. For example, for a given period of production, these estimations may be used to determine hanging-wall and foot-wall outlines in a subhorizontal sedimentary deposit, or to carry out selection that satisfies the various constraints on tonnage and quality. Mine–mill production schedules can be made more reliable by specifying confidence intervals for the mean grades of ore sent for milling.

II

The Theory of Regionalized Variables

SUMMARY

This fundamental theoretical chapter gives the conceptual basis of geostatistics, then introduces, successively, the three main geostatistical tools: the variogram, the estimation variance and the dispersion variance.

Section II.A gives the interpretation of a regionalized variable $z(x)$ as a realization of a random function $Z(x)$. The spatial correlations of this random function, i.e., the variability in space of the variable being studied, are characterized by the covariance function $C(h)$ or by its equivalent, the semi-variogram function defined by relation (II.4):

$$\gamma(h) = \tfrac{1}{2} E\{[Z(x+h) - Z(x)]^2\}.$$

Section II.B introduces the distribution function of the error of estimation. If a normal distribution is assumed, this error is then completely characterized by its expectation (or estimation bias) and its variance (or estimation variance). For the important class of linear estimators, the estimation variance of any domain V by any other domain v is given by a combination of mean values of the semi-variogram $\gamma(h)$, cf. formula (II.27):

$$\sigma_E^2(v, V) = 2\bar{\gamma}(V, v) - \bar{\gamma}(V, V) - \bar{\gamma}(v, v).$$

Section II.C introduces a second variance, the dispersion variance $D^2(v/V)$, which characterizes the dispersion of the grades of mining units v within a stope V, for example. This dispersion variance is written as a combination of mean values of the semi-variogram $\gamma(h)$, cf. formula (II.36):

$$D^2(v/V) = \bar{\gamma}(V, V) - \bar{\gamma}(v, v).$$

Section II.D introduces the important notion of regularization (taking the average value in a certain support v). If $Z(x)$ is the point grade with the point semi-variogram $\gamma(h)$, the mean value $Z_v(x)$ in the volume v is then characterized by a regularized semi-variogram $\gamma_v(h)$, which is written as a combination of mean values of the point semi-variogram, cf. formula (II.41):

$$\gamma_v(h) = \bar{\gamma}(v, v_h) - \bar{\gamma}(v, v).$$

26

Two sets of information defined on two different supports v and v' and characterized by the two regularized semi-variograms γ_v and $\gamma_{v'}$ can then be compared.

As all the previous geostatistical operations make continuous use of mean values $\bar{\gamma}$ of the point semi-variogram $\gamma(h)$, section II.E is devoted to the practical calculation procedures of those mean values $\bar{\gamma}$. Either a direct numerical calculation or the use of appropriate auxiliary functions and charts is possible.

II.A. REGIONALIZED VARIABLES AND RANDOM FUNCTIONS

II.A.1. Regionalized variables and their probabilistic representation

When a variable is distributed in space, it is said to be *"regionalized"*. Such a variable is usually a characteristic of a certain phenomenon, as metal grades, for example, are characteristics of a mineralization. The phenomenon that the regionalized variable is used to represent is called a "regionalization", examples of which are:

(i) the market price of a metal which can be seen as the distribution of the variable price in time (one-dimensional space);

(ii) a geological phenomenon such as the thickness of a subhorizontal bed which can be regarded as the distribution in the two-dimensional space of the variable thickness.

(iii) a mineralized phenomenon can be characterized by the distribution in the three-dimensional space of variables such as grades, densities, recoveries, granulometries etc.

Regionalized variables are not restricted to mining, and examples from other fields include population density in demography, rainfall measurements in pluviometry and harvest yields in agronomy. In fact, almost all variables encountered in the earth sciences can be regarded as regionalized variables.

The definition of a regionalized variable (abbreviated to ReV) as a variable distributed in space is purely descriptive and does not involve any probabilistic interpretation. From a mathematical point of view, a ReV is simply a function $f(x)$ which takes a value at every point x of coordinates (x_u, x_v, x_w) in the three-dimensional space. However, more often than not, this function varies so irregularly in space as to preclude any direct mathematical study of it. Even a curve as irregular as that of the nickel grades of core samples shown in Fig. II.1 underestimates the true variability of the point grades due to the smoothing effect of taking average values over the volume of the core. However, the profiles of the nickel grades from three neighbouring vertical bore-holes have a common general behaviour: a relatively slow increase of the grade values followed by a rapid decrease, and, then, at the contact with the bedrock, a further increase. This behaviour is characteristic of the vertical variability of nickel grades in the lateritic type deposits of New Caledonia, (cf. Ch. Huijbregts and A. Journel, 1972).

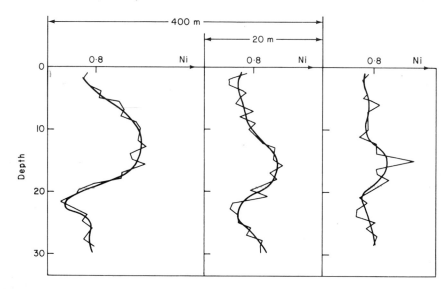

FIG. II.1. Vertical variability of Ni grades.

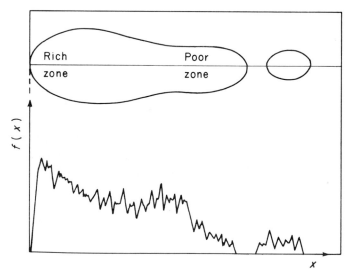

FIG. II.2. Random and structured aspect of a ReV.

In practically all deposits, a characteristic behaviour or structure of the spatial variability of the ReV under study can be discerned behind a locally erratic aspect. In almost all deposits, there exist zones which are richer than others; samples taken in a rich zone will be, on average, richer than those taken in a poorer zone, i.e., the value of the ReV $f(x)$ depends on the spatial position x, see Fig. II.2.

Thus a ReV possesses two apparently contradictory characteristics:

(i) a local, random, erratic aspect which calls to mind the notion of random variable;

(ii) a general (or average) structured aspect which requires a certain functional representation.

A proper formulation must take this double aspect of randomness and structure into account in such a way as to provide a simple representation of the spatial variability and lead to a consistent and operational approach to the solution of problems. One such formulation is the probabilistic interpretation as provided by *random functions*.

The concept of a random function

A random variable (abbreviated to RV) is a variable which takes a certain number of numerical values according to a certain probability distribution. For instance, the result of casting an unbiased die can be considered as a RV which can take one of six equally probable values. If one particular cast result is 5, then, by definition, we shall say that this value 5 is a particular *realization* of the RV "result of casting the die".

Similarly, let us consider the grade $z(x_1) = 1 \cdot 5\%$ Cu at a particular point x_1 in a copper deposit. This grade can be considered as a particular realization of a certain RV $Z(x_1)$ defined at point x_1. Thus, the set of grades $z(x)$ for all points x inside the deposit, i.e., the regionalized variable $z(x)$, can be considered as a particular realization of the set of RV $\{Z(x), x \in \text{deposit}\}$. This set of RV is called a random function and will be written $Z(x)$.[†]

This definition of a random function (abbreviated to RF) expresses the random and structured aspects of a regionalized variable:

(i) locally, at a point x_1, $Z(x_1)$ is a random variable;

(ii) $Z(x)$ is also a RF in the sense that for each pair of points x_1 and $x_1 + h$, the corresponding RV's $Z(x_1)$ and $Z(x_1 + h)$ are not, in

---

[†] *Notation* In order to distinguish clearly between observations and a probabilistic model, the regionalized variable or the realizations of the RV or RF are written in lower-case letters, thus: $y(x)$, $z(x)$, $t(x)$, while the RV or RF themselves are written in upper-case letters, thus: $Y(x)$, $Z(x)$, $T(x)$.

general, independent but are related by a correlation expressing the spatial structure of the initial regionalized variable $z(x)$.

Statistical inference

The probabilistic interpretation of a ReV $z(x)$ as a particular realization of a certain RF $Z(x)$ has an operative sense only when it is possible to infer all or part of the probability law which defines this RF in its entirety.

Obviously, it is not rigorously possible to infer the probability law of a RF $Z(x)$ from a single realization $z(x)$ which is, in addition, limited to a finite number of sample points x_i. It is impossible, for instance, to determine the law of the RV "the result of casting the die" from the single numerical result 5 resulting from a single cast of the die, and, in particular, it is not possible to determine whether or not the die is biased. Many casts of the die are necessary to answer such a question. Similarly, many realizations $z_1(x)$, $z_2(x)$, ..., $z_k(x)$ of the RF $Z(x)$ are required in order to infer the probability law of $Z(x)$. Since, in practice we shall be limited to a single realization $\{z(x_i)\}$ of the RF at the positions x_i, we appear to have come to a dead end and, in order to solve it, certain assumptions are necessary. These assumptions involve various degrees of spatial homogeneity and are introduced under the general heading of the hypothesis of stationarity.

In practice, even if only over a certain region, the phenomenon under study can very often be considered as homogeneous, the ReV repeating itself in space. This homogeneity or repetition provides the equivalent of many realizations of the same RF $Z(x)$ and permits a certain amount of statistical inference. Two experimental values $z(x_0)$ and $z(x_0+h)$ at two different points x_0 and x_0+h can, thus, be considered as two different realizations of the same RV $Z(x_0)$. This type of approach is not peculiar to geostatistics, it is used to infer the distribution law of the RV $Z(x)$ from a histogram of data values $\{z(x_i)\}$, or more simply, to infer the mathematical expectation $E\{Z(x)\}$ from the arithmetic mean of the data.

II.A.2. Moments and stationarity

Consider the RF $Z(x)$. For every set of k points in R^n (n-dimensional space), $x_1, x_2, \ldots, x_k$, called support points, there corresponds a k-component vectorial random variable

$$\{Z(x_1), Z(x_2), \ldots, Z(x_k)\}.$$

This vectorial RV is characterized by the k-variable distribution function

$$F_{x_1 x_2 \ldots x_k}(z_1, z_2, \ldots, z_k) = \text{Prob}\{Z(x_1) < z_1, \ldots, Z(x_k) < z_k\}.$$

The set of all these distribution functions, for all positive integers k and for every possible choice of support points in R^n, constitutes the "spatial law" of the RF $Z(x)$.

In mining applications, the entire spatial law is never required, mainly because the first two moments of the law are sufficient to provide an acceptable approximate solution to most of the problems encountered. Moreover, the amount of data generally available is insufficient to infer the entire spatial law. In geostatistics (more precisely, *linear* geostatistics), only the first two moments of the RF are used; in other words, no distinction is made between two RF's $Z_1(x)$ and $Z_2(x)$ which have the same first- and second-order moments and both functions are considered as comprising the same model.

Mathematical expectation, or first-order moment

Consider a RV $Z(x)$ at point x. If the distribution function of $Z(x)$ has an expectation (and we shall suppose that it has), then this expectation is generally a function of x, and is written

$$E\{Z(x)\} = m(x).$$

Second-order moments

The three second-order moments considered in geostatistics are as follows.

(i) The *variance*, or more precisely the "*a priori*" variance of $Z(x)$. When this variance exists, it is defined as the second-order moment about the expectation $m(x)$ of the RV $Z(x)$, i.e.,

$$\text{Var}\{Z(x)\} = E\{[Z(x) - m(x)]^2\}.$$

As with the expectation $m(x)$, the variance is generally a function of x.

(ii) The *covariance*. It can be shown that if the two RV's $Z(x_1)$ and $Z(x_2)$ have variances at the points x_1 and x_2, then they also have a covariance which is a function of the two locations x_1 and x_2 and is written as

$$C(x_1, x_2) = E\{[Z(x_1) - m(x_1)][Z(x_2) - m(x_2)]\}.$$

(iii) The *variogram*. The variogram function is defined as the variance of the increment $[Z(x_1) - Z(x_2)]$, and is written as

$$2\gamma(x_1, x_2) = \text{Var}\{Z(x_1) - Z(x_2)\}.$$

The function $\gamma(x_1, x_2)$ is then the "semi-variogram".

The hypothesis of stationarity

It appears from the definitions that the covariance and variogram functions depend simultaneously on the two support points x_1 and x_2. If this is indeed the case, then many realizations of the pair of RV's $\{Z(x_1), Z(x_2)\}$ must be available for any statistical inference to be possible.

On the other hand, if these functions depend only on the distance between the two support points (i.e., on the vector $h = x_1 - x_2$ separating x_1 and x_2), then statistical inference becomes possible: each pair of data $\{z(x_k), z(x_{k'})\}$ separated by the distance $(x_k - x_{k'})$, equal to the vector h, can be considered as a different realization of the pair of RV's $\{Z(x_1), Z(x_2)\}$.

It is intuitively clear that, in a zone of homogeneous mineralization, the correlation that exists between two data values $z(x_k)$ and $z(x_{k'})$ does not depend on their particular positions within the zone but rather on the distance which separates them.

Strict stationarity   A RF is said to be stationary, in the strict sense, when its spatial law is invariant under translation. More precisely, the two k-component vectorial RV's $\{Z(x_1), \ldots, Z(x_k)\}$ and $\{Z(z_1 + h), \ldots, Z(x_k + h)\}$ are identical in law (have the same k-variable distribution law) whatever the translation vector h.

However, in linear geostatistics, as we are only interested in the previously defined two first-order moments, it will be enough to assume first that these moments exist, and then to limit the stationarity assumptions to them.

Stationarity of order 2   A RF is said to be stationary of order 2, when:

(i) the mathematical expectation $E\{Z(x)\}$ exists and does not depend on the support point x; thus,

$$E\{Z(x)\} = m, \qquad \forall x; \tag{II.1}$$

(ii) for each pair of RV's $\{Z(x), Z(x+h)\}$ the covariance exists and depends on the separation distance h,

$$C(h) = E\{Z(x+h) \cdot Z(x)\} - m^2, \qquad \forall x, \tag{II.2}$$

where h represents a vector of coordinates (h_u, h_v, h_w) in the three-dimensional space.

The stationarity of the covariance implies the stationarity of the variance and the variogram. The following relations are immediately evident:

$$\text{Var } Z(x)\} = E\{[Z(x) - m]^2\} = C(0), \qquad \forall x \tag{II.3}$$

$$\gamma(h) = \tfrac{1}{2}E\{[Z(x+h) - Z(x)]^2\} = C(0) - C(h), \qquad \forall x. \tag{II.4}$$

Relation (II.4) indicates that, under the hypothesis of second-order stationarity, the covariance and the variogram are two equivalent tools for characterizing the auto-correlations between two variables $Z(x+h)$ and $Z(x)$ separated by a distance h. We can also define a third tool, the *correlogram*:

$$\rho(h) = \frac{C(h)}{C(0)} = 1 - \frac{\gamma(h)}{C(0)} \qquad (\text{II.5})$$

The hypothesis of second-order stationarity assumes the existence of a covariance and, thus, of a finite *a priori* variance, $\text{Var}\{Z(x)\} = C(0)$. Now, the existence of the variogram function represents a weaker hypothesis than the existence of the covariance; moreover, there are many physical phenomena and RF's which have an infinite capacity for dispersion (cf. section II.A.5, Case Study 1), i.e., which have neither an *a priori* variance nor a covariance, but for which a variogram can be defined. As a consequence, the second-order stationarity hypothesis can be slightly reduced when assuming only the existence and stationarity of the variogram.

Intrinsic hypothesis   A RF $Z(x)$ is said to be intrinsic when:

(i) the mathematical expectation exists and does not depend on the support point x,

$$E\{Z(x)\} = m, \qquad \forall x;$$

(ii) for all vectors h the increment $[Z(x+h) - Z(x)]$ has a finite variance which does not depend on x,

$$\text{Var}\{Z(x+h) - Z(x)\} = E\{[Z(x+h) - Z(x)]^2\} = 2\gamma(h), \qquad \forall x. \quad (\text{II.6})$$

Thus, second-order stationarity implies the intrinsic hypothesis but the converse is not true: the intrinsic hypothesis can also be seen as the limitation of the second-order stationarity to the increments of the RF $Z(x)$.

Quasi-stationarity   In practice, the structural function, covariance or variogram, is only used for limited distances $|h| \leqslant b$. The limit b represents, for example, the diameter of the neighbourhood of estimation (i.e., the zone which contains the information to be used, cf. Fig. II.3). In other cases, b may be the extent of an homogeneous zone and two variables $Z(x_k)$ and $Z(x_k+h)$ cannot be considered as coming from the same homogeneous mineralization if $|h| > b$.

In such cases, we can, and indeed we must, be content with a structural function $C(x, x+h)$ or $\gamma(x, x+h)$, which is no more than locally stationary (for distances $|h|$ less than the limit b). This limitation of the hypothesis of second-order stationarity (or the intrinsic hypothesis if only the variogram is assumed) to only those distances $|h| \leqslant b$ corresponds to an hypothesis of quasi-stationarity (or to a quasi-intrinsic hypothesis).

In practical terms, we can define sliding neighbourhoods inside of which the expectation and the covariance can be considered as stationary and the data are sufficient to make statistical inference possible.

The hypothesis of quasi-stationarity is really a compromise between the scale of homogeneity of the phenomenon and the amount of available data. In fact, it is always possible to produce stationarity by considerably reducing the dimension b of the zones of quasi-stationarity, but then most of these zones would not have any data in them, and, thus, it would be impossible to infer the quasi-stationary moments in these zones.

From a data grid of 1 to 3 m, a 200-m-long hillside would appear as a quasi-stationary model, but would require a non-stationary model (because of the increasing altitude) for a grid of 20 to 30 m, cf. Fig. II.3. If the data grid is of the order of 200 m, then a certain stationarity again becomes evident: at the scale of the peneplanar basin the entire hillside appears as a local variability of the stationary RF "altitude". This example shows, incidentally, that it is not possible to devise a theoretical test for the *a priori* verification of the hypothesis of stationarity using a single realization of the RF. It is always possible to find a particular realization of a stationary RF which would display a systematic and continuous change in value (apparent non-stationarity) over a limited distance, so that a theoretical test could never refute the hypothesis of stationarity.

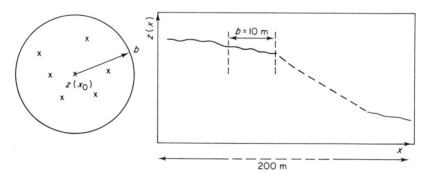

FIG. II.3. Quasi-stationarity neighbourhood.

II.A.3. Properties of the covariance and the variogram

The following properties are classical results of the theory of stochastic processes, cf., in particular, J. Bass (1974) or J. L. Doob (1953).

Their proof is of no interest in the application of geostatistics, we shall simply state the properties and comment on them.

Positive definite conditions

Let $Z(x)$ be a stationary RF of expectation m and covariance $C(h)$ or semi-variogram $\gamma(h)$. Let Y be any finite linear combination of the type

$$Y = \sum_{i=1}^{n} \lambda_i Z(x_i) \tag{II.7}$$

for any weights λ_i.

This linear combination is a RV and its variance must never be negative, $\text{Var}\{Y\} \geqslant 0$. Explicitly, this variance is written

$$\text{Var}\{Y\} = \sum_i \sum_j \lambda_i \lambda_j C(x_i - x_j) \geqslant 0. \tag{II.8}$$

The covariance function must be such that it insures that the previous variance will always be positive or zero. By definition then, the function $C(h)$ is said to be "positive definite".

Thus, not any function $g(h)$ can be considered as the covariance of a stationary RF: it is imperative that it be positive definite.

Taking account of relation (II.4), expression (II.8) can be written in terms of the semi-variogram:

$$\text{Var}\{Y\} = C(0) \sum_i \lambda_i \sum_j \lambda_j - \sum_i \sum_j \lambda_i \lambda_j \gamma(x_i - x_j). \tag{II.9}$$

In the case where the variance $C(0)$ does not exist, only the intrinsic hypothesis is assumed, and the variance of Y is defined on the condition that

$$\sum_{i=1}^{n} \lambda_i = 0$$

and the term $C(0)$ is thus eliminated, leaving

$$\text{Var}\{Y\} = -\sum_i \sum_j \lambda_i \lambda_j \gamma(x_i - x_j) \tag{II.10}$$

The variogram function must be such that $\text{Var}\{Y\}$ is positive or zero, with the condition that $\sum_i \lambda_i = 0$. By definition, $\gamma(h)$ is said to be a "conditional positive definite" function.

Authorized linear combinations

When only the intrinsic hypothesis is assumed, $\gamma(h)$ exists but $C(h)$ may not exist; the only linear combinations that have a finite variance are those verifying the weights condition $\sum_i \lambda_i = 0$. Thus, if only these authorized linear combinations are considered (which is the case in linear geostatistics), there is no need to calculate the *a priori* variance $C(0)$, nor to know the expectation $m = E\{Z(x)\}$ or the covariance $C(h)$; it is enough to know the semi-variogram.

Note that, for the authorized linear combinations, formula (II.10) is simply deduced from formula (II.8) by substituting $(-\gamma)$ for C. This rule for passing from a formula written in terms of covariance to the corresponding formula written in terms of semi-variogram is general in linear geostatistics.

Properties of the covariance

The positive definite character of the covariance function $C(h)$ entails the following properties:

$C(0) = \text{Var}\{Z(x)\} \geq 0$,   an *a priori* variance cannot be negative;

$C(h) = C(-h)$,   the covariance is an even function; $\qquad$ (II.11)

$|C(h)| \leq C(0)$,   Schwarz's inequality.

Since the degree of correlation between two variables $Z(x)$ and $Z(x+h)$ generally decreases as the distance h between them increases, so, in general, does the covariance function, which decreases from its value at the origin $C(0)$. Correspondingly, the semivariogram $\gamma(h) = C(0) - C(h)$ increases from its value at the origin, $\gamma(0) = 0$, see Fig. II.4.

Absence of correlation

Very often, in practice, the correlation between two variables $Z(x)$ and $Z(x+h)$ disappears when the distance h becomes too large:

$$C(h) \to 0, \quad \text{when } |h| \to \infty,$$

and, in practice, we can put $C(h) = 0$, once $|h| \geq a$. The distance a beyond which $C(h)$ can be considered to be equal to zero is called the "range" and it represents the transition from the state in which a spatial correlation exists $(|h| < a)$ to the state in which there is absence of correlation $(|h| > a)$, cf. Fig. II.4.

Properties of the variogram

The definition of the variogram as the variance of increments entails the following properties:

$$\gamma(0)=0, \qquad \gamma(h)=\gamma(-h)\geqslant 0. \qquad (II.12)$$

In general, but not always, as h increases, the mean quadratic deviation between the two variables $Z(x)$ and $Z(x+h)$ tends to increase and so $\gamma(h)$ increases from its initial zero value.

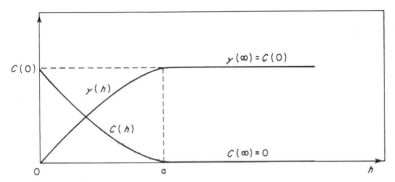

FIG. II.4. Covariance and semi-variogram.

Transition phenomena   Very often, in practice, the semi-variogram stops increasing beyond a certain distance and becomes more or less stable around a limit value $\gamma(\infty)$ called a "*sill*" value, which is simply the *a priori* variance of the RF

$$\gamma(\infty)= \text{Var}\,\{Z(x)\}= C(0). \qquad (II.13)$$

In such cases, the *a priori* variance exists as does the covariance. Such variograms, which are characterized by a sill value and a range, are called "transition" models, and correspond to a RF which is not only intrinsic but also second-order stationary, cf. Fig. II.4.

Zone of influence   In a transition phenomenon, any data value $z(x)$ will be correlated with any other value falling within a radius a of x. This correlation and, hence, the influence of one value on the other will decrease as the distance between the two points increases. Thus, the range corresponds to the intuitive idea of a zone of influence of a RV: beyond the distance $|h|=a$, the RV's $Z(x)$ and $Z(x+h)$ are no longer correlated.

Anisotropies   There is no reason to expect that the mineralization will exhibit the same behaviour in every direction, i.e., that the mineralization will be *isotropic*. In the three-dimensional space, x represents the coordinates (x_u, x_v, x_w) and h represents a vector of modulus $|h|$ and direction α. Thus, in condensed form, $\gamma(h)$ represents the set of semi-variograms $\gamma(|h|, \alpha)$ for each direction α. By studying $\gamma(h)$ in various directions α, it would be possible to determine any possible anisotropies, such as the variability of the range $a(\alpha)$ with the direction α.

In the example of Fig. II.5, the semi-variogram for the vertical direction reveals a short range a_1 corresponding to the mean vertical width of the mineralized lenses, while the semi-variogram for the horizontal direction has a larger range a_2 corresponding to the mean horizontal dimensions of the same lenses. The directional graph of ranges $a(\alpha)$ thus represents the average morphology of the mineralized lenses.

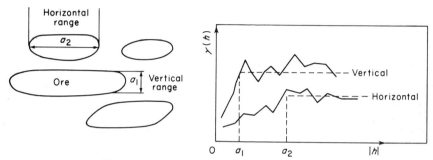

FIG. II.5. The variogram as a structural tool.

Behaviour of the variogram near the origin

The continuity and regularity in space of the RF $Z(x)$ and, thus, of the ReV $z(x)$ that it represents, are related to the behaviour of the variogram near the origin. In order of decreasing regularity, four main types of behaviour can be distinguished, see Fig. II.6 and section II.A.5, Case Study 3.

   (a) *Parabolic*: $\gamma(h) \sim A|h|^2$ when $h \to 0$. $\gamma(h)$ is twice differentiable at the origin and the RF is itself differentiable in the mean square sense. This type of behaviour is characteristic of a highly regular spatial variability.

   (b) *Linear behaviour*: $\gamma(h) \sim A|h|$ when $h \to 0$. $\gamma(h)$ is no longer differentiable at the origin† but remains continuous at $h = 0$, and, thus, for all $|h|$. The RF $Z(x)$ is mean-square continuous‡ but no longer differentiable.

---

† In fact, the right and left derivative, $\gamma'(0^+)$ and $\gamma'(0^-)$ exist but have different signs.
‡ A RF is said to be mean-square continuous if
$$\lim E\{[Z(x+h)-Z(x)]^2\} = 0, \qquad \text{when } h \to 0.$$

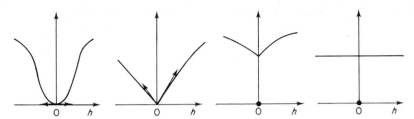

FIG. II.6. Behaviour near the origin of the variogram. (a) Parabolic behaviour; (b) linear behaviour; (c) nugget effect; (d) pure nugget effect.

(c) *Discontinuity at the origin*: $\gamma(h)$ does not tend towards zero when h tends towards zero, although by definition $\gamma(0)=0$. The RF is no longer even mean-square continuous: the variability between two values $z(x)$ and $z(x+h)$ taken at two very close points may be quite high and increases as the size of the discontinuity at the origin of $\gamma(h)$ increases. This local variability can be compared to the random phenomenon of white noise known to physicists. As the distance h increases, the variability often becomes more continuous and this is reflected in the continuity of $\gamma(h)$ for $h>0$.

The discontinuity of the variogram at the origin is called a "*nugget effect*" and is due both to measurement errors and to micro-variabilities of the mineralization; since the structure of these micro-variabilities is not accessible at the scale at which the data are available, they appear in the form of a white noise. A whole section (III.A) will be devoted to the study of this nugget effect.

(d) *Pure nugget effect.* This is the limit case when $\gamma(h)$ appears solely as a discontinuity at the origin:

$$\gamma(0)=0 \quad \text{and} \quad \gamma(h)=C_0 \quad \text{when } h>\varepsilon.$$

In practice, such a variogram can be represented as a transition phenomenon with a sill value C_0 and a range $a=\varepsilon$, where ε is very small in relation to the distances of experimental observation. For all experimental distances, small as they may be, the two RV's $Z(x)$ and $Z(x+h)$ are uncorrelated. The pure nugget effect thus corresponds to the total absence of auto-correlation. It should be pointed out that such a case is very rare in mining.

Behaviour at infinity

Using the property that $-\gamma$ is a conditional positive definite function, it can be shown that the variogram necessarily increases more slowly at infinity than does $|h|^2$:

$$\lim \frac{\gamma(h)}{|h|^2}=0 \quad \text{when } |h|\to\infty. \tag{II.14}$$

As a consequence, an experimental variogram which increases at least as rapidly as $|h|^2$ for large distances h is incompatible with the intrinsic hypothesis. Such an increase in the variogram most often indicates the presence of a trend or *drift*,† i.e., a non-stationary mathematical expectation:

$$E\{Z(x)\} = m(x),$$

where $m(x)$ depends on x.

II.A.4. Coregionalization

A regionalized phenomenon can be represented by several inter-correlated variables and, in certain cases, it may be useful to study them simultaneously.

In a lead–zinc deposit, for example, the two mineralizations are closely related and it may be of interest to know if, on average, zones that are rich in Pb are equally rich in Zn. In other polymetallic deposits, it would be useful to be able to use the information provided by one variable to make up for missing information on another variable.

The probabilistic approach to a coregionalization is similar to that of the regionalization of a single variable.

The simultaneous regionalizations (or the coregionalization) of K ReV's $\{z_1(x), \ldots, z_K(x)\}$ are interpreted as a particular realization of the set of K inter-correlated RF's $\{Z_1(x), \ldots, Z_K(x)\}$.

Under the hypothesis of second-order stationarity we define:

(i) for each RF $Z_k(x)$, the mathematical expectation

$$E\{Z_k(x)\} = m_k = \text{constant}, \qquad \forall x;$$

(ii) for each pair of RF's $Z_k(x)$, $Z_{k'}(x)$, the *cross-covariance*

$$E\{Z_{k'}(x+h) . Z_k(x)\} - m_{k'} m_k = C_{k'k}(h), \qquad \forall x; \qquad (\text{II.15})$$

(iii) and the *cross-variogram*

$$E\{[Z_{k'}(x+h) - Z_{k'}(x)][Z_k(x+h) - Z_k(x)]\} = 2\gamma_{k'k}(h), \qquad \forall x. \qquad (\text{II.16})$$

Properties of cross-moments

1. When $k' = k$, the formulae (II.15) and (II.16) yield the definitions of the covariance and variogram, $C_{kk}(h)$ and $2\gamma_{kk}(h)$.

---

† The term "drift" is preferred in geostatistical applications as it avoids the subjective interpretation that is often associated with the word "trend".

2. A cross-semi-variogram $\gamma_{k'k}(h)$ can take negative values, whereas a direct semi-variogram is always positive. A negative value of the cross-semi-variogram indicates that a positive increase in one of the variables (k') corresponds, on average, to a decrease in the other (k). This could be explained, for example, by a phenomenon in which element k is substituted for element k'.

3. Under the second-order stationary hypothesis, the existence of the cross-covariances entails that of the cross-variograms and these two groups of structural tools are related by the expression

$$2\gamma_{k'k}(h) = 2C_{k'k}(0) - C_{k'k}(h) - C_{kk'}(h). \qquad (II.17)$$

4. The cross-variogram is symmetric in (k', k) and $(h, -h)$ while this is not necessarily so for the cross-covariance:

$$\left. \begin{array}{ll} \gamma_{k'k}(h) = \gamma_{kk'}(h) & \text{and} \quad \gamma_{k'k}(h) = \gamma_{k'k}(-h), \\[2mm] C_{kk'}(h) = C_{k'k}(-h) & \text{and} \quad C_{k'k}(-h) \neq C_{k'k}(h). \end{array} \right\} \qquad (II.18)$$

A lag effect in one variable in relation to another (e.g., rich Pb grades lagging behind rich Zn grades in a given direction) is represented by a dissymetric cross-covariance in $(h, -h)$. Such a lag effect would not be visible on the cross-variogram, and, where it is suspected, the cross-covariance should be used. Otherwise, for reason of consistency with the study of the direct variograms, it is preferable to use cross-variograms.

5. *Correlation coefficient.* The point to point correlation coefficient between two variables $Z_{k'}(x)$ and $Z_k(x)$ located at the same point x is written

$$\rho_{k'k} = \frac{C_{k'k}(0)}{\sqrt{[C_{kk}(0) \cdot C_{k'k'}(0)]}} = \rho_{kk'}. \qquad (II.19)$$

6. *Linear combination.* It may be useful in certain studies to define a finite linear combination of ReV's $\{z_k(x), k = 1 \text{ to } K\}$ such as

$$f(x, \lambda_k) = \sum_{k=1}^{K} \lambda_k z_k(x), \qquad (II.20)$$

with, for example,

$$\begin{cases} \lambda_k, & \text{the profit or loss associated with a unit quantity of the metal } k; \\ z_k(x), & \text{the grade of metal } k \text{ on a given production unit;} \end{cases}$$

or

$$\begin{cases} \lambda_k, & \text{costs associated with the extraction of waste and mineral;} \\ z_k(x), & \text{thickness of overburden and mineralized thickness.} \end{cases}$$

The ReV $f(x, \lambda_k)$ can then be interpreted as a particular realization of the RF $F(x, \lambda_k) = \sum_k \lambda_k Z_k(x)$.

If the coregionalization $\{Z_1(x), \ldots, Z_K(x)\}$ verifies the intrinsic hypothesis, then the RF $F(x, \lambda_k)$ is also intrinsic and has the following stationary semi-variogram:

$$\Gamma(h) = \tfrac{1}{2}E\{[F(x+h, \lambda_k) - F(x, \lambda_k)]^2\} = \sum_k \sum_{k'} \lambda_k \lambda_{k'} \gamma_{kk'}(h),$$

and this single direct semi-variogram $\Gamma(h)$ will be used in subsequent operations. In particular, the economic function $f(x, \lambda_k)$ can be kriged (estimated optimally).

II.A.5. Case studies

1. *Infinite* a priori *variance (Witwatersrand gold province)*

Under the hypothesis of second-order stationarity, both a covariance and, thus, a finite *a priori* variance, Var $\{Z(x)\} = C(0)$, are assumed. But there are theoretical models of RF's and also physical phenomena that display an unlimited capacity for dispersion and for which neither an *a priori* variance nor a covariance can be defined.

Brownian motion is an example of such a phenomenon and is well known to physicists. It can be described by the Wiener–Lévy process, which has an infinite variance and, thus, no covariance. However, the increments of this process do have a finite variance:

$$\text{Var}\{Z(x+h) - Z(x)\} = 2\gamma(h) = A \cdot |h|.$$

The variogram of the Wiener–Lévy process is, thus, defined and it can be shown that it increases linearly with the distance $|h|$.

The dispersion of gold grades in the Witwatersrand provinces of South Africa provides a more practical mining example. South African authors (in particular D. G. Krige, 1951, 1976) have shown that the experimental variance $\sigma^2(V)$ of gold grades within a field of size V increases continuously with the dimensions of the field.

The essence of Krige's original results is shown in Fig. II.7. The variance of the logarithms of the gold grades increases without interruption as a function of the logarithm of the area of the field V over which the variance is calculated. A complete range of field sizes was considered, beginning with the cross-sectional area of analysed core samples ($0 \cdot 0075$ ft^2) and followed in turn by the horizontal cross-sectional area of a mining panel ($63\,000$ ft^2), the mean horizontal surface of a mine (about $3 \cdot 4 \times 10^8$ ft^2) and, finally, the total surface area of the gold bearing region of the Rand (about $3 \cdot 4 \times 10^9$ ft^2).

The experimental curve of Fig. II.7 can be approximated by a straight line representing the uninterrupted logarithmic growth of the variance with the increase in field size V:

$$\sigma^2(V) = \alpha \log (V/v),$$

where $v = 0.0075 \text{ ft}^2$ is the area of cross-section of the analysed core samples, and V is the horizontal surface area of the field. The *a priori* variance of the logarithms of the gold grades should be the limit of $\sigma^2(V)$ as the field size V tends towards infinity. Experimentally, such a limit does not exist and, as a consequence, a second-order stationary model of RF cannot be used. It then becomes necessary to use a model that is only intrinsic, having a variogram but no covariance.

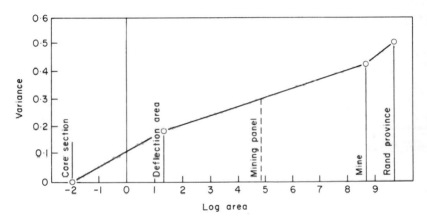

FIG. II.7. Variance versus area of dispersion.

2.   Quasi-stationarity (on a copper deposit)

A quasi-stationary type of mineralization is illustrated by a massive copper deposit explored by vertical bore-holes. The copper grade decreases regularly with the depth, which indicates the presence of a trend (non-stationarity) in the vertical direction.

The experimental variogram of the copper grades calculated using all the vertical drill cores is shown on Fig. II.8. The characteristics of this variogram are:

   (i) a nugget effect, the amplitude C_0 of the discontinuity at the origin equals about $0.4 (\% \text{ Cu})^2$;
   (ii) a transition phenomenon between the origin and a distance of about 100 ft with a sill value of 1 and a range of about 50 ft;

(iii) beyond 100 ft there is a sharp increase in the variogram values, indicating the presence of a trend due to Cu grades decreasing with depth.

Thus, over neighbourhoods of less than 100 ft in the vertical direction, the mineralization can be considered as stationary and characterized by a semi-variogram with a finite range, i.e., we accept an hypothesis of quasi-stationarity. More precisely, the theoretical model adopted is known as a spherical scheme with a nugget effect and is represented by the equation

$$\gamma(h) = C_0 + C[1 \cdot 5h/a - 0.5h^3/a^3], \qquad \forall h \in [0, 100] \text{ ft}$$

with

$$C_0 = 0 \cdot 4, \qquad C = 0 \cdot 6 (\% \text{ Cu})^2, \qquad a = 50 \text{ ft}.$$

This model is shown on Fig. II.8 as the broken-line curve.

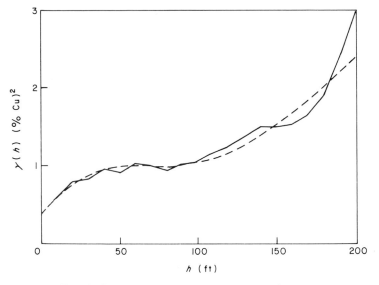

FIG. II.8. Trend effect for large vertical distances.

3. Behaviour near the origin (pluviometry at Korhogo)

The example given here has the advantage of demonstrating the four different types of behaviour of the variogram near the origin cited in section II.A.3 and Fig. II.6.

This variography is taken from a prominent unpublished study carried out by J. P. Delhomme (1976) on Korhogo (Ivory Coast) piezometric data. The data $z(x, t)$ at each position x give the piezo height (height of the top of the water table) as a function of time t. The measurements are carried out daily. The piezometers are placed at intervals of about 500 m, see Fig. II.9.

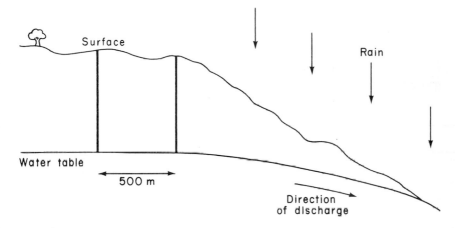

FIG. II.9. Piezometric survey at Korhogo.

At every location x of a piezometer, a profile of the piezo height is available over the rainy season (t varying from August to November). Figure II.10 shows four of these profiles among the most characteristic ones. The following observations can be made.

At piezometer no. 3   The water table is very close to the surface and the water level reacts each time it rains (the amount of water retained by the ground is proportional to the thickness of the ground through which the water passes before reaching the water table). The profile from piezo no. 3 is very erratic and is directly related to the daily rainfall. The variogram along this profile is linear with a nugget effect (Brownian motion for small lags t).

At piezometer no. 4   The water table is deeper and the absorption effect of the ground begins to take effect. The corresponding variogram has a smaller nugget effect and a transition phenomenon is becoming evident (appearance of a sill).

At piezometer no. 33   The water table is deeper still and the regularizing effect of the ground absorption is almost total. There are, however, a certain number of discontinuities evident on the profile and these are due to abrupt refilling of the table. The corresponding variogram has a small nugget effect representing the discontinuous refilling of the table followed by a parabolic shape which is characteristic of the high continuity of the ReV $z(t)$. More precisely, the RF $Z(t)$ is mean-square differentiable except during the periods of refilling.

At piezometer no. 18   The water table is even deeper and the effects of the discontinuous refilling are no longer evident. The nugget effect has completely disappeared and the variogram presents an almost perfect parabolic behaviour near the origin, characteristic of the differentiability of the profile $z(t)$ at any time t.

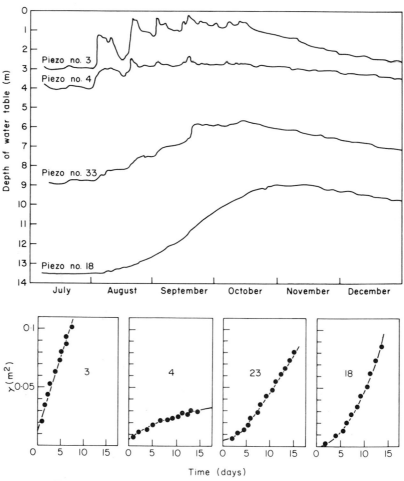

FIG. II.10. Variography of the piezo heights at Korhogo.

4. *Coregionalizations at Exotica*

The experimental direct and cross-semi-variograms for Fe, soluble Cu and total Cu, shown on Fig. II.11, were calculated from the grades defined on 13-m vertical bench heights in the Exotica copper deposit (Chile) and represent the horizontal coregionalization of Fe, Cu S, Cu T grades.

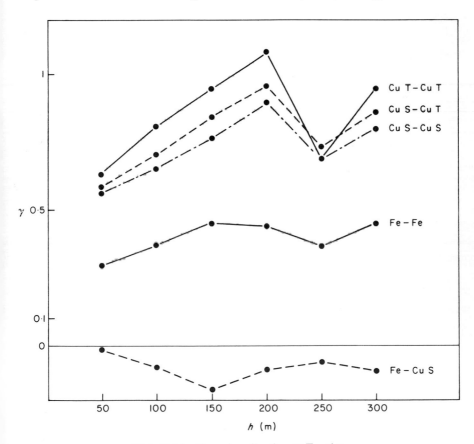

FIG. II.11. Coregionalization at Exotica.

Note that the cross-semi-variogram for Cu T–Cu S is positive while that of Fe–Cu S is negative. The negative Fe–Cu S correlation can be explained by a superficial alteration phenomenon in which water liberates the initial sulphuretted copper elements (chalcopyrite in particular), and the almost insoluble Fe remains close to the surface while the soluble copper Cu S is washed away. As a consequence, a positive increment in iron grades $[z_k(x+h)-z_k(x)]>0$ corresponds, on average, to a negative increment in soluble copper $[z_{k'}(x+h)-z_{k'}(x)]<0$, and the resulting cross-variogram is negative.

II.B. ESTIMATION VARIANCE

II.B.1. The estimation error and its distribution

Every estimation method involves an estimation error, arising from the simple fact that the quantity to be estimated generally differs from its estimator z^*, thus implying an error of estimation $z-z^*$.

When the mean grade $z(x_i)$ of a vertical bore-hole through the middle of a block V is used to estimate the true mean grade $z_V(x_i)$ of the block, the error involved is $r(x_i)=z_v(x_i)-z(x_i)$, cf. Fig. II.12. Just as the ReV $z(x)$ has been interpreted as a particular realization of the RF $Z(x)$, so the error $r(x_i)$ also appears as a particular realization of the RV $R(x_i)=Z_V(x_i)-Z(x_i)$ at the point x_i.

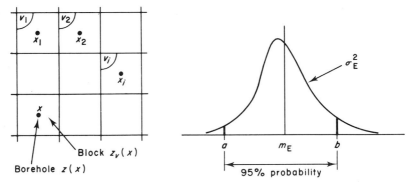

FIG. II.12. Distribution of the error.

Suppose now that the entire deposit is divided into blocks of equal size V, and that each block is intersected by a vertical bore-hole passing through its centre. If the deposit is an homogeneous mineralization, i.e., if the RF $Z(x)$ can be considered as stationary, then the error $R(x)$ is also stationary and any two errors $r(x_i)$ and $r(x_j)$ can be considered as two different realizations of the same stationary RF $R(x)=Z_V(x)-Z(x)$.

If the histogram of experimental errors is available in a control zone, then it will be possible, because of the stationary assumption, to infer the complete distribution function of $R(x)$. Even if such a histogram is not available, it will still be possible to calculate the stationary expectation $m_E=E\{R(x)\}$ and variance $\sigma_E^2=\mathrm{Var}\{R(x)\}$ of the distribution function of the error.

The particular error $r(x_i) = z_V(x_i) - z(x_i)$ involved when estimating the block $V(x_i)$ remains unknown, but the mean and variance of the errors (or the complete distribution function if it is known) will provide a measure of the quality of the estimation.

The distribution of the estimation error

As with all RV's, the error $R(x)$ is characterized by its distribution function:

$$F(u) = \text{Prob } \{R(x) < u\}.$$

If the RF $R(x)$ is stationary, then this function does not depend on the location x. Given the function $F(u)$, it is possible to calculate the probability that the error will fall within any interval $[a, b]$:

$$\text{Prob } \{a \leqslant R(x) < b\} = F(b) - F(a), \qquad \forall x.$$

In practice, the only error distributions considered will be those with finite moments, at least the first two moments. This amounts to the assumption of second-order stationarity of the RF $R(x)$ with

(i) a mathematical expectation,

$$E\{R(x)\} = m_E = \text{Constant}, \qquad \forall x;$$

(ii) a variance, called the *"estimation variance"*,

$$\text{Var } \{R(x)\} = E\{R(x)]^2\} - m_E^2 = \sigma_E^2 = \text{constant}, \qquad \forall x.$$

The expectation m_E characterizes the mean error, while the variance σ_E^2 is a dispersion index of the error, cf. Fig. II.12. Thus, a good estimation procedure must be such that it ensures

(i) a mean error close to zero, this property of the estimator is known as *unbiasedness*;
(ii) a dispersion of errors very concentrated around this mean zero value, this second criterion being expressed by a small value of the estimation variance σ_E^2.

At the estimation stage, the type of the distribution function of the errors is unknown in most cases. But, since the two most important characteristics of this function – its expectation and variance – can be calculated, we shall refer to a standard two-parameter (m_E and σ_E^2) function which will provide an order of magnitude of the required confidence interval.

Confidence intervals for the normal distribution

Among all the two-parameter distribution functions, the one most often used to characterize an error is the normal distribution. The main justification for using this distribution is that it is the one most often observed in practice, particularly in mining practice, cf. section II.B.3, Case Studies 1 and 2.

In mining applications, the error distribution functions are generally symmetric with a slightly more pronounced mode and larger tails than a normal distribution with the same expectation and variance, cf. Fig. II.13. Thus, in relation to the standard normal distribution, there are more small errors (in the region of $m_E = 0$) and more large errors (in the tails of the distribution). However, the classic confidence interval $[m_E \pm 2\sigma_E]$ contains approximately 95% of the observed errors. The 95% normal confidence interval is then a good criterion for judging the quality of a geostatistical estimation, but it should always, as far as possible, be verified experimentally through test or control zones.

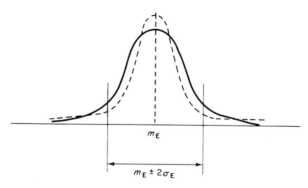

FIG. II.13. Normal standard for the error distribution. ———, Gaussian standard; ––––, experimental distribution of errors.

Remark Under the non-bias condition $(m_E = 0)$ the standard 95% confidence interval can be expressed as $[z^* \pm 2\sigma_E]$ for the true value z estimated by z^*. It sometimes happens that the value of the estimation standard deviation σ_E is such that the lower bound of the 95% confidence interval $[z^* \pm 2\sigma_E]$ is negative, e.g., with $z^* = 0.5\%$ Cu and $\sigma_E = 1\%$, so that $z^* - 2\sigma_E = -1.5\%$. A practical solution to this problem is to define the relative standard deviation $\sigma_E/\bar{z}^*$, i.e., the ratio of the standard deviation to the experimental mean

$$\bar{z}^* = \frac{1}{n} \sum_{i=1}^{n} z_i^*$$

of the n available estimators. The 95% confidence interval of a particular true value z_i estimated by z_i^* will then be

$$z_i \in \left[z_i^* \pm 2\left(\frac{\sigma_E}{\bar{z}^*}\right) z_i^* \right]. \tag{II.21}$$

This formula will not ensure that the lower bound is always positive, since it becomes negative when the relative standard deviation is greater than 50%, $\sigma_E/\bar{z}^* > 0.5$, but it is sufficient for practical applications. The case where $\sigma_E/\bar{z}^* > 0.5$ corresponds to poor estimations and, in such cases, it would be better not to use confidence limits at all.

Formula (II.21) corresponds to a non-stationary standard deviation of estimation, since it depends on each estimator z_i^*. It can be shown (cf. G. Matheron, 1974, and section III.B.6, "Remark 3. Proportional effect in m^{*2}") that this formula corresponds to a conditioning of the RF $Z(x)$ to the estimator z_i^*, the distribution of $Z(x)$ being lognormal.

II.B.2. Linear estimators

Consider the problem of estimating the unknown value Z_V (e.g., the mean grade $Z_V(x)$ of a block of size V centred on point x) from a set of n data values $\{Z(x_i), i = 1 \text{ to } n\}$.

The estimator Z^* will be a function of the available data:

$$Z^* = f\{Z(x_1), \ldots, Z(x_n)\},$$

but not any function of this data. It must be a function such that

(i) it satisfies the non-bias condition, $E\{Z_V - Z^*\} = 0$;
(ii) it is reasonably simple, so as to permit the calculation of the estimation variance,

$$\sigma_E^2 = E\{[Z_V - Z^*]^2\} = E\{Z_V^2\} + E\{Z^{*2}\} - 2E\{Z_V Z^*\}.$$

For any function f, the calculation of the first-order moment $E\{Z_V - Z^*\}$ and the various second-order moments of the expression of σ_E^2 requires that the n-variables distribution $\{Z(x_1), \ldots, Z(x_n)\}$ be known. However, since it is generally not possible to infer this distribution from a unique realization of the RF $Z(x)$, we are limited to the *class of linear estimators* (at least in linear geostatistics). It will then always be possible to calculate the mean and variance of the error from the variogram or the covariance.

Research into non-linear geostatistics is being carried out and has shown that restrictions imposed by linear estimators can be removed, cf. disjunctive kriging and transfer functions, G. Matheron, (1975a, b, c) and section VIII.B. Such non-linear geostatistics requires a much more complete knowledge of the spatial structure of the RF $Z(x)$ than that provided by the variogram.

With the exception of Chapter VIII, non-linear methods are outside the scope of this book and we shall limit all further discussions to estimators Z^* which are linear combinations of the n available data:

$$Z^* = \sum_{i=1}^{n} \lambda_i Z(x_i). \tag{II.22}$$

Estimation variance

Let $Z(x)$ be a second-order stationary RF with expectation m, covariance $C(h)$ and semi-variogram $\gamma(h)$, cf. definition formulae (II.1) to (II.4).

Consider first the simple case of the estimation of the arithmetic mean z_K of the K unknown values $\{z(x_k), k = 1 \text{ to } K\}$:

$$z_K = \frac{1}{K} \sum_{k=1}^{K} z(x_k).$$

The linear estimator z_K^* is the arithmetic mean of the n available data values $\{z(x_i), i = 1 \text{ to } n\}$:

$$z_K^* = \frac{1}{n} \sum_{i=1}^{n} z(x_i).$$

The value z_K^* is interpreted as a particular realization of the RV Z_K^* and the unknown error is, thus, a particular realization of the RV $Z_K - Z_K^*$.

Under the second-order stationary hypothesis, the non-bias is automatically verified, since

$$\left. \begin{array}{l} E\{Z_K\} = \dfrac{1}{K} \sum_{k} E\{Z(x_k)\} = m \\[2mm] E\{Z_K^*\} = \dfrac{1}{n} \sum_{i} E\{Z(x_i)\} = m \end{array} \right\} \Rightarrow E\{Z_K - Z_K^*\} = 0.$$

The estimation variance is written

$$\sigma_E^2 = E\{[Z_K - Z_K^*]^2\} = E\{Z_K^2\} + E\{Z_K^{*2}\} - 2E\{Z_K Z_K^*\},$$

with

$$E\{Z_K^2\} = \frac{1}{K^2} E\{\sum_{k} \sum_{k'} Z(x_k) Z(x_{k'})\}.$$

We can invert the summations $\sum$ and the expectation operator E:

$$E\{Z_K^2\} = \frac{1}{K^2} \sum_{k} \sum_{k'} E\{Z(x_k) Z(x_{k'})\} = \frac{1}{K^2} \sum_{k} \sum_{k'} [C(x_k - x_{k'}) + m^2].$$

Similarly,

$$E\{Z_K^{*2}\} = \frac{1}{n^2} \sum_i \sum_j [C(x_i - x_j) + m^2],$$

$$E\{Z_K Z_K^*\} = \frac{1}{Kn} \sum_k \sum_i [C(x_k - x_i) + m^2].$$

the constant term m^2 is then eliminated from the expression of σ_E^2, leaving

$$\sigma_E^2 = E\{[Z_K - Z_K^*]^2\}$$

$$= \frac{1}{K^2} \sum_k \sum_{k'} C(x_k - x_{k'}) + \frac{1}{n^2} \sum_i \sum_j C(x_i - x_j) - \frac{2}{Kn} \sum_k \sum_i C(x_k - x_i). \quad \text{(II.23)}$$

We will denote by $\bar{C}((K), (n))$ the mean value of the covariance $C(h)$ when one extremity of the vector h describes the set $\{x_k, k = 1 \text{ to } K\}$ and the other extremity independently describes the set $\{x_i, i = 1 \text{ to } n\}$:

$$\bar{C}((K), (n)) = \frac{1}{Kn} \sum_{k \in (K)} \sum_{i \in (n)} C(x_k - x_i).$$

The previous estimation variance can then be expressed as

$$\sigma_E^2 = \bar{C}((K), (K)) + \bar{C}((n), (n)) - 2\bar{C}((K), (n)). \quad \text{(II.24)}$$

Generalization to the continuous case

Suppose now that the K points x_k are located within a volume V centred on point x and the n points x_i within another volume v centred on point x'. As K and n tend towards infinity, the preceding arithmetic means z_K and z_K^* tend towards the mean values in V and v of the point variable $z(y)$, i.e.,

$$z_K \to z_V(x) = \frac{1}{V} \int_{V(x)} z(y) \, dy \quad \text{and} \quad z_K^* \to z_v(x') = \frac{1}{v} \int_{v(x')} z(y) \, dy.$$

These mean values $z_V(x)$ and $z_v(x')$ are interpreted as particular realizations of two RV's $Z_V(x)$ and $Z_v(x')$:

$$Z_V(x) = \frac{1}{V} \int_{V(x)} Z(y) \, dy \quad \text{and} \quad Z_v(x') = \frac{1}{v} \int_{v(x')} Z(y) \, dy.$$

(Recall here that the preceding unidimensional integrals in fact represent double or triple integrals according to whether V and v are two-dimensional or three-dimensional, cf. notations in Index D.)

Under the hypothesis of second-order stationarity, the expectations of $Z_V(x)$ and $Z_v(x')$ are equal to m, and the non-bias condition is satisfied.

For the estimation variance, the formulae (II.23) and (II.24) for the discrete case give, for the continuous case of estimating the mean grade of V by that of v,

$$\sigma_E^2 = E\{[Z_V(x) - Z_v(x')]^2$$

$$= \frac{1}{V^2} \int_{V(x)} dy \int_{V(x)} C(y - y') \, dy' + \frac{1}{v^2} \int_{v(x')} dy \int_{v(x')} C(y - y') \, dy'$$

$$- \frac{2}{Vv} \int_{V(x)} dy \int_{v(x')} C(y - y') \, dy. \tag{II.25}$$

If the domains $V(x)$ and $v(x')$ are three-dimensional, each of the integral signs of formula (II.25) represents a triple integral and the variance σ_E^2 is a sum of sextuple integrals. To simplify notations, we shall denote by $\bar{C}(V, v)$ the mean value of the covariance $C(h)$ when one extremity of the distance vector h describes the domain $V(x)$ and the other extremity independently describes the domain $v(x')$. It is then written

$$\sigma_E^2 = \bar{C}(V, V) + \bar{C}(v, v) - 2\bar{C}(V, v). \tag{II.26}$$

When the covariance $C(h)$ exists, the semi-variogram $\gamma(h)$ also exists and these two structural functions are related by formula (II.4):

$$C(h) = C(0) - \gamma(h).$$

Formulae (II.23) to (II.26) can then be rewritten in terms of semi-variograms. Since the constant term $C(0)$ disappears, it is enough to replace C or $\bar{C}$ by $-\gamma$ or $-\bar{\gamma}$ and the estimation variance becomes

$$\boxed{\sigma_E^2 = 2\bar{\gamma}(V, v) - \bar{\gamma}(V, V) - \bar{\gamma}(v, v),} \tag{II.27}$$

where $\bar{\gamma}(V, v)$, for instance, represents the mean value of $\gamma(h)$ when one extremity of the vector h describes the domain $V(x)$ and the other extremity independently describes the domain $v(x')$.

It can be shown that this formula (II.27) remains valid even when the covariance $C(h)$ does not exist, provided of course that the semi-variogram is defined (the RF $Z(x)$ is then only intrinsic).

Remark 1   The estimation variance of V by v is sometimes referred to as the variance of extending the grade of v to V or simply the *extension variance* of v to V and is then denoted by $\sigma_E^2(v, V)$.

Remark 2   The two formulae (II.26) and (II.27) are completely general whatever the domains v and V. In particular, the domains need not necessarily be connex (single continuous regions); the domain V to be estimated may be, for example, two distinct blocks $V = V_1 + V_2$ and the information domain v may consist of several samples $v = \sum_i S_i$, some of which may be situated within the domain V.

Remark 3 The variogram $2\gamma(h)$ can itself be interpreted as the elementary estimation variance of a variable $Z(x)$ by another variable $Z(x+h)$ at a distance h from $Z(x)$:

$$E\{[Z(x+h)-Z(x)]^2\} = 2\gamma(h),$$

cf. definition formula (II.6).

Remark 4 Formulae (II.26) and (II.27) express four essential and intuitive facts which condition all estimations. The quality of the estimation of V by v is a function of the following.

(i) The geometry of the domain V to be estimated: term $-\bar{\gamma}(V, V)$ appearing in formula (II.27). As the semi-variogram $\gamma(h)$ is an increasing function of h, $\bar{\gamma}(V, V)$ increases with the size of V. Hence, if v and the interdistance (v, V) remain fixed, it is easier to estimate the mean grade of a large block V than the grade of an unknown point $(V = x)$. Finally, if the size V is fixed, the term $\bar{\gamma}(V, V)$, and, consequently, the estimation variance σ_E^2, depend also on the geometry of V.

(ii) The distances between the domain V to be estimated and the support v of the estimator: term $+\bar{\gamma}(V, v)$ appearing in formula (II.27). As the distance between V and v increases, so does the value of $\bar{\gamma}(V, v)$ and the value of the estimation variance σ_E^2.

(iii) The geometry of the estimating domain v: term $-\bar{\gamma}(v, v)$. As the dimension of the domain v increases, the value of $\bar{\gamma}(v, v)$ also increases and, consequently, σ_E^2 will decrease. But, for fixed volumes V, v and fixed interdistance (v, V), the estimation variance will also depend on the configuration of the information v. The estimation of the same block V on Fig. II.14 by the two distinct samples v_1 and v_2 will be better than that provided by the two closely spaced samples v_1' and v_2'; indeed, $\bar{\gamma}(v_1' + v_2', v_1' + v_2')$ will be less than $\bar{\gamma}(v_1 + v_2, v_1 + v_2)$. This intuitive notion of the importance

FIG II.14. Influence of the information configuration.

of the sample configuration, which is formalized in geostatistics through the term $\bar{\gamma}(v, v)$, is ignored by the more usual estimation methods such as weighting by inverse distances or their squares.

(iv) The structural function, semi-variogram or covariance. The quality of the estimation should depend on the structural characteristics (anisotropies, degree of regularity) of the regionalized phenomenon under study. The semi-variogram γ may depend on the direction α of the vector h, i.e., the semi-variogram will be a function of distance and direction, $\gamma(|h|, \alpha)$. In horizontally stratified mineralizations, for example, the variation in grade values will be much more continuous in an horizontal direction than they will be vertically, and, by taking the anisotropic semi-variogram into account, more weight would be given to sample v_1 in estimating block V in the same stratum than to sample v_2 in a different stratum, cf. Fig. II.15.

This fundamental geological notion of the spatial continuity of a mineralization as a function of direction is quite often overlooked or simply ignored in the application of many estimation methods such as that of weighting by inverse distances.

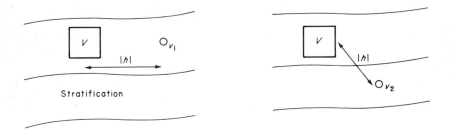

FIG. II.15. Influence of the mineralization structure.

The estimation of a mean value by a weighted average

The general formulae (II.26) and (II.27) given above also cover the particular case of estimating the mean value Z_V of a block V by a linear combination Z^* of n available data values $Z(x_i)$ taken at the points $\{x_i, i = 1 \text{ to } n\}$:

$$Z_V = \frac{1}{V} \int_{V(x)} Z(y) \, dy \quad \text{and} \quad Z_V^* = \sum_{i=1}^{n} \lambda_i Z(x_i).$$

The non-bias condition entails that $E\{Z_V - Z_V^*\} = 0$, i.e.,

$$E\{Z_V^*\} = \sum_i \lambda_i E\{Z(x_i)\} = m \sum_i \lambda_i = E\{Z_V\} = m.$$

So the non-bias is ensured if the sum of the n weights equals 1:

$$\sum_{i=1}^{n} \lambda_i = 1.$$

The estimation variance can then be calculated from formula (II.27):

$$\sigma_E^2 = E\{[Z_V - Z_V^*]^2\}$$

$$= 2\sum_i \lambda_i \bar{\gamma}(x_i, V) - \bar{\gamma}(V, V) - \sum_i \sum_j \lambda_i \lambda_j \gamma(x_i - x_j), \qquad \text{(II.28)}$$

where $\bar{\gamma}(x_i, V)$ denotes the mean value of $\gamma(h)$ when one extremity of the distance vector h is fixed at point x_i and the other extremity independently describes volume V.

Remark 5 Formula (II.28) expresses the estimation variance σ_E^2 as a linear function of the n weights λ_i. The estimation procedure known as *kriging* (cf. Chapter V) determines the optimal set of weights $\{\lambda_i, i = 1$ to $n\}$, i.e., the weights λ_i which minimize the variance σ_E^2 subject to the non-bias condition $\sum_i \lambda_i = 1$. Hence, kriging appears as the best linear unbiased estimator (in short a "BLUE" estimator).

Remark 6 Formula (II.28) is completely general, whatever the geometry of the block V and the choice of the weights λ_i, subject of course to the condition $\sum_i \lambda_i = 1$. Thus, the formula can be used to calculate the estimation variance of any linear weighted average type of estimator, such as inverse distance weighting or weighting by polygons of influence. This is sufficient justification for the calculation of the semi-variogram (i.e., a geostatistical approach), even if for reasons of cost or speed we are limited to one of the usual types of estimator such as polygons of influence.

II.B.3. Case studies

1. *Block kriging at Chuquicamata after I. Ugarte (1972)*

Chuquicamata is an open-pit porphyry-copper deposit in Chile where short-term planning is based on data from blast-holes. The mean copper grade of each block $V(20\,\text{m} \times 20\,\text{m} \times 13\,\text{m})$ is estimated by kriging from the blast-hole data available from the nine neighbouring blocks which have already been mined out, as shown on Fig. II.16, i.e., each block V is assigned a weighted average of the blast-hole values, the weights being chosen to ensure the minimum estimation variance. The kriged estimator of block V will be written Z_V^*. Next, the block V is put into production and blast-hole samples are taken from it, an average of six being taken from each block. Experimental observation has shown that the arithmetic mean of these six holes, Z_V, can be taken as the true grade of block V.

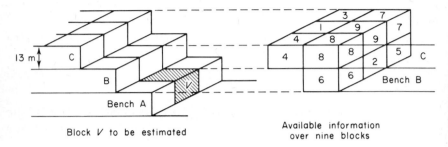

Block V to be estimated

Available information over nine blocks

FIG. II.16. Short-term kriging at Chuquicamata.

A test zone of homogeneous mineralization which comprised 397 blocks gave the histogram of the observed errors $(Z_V - Z_V^*)$, shown on Fig. II.17.

The arithmetic mean of the 397 errors was $+0\cdot01\%$ Cu and, given that the mean copper grade over the test zone was 2% Cu, it is obvious that the non-bias condition has been met.

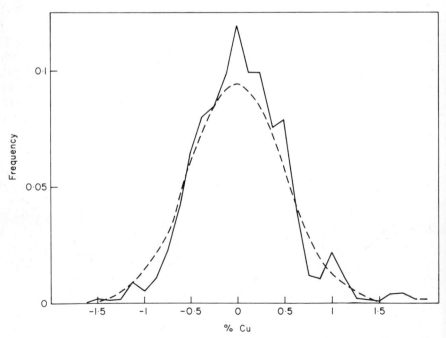

FIG. II.17. Histogram of kriging errors. ——, Histogram, $m = 0\cdot01$, $\sigma_E^2 = 0\cdot273$; – – –, Gaussian distribution, $m = 0$, $\sigma_E^2 = 0\cdot273$.

The experimental variance of the 397 errors was $0·273$ (% Cu)^2 while the predicted theoretical estimation variance calculated from formula (II.28) was $0·261$ (% Cu)^2. Therefore, the estimation variance was predicted with the very acceptable relative precision of 4%: $(0·273 - 0·261)/0·273 = 0·04$.

A normal distribution with zero mean and a variance equal to that of the experimentally observed errors ($\sigma_E^2 = 0·273$) is also shown on Fig. II.17. It fits reasonably well the experimental distribution of errors with the following remarks.

(i) The normal distribution underestimates the proportion of both small errors (between $\pm 0·5$% Cu) and large errors (greater than $1·5$% Cu).

(ii) The standard normal interval ($\pm 2\sigma_E$) correctly estimates the experimental 95% confidence interval: 96% of all the observed errors fall within $\pm 2\sigma_E = \pm 1·04$% Cu. If the predicted theoretical variance, $\sigma_E^2 = 0·261$ had been used, the result would have made little difference: $\pm 2\sigma_E = \pm 1·02$% Cu.

2. *Cartography at Mea after J. Deraisme (1976, unpublished)*

The nickel silicate deposit of Mea (New Caledonia) was sampled by vertical bore-holes on a pseudo-regular square grid of 80 m, cf. Fig. II.18. One of the problems in the exploration of this deposit was to map the hanging wall and foot wall of the mineralized bed. Let $Z(x)$ be the height of the hanging wall at the horizontal coordinate point x. At each point x_i where a drill intersected the bed, the height $z(x_i)$ of the hanging wall was measured. In order to map the hanging wall, the unknown value $z(x_0)$ of the height at

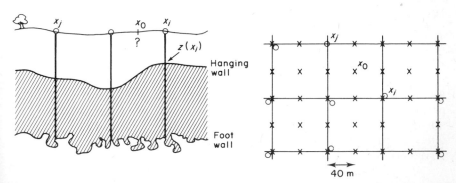

FIG. II.18. Cartography of the hanging wall at Mea. $\bigcirc$, x_i bore-holes; $\times$, x_0 node of the estimated grid.

each node of a regular square grid of 40-m side was estimated. The estimator used was the BLUE $Z^*(x_0)$ provided by kriging.

Isobath contours of the hanging wall were plotted, using the estimated 40-m grid pattern. Similarly, the foot-wall depth and the mineralized thickness (hanging-wall height minus foot-wall depth) were plotted.

The kriging estimator $z^*(x_0)$ used was a linear combination of the nine bore-hole data closest to x_0:

$$z^*(x_0) = \sum_{j=1}^{9} \lambda_j z(x_j).$$

Kriging provides the system of weights which minimizes the estimation variance:

$$\sigma_E^2(x_0) = E\{[Z(x_0) - Z^*(x_0)]^2\}.$$

Since the configuration of the available data was not strictly regular, the estimation variance depends on the position of each point x_0 to be estimated.

In order to verify the procedure, each one of the known values $z(x_i)$ was estimated by kriging, i.e., by the optimal linear combination of the nine closest data values $\{x_j, j = 1$ to 9, and $j \neq i\}$. Using the theoretical estimation variance $\sigma_E^2(x_i)$ and the known experimental error $z(x_i) - z^*(x_i)$, the deviation ratio $e(x_i) = [z(x_i) - z^*(x_i)]/\sigma_E(x_i)$ was calculated for each data point x_i. For the theoretical estimation variance $\sigma_E^2(x_i)$ to be an effective estimator of the true estimation variance, the variance of the n values of $e(x_i)$ should be approximately equal to 1.

Figure II.19 shows the histogram of 167 $e(x_i)$ values along with a normal distribution with the same variance and a zero mean. The arithmetic mean of the 167 experimental errors $[z(x_i) - z^*(x_i)]$ is 0.15 m and, given that the mean hanging-wall height is 840 m and the mean bed thickness is 20 m, the non-bias can be accepted as having been verified.

The variance of the deviation ratio is 0.863, which is reasonably close to 1, and, thus, the theoretical estimation variance $\sigma_E^2(x)$ provided by formula (II.15) is an effective predictor of the experimental variance of the errors.

As in the preceding Chuquicamata case, the normal distribution underestimates both the proportion of small errors (between ± 0.5 m) and the proportion of large errors (greater than ± 2 m). Again, the usual normal interval of $\pm \sigma_E^2$ correctly estimates the 95% experimental confidence interval: 92% of the deviation ratios fall within this interval ($\pm 2\sigma_E = \pm 1.86$ m).

Remark The verification of the estimation method by constructing the histogram of the experimental deviation ratios, $e(x) = [z(x) - z^*(x)]/\sigma_E(x)$

is sufficiently simple to be carried out systematically, provided that the support of the elements to be estimated is the same as that of the data used in the estimation. This is often the case in cartography.

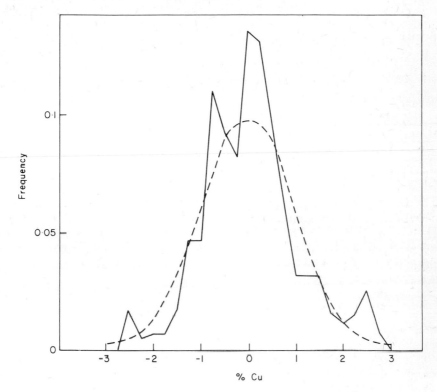

FIG. II.19. Histogram of the experimental deviation ratio. ———, Histogram, $m = 0.01$, $\sigma_E^2 = 0.863$; Gaussian distribution, $m = 0$, $\sigma_E^2 = 0.863$.

II.C. DISPERSION VARIANCE

II.C.1. Definition of the dispersion variance

In selection problems, it is of little use to know the mean grade of a working stope if we don't have some measure of the dispersion (or variability) of the grades of production size units within the stope.

Let V be a production stope centred on point x and divided into N equally sized production units $v(x_i)$ centred on points x_i:

$$V = \sum_{i=1}^{N} v_i = Nv.$$

Let $z(y)$ be the ReV grade at point y. The mean grade of each unit $v(x_i)$ centred on point x_i is

$$z_v(x_i) = \frac{1}{v} \int_{v(x_i)} z(y)\, dy.$$

Similarly, the mean grade of stope V centred on x is

$$z_V(x) = \frac{1}{V} \int_{V(x)} z(y)\, dy = \frac{1}{N} \sum_{i=1}^{N} z_v(x_i).$$

To each of the N positions x_i of the units $v(x_i)$ inside stope V corresponds a deviation $[z_V(x) - z_v(x_i)]$. The dispersion of the N grades $z_v(x_i)$ about their mean value $z_V(x)$ can be characterized by the mean square deviation:

$$s^2(x) = \frac{1}{N} \sum_{i=1}^{N} [z_V(x) - z_v(x_i)]^2. \tag{II.29}$$

The dispersion can also be represented by the histogram of the N values $z_v(x_i)$, i.e., by the frequency of occurence of each value z as shown on Fig. II.20. In practice, however, at the estimation stage the true grades $z_v(x_i)$ of the production units v are not known and neither is the true grade $z_V(x)$ of the stope V; thus, the histogram of Fig. II.20 is not available.

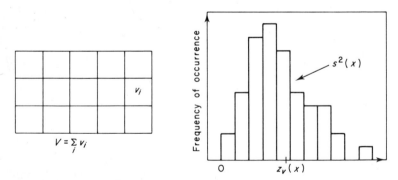

FIG. II.20. Dispersion of units v grades within stope V.

The problem is then to estimate, if not the entire histogram, at least its two main characteristics, mean $z_V(x)$ and variance $s^2(x)$.

Probabilistic interpretation

When the point ReV $z(y)$ is interpreted as a particular realization of a RF $Z(y)$, the unit grades $z_v(x_i)$, as well as the stope grade $z_V(x)$, appear as realizations of RV's $Z_v(x_i)$ and $Z_V(x)$; the preceding mean square deviation, $s^2(x)$, also appears as a particular realization of a RV $S^2(x)$ defined at point x, i.e., corresponding to the stope V centred on point x:

$$S^2(x) = \frac{1}{N} \sum_i [Z_V(x) - Z_v(x_i)]^2. \tag{II.30}$$

Variance of dispersion

Under the hypothesis of stationarity of the point RF $Z(y)$, the stationary expectation of this RV $S^2(x)$ is, by definition, the dispersion variance of the units v within V. The following notation is used:

$$D^2(v/V) = E\{S^2(x)\} = E\left\{\frac{1}{N} \sum_i [Z_V(x) - Z_v(x_i)]^2\right\}. \tag{II.31}$$

Remark   It must be stressed that the RV $S^2(x)$ and its particular realization $s^2(x)$ depend on the particular position x of stope $V(x)$, while, under the hypothesis of stationarity, $D^2(v/V)$ no longer depends on the position x but only on the geometries of v and V and the covariance $C(h)$, as will be demonstrated.

For each one of K stopes, identical in shape and size, $\{V(x_k), k = 1$ to $K\}$, there is an experimental mean square deviation $s^2(x_k)$ calculated from (II.29). As $K \to \infty$, the arithmetic mean of these K experimental deviations tends towards the stationary expectation of $S^2(x)$, i.e., the variance of dispersion $D^2(v/V)$.

If we are only interested in one particular stope $V(x)$, then $D^2(v/V)$ will be an estimator of the experimental mean square deviation $s^2(x)$, just as the expectation of a RV is an estimator of a particular realization of this RV.

Example   Suppose that a stope $V(x_k)$ is divided into four equal units $\{v(x_i), i = 1$ to $4\}$ and that grades $Z_v(x_i)$ are assigned to each unit according to the result of casting an unbiased six-sided die.

The distribution resulting from casting the die is uniform with a mean equal to $(1+2+3+4+5+6)/6 = 3\cdot5$ and a variance $\sigma^2 = 2\cdot917$.

Consider three stopes $\{V(x_k), k = 1$ to $3\}$ having the same mean grade $z_V = 3\cdot5$ and consisting of the following three series of four units:

$$k = 1, \qquad z_v = 6, 3, 2, 3, \qquad z_V = 3\cdot5, \qquad s^2 = 2\cdot25;$$

$$k = 2, \qquad z_v = 6, 5, 1, 2, \qquad z_V = 3\cdot5, \qquad s^2 = 4\cdot25;$$

$$k = 3, \qquad z_v = 6, 2, 2, 4, \qquad z_V = 3\cdot5, \qquad s^2 = 2\cdot75.$$

The variance of dispersion $D^2(v/V)$ of these units v within V is given by the definition formula (II.31):

$$D^2(v/V) = E\left\{\frac{1}{4} \sum_{i=1}^{4} [Z_V(x) - Z_v(x_i)]^2\right\},$$

which is equivalent to

$$D^2(v/V) = \frac{1}{4} \sum_{i=1}^{4} E\{[Z_V(x) - Z_v(x_i)]^2\} = \frac{1}{4} \sum_i \sigma^2 = \sigma^2 = 2 \cdot 917.$$

Thus, the three experimental deviations, $s^2 = 2 \cdot 25$, $4 \cdot 25$, $2 \cdot 75$ fluctuate around their theoretical expectation $D^2(v/V) = E\{S^2(x)\} = 2 \cdot 917$.

Generalization of the definition of the dispersion variance

The definition (II.31) corresponds to a domain V divided into an exact number of units v.

When v is very small in relation to the field of dispersion V, $v \ll V$, every unit v centred on a point y inside V can be considered as being wholly within V. This means that the border effect (units v overlapping the border of V) can be ignored, cf. Fig. II.21. The mean square deviation $s^2(x)$ is then an integral over the field of dispersion V:

$$s^2(x) = \frac{1}{V} \int_{V(x)} [z_V(x) - z_v(y)]^2 \, dy.$$

Under the stationarity hypothesis, the dispersion variance of units v within V is defined as the stationary expectation of $S^2(x)$:

$$D^2(v/V) = E\left\{\frac{1}{V} \int_{V(x)} [Z_V(x) - Z_v(y)]^2 \, dy\right\}, \qquad v \ll V. \qquad \text{(II.32)}$$

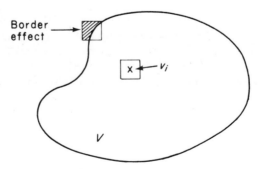

FIG. II.21. Border effect.

It can be shown that the order of the integral sign and the expectation symbol can be reversed to give

$$D^2(v/V) = \frac{1}{V}\int_{V(x)} E\{[Z_V(x) - Z_v(y)]^2\}\, dy$$

$$= \frac{1}{V}\int_{V(x)} \sigma_E^2(V(x),\, v(y))\, dy, \qquad v \ll V.$$

(II.33)

The variance of dispersion $D^2(v/V)$ thus appears as the mean value over V of the estimation variance of grade $Z_V(x)$ by grade $Z_v(y)$ of a unit v located inside V.

(In practical applications, care must always be taken to distinguish between the dispersion variance, denoted by D^2, and the estimation variance, denoted by σ_E^2.)

Remark   For $v = 0$, the unit v becomes a point and is said to have a point support. In this case, the restriction $v \ll V$ is verified and formula (II.33) gives the variance of dispersion of a point variable within V, which will be denoted by

$$D^2(0/V) = \frac{1}{V}\int_{V(x)} E\{[Z_V(x) - Z(y)]^2\}\, dy$$

(II.34)

with

$$Z_V(x) = \frac{1}{V}\int_{V(x)} Z(y)\, dy.$$

Experimental dispersion variance

Consider again the particular stope of Fig. II.20, V consisting of N units v. Now, a limited number n of the N unit grades $(n \leqslant N)$ may be known and the mean square deviation of these n values can be calculated according to formula (II.29):

$$s_n^2 = \frac{1}{n}\sum_{i=1}^{n} [\bar{z} - z_v(x_i)]^2$$

with

$$\bar{z} = \frac{1}{n}\sum_i z_v(x_i).$$

If n is not too small in relation to N and if the n known values $z_v(x_i)$ are distributed approximately *uniformly* over the field V, then s_n^2 can be considered as an estimator of the dispersion characteristic $s^2(x)$ given by formula (II.29) and we say that s_n^2 is the experimental dispersion variance of v within V.

II.C.2. Calculation of the dispersion variance

Let $Z(x)$ be a stationary point RF with expectation m, covariance $C(h)$ and semi-variogram $\gamma(h)$.

Expression (II.33) gives the dispersion variance $D^2(v/V)$ in terms of the mean value of the estimation variance over the field V:

$$D^2(v/V) = \frac{1}{V}\int_{V(x)} \sigma_E^2(V(x), v(y))\,dy, \qquad v \ll V.$$

The general formula (II.26) expresses the estimation variance as

$$\sigma_E^2(V(x), v(y)) = \bar{C}(V(x), V(x)) + \bar{C}(v(y), v(y)) - 2\bar{C}(V(x), v(y)).$$

As the covariance $C(h)$ is stationary, the first two terms of this last expression do not depend on the positions x and y of the volumes V and v but only on their respective geometries; thus, we can write

$$\bar{C}(V(x), V(x)) = \bar{C}(V, V), \qquad \forall x,$$

and

$$\bar{C}(v(y), v(y)) = \bar{C}(v, v), \qquad \forall y,$$

and, as a consequence, these two terms remain invariant when being taken on average over the field $V(x)$:

$$\frac{1}{V}\int_{V(x)} [\bar{C}(V(x), V(x)) + \bar{C}(v(y), v(y))]\,dy = \bar{C}(V, V) + \bar{C}(v, v).$$

The third term becomes

$$\frac{1}{V}\int_{V(x)} \bar{C}(V(x), v(y))\,dy = \bar{C}(V(x), V(x)) = \bar{C}(V, V), \qquad \text{if } v \ll V.$$

Because of the condition $v \ll V$, the border effect can be neglected and the displacement of $v(y)$ within V amounts to displacing a point y within V. When the field $V(x)$ consists of an exact number of units $v(y)$, the condition $v \ll V$ is no longer required, as the displacement of $v(y)$ within V amounts exactly to displacing a point within V.

The general formula giving the dispersion variance $D^2(v/V)$ as a function of the covariance is then

$$D^2(v/V) = \bar{C}(v, v) - \bar{C}(V, V). \qquad (II.35)$$

The dispersion variance $D^2(v/V)$ appears as the difference of the two mean values $\bar{C}$ over the two volumes v and V.

This formula can be rewritten in terms of semi-variogram by replacing C by $-\gamma$:

$$\boxed{D^2(v/V) = \bar{\gamma}(V, V) - \bar{\gamma}(v, v).}$$ (II.36)

It can be shown that this last expression remains valid even if the covariance $C(h)$ does not exist, provided that the semi-variogram $\gamma(h)$ is defined (the RF $Z(x)$ is then only intrinsic).

Remark 1   Krige's relation. The additivity property of dispersion variances can be established as a consequence of the linearity of expressions (II.35) and (II.36). This relation was found experimentally by D. G. Krige using data from the gold deposits of Witwatersrand and is thus called "Krige's relation":

$$D^2(v/G) = D^2(v/V) + D^2(V/G), \qquad \text{if } v \subset V \subset G.$$ (II.37)

The dispersion of the unit v within the deposit G is equal to the sum of the dispersion of v in the stope V and the dispersion of these stopes V within the deposit G.

Note that $D^2(v/G) \geq D^2(V/G)$ if $v < V$.

The dispersion variance decreases as the support increases; it is well known that there is much more variability among the grades of core samples (v of several kilograms) than there is among the mean grades of blocks (V of several thousand tons).

Remark 2   A priori variance. Consider a very large field of dispersion ($V \to \infty$). Under the hypotheses of stationarity and ergodicity, the mean value Z_V over this field tends towards the stationary expectation of the point RF $Z(x)$:

$$Z_V(x) \to E\{Z(x)\} = m \quad \text{when } V \to \infty.$$

Thus, the dispersion variance $D^2(0/\infty)$ of the point variable in a very large field V is given, according to formula (II.34), by

$$D^2(0/\infty) = \lim_{V \to \infty} \frac{1}{V} \int_V E\{[Z(y) - m]^2\} \, dy.$$

If it exists, the *a priori* variance of the point RF $Z(y)$ is precisely equal to the term under the integral sign, i.e.,

$$\text{Var}\{Z(y)\} = E\{[Z(y) - m]^2\} = C(0)$$

(cf. relation (II.3)), and, thus, the preceding limit exists and is written

$$D^2(0/\infty) = \lim_{V \to \infty} D^2(0/V) = \text{Var}\{Z(y)\} = C(0). \qquad (II.38)$$

When the RF $Z(y)$ has a stationary and finite *a priori* variance, this variance is the limit of the dispersion variance of the point variable $Z(y)$ in a very large field.

Under the hypothesis that this *a priori* variance exists and, thus, the covariance $C(h)$ exists, expressions (II.35) and (II.36) give

$$D^2(0/\infty) = C(0) - \bar{C}(\infty, \infty) = C(0),$$

$$D^2(0/\infty) = \bar{\gamma}(\infty, \infty) - \gamma(0) = \bar{\gamma}(\infty, \infty) = C(0),$$

since $\gamma(0) = 0$. Thus, for the finite *a priori* variance RF, we have

$$\bar{C}(\infty, \infty) = C(\infty) = 0 \qquad (II.39)$$

and

$$\bar{\gamma}(\infty, \infty) = \gamma(\infty) = C(0) = \text{Var}\{Z(y)\}.$$

However, there are RF's which have an unlimited capacity of dispersion, at least over an entire deposit or a metallogenic province. In such cases, neither the *a priori* variance nor the covariance are defined and the dispersion variance $D^2(0/V)$ increases indefinitely with V, cf. Fig. II.7.

II.C.3. Case studies

1. *Grade dispersions at Chuquicamata after I. Ugarte (1972)*

This example is taken from the study of the Chuquicamata (Chile) copper deposit already cited in section II.B.3. This porphyry-copper deposit is mined by open-pit and the mining unit is a block V of dimensions $20\,\text{m} \times 20\,\text{m} \times 13\,\text{m}$, cf. Fig. II.16. On average, each block V contains six blast holes and experience has shown that the arithmetic mean Z_V of the grades of these six blast-holes can be taken as the mean copper grade of a block.

On a particular mined out bench of area $410\,\text{m} \times 160\,\text{m}$, there are approximately 1000 blast-holes which are used to provide the mean grades Z_V of 160 blocks V. Using this set of data the following calculations were made.

(i) The histogram of the dispersion of the blast hole grades over the field (bench) B, cf. Fig. II.22. The horizontal cross-section of the blast-holes can be considered as zero when compared to the dimensions of the field B or of the block V and, thus, this histogram corresponds to a realization of the dispersion law $F_{0/B}$ of the

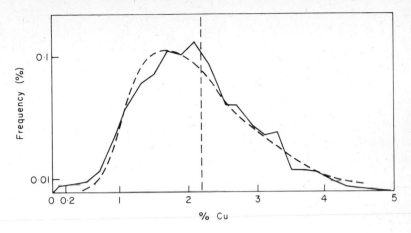

FIG. II.22. Dispersion of blast-hole grades in the bench. 13-m blast-hole, $m = 2 \cdot 12\%$ Cu, $s^2 = 0 \cdot 939$ (% Cu)^2.

quasipoint grades in the field B. The theoretical dispersion variance of this law is $D^2(0/B)$.

A lognormal distribution with the same mean (2·12% Cu) and same variance ($s^2 = 0·939$ (% Cu)^2) has been fitted. The fit is not perfect, especially at the extremities of the curve (the quality of this fit can be better appreciated by drawing the experimental and fitted cumulative distributions on logarithmic probability paper).

The theoretical value $D^2(0/B)$ calculated with the semi-variogram through formula (II.36) is $D^2(0/B) = 0·92$, which is close enough to the experimental dispersion variance $s^2 = 0·939$.

(ii) The histogram of dispersion of the block grades Z_V in the field B, cf. Fig. II.23. This histogram is a realization of the dispersion law $F_{V/B}$ of the block grades in B.

The theoretical dispersion variance $D^2(V/B)$ calculated from formula (II.36) is $D^2(V/B) = 0·37$, which is close enough to the experimental dispersion $s^2 = 0·346$ (% Cu)^2.

The adjustment of this histogram to a lognormal distribution with the same mean (2·16% Cu) and same variance s^2 is quite satisfactory.

(iii) The histogram of dispersion of the $2V$-block grades in the field B, cf. Fig. II.24. These blocks $2V$ (40 m × 20 m × 13 m) were obtained by grouping the blocks V in pairs in the N–S direction.

The calculated theoretical dispersion variance is $D^2(2V/B) = 0·287$, which is close enough to the experimental dispersion variance $s^2 = 0·269$ (% Cu)^2.

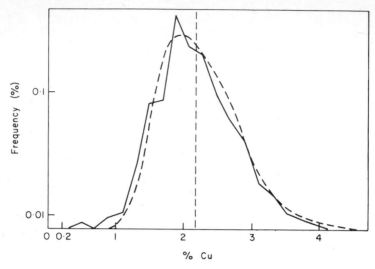

FIG. II.23. Dispersion of V-block grades in the bench. Block V, $20 \text{ m} \times 20 \text{ m} \times 13 \text{ m}$; $m = 2 \cdot 16$, $s^2 = 0 \cdot 346\%$ Cu^2.

The adjustment of this histogram to a lognormal distribution with the same mean ($2 \cdot 17\%$ Cu) and same variance s^2 is not satisfactory at all.

(The difference between the experimental means of the blast-holes and those of the blocks V and $2V$ is explained by the fact that the grid of blast-holes is not regular and that the field B is not divided into an exact number of blocks V and $2V$).

As a conclusion concerning the type of the dispersion laws, it can be seen that, if all three histograms of Figs II.22 to II.24 present the same asymmetry, only the histogram of the dispersion of the V-block can be correctly fitted to a two-parameter (m and s^2) lognormal distribution.

2. *Grades dispersion in the Nimba Range*

The iron deposits of the Nimba range (Liberia and Guinea) are of itabirite type concentrated in iron oxides by leaching of the silica.

Figure II.25(a) and (b) gives the histograms of the dispersion of the grades in Fe and SiO_2 impurity measured in core samples of constant length 5 m. The field of dispersion is that of a particularly enriched surface ore essentially made up of hematite.

The extreme homogeneity of this ore is demonstrated by a clearly unimodal Fe histogram of inverse-lognormal type. If $Z(x)$ is the Fe grade

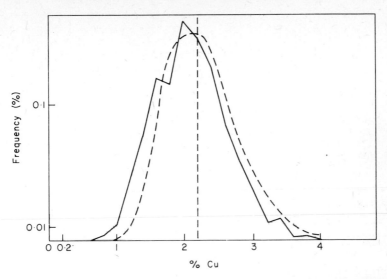

FIG. II.24. Dispersion of $2V$-block grades in the bench. $2V$ block, $40\,m \times 20\,m \times 13\,m$; $m = 2 \cdot 17\%$ Cu, $s^2 = 0 \cdot 269$ (% Cu)^2.

expressed as a percentage (%), the variable $\log[70 - Z(x)]$ is distributed according to a normal law. The value 70% corresponds exactly to the maximum iron grade of a sample composed entirely of hematite Fe_2O_3. The adjustment of an inverse-lognormal distribution with the same mean (66·6% Fe) and same variance ($s^2 = 2 \cdot 3$ (% Fe)^2) is satisfactory.

The histogram of the SiO_2 grades is of straightforward lognormal type and can be adjusted to a lognormal distribution with the same mean (0·58% SiO_2) and same variance ($s^2 = 0 \cdot 20$ (% SiO_2)^2). The adjustment is satisfactory at least for SiO_2 grades greater than the mean 0·58.

However, the histograms are not always unimodal. Figure II.26 shows the histogram of the Fe grades over a much larger field, covering not only the surface hematite ore but also the deeper and poorer ore (little altered itabirite). Note that this global histogram is multimodal with a first mode of 67% Fe corresponding to the hematite ore and several other modes around 50% Fe corresponding to the various poorer ores.

Figure II.27 shows the histogram of the dispersion of Fe grades within the poor ores. This histogram appears as a mere development of the (30 to 60% Fe) part of the preceding global histogram.

Note again the extreme homogeneity ($s^2 = 2 \cdot 3$ (% Fe)^2) of the hematite ore in relation to the variability of the grades of the deeper, poorer ores ($s^2 = 50 \cdot 1\%$).

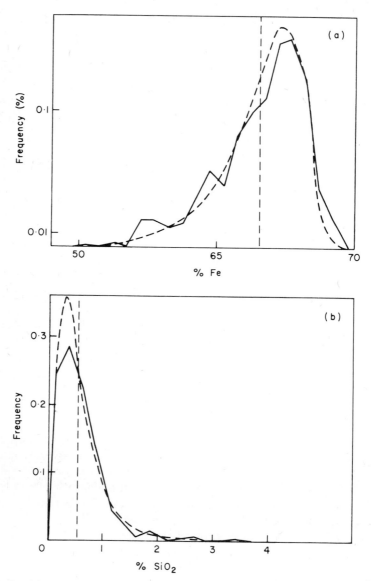

FIG. II.25 (a). Dispersion of Fe grades (hematite ore). $m = 66 \cdot 6\%$ Fe, $s^2 = 2 \cdot 3$ (% Fe)^2. (b) Dispersion of SiO$_2$ grades (hematite ore). $m = 0.58\%$ Fe, $s^2 = 0 \cdot 20$ (% SiO$_2$)^2.

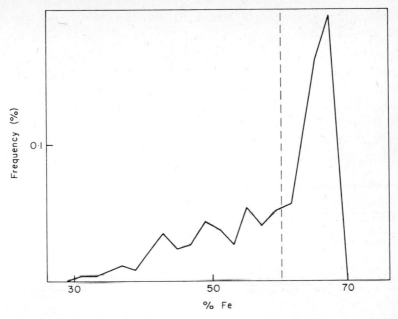

FIG. II.26. Dispersion of Fe grades (global ore). $m = 58 \cdot 8\%$ Fe, $s^2 = 84 \cdot 9\%^2$.

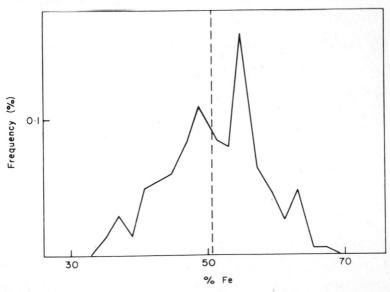

FIG. II.27. Dispersion of Fe grades (deep poor ores). $m = 50 \cdot 5\%$ Fe, $s^2 = 50 \cdot 1\%^2$.

3. *Grade distribution at Haartebeestfontein after D. G. Krige (1976) and J.*
   M. Rendu (1976)

The field of dispersion B is a section (3000 m × 900 m) of the Haartebeestfontein gold mine in South Africa. Each unit v (7·5 m × 7·5 m) contains
between zero and four samples (two on average) and the arithmetic mean
of the samples, Z_v, in each unit v is said to represent the cluster of samples
inside the unit v.

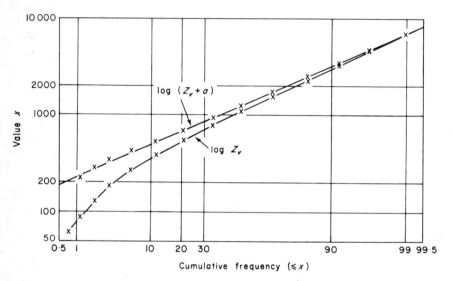

FIG. II.28. Dispersion of cluster grades at Haartebeestfontein.

Figure II.28 shows the cumulative histogram of the dispersion of the
grades Z_v in the field B in normal logarithmic coordinates. Note that for
the low grades Z_v there is a significant deviation compared with the line
corresponding to the lognormal model.

The two-parameter lognormal model can be improved by introducing a
third parameter a, so that $\log (Z_v + a)$, rather than $\log Z_v$ is assumed to be
distributed normally. With $a = 150$ g cm/ton, a lognormal model can
be fitted remarkably well to the experimental distribution of the grades
$(Z_v + a)$.

II.C.4. Type of the dispersion laws

It would not be possible within the bounds of this book to present all the
available examples of laws of dispersion of grades (grade histograms) but
the following remarks, drawn from practical experience, can be made.

(i) When the dimensions of support v are small compared with those of the field of dispersion V (e.g., v representing a core sample and V a mining unit or a stope), then (a) the histogram of the grades of elements of weak concentration (Au, Ag, precious metals, U, Cu, Ni, impurities in weak concentration such as SiO_2, Al_2O_3, Fe, ...) are asymmetric with a tail towards the higher grades (direct lognormal type); (b) the histogram of the grades of macro-elements of strong concentration (Fe in deposit, Mn in deposit, phosphate P_2O_5, ...) are asymmetric with a tail towards the lower grades (inverse lognormal type).

(ii) As the support v increases in size (e.g., v representing a mining unit) a blending effect is produced, which tends to slowly tone down the previous asymmetries. Lognormality is not preserved when the support v becomes too large.

(iii) A cleary multimodal histogram reveals, in general,[†] the existence of heterogeneous mineralizations (the presence of several types of ore or the influence of poorly mineralized areas on the fringes of the deposit). The interpretation of such a histogram must be made with due regard to statistics (e.g., fluctuation due to an insufficient number of samples), geometry (e.g., position of the field of dispersion in the deposit) and geology (e.g., mineralogy, faults, directions of enrichment of the ore).

(iv) The preceding remarks concern the grades of elements. For geometric-type regionalized variables, such as mineralized or overburden thickness, or mining type ReV such as percentages of ore recovery or granulometric ratios, the histograms are of a more symmetrical type, as shown on Fig. II.29(a) and (b).

Figure II.29(a) shows the histogram of the mineralized Ni-laterites thickness of the Prony ore body (New Caledonia). The histogram was constructed from 176 values measured on the 176 vertical bore-holes of a regular square grid of 100-m side.

Figure II.29(b) shows the histogram of the percentage of ore (mesh >6 mm) recovered on mining units at Moanda (Gabon) manganese deposit. This histogram was constructed from the results of mining 1400 unit blocks corresponding to an area of about 42 ha.

---

† But the presence of several modes on a histogram does not imply that each of these modes necessarily corresponds to one type of mineral that can be distinguished at the mining stage. For example, a banded hematite quartzite ore surveyed by drill cores (of a metre or so in length) would show a multimodal histogram of the type shown on Fig. II.27 and still be considered as an homogeneous ore at the scale of mining (on blocks of 10 or 20 m in dimension).

Of course, non-symmetrical histograms of thicknesses can be found just as well as symmetrical histograms of grades. The previous remarks are not to be taken as laws. As an example, Fig. II.30 shows a histogram of exponential type corresponding to the vertical thicknesses of bauxite lenses within a zone of the Delphi deposit (Greece). Only the positive drill-holes that actually meet a mineralized lens (i.e., with a non-zero mineralized thickness) were retained for this histogram. Indeed 60% of the drill-holes

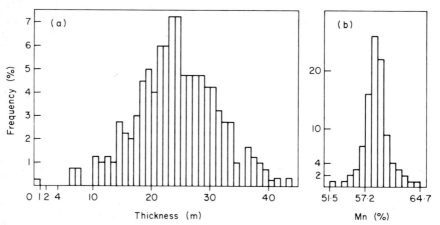

FIG. II.29. Histograms of symmetrical type. (a) Mineralized thickness at Prony. (b) Ore recovery at Moanda.

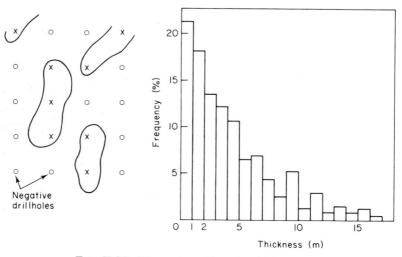

FIG. II.30. Dispersion of bauxite thicknesses.

were negative and would have appeared on a global histogram as a strong atom at the origin (zero thickness).

II.D. REGULARIZATION

II.D.1. Regularization and estimation variance

Very rarely, in practice, will point data $z(x)$ be available. Most often, the available data $z_v(x)$ will be defined on a certain *support* $v(x)$ centred on a point x, v being, for example, a core sample or, more generally, the volume of a sample. The grade $z_v(x)$ of this sample is the mean value of the point grade $z(y)$ in $v(x)$, i.e.,

$$z_v(x) = \frac{1}{v} \int_{v(x)} z(y)\, dy.$$

The mean value $z_v(x)$ is said to be the *regularization* of the point variable $z(y)$ over the volume $v(x)$.

Let the point-regionalized variable $z(y)$ be a particular realization of a second-order stationary RF $Z(y)$, with expectation m and covariance $C(h)$ or variogram $2\gamma(h)$. The regularization of the point RF $Z(y)$ over the volume $v(x)$ is, likewise, a RF denoted by $Z_v(x)$ and written as

$$Z_v(x) = \frac{1}{v} \int_{v(x)} Z(y)\, dy.$$

It can be shown that the regularized RF $Z_v(x)$ of a second-order stationary point RF is also second-order stationary, with

(i) an expectation identical to the point expectation m – indeed,

$$E\{Z_v(x)\} = \frac{1}{v} \int_v E\{Z(y)\}\, dy = \frac{1}{v} \int_{v(x)} m \cdot dy = m, \qquad \forall x;$$

(ii) a variogram $2\gamma_v(h)$ defined, according to formula (II.4), as

$$2\gamma_v(h) = E\{[Z_v(x+h) - Z_v(x)]^2\}. \tag{II.40}$$

The problem is then to derive this regularized variogram $2\gamma_v(h)$ from the point variogram $2\gamma(h)$. To do so, an easy way is to consider the expression (II.40) of the regularized variogram as the variance of the estimation of the mean grade $Z_v(x)$ by the mean grade $Z_v(x+h)$ separated by the vector h, cf. section II.B.2, Remark 3. This estimation variance is then given by the general formula (II.27):

$$2\gamma_v(h) = 2\bar{\gamma}(v(x), v(x+h)) - \bar{\gamma}(v(x), v(x)) - \bar{\gamma}(v(x+h), v(x+h)).$$

Since the point semi-variogram $\gamma(h)$ is stationary, the last two terms of
the previous expression are equal and, thus,

$$\boxed{\gamma_v(h) = \bar{\gamma}(v, v_h) - \bar{\gamma}(v, v),}$$ \hfill (II.41)

v_h denoting the support v translated from v by the vector h, $\bar{\gamma}(v, v_h)$
represents classically the mean value of the point semi-variogram $\gamma(u)$
when one of the extremities of the vector u describes the support v and the
other extremity independently describes the translated support v_h.

Remark 1   For distances h which are very large in comparison with the
dimension of support v, the mean value $\bar{\gamma}(v, v_h)$ is approximately equal to
the value $\gamma(h)$ of the point semi-variogram and we obtain the very useful
practical approximation

$$\gamma_v(h) \simeq \gamma(h) - \bar{\gamma}(v, v) \quad \text{for } h \gg v.$$ \hfill (II.42)

At large distances, $h \gg v$, the regularized semi-variogram is simply derived
from the point semi-variogram by subtracting a constant term $\bar{\gamma}(v, v)$
related to the dimensions and geometry of the support v of the regularization, cf. Fig. II.31.

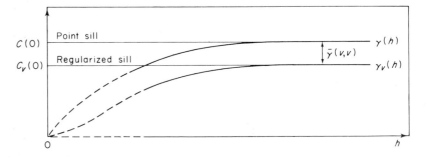

FIG. II.31. Point and regularized semi-variograms at large distances.

Remark 2   Regularized covariance and a priori *variance.* It can be shown
that if the point *a priori* variance $\text{Var}\{Z(y)\} = C(0)$ exists, so do the
regularized *a priori* variance $C_v(0)$ and the regularized covariance $C_v(h)$,
which are written in terms of mean values $\bar{C}$ of the point covariance $C(h)$:

$$\text{Var}\{Z_v(x)\} = C_v(0) = \bar{C}(v, v),$$ \hfill (II.43)

$$C_v(h) = C_v(0) - \gamma_v(h) = \bar{C}(v, v_h).$$ \hfill (II.44)

Hence, if the point semi-variogram $\gamma(h)$ is of a transition type with a sill
value equal to the point *a priori* variance, $\gamma(\infty) = C(0) = \text{Var}\{Z(y)\}$, cf.

formula (II.39), the regularized semi-variogram $\gamma_v(h)$ is also of a transition, type with a sill value equal to the regularized *a priori* variance:

$$\gamma_v(\infty) = C_v(0) = \text{Var}\{Z_v(x)\}, \qquad (II.45)$$

cf. Fig. II.31.

II.D.2. Regularization by cores along a bore-hole

This type of regularization, very frequent in practice, corresponds to the construction of the variogram of the mean grades of core samples along the length of a bore-hole, cf. Fig. II.32. It is assumed that all the core samples have the same length l and the same cross-sectional area s.

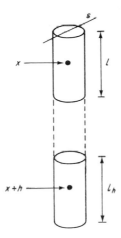

FIG. II.32. Core samples aligned along a bore-hole.

Let $Z(x)$ be a point RF which is either second-order stationary or at least intrinsic with a semi-variogram $\gamma(h)$.

The RF regularized on the support $v = s \times l$ of the core sample is written

$$Z_v(x) = \frac{1}{v} \int_{v(x)} Z(y)\, dy,$$

where the sign $\int_{v(x)} dy$ represents, in fact, a triple integral over the volume v. When the diameter of the core is small† compared to length l, the regularization effect of the cross-sectional area s of the core sample can be

---

† This approximation will not be valid for samples taken from sampling pits, for example, because the diameter of the pits will be of the order of the distance between the samples (0·5 to 1 m).

neglected. The mean value over the length l of the core sample can then be written in terms of a single integral in the direction of the length of the core as

$$Z_v(x) \simeq Z_l(x) = \frac{1}{l} \int_{l(x)} Z(y)\, dy. \qquad (II.46)$$

Regularized semi-variogram

When the cross-sectional area s of the core sample is negligible, two core samples can be considered as two *aligned* segments l and l_h of the same length l and separated by a distance h, cf. Fig. II.32.

According to formula (II.41), the regularized semi-variogram can then be written

$$\gamma_l(h) = \tfrac{1}{2} E\{[Z_l(x+h) - Z_l(x)]^2\} = \bar{\gamma}(l, l_h) - \bar{\gamma}(l, l). \qquad (II.47)$$

Application to point models in current use

This regularization by cores assimilated to a segment of length l will be illustrated on the four theoretical point variogram models in current use (these models are precisely defined in section III.B.2).

This part is not essential for a complete understanding of the problem of regularization, and the reader may go directly to the next section, II.D.3, "Grading over a constant thickness", which provides another example of regularization.

1. The linear model

$$\gamma(h) = r \quad \text{with } r = |h|.$$

If u denotes the coordinate along the length of the bore-hole, the first term $\bar{\gamma}(l, l_h)$ of the expression (II.47) for the regularized semi-variogram is written

$$\bar{\gamma}(l, l_h) = \frac{1}{l^2} \int_0^l du \int_r^{r+l} |u - u'|\, du'.$$

(Recall here that vector h is parallel to the direction of the bore-hole length.)

According to whether or not $r = |h|$ is less than l, i.e., whether or not the two segments l and l_h overlap (cf. Fig. II.33), we have the following two expressions.

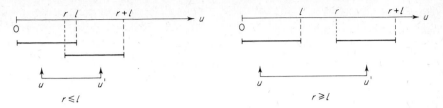

FIG. II.33. Calculation of the term $\bar{\gamma}(l, l_h)$.

(i) $|h| \leqslant l$:

$$l^2 \bar{\gamma}(l, l_h) = \int_0^r du \int_r^{r+l} (u' - u) du + \int_r^l du \left[\int_r^u (u - u') du' \right.$$

$$\left. + \int_u^{r+l} (u' - u) du' \right].$$

These integrals can be evaluated quite simply to give

$$\bar{\gamma}(l, l_h) = \frac{1}{3} \frac{r^2}{l^2} (3l - r) + \frac{l}{3} \quad \text{for } r = |h| \leqslant l$$

and, when $h = 0$, $\bar{\gamma}(l, l) = l/3$.

(ii) $|h| \geqslant l$:

$$l^2 \bar{\gamma}(l, l_h) = \int_0^l du \int_r^{r+l} (u' - u) du' = rl^2$$

and, thus, $\bar{\gamma}(l, l_h) = r$ for $r = |h| \geqslant l$.

Applying formula (II.47), the regularized semi-variogram is

$$\gamma_l(h) = \begin{cases} \dfrac{r^2}{3l^2} (3l - r), & \forall r = |h| \in [0, l] \\ r - l/3, & \forall r \geqslant l, \end{cases} \tag{II.48}$$

cf. Fig. II.34.

For distances $|h| \leqslant l$, the regularized semi-variogram is parabolic. The cores of length l overlap each other and the regularizing effect is strong enough to change the behaviour of the model at the origin from linear to parabolic.

For distances $|h| \geqslant l$, which, in practice, are the only distances observable, the linear behaviour of the point model is preserved and the regularized model differs from the point model by a constant value

$$\gamma_l(h) = \gamma(h) - \bar{\gamma}(l, l) = r - l/3, \quad r = |h| \geqslant l.$$

(Note that for the linear point model the approximation of formula (II.42) becomes a strict equality.)

If the regularized curve for $|h| \geq l$ is projected towards $h = 0$, the ordinate axis is intercepted at $-l/3$ and this negative value is called a "pseudo-negative nugget effect" due to regularization.

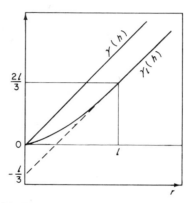

FIG. II.34. The linear model and its regularization.

2. *The logarithmic model*

$$\gamma(h) = \log r \quad \text{with } r = |h|.$$

Note that for $h = 0$, $\log |h| = -\infty \neq 0$, so that this logarithmic model cannot be used for point samples, only models $\gamma_v(h)$, regularized on non-point supports, are defined.

For distances observed in practice, $|h| \geq l$, the semi-variogram regularized by cores is written as

$$\gamma_l(h) = \frac{(r+l)^2}{2l^2} \log (r+l) + \frac{(r-l)^2}{2l^2} \log (r-l)$$

$$- \frac{r^2}{l^2} \log l - \log l, \qquad \forall r \geq l. \tag{II.49}$$

In *practice*, this exact formula is rather difficult to use and the approximation of formula (II.42) is preferred:

$$\gamma_l(h) \simeq \gamma(h) - \bar{\gamma}(l, l) = \log \frac{r}{l} + \frac{3}{2}, \qquad \forall r \geq l, \tag{II.50}$$

with

$$\bar{\gamma}(l, l) = \log l - \tfrac{3}{2}.$$

The relative error involved in this approximation is 8% for $r = l$ and 0.1% for $r = 5l$.

3. The spherical model

$$\gamma(h) = \begin{cases} \dfrac{3}{2}\dfrac{r}{a} - \dfrac{1}{2}\dfrac{r^2}{a^3}, & \forall r = |h| \in [0, a], \\ 1 = \text{sill value}, & \forall r \geqslant a = \text{range}. \end{cases}$$

This model reaches a sill value, $\gamma(\infty) = 1$, for a distance equal to the range a, cf. Fig. III.8.

The expressions for the semi-variogram $\gamma_l(h)$ regularized by cores are long and awkward. They can be found along with those for the exponential model in Ch. Huijbregts (1971b).

In *practice* we can make use of the following charts, given in section II.E.5:

(i) chart no. 21, which gives the values of the first experimentally accessible points of $\gamma_l(h)$ and also of the sill value $\gamma_l(\infty)$ as functions of a/l;

(ii) charts no. 22(a) and 22(b), which give the values of $\gamma_l(h)$ for different values of a/l as a function of $|h|/l$.

Furthermore, for the practical distances $|h| \geqslant l$, the regularized semi-variogram can be derived from the point model by the approximate formula (II.42):

$$\gamma_l(h) \simeq \gamma(h) - \bar{\gamma}(l, l), \quad \text{for } |h| \geqslant l,$$

the mean value $\bar{\gamma}(l, l)$ for $l \leqslant a$ (the length is not greater than the range), being given by the expression

$$\bar{\gamma}(l, l) = \frac{l}{2a} - \frac{l^3}{20a^3}, \qquad \forall a \geqslant l. \tag{II.51}$$

Given the conditions, $|h| \geqslant l$ and $a \geqslant 3l$, which cover almost all practical cases, this approximation is excellent.

4. The exponential model

$$\gamma(h) = 1 - \exp(-r/a), \qquad \forall r = |h|.$$

This model reaches its sill value, $\gamma(\infty) = 1$, asymptotically. In practice, a practical range $a' = 3a$ is taken, for which

$$\gamma(a') = 0.95 \simeq \text{sill } 1,$$

cf. Fig. III.8.

The semi-variogram regularized by cores of length l takes the form

$$\gamma_l(h) = \begin{cases} \dfrac{2a^2}{l^2}(e^{-l/a}-1)+\dfrac{2ar}{l^2}+\dfrac{a^2}{l^2}\,e^{-r/a}(2-e^{-l/a})-\dfrac{a^2}{l^2}\,e^{(r-l)/a}, & \forall r \leqslant l, \\[2mm] \dfrac{a^2}{l^2}\Big[e^{-l/a}-e^{l/a}+\dfrac{2l}{a}+(e^{-l/a}+e^{l/a}-2)(1-e^{-r/a})\Big], & \forall r \geqslant l. \end{cases} \quad \text{(II.52)}$$

In *practice*, the regularized semi-variogram can be deduced from the point model by the approximate formula (II.42):

$$\gamma_l(h) \simeq \gamma(h) - \bar{\gamma}(l, l) \quad \text{for } |h| \geqslant l,$$

the mean value $\bar{\gamma}(l, l)$ being given by the expression

$$\bar{\gamma}(l, l) = 1 - \frac{a^2}{l^2}\Big[2\,e^{-l/a} + \frac{2l}{a} - 2\Big]. \quad \text{(II.53)}$$

Given the conditions, $|h| \geqslant l$ and $a' = 3a \geqslant 3l$, which cover almost all practical cases, this approximation is excellent.

Regularized range For a transition point model, i.e., for a point model with a sill and a (practical or actual) range a, two point variables $Z(x)$ and $Z(x+h)$ are uncorrelated when the distance h separating them is greater than the range ($|h| > a$.)

As a consequence, the two variables regularized by cores of length l, $Z_l(x)$ and $Z_l(x+h)$ are uncorrelated when the distance between each point of $l(x)$ and each point of $l(x+h)$ is greater than the range a, cf. Fig. II.35; for which it is necessary and sufficient that $|h| > a + l$.

The regularized range is, thus, $(a + l)$ and this is the distance at which the regularized model $\gamma_l(h)$ reaches (actually or practically) its sill $\gamma_l(\infty)$. Note that the previous practical approximation, $\gamma_l(h) \simeq \gamma(h) - \bar{\gamma}(l, l)$ amounts to assigning to the regularized model the same range as that of the point

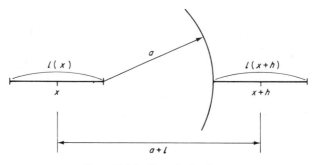

FIG. II.35. Regularized range.

model. In fact, for the two transition models in current use (the spherical and the exponential), when $|h| > a$ and $a \geqslant 3l$, $\gamma_l(h)$ can be approximated with excellent precision by its sill value $\gamma_l(\infty)$.

II.D.3. Grading over a constant thickness

This second type of regularization is also very common in practice and corresponds to taking the mean over a constant thickness l of a horizontal bench or of a vein of more or less constant thickness, cf. Fig. II.36. Let (x_u, x_v, x_w) be the three coordinates defining the position x of the point RF $Z(x)$ in the three-dimensional space. The point RF is assumed to be second-order stationary or at least intrinsic and to have the semi-variogram $\gamma(h_u, h_v, h_w)$.

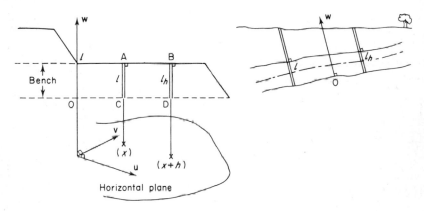

FIG. II.36. Grading over a constant thickness.

If x_w and h_w represent the coordinates in the vertical direction, then the mean value over the constant thickness l of a horizontal bench is given by

$$Z_G(x_u, x_v) = \frac{1}{l} \int_0^l Z(x_u, x_v, x_w)\, dx_w. \tag{II.54}$$

We thus define a new RF, $Z_G(x_u, x_v)$, which is a regularization of the point RF $Z(x_u, x_v, x_w)$ and is studied in the horizontal two-dimensional plane (x_u, x_v) or, more generally, in the plane which is orthogonal to the direction x_w of regularization. $Z_G(x_u, x_v)$ is called the "*grading of the point RF over the constant thickness l*".

(There is some difficulty here with terminology. The original French word "*montée*" has been translated as "grading" in the sense of a regularization over an interval.)

Example If $Z(x_u, x_v, x_w)$ is the point grade in a horizontal stratiform formation of quasi-constant thickness l, then $Z_G(x_u, x_v)$ is the mean† grade of a vertical bore-hole positioned at the point of horizontal coordinates (x_u, x_v) on the topographical surface.

Remark Even though the support of the regularization (l) may be the same, care should always be taken to distinguish between grading over a constant thickness l and regularization by core samples of length l along the length of a bore-hole, cf. Figs II.36 and II.32. In the case of grading, the two variables $Z_G(x_u, x_v)$ and $Z_G(x_u + h_u, x_v + h_w)$ are defined on two segments l and l_h which are parallel and separated by a distance vector $h = (h_u, h_v)$ orthogonal to the direction of the length l, while, in the case of regularization by cores, the two segments l and l_h are *aligned* and separated by a distance vector h parallel to the direction of length l.

Graded semi-variogram

The semi-variogram of the graded RF over the constant thickness l is written in the standard form of formula (II.41):

$$\gamma_G(h) = \tfrac{1}{2}E\{[Z_G(x+h) - Z_G(x)]^2\} = \bar{\gamma}(l, l_h) - \bar{\gamma}(l, l), \qquad (\text{II.55})$$

where $x = (x_u, x_v)$ and $h = (h_u, h_v)$ denote the coordinates in the plane orthogonal to the direction of the grading thickness, cf. Fig. II.36.

In calculating the mean value $\bar{\gamma}(l, l_h)$, three directions in space are involved and the calculation is simple only when the point model $\gamma(h_u, h_v, h_w)$ is isotropic, i.e., if $\gamma(h) = \gamma(r)$ with

$$r = |h| = \sqrt{(h_u^2 + h_v^2 + h_w^2)}$$

Isotropic expression of the graded semi-variogram

If the point model is isotropic in the three-dimensional space (h_u, h_v, h_w), the graded semi-variogram $\gamma_G(h_u, h_v)$ is also isotropic in the two-dimensional space (h_u, h_v) and is written in terms of simple integrals, cf. Fig. II.37, as

$$\gamma_G(h) = \frac{2}{l^2}\int_0^l (l-y)\gamma(\sqrt{(r^2+y^2)})\,dy - \frac{2}{l^2}\int_0^l (l-y)\gamma(y)\,dy \qquad (\text{II.56})$$

with $r^2 = |h|^2 = h_u^2 + h_v^2$.

---

† The regularization effect of the cross-sectional area s of the core can be neglected in relation to the thickness l of the formation when integrating over the support $(v = s \times l)$ of the intersection of the bore-hole with the formation.

Application to point models in current use

This regularization by grading will be illustrated for the four point models in current use. These point models will be supposed to be isotropic in the three-dimensional space.

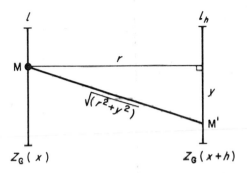

FIG. II.37. Calculation of the term $\bar{\gamma}(l, l_h)$.

This part is not essential and the reader may go directly to the next section, II.D.4.

1. *The linear model*

$$\gamma(h) = r \quad \text{with } r = |h|.$$

The exact isotropic expression for the semi-variogram graded over a constant thickness l can be found by expanding formula (II.56):

$$\gamma_G(h) = \frac{1}{3}\sqrt{(r^2+l^2)} + \frac{r^2}{l}\log\frac{l+\sqrt{(r^2+l^2)}}{r} + \frac{2}{3}\frac{r^2}{l^2}(r-\sqrt{(r^2-l^2)}) - \frac{l}{3} \qquad \text{(II.57)}$$

with $\bar{\gamma}(l, l) = l/3$.

In practice, this exact expression will be used only for small distances $|h| < l$. For $|h| \geq l$, the standard approximation given in (II.42) is used henceforth:

$$\gamma_G(h) \simeq \gamma(h) - \bar{\gamma}(l, l) = r - l/3, \qquad \forall r = |h| \geq l. \qquad \text{(II.58)}$$

The relative error corresponding to this approximation is 10% for $|h| = l$, 1% for $|h| = 3l$ and less than 0·1% for $|h| = 10l$.

Remark For distances $h \geq l$, the model $\gamma_G(h)$ graded over the length l is practically indistinguishable from the model $\gamma_l(h)$ regularized by cores of length l, cf. formulae (II.58) and (II.48). On the other hand, for distances $|h| \leq l$, the effect of regularization by cores of length l is greater than that of regularization by grading over l and we have $\gamma_l(h) \le \gamma_G(h)$. Such distances $|h| \le l$ are scarcely considered on the core-regularized semi-variogram $\gamma_l(h)$.

2. The logarithmic model

$$\gamma(h) = \log r \quad \text{with } r = |h|.$$

(Recall that this logarithmic model should only be used when regularized over non-point supports.)

The exact isotropic expression of the semi-variogram graded over l is found by expanding formula (II.56):

$$\gamma_G(h) = \frac{2\theta}{\tan \theta} + \frac{\log \cos \theta}{\tan^2 \theta} - \log \sin \theta \tag{II.59}$$

with $\theta = \arctan r/l$ and $\bar{\gamma}(l, l) = \log l - 3/2$.

This formula gives rise to the following limited expansions:

$$\left.\begin{aligned}
\text{for } r \le l, \quad & \gamma_G(h) = \pi \frac{r}{l} + \frac{r^2}{l^2} \log \frac{r}{l} - \frac{3}{2} \frac{r^2}{l^2} - \frac{1}{12} \frac{r^4}{l^4} + \cdots; \\
\text{for } r \ge l, \quad & \gamma_G(h) = \frac{3}{2} + \log \frac{r}{l} + \frac{1}{12} \frac{l^2}{r^2} - \frac{1}{60} \frac{l^4}{r^4} + \cdots.
\end{aligned}\right\} \tag{II.60}$$

In practice, for distances $|h| \ge l$, as a first-order approximation we can use the standard expression

$$\gamma_G(h) \simeq \gamma(h) - \bar{\gamma}(l, l) = \log \frac{r}{l} + \frac{3}{2}. \tag{II.61}$$

The relative error involved in this approximation is 6% for $r = l$, 1% for $r = 2l$, 0·1% for $r = 5l$.

3, 4. The spherical and exponential models
The analytical expressions for these graded semi-variograms are long and awkward to handle and for this reason the following charts are used:

  (i) chart no. 1, which gives the graded semi-variogram of the point spherical model (with sill 1 and range a) as a function of $|h|/a$ and for different values of l/a;

  (ii) chart no. 11, which gives the graded semi-variogram of the point exponential model (with sill 1 and parameter a and, thus, practical range $a' = 3a$) as a function of $|h|/a$ and for different values of l/a.

The *graded sill* is defined as follows. If the sill of the point model is 1, then, according to formula (II.42), the sill of the model graded over l is

$$\gamma_G(\infty) = 1 - \bar{\gamma}(l, l),$$

the mean values $\bar{\gamma}(l, l)$ being given by formulae (II.51) and (II.53).

The model graded over l and the model regularized by cores of length l thus have the same sill.

The *graded range* may also be defined. When the distance $|h|$ becomes greater than the (actual or practical) range, every point of l is independent of every point of l_h and, as a consequence, the range of the graded model is equal to the range of the point model, whatever the grading thickness l may be.

Recall that the range of the model regularized by cores of length l is, strictly, equal to $(a + l)$, where a is the range of the point model.

In practice, for both the spherical and the exponential models we have the following.

(i) For distances $|h| \geq l$, we will use the standard approximation of (II.42):

$$\gamma_G(h) \simeq \gamma(h) - \bar{\gamma}(l, l). \qquad \text{(II.62)}$$

The relative error involved in this approximation never exceeds 10% in the most unfavourable cases (e.g., $|h| = l$ and $l/a = 0 \cdot 1$) and is less than 1% in the majority of cases encountered in practice ($|h| > 3l$ whatever the ratio l/a may be).

(ii) For distances $|h| < l$, we can either (a) use the exact formulae of the graded models given in J. Serra, (1967a), Vol. 2, pp. 446 and 461) or (b) if $a \gg l$ (in practice if the actual or practical range is greater than $3l$), we can assimilate the point models to their tangents at the origin (linear model of slope w); the graded models can then be deduced from formulae (II.57) or (II.58) corresponding to the linear model.

II.D.4. Practical rule for regularization

The two types of regularization considered in the two previous sections are particularly useful, for they correspond to the two main data configurations encountered in practice: core samples aligned along a bore-hole and data defined over the constant thickness of a mining bench.

Although these two types of regularization are different in their principle, they lead to a *common* approximate formula giving the regularized semi-variogram

$$\boxed{\gamma_l(h) \simeq \gamma(h) - \bar{\gamma}(l, l), \qquad \forall |h| \geq l.} \qquad \text{(II.63)}$$

The regularized model is deduced from the point model by subtracting a constant term $\bar{\gamma}(l, l)$ which depends only on the length l of the support of the regularization.

Remark 1   For the regularization by cores of length l of transition-type point model with a (practical or actual) range a, the additional condition $a \gg 3l$ should also be satisfied before applying the approximation (II.63). In practice, this additional condition is often met, since the length l of the data support should, generally, be much smaller than the range a of the sought after structure.

Remark 2   *Positive definite-type functions for regularized model.* As it is an approximation, the expression (II.63) for the regularized model is not strictly a conditional positive definite function; hence, we are not sure that all variances calculated from this approximate expression of $\gamma_l(h)$ will be strictly positive or zero, cf. the condition (II.10) in section II.A.3.

Strictly speaking, if the point semi-variogram $\gamma(h)$ is of a certain model – spherical for instance – the regularized semi-variogram $\gamma_l(h)$ is not of the same model. However, for practical applications, it is an acceptable approximation to consider that the model is the same for the regularized $\gamma_l(h)$ – i.e., spherical – with, of course, parameters (regularized range and sill) different from those of the point model $\gamma(h)$. This model for $\gamma_l(h)$ would then be a conditional positive definite function and ensures that all variances calculated from it are non-negative.

Thus, if the point semi-variogram $\gamma(h)$ is a spherical model with range a and sill C, the regularized semi-variogram $\gamma_l(h)$ can be assimilated (for distances $|h| \geq l$) to a spherical model with

   (i) a regularized sill, $C_l = C - \bar{\gamma}(l, l)$;
  (ii) a regularized range a_l equal to $(a + l)$ if the regularization is done by cores of length l, and equal to a if the regularization is a grading over the constant thickness l.

Remark 3   *Deconvolution (passing from γ_v to γ).* In practice, since the available data are most often defined on a non-point support v, it is often required to work out a theoretical point model $\gamma(h)$ which is coherent with the observed regularized semi-variogram $\gamma_v^*(h)$. Deducing a point model $\gamma(h)$ from a regularized model $\gamma_v(h)$ amounts to a "deregularization" or a "deconvolution" of the model $\gamma_v(h)$. In practice, this deconvolution is made through the following identification procedures.

   (i) A point model $\gamma(h)$ is defined from an inspection of the regularized curve $\gamma_v^*(h)$ experimentally available.

(ii) The regularized theoretical expression $\gamma_v(h)$ (or its approximation given by formula (II.63)) of this point model $\gamma(h)$ is then calculated and compared with the experimental curve $\gamma_v^*(h)$. The various parameters (ranges, sills, ratios of anisotropies, etc.) of the point model are then adjusted in such a way as to bring $\gamma_v(h)$ in line with $\gamma_v^*(h)$.

(iii) Once the point model $\gamma(h)$ is determined, the theoretical expression $\gamma_{v'}(h)$ for the regularization over a possible second support v' can be deduced and checked with the corresponding experimental curve $\gamma_{v'}^*(h)$.

The calculation of the regularized theoretical expressions $\gamma_v(h)$ and $\gamma_{v'}(h)$ is simple only when the point model consists of the established models which are in current use, and for which formulae and graphs are available, e.g., if the point semi-variogram $\gamma(h)$ is a spherical model.

II.D.5. Case studies

1. *Variograms of grades at Tazadit*

The iron (banded hematite quartzite) deposit of Tazadit (Mauritania) was sampled by vertical bore-holes. The particular bore-hole considered for this example was 177 m (L) long and was cut into core samples of constant length $l = 3$ m.

Let $Z(x)$ be the point RF representing the iron grade regionalization, $Z_l(x)$ the RF regularized by cores of length $l = 3$ m and $Z_{4l}(x)$ the RF regularized by cores of length $4l = 12$ m (obtained by grouping the available core samples in fours).

Figure II.38 gives the two experimental semi-variograms of the two variables $Z_l(x)$ and $Z_{4l}(x)$ (the circles ● correspond to the experimental points of γ_l and the crosses × to those of γ_{4l}).

These semi-variograms reach a sill value for a range a of approximately 40 m. The sill C_l is approximately 133 (% Fe)^2. If a spherical model with range $a = 40$ m and sill C is assumed for the theoretical point model $\gamma(h)$, then the theoretical value of the sill C_l is written, according to formula (II.51), as

$$C_l = \gamma_l(\infty) = C - \bar{\gamma}(l, l) = C\left[1 - \frac{l}{2a} - \frac{l^3}{20a^3}\right],$$

with $l = 3$ m, $a = 40$ m and $C_l = 133$. Thus, the sill of the point spherical model is $C = 138$ (% Fe)^2.

In Fig. II.38, the theoretical point spherical model $\gamma(h)$ with range $a = 40$ m and sill $C = 138$ is shown as a dashed line. The theoretical regularized model γ_l deduced from the previous point model by the practical approximation (II.63),

$$\gamma_l(h) \simeq \gamma(h) - \bar{\gamma}(l, l), \qquad \forall h \geqslant l, \qquad a > 3l = 12 \text{ m},$$

is shown as a first solid line. The regularized model γ_{4l} deduced by the same approximation is shown as a second solid line.

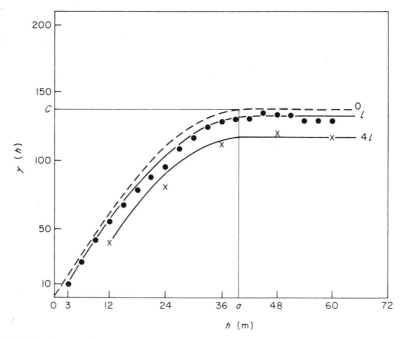

FIG. II.38. Regularized semi-variograms of Fe grades at Tazadit. Experimental points: ●, γ_l; ×, γ_{4l}. $L = 177$ m, $m = 49.2\%$ Fe, $s = 120$ (% Fe)^2.

The theoretical value of the sill C_{4l} is found from formula (II.51) to be

$$C_{4l} = \gamma_{4l}(\infty) = C\left[1 - \frac{4l}{2a} + \frac{(4l)^3}{20a^3}\right] = 117.5,$$

and correctly fits the experimentally observed sill.

Thus, the practical rule (II.63) for the regularization by cores of length l provides models γ_l and γ_{4l} which are very good fits to the corresponding experimental semi-variograms.

Variance of dispersion within L   The experimental variance of dispersion of the grades of the 59 cores of length l along the total length L of the bore-hole was found to be $s^2 = 120$ (% Fe)^2 and the mean grade $m = 49\cdot2\%$ Fe.

The corresponding theoretical dispersion variance $D^2(l/L)$ can be calculated from formula (II.36) either (i) by using the previous point spherical model $\gamma(h)$,

$$D^2(l/L) = \bar{\gamma}(L, L) - \bar{\gamma}(l, l) = C[0\cdot841 - 0\cdot037] = 111\%^2,$$

or (ii) by assimilating the core length l to a quasi-point support and by considering a point spherical model with the same range $a = 40$ m and a sill $C_l = 133$,

$$D^2(l/L) \simeq D^2(0/L) = \bar{\gamma}(L, L) = C_l \times 0\cdot841 = 111\cdot9\%^2.$$

Both these theoretical values are close to the experimental value $s^2 = 120$.

2. *Regularized and graded variograms at Mounana after G. Matheron (1962)*

The massive Mounana (Gabon) uranium ore body was sampled by:

(i) vertical bore-holes cut into cores of length $l = 0\cdot5$ m and analysed for uranium grade;
(ii) several horizontal mine workings along which vertical channel samples of length $2l = 1$ m were taken at regular horizontal intervals of 1 m, cf. Fig. II.39.

As the histogram of the dispersion of the uranium grades in the deposit is clearly lognormal, the ReV studied is the logarithm of the grade which is then normally distributed.

Figure II.39 also shows, in semi-logarithmic coordinates:

(i) the experimental semi-variogram $\gamma_l^*(h)$ regularized by cores of length l and computed from the vertical information (bore-holes);
(ii) the experimental semi-variogram $\gamma_G^*(h)$ graded over the length $2l$ and computed from the horizontal information (channel samples).

As the experimental points of $\gamma_l^*(h)$ are reasonably well aligned, a logarithmic model regularized by cores of length $l = 0\cdot5$ was fitted according to the approximation (II.50):

$$\gamma_l(h) \simeq C\left[\log\frac{h}{l} + \frac{3}{2}\right] \quad \text{for } h \geqslant l,$$

h being a vertical distance, with $C = 0\cdot27$.

If the mineralization is isotropic in the three-dimensional space, it should be possible to fit the experimental horizontal semi-variogram $\gamma_G^*(h)$ to a logarithmic model graded over $2l$. This graded model is given by the same approximation as

$$\gamma_G(h) \simeq C\left[\log\frac{h}{2l} + \frac{3}{2}\right] = \gamma_l(h) - C\log 2 = \gamma_l(h) - 0\cdot187.$$

But the experimental curve $\gamma_G^*(h)$ is clearly above the theoretical straight line $\gamma_G(h)$ derived from the hypothesis of isotropy, and we can, thus, conclude that the phenomenon is anisotropic: the horizontal variability is greater than the vertical variability.

(Note that this study dealt only with small distances, h being less than 10 m in Fig. II.39.)

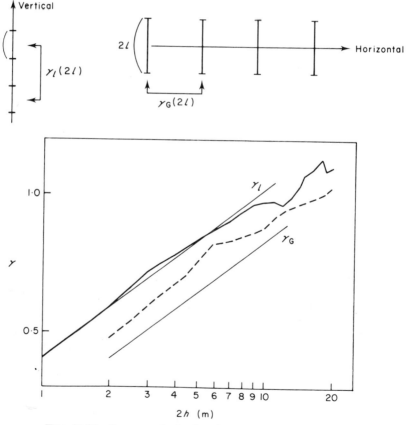

FIG. II.39. Core-regularized and graded semi-variograms.

II.E. CALCULATION OF THE MEAN VALUES $\bar{\gamma}$

The main geostatistical operations (estimation and dispersion variances, regularization, kriging) make continuous use of mean values $\bar{\gamma}$ of the point semi-variogram $\gamma(h)$ such as

$$\bar{\gamma}(v, v') = \frac{1}{vv'} \int_v dx \int_{v'} \gamma(x - x') \, dx'. \qquad \text{(II.65)}$$

$\bar{\gamma}(v, v')$ denotes the mean value of the function $\gamma(h)$ when one extremity of the vector h describes the domain v and the other extremity independently describes the domain v'.

It is recalled that if the domain v is three-dimensional, then the notation $\int_v dx$ in fact represents the triple integral $\iiint_v dx_u \, dx_v \, dx_w$, (x_u, x_v, x_w) being the three coordinates of the point x.

Thus, in the three-dimensional space, the mean value $\bar{\gamma}(v, v')$ between the two volumes v and v' involves a sextuple integral. The advantage of avoiding a direct analytical computation of these sextuple integrals is obvious.

Two solutions are possible. The first one is to calculate the mean values $\bar{\gamma}$ numerically using a computer. The second one is to break the analytical resolution of the multiple integrals into successive stages, some of them being calculated in advance once for all. These stages or intermediate integrations correspond to the definition of auxiliary functions which can be presented either as charts or in their exact analytical form.

II.E.1. Numerical calculation

This solution is completely general whatever the domains of integration v and v', and is often much more rapid than might be supposed at first, providing that there is no unnecessary generation of discrete points to represent the volumes v and v'.

Centred, regular grids of points $(x_i, i = 1$ to $n)$ and $(x_j, j = 1$ to $n')$ are placed over the domains v and v'. The multiple integral $\bar{\gamma}(v, v')$ is then approximated by the double discrete sum

$$\bar{\gamma}(v, v') \simeq \frac{1}{nn'} \sum_{i=1}^{n} \sum_{j=1}^{n'} \gamma(x_i - x'_j), \qquad \text{(II.66)}$$

cf. Fig. II.40, which represents a parallelepiped-shaped domain v and a domain v' consisting of a section of drill core.

Remark 1   The numerical integration formula (II.66) is the simplest discrete approximation method applying the same weight $(1/n)$ to each of the

n discrete points x_i which form a regular partition of the domain v. There are, of course, other methods of discrete approximation (the methods of Newton, Gauss, etc., cf. R. W. Hammings, (1971)) which sometimes provide more precise numerical estimation of the multiple integral $\bar{\gamma}$.

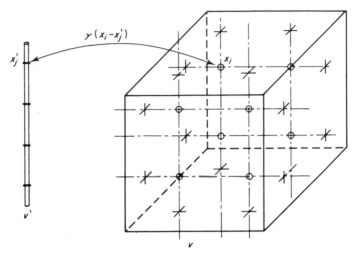

FIG. II.40. Discrete representation of the domains v and v'.

In practice, in mining geostatistics, taking account of

(i) the rapidity of calculation provided by the various micro-computers now available,

(ii) the small number n of discrete points required (see remark 3 hereafter),

(iii) the simplicity and generality of the method of centred regular discrete approximation with uniform weighting,

the numerical expression of formula (II.66) will be used, provided that care is taken to avoid any "zero effect" which may be produced when the two domains v and v' are not disjoint.

Remark 2   The zero effect. Whatever the adopted structural model, $\gamma(0) = 0$ by definition.

As a consequence, the two centred regular grids should be chosen so that none of the discrete approximation points (x_i, $i = 1$ to n) and (x'_j, $j = 1$ to n') coincide. Each time two points coincide ($x_i = x'_j$) in the numerical estimation of $\bar{\gamma}$, a greater importance is attached to the zero value $\gamma(0) = 0$. This zero effect will cause the mean value of the semi-variogram to be systematically underestimated and, correlatively, the mean value $\bar{C}$ of the corresponding covariance will be systematically overestimated as $C(h) = C(0) - \gamma(h)$.

Note that the zero effect can only be produced when the two domains v and v' are not disjoint, i.e., if there is a non-empty intersection of the two domains within which points may coincide. It is particularly dangerous when one of the two domains contains the other, e.g., when a mean value of the type $\bar{\gamma}(v, v)$ is to be estimated.

For such mean values $\bar{\gamma}(v, v)$, the zero effect can be avoided either by slightly shifting each point of the second grid $(x'_j, j = 1$ to $n')$ independently with respect to the corresponding point of the first grid $(x_i, i = 1$ to $n)$, or by using a more sophisticated weighting method for the calculation of the multiple integral $\bar{\gamma}(v, v)$ such as the Cauchy–Gauss method proposed in remark 5.

Remark 3   Required density of discrete approximation. The main rule is that the density of discrete approximation must be such that all the theoretical-order relations that are satisfied by the true mean values $\bar{\gamma}$ must also be satisfied by the numerical estimators $\bar{\gamma}^*$. Thus, with $v > v'$, it follows that $\bar{\gamma}(v, v) \geqslant \bar{\gamma}(v', v')$ and the two numerical estimators must satisfy the inequality $\bar{\gamma}^*(v, v) \geqslant \bar{\gamma}^*(v', v')$, otherwise the dispersion variance $D^2(v'/v)$ of v' within v will be negative.

The required density of discrete approximation can be found iteratively by progressively increasing the numbers n and n' until any further increase produces no significant improvement.

Experience has shown that in order to obtain an often unnecessary accuracy, too high densities were considered, leading to a substantial increase in computing time.

The densities of discrete approximation required for kriging have been calculated as to ensure a loss of less than 1% of the kriging variances (estimation variances minimized by the kriging procedure). In practice, these values have proved to be largely sufficient for most of the other purposes of the $\bar{\gamma}$ calculation.

In practical applications of mining geostatistics, the following maximum values are used:

$$\left.\begin{array}{l} \text{10 points for a one-dimensional domain,} \\ 6 \times 6 \text{ points for a two-dimensional domain,} \\ 4 \times 4 \times 4 \text{ points for a three-dimensional domain.} \end{array}\right\} \quad \text{(II.67)}$$

These maximum values are given only as indications. In many cases, common sense will allow the number of points to be considerably reduced. For example, if two domains v and v' are separated by a distance which is greater than the largest range of the model $\gamma(h)$, then $\bar{\gamma}(v, v') =$ sill value. Again, if the dimensions of the two domains v and v', separated by a distance h, are very small with respect to the range of the transition model $\gamma(h)$, then $\bar{\gamma}(v, v') \simeq \gamma(h)$. Now, if only the dimensions of v' are small with respect to this range, then v' can be approximated by its centre of gravity ($n' = 1$) and v can be approximated by n discrete points in the usual manner. Still again, if the domain v is a parallelepiped very much flattened in the vertical direction, and if the model $\gamma(h)$ is isotropic, then v can be approximated by its median horizontal plane – which is then approximated by discrete points over two dimensions ($n \leqslant 6 \times 6$ points); on the other hand, if there is a strong vertical variability represented by an anisotropic model $\gamma(h)$, it is advisable to approximate the vertical dimension of the parallelepiped v also.

Remark 4 *Mean values $\bar{\gamma}$ of linear combinations.* As the operation of taking the mean value $\bar{\gamma}(v, v')$ is linear in γ, more precisely, if $\gamma(h) = \sum_k C_k \gamma_k(h)$, then $\bar{\gamma}(v, v') = \sum_k C_k \bar{\gamma}_k(v, v')$, each of the terms $\bar{\gamma}_k(v, v')$ can be treated separately. Some of these terms can be estimated numerically, while others can be determined directly either from charts or auxiliary functions.

Remark 5 *Cauchy–Gauss method for calculating $\bar{\gamma}(v, v)$ and $\bar{\gamma}(v, v_h)$.* As the following developments (rather tedious) are simply exercises in numerical calculus and do not result in any new geostatistical concepts, the reader may go directly to the next section, "Computer subroutines", which will give, among others, the subroutine corresponding to this Cauchy–Gauss method.

Let v be any domain in one, two or three dimensions. The mean value $\bar{\gamma}(v, v)$ can be written in the form of a weighted mean in which the weighting function, $(1/v)k(x)$, is proportional to the indicator function $k(x)$ of the domain v:

$$\bar{\gamma}(v, v) = \frac{1}{v^2} \int_v dx \int_v \gamma(x - x') dx' = \frac{1}{v^2} \int_{-\infty}^{+\infty} k(x) dx \int_{-\infty}^{+\infty} \gamma(x - x')k(x') dx'$$

with

$$k(x) = \begin{cases} 1 & \text{if } x \in v \\ 0 & \text{if not} \end{cases} \quad \text{and} \quad \frac{1}{v} \int_{-\infty}^{+\infty} k(x) dx = 1.$$

By using the function $K = k * \check{k}$, auto-convolution of the indicator function k, the preceding multiple integral can be reduced to

$$\bar{\gamma}(v, v) = \frac{1}{v^2} \int_{-\infty}^{+\infty} \gamma(u)K(u) du \tag{II.68}$$

with

$$K(u) = k * \check{k} = \int_{-\infty}^{+\infty} k(u+x')k(x')\,dx'.$$

Formula (II.68) is merely the Cauchy algorithm for reducing a multiple integral of order $2n$ to a multiple integral of order n. This algorithm applies equally well to the calculation of the mean values $\bar{\gamma}(v, v_k)$ between two equal domains v and v_h translated from each other by the vector h. The formula then becomes

$$\bar{\gamma}(v, v_h) = \frac{1}{v^2} \int_v dx \int_{v_h} \gamma(x-x')\,dx' = \frac{1}{v^2} \int_{-\infty}^{+\infty} \gamma(h+u)K(u)\,du.$$
$$(\text{II.69})$$

The Cauchy algorithm for reducing the order of the integral is of interest only if the function $K(u)$ is known *a priori*. Now, for rectangular domains v with sides parallel to the system of axes, the corresponding function $K(u)$, also called the "geometric covariogram of v", can be readily determined.

(a) *In one dimension*, v is a segment of length l and its geometric covariogram is a function of one variable u:

$$K(u) = \begin{cases} l - |u| & \text{if } |u| \le l, \\ 0 & \text{if not.} \end{cases}$$

Consider now the symmetric function of $\gamma(h)$:

$$\Gamma(h+u) = \tfrac{1}{2}[\gamma(h+u)+\gamma(h-u)].$$

Then, for aligned segments of identical length l translated one from the other by the distance h, cf. Fig. II.41, the Cauchy algorithm (II.69) leads to the following expression for the mean value $\bar{\gamma}(l, l_h)$:

$$\bar{\gamma}(l, l_h) = 2 \int_0^1 (1-u)\Gamma(h+lu)\,du. \qquad (\text{II.70})$$

The mean value $\bar{\gamma}(l, l)$ is obtained by setting h to zero in formula (II.70).

FIG. II.41. Two aligned segments

(b) *In two dimensions*, v is a rectangle S with sides l_1 and l_2 parallel to the two-coordinate axes (u_1, u_2) and its geometric covariogram is a function of the two variables u_1 and u_2:

$$K(u_1, u_2) = \begin{cases} (l_1 - |u_1|)(l_2 - |u_2|) & \text{if } |u_1| \leq l_1 \quad \text{and} \quad |u_2| \leq l_2, \\ 0 & \text{if not.} \end{cases}$$

Consider now the symmetric function of the two-dimensional semi-variogram $\gamma(h_1, h_2)$:

$$\Gamma(h_1 + u_1, h_2 + u_2) = \frac{1}{4} \sum_{\varepsilon_1 = -1}^{\varepsilon_1 = +1} \sum_{\varepsilon_2 = -1}^{\varepsilon_2 = +1} \gamma(h_1 + \varepsilon_1 u_1, h_2 + \varepsilon_2 u_2).$$

Then, for two coplanar rectangles of same dimensions (l_1, l_2) and translated from each other by a vector h of coordinates (h_1, h_2) in the directions of the sides (l_1, l_2) of the rectangles, cf. Fig. II.42, Cauchy's algorithm (II.69) leads to the following expression for the mean value $\bar{\gamma}(S, S_h)$:

$$\bar{\gamma}(S, S_h) = 4 \int_0^1 \int_0^1 (1 - u_1)(1 - u_2)\Gamma(h_1 + l_1 u_1, h_2 + l_2 u_2) \, du_1 \, du_2.$$

$$(\text{II.71})$$

The mean value $\bar{\gamma}(S, S)$ is obtained by setting the vector h to zero in formula (II.71).

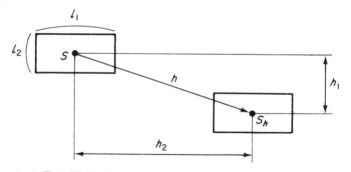

FIG. II.42. Two coplanar translated rectangles.

(c) *In three dimensions*, v is a rectangular parallelepiped with sides l_1, l_2, l_3 parallel to the three coordinate axes (u_1, u_2, u_3) and its geometric covariogram is a function of the three variables u_1, u_2, u_3:

$$K(u_1, u_2, u_3) =$$
$$\begin{cases} (l_1 - |u_1|)(l_2 - |u_2|)(l_3 - |u_3|) & \text{if } |u_1| \leq l_1, \quad |u_2| \leq l_2, \quad |u_3| \leq l_3, \\ 0 & \text{if not.} \end{cases}$$

Consider now the symmetric function of the three-dimensional semi-variogram $\gamma(h_1, h_2, h_3)$:

$$\Gamma(h_1+u_1, h_2+u_2, h_3+u_3) = \frac{1}{8} \sum_{\varepsilon_1=-1}^{\varepsilon_1=+1} \sum_{\varepsilon_2} \sum_{\varepsilon_3} \gamma(h_1+\varepsilon_1 u_1, h_2+\varepsilon_2 u_2, h_3$$
$$+\varepsilon_3 u_3).$$

Then, for two rectangular parallelepipeds with the same dimensions (l_1, l_2, l_3) and translated from each other by the vector h of coordinates (h_1, h_2, h_3) in the directions of the sides (l_1, l_2, l_3) of the parallelepipeds, cf. Fig. II.43, Cauchy's algorithm leads to the following expression for the mean value $\bar{\gamma}(v, v_h)$:

$$\bar{\gamma}(v, v_h) = 8 \int_0^1 \int_0^1 \int_0^1 (1-u_1)(1-u_2)(1-u_3)$$

$$\times \Gamma(h_1+l_1 u_1, h_2+l_2 u_2, h_3+l_3 u_3) \, du_1 \, du_2 \, du_3.$$

(II.72)

The mean value $\bar{\gamma}(v, v)$ is obtained by setting the vector h to zero in formula (II.72).

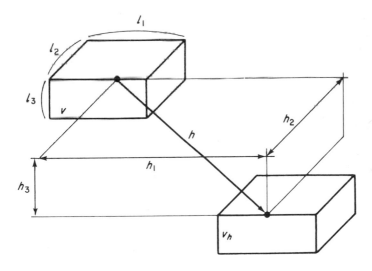

FIG. II.43. Two translated rectangular parallelepipeds.

Thus, the calculation of a mean value $\bar{\gamma}(v, v)$ over rectangular domains v amounts to the integration over the domains $(0, 1)$ of the nucleus $2(1-u)$, $4(1-u_1)(1-u_2)$ and $8(1-u_1)(1-u_2)(1-u_3)$ in one, two and three dimensions, respectively.

This integration can be carried out by choosing a regular centred grid of points within the intervals $(0, 1)$ and considering uniform weights ($1/n$ if n points are used). The Gauss numerical integration method provides much better precision for the same number n of points. For example, four abscissae points (x_i, $i = 1$ to 4) were chosen in the interval $(0, 1)$ with the corresponding four Gauss weights (λ_i, $i = 1$ to 4). The three integrals of formulae (II.70) to (II.72) are then estimated by the following discrete sums.

(a) *In one dimension*, the domain being a segment of length l:

$$\bar{\gamma}(l, l_h) \simeq \sum_{i=1}^{4} \lambda_i \Gamma(h + lx_i). \qquad (II.73)$$

(b) *In two dimensions*, the domain being a rectangle S of sides (l_1, l_2) parallel to the coordinate (h_1, h_2) directions of vector h:

$$\bar{\gamma}(S, S_h) \simeq \sum_{i=1}^{4} \sum_{j=1}^{4} \lambda_i \lambda_j \Gamma(h_1 + l_1 x_i, h_2 + l_2 x_j). \qquad (II.74)$$

(c) *In three dimensions*, the domain being a rectangular parallelepiped v of sides (l_1, l_2, l_3) parallel to the coordinate (h_1, h_2, h_3) directions of vector h:

$$\bar{\gamma}(v, v_h) \simeq \sum_{i=1}^{4} \sum_{j=1}^{4} \sum_{k=1}^{4} \lambda_i \lambda_j \lambda_k \Gamma(h_1 + l_1 x_i, h_2 + l_2 x_j, h_3 + l_3 x_k). \qquad (II.75)$$

The four abscissae x_i, x_j, x_k of the discrete approximation and the corresponding weights λ_i, λ_j, λ_k have been calculated by A. Marechal (1976, internal report) and are given in Table II.1.

TABLE II.1. *Parameters of the Gauss numerical integration* ($n = 4$ *points*)

	1	2	i 3	4
x_i	0·0571	0·2768	0·5836	0·860 25
λ_i	0·2710	0·4069	0·2597	0·062 35

$\sum_{i=1}^{4} \lambda_i = 0·999\ 94 \simeq 1$

Thus, *in practice*, the Cauchy–Gauss procedure reduces the numerical calculation of the mean values $\bar{\gamma}(v, v_h)$ and $\bar{\gamma}(v, v)$ to that of 4^p values of the symmetric semi-variogram function $\Gamma(u)$, where p is the dimension of the domain v considered (in practice, $p = 1$, 2 or 3 dimensions).

For a spherical or exponential model $\gamma(h)$ and for practical values of the range, the relative precision is always better than 1%.

Computer subroutines

The following subroutines calculate the mean value $\bar{\gamma}(v, v')$ between any two rectangular domains v and v', whatever their respective positions in relation to the system of axes on which the model $\gamma(h)$ is defined.

The first subroutine GBAR is completely general and considers a discrete approximation of each of the domains v and v' by a centred regular grid of n points.

The second subroutine F is more particular but much less time-consuming, and considers the Cauchy–Gauss discrete approximation to calculate the mean value $\bar{\gamma}(v, v)$ within a single domain v.

FUNCTION GBAR(– – –)

v and v' can be any domain, provided that they are of rectangular form. The dimensions of the two domains do not have to be the same, e.g., v can be a three-dimensional parallelepiped (a mining block) while v' is a segment (core samples along a bore-hole), cf. Fig. II.44.

Each of the two domains is approximated by a centred regular grid of n points, n taking the values given in (II.67).

When the two domains v and v' have a non-empty intersection, and particularly when one contains the other, the index INTER $= 1$ allows

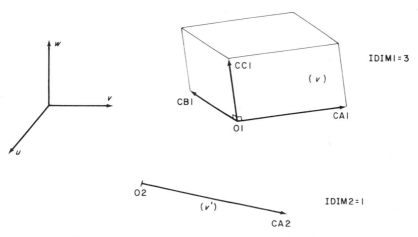

FIG. II.44. Coding the domains v and v' in GBAR.

avoiding the zero effect: each point of the discrete approximation of one domain is shifted randomly from its corresponding node on the regular centred grid.

The three-dimensional point semi-variogram $\gamma(h)$ is read as a sub-routine FUNCTION GAM(HX, HY, HZ). The three coordinates (HX, HY, HZ) of the vector h are relative to the reference system of axes. (In section III.B.5, an example of FUNCTION GAM corresponding to a very general structural model $\gamma(h)$ is given.)

Relative to the reference system of axes, each domain v or v' is defined by one apex and one, two or three side vectors according to whether the domain is one-, two- or three-dimensional. Thus, in the example of Fig. II.44,

   (i) the rectangular parallelepiped v is defined by the coordinates of one of its apex 01 and the coordinates of the three side vectors CA1, CB1 and CC1.
   (ii) the segment v' is defined by the coordinates of one of its extremities 02 and the coordinates of the vector CA2.

```
      FUNCTICN GBAR(CA1,CB1,CC1,CC1,ICIM1,CA2,CB2,
     1CC2,CC2,ICIM2,INTER)
C
C  ........................................................
C      COMPLTES THE AVERAGE VALLE CF GAMMA
C BETWEEN TWC COMAINS WITH ANY CRIENTATICN WITH
C RESPECT TC COORDINATES SYSTEM..EITHER CCMAIN MIGHT
C BE A SEGMENT,RECTANGLE OR PARALLELEPIPED..DOMAINS
C MIGHT BE CF A CIFFERENT CIMENSION..
C
C  ....PARAMETERS
C   CA1(3),CB1(3),CC1(3)  VECTCRS DEFINING CCMAIN 1
C               VECTORS HAVE CCMMCN CRIGIN LCCATED
C               ON CNE OF THE APEXES OF THE CCMAIN
C   CO1(3)  CCORDINATES CF THIS REFERENCE AFEX
C   IDIM1  NUMBER OF CIMENSIONS CF DOMAIN 1
C      =1  SEGMENT       VECTCR CA1
C      =2  RECTANGLE     VECTCRS CA1,CB1
C      =3  PARALLELEPIPED VECTCRS CA1,CB1,CC1
C      UNUSEC VECTCRS SHOULC BE PUT TC C.
C   CA2(3),CB2(3),CC2(3),CO2(3),ICIM2  CEFINITICN CF
C                       CCMAIN 2
C
C  ....SUBRCLTINES CALLEC
C        GAM(HX,HY,HZ) SEMI-VARIOGRAM FLNCTICN
C        RANCOM(K) RANDCM NLMBERS GENERATCR (C,1)
C
C  ....CPTICNS
```

```
C   INTER=1 CORRECTION OF INTEGRATION ERROR BY
C            GENERATION OF A RANDOM STRATIFIED GRID
C            ON SECOND DOMAIN
C   ............................................
C
      DIMENSION CA1(3),CB1(3),CC1(3),CA2(3),CB2(3),
     1CC2(3),CO1(3),CO2(3)
      DIMENSION L1(3),L2(3),L3(3)
      DIMENSION X1(3),X2(64,3),AL(3)
      DATA L1/10,6,4/,L2/1,6,4/,L3/1,1,4/
      GB=0.
      N1=L1(IDIM1)
      N2=L2(IDIM1)
      N3=L3(IDIM1)
      M1=L1(IDIM2)
      M2=L2(IDIM2)
      M3=L3(IDIM2)
      IJK=0
      IF(INTER.EQ.1)GO TO 200
C
C                   GRIDDING OF SECOND DOMAIN
C
      DO 100 I2=1,M1
      DO 100 J2=1,M2
      DO 100 K2=1,M3
      IJK=IJK+1
      DO 110 IC=1,3
      X2(IJK,IC)=CO2(IC)+(I2-0.5)*CA2(IC)/M1+
     1(J2-0.5)*CB2(IC)/M2+(K2-0.5)*CC2(IC)/M3
  110 CONTINUE
  100 CONTINUE
      GO TO 300
C
  200 DO 210 I2=1,M1
      DO 210 J2=1,M2
      DO 210 K2=1,M3
      IJK=IJK+1
      DO 220 IL=1,3
  220 AL(IL)=-RANDOM(K)
      DO 230 IC=1,3
      X2(IJK,IC)=CO2(IC)+(I2+AL(1))*CA2(IC)/M1+
     1(J2+AL(2))*CB2(IC)/M2+(K2+AL(3))*CC2(IC)/M3
  230 CONTINUE
  210 CONTINUE.
C
C                   COMPUTE GBAR
C
  300 DO 1 I1=1,N1
      DO 1 J1=1,N2
      DO 1 K1=1,N3
      DO 10 IC=1,3
      X1(IC)=CO1(IC)+(I1-0.5)*CA1(IC)/N1+
```

```
     1(J1-0.5)*CB1(IC)/N2+(K1-C.5)*CC1(IC)/N3
 10  CONTINUE
     DO 2  IJ=1,IJK
     HX=X1(1)-X2(IJK,1)
     HY=X1(2)-X2(IJK,2)
     HZ=X1(3)-X2(IJK,3)
     GB=GB+GAM(HX,HY,HZ)
  2  CONTINUE
  1  CCNTINUE
     GBAR=CB/(N1*N2*N3)/IJK
     RETURN
     END
```

FUNCTION F(– – –)

This subroutine calculates the mean value $\bar{\gamma}(v, v)$ over a rectangular domain of any dimension (segment, rectangle or parallelepiped). The subroutine considers the Cauchy–Gauss discrete approximation method and the corresponding formulae (II.73) to (II.75).

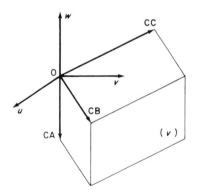

FIG. II.45. Coding the domain v in FUNCTION F.

The three-dimensionnal point semi-variogram $\gamma(h)$ is read as a subroutine FUNCTION GAM(HX, HY, HZ), the three coordinates (HX, HY, HZ) of the vector h are relative to the reference system of axes.

Relative to this reference system of axes, the domain v is defined by one apex considered as being the origin O of the coordinates and by one, two or three side vectors. Thus, on the example of Fig. II.45, the rectangular parallelepiped v is defined by the coordinates of its three side vectors CA, CB and CC.

```
      FUNCTION F(CA,CB,CC, IDIM)
C
C  ................................................
C       COMPUTES THE AVERAGE VALUE OF GAMMA
C WITHIN A SEGMENT,RECTANGLE OR PARALLELEPIPED
C FAVING ANY ORIENTATION WITH RESPECT TO COORDI-
C NATES SYSTEM..DOMAIN IS DEFINED BY VECTORS..ALGO-
C RITHM IS CAUCHY-GAUSS:POINTS P(4),WEIGHTS W(4)..
C
C ....PARAMETERS
C   CA(3),CB(3),CC(3)   VECTORS DEFINING DOMAIN
C               VECTORS HAVE COMMON ORIGIN LOCATED
C               ON ONE OF THE APEXES OF THE DOMAIN
C    IDIM   NUMBER OF DIMENSIONS OF DOMAIN
C       =1   SEGMENT VECTOR  CA
C       =2   RECTANGLE VECTORS CA,CB
C       =3   PARALLELEPIPED VECTORS CA,CB,CC
C
C ....SUBROUTINE CALLED
C        GAM(HX,HY,HZ) SEMI-VARIOGRAM FUNCTION
C  ................................................
C
      DIMENSION CA(3),CB(3),CC(3),P(4),W(4)
      DATA P/.0571,.2766,.5836,.86025/
      DATA W/.2710,.4069,.25969,.06235/
C
      F=0.
      GO TO (1,2,3),IDIM
C
C                      SEGMENT
C
    1 DO 10 I=1,4
      HX=P(I)*CA(1)
      HY=P(I)*CA(2)
      HZ=P(I)*CA(3)
   10 F=F+W(I)*GAM(HX,HY,HZ)
      RETURN
C
C                      RECTANGLE
C
    2 DO 20 I=1,4
      DO 20 J=1,4
      G=0.
      J1=-1
      J2=-1
      DO 21 IN=1,2
      J1=-J1
      DO 21 IM=1,2
      J2=-J2
      HX=J1*P(I)*CA(1)+J2*P(J)*CB(1)
      HY=J1*P(I)*CA(2)+J2*P(J)*CB(2)
      HZ=J1*P(I)*CA(3)+J2*P(J)*CB(3)
```

```
 21  G=G+GAM(HX,HY,HZ)
     F=F+W(I)*W(J)*G/4.
 20  CONTINUE
     RETURN
C
C                                        PARALLELEPIPED
C
  3  DO 30 I=1,4
     DO 30 J=1,4
     DO 30 K=1,4
     G=0.
     J1=-1
     J2=-1
     J3=-1
     DO 31 IN=1,2
     J1=-J1
     DO 31 IM=1,2
     J2=-J2
     DO 31 IL=1,2
     J3=-J3
     HX=J1*P(I)*CA(1)+J2*P(J)*CE(1)+J3*F(K)*CC(1)
     HY=J1*P(I)*CA(2)+J2*P(J)*CE(2)+J3*F(K)*CC(2)
     HZ=J1*P(I)*CA(3)+J2*P(J)*CE(3)+J3*F(K)*CC(3)
 31  G=G+GAM(HX,HY,HZ)
     F=F+W(I)*W(J)*W(K)*G/8.
 30  CONTINUE
     RETURN
     END
```

II.E.2. Auxiliary functions

An auxiliary function is a pre-calculated mean value $\bar{\gamma}(v, v)$ corresponding to particularly simple geometries of v and v', which are frequently found in practice. There are four basic auxiliary functions, denoted by α, χ, F and H, defined for rectangular shaped domains in one or two dimensions.

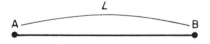

FIG. II.46. Segment AB of length L.

Let the point variogram $\gamma(h)$ be *isotropic*:

$$\gamma(h) = \gamma(r) \quad \text{with } r = |h|.$$

Now, *in one dimension*, let AB be a segment of length L as in Fig. II.46.

The auxiliary function $\chi(L)$ is defined as the mean value of $\gamma(h)$ when the extremity of the vector h is fixed at the point A and the other extremity describes the segment AB of length L:

$$\chi(L) = \bar{\gamma}(A, AB) = \frac{1}{L} \int_0^L \gamma(u) \, du. \qquad (II.76)$$

The auxiliary function $\chi(L)$ thus represents a simple integral of the semi-variogram $\gamma(h)$.

The auxiliary function $F(L)$ is defined as the mean value of $\gamma(h)$ when the two extremities of the vector h describe, independently of each other, the segment AB of length L:

$$F(L) = \bar{\gamma}(AB, AB) = \frac{1}{L^2} \int_0^L du \int_0^L \gamma(u - u') \, du'. \qquad (II.77)$$

The auxiliary function $F(L)$ thus represents a double integral of $\gamma(h)$. By means of the Cauchy algorithm of formula (II.70), this double integral can be reduced to the following simple integrals:

$$F(L) = \frac{2}{L^2} \int_0^L (L - u)\gamma(u) \, du = \frac{2}{L^2} \int_0^L u\chi(u) \, du. \qquad (II.78)$$

In two dimensions, let ABCD be a rectangle with sides (L, l) as in Fig. II.47.

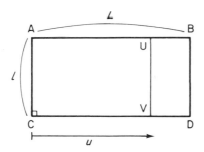

FIG. II.47. Rectangle ABCD of sides (L, l).

The two-variable auxiliary function $\alpha(L; l)$ is defined as the mean value of $\gamma(h)$ when one extremity of the vector h describes a side of length l and the other extremity independently describes the opposite side at a distance L:

$$\alpha(L; l) = \bar{\gamma}(AC, BD).$$

It can be shown that this double integral of the isotropic function $\gamma(r)$ can be reduced to the following simple integral:

$$\alpha(L;l) = \bar{\gamma}(AC, BD) = \frac{2}{l^2} \int_0^l (l-u)\gamma(\sqrt{(L^2+u^2)})\,du. \qquad (II.79)$$

Note that the auxiliary function $\alpha(L;l)$ is not symmetric in $(L;l)$:

$$\alpha(l;L) = \bar{\gamma}(AB, CD) \neq \bar{\gamma}(AC, BD).$$

The two-variable auxiliary function $\chi(L;l)$ is defined as the mean value of $\gamma(h)$ when one extremity of the vector h describes a side of length l and the other extremity independently describes the entire rectangle ABCD:

$$\chi(L;l) = \bar{\gamma}(AC, ABCD) = \bar{\gamma}(BD, ABCD).$$

The auxiliary function $\chi(L;l)$ thus represents a triple integral of the isotropic function $\gamma(r)$, and appears as the mean value of the auxiliary function $\alpha(u;l) = \bar{\gamma}(AC, UV)$ when the side UV describes the entire rectangle ABCD, as in Fig. II.47. The triple integral can then be reduced to the following simple integral:

$$\chi(L;l) = \bar{\gamma}(AC, ABCD) = \frac{1}{L} \int_0^L \alpha(u;l)\,du. \qquad (II.80)$$

Note that the auxiliary function $\chi(L;l)$ is not symmetric in $(L;l)$:

$$\chi(l;L) = \bar{\gamma}(AB, ABCD) \neq \bar{\gamma}(AC, ABCD).$$

The two-variable auxiliary function $F(L;l)$ is defined as the mean value of $\gamma(h)$ when both extremities of the vector h describe, independently of each other, the entire rectangle ABCD:

$$F(L;l) = \bar{\gamma}(ABCD, ABCD) \qquad (II.81)$$

This auxiliary function is symmetrical in $(L;l)$. It represents a quadruple integral of the isotropic function $\gamma(r)$. The Cauchy algorithm of formula (II.70) can be applied to the auxiliary functions $\alpha(L;l)$ and $\chi(L;l)$ to reduce this quadruple integral to the following simple integrals:

$$F(L;l) = \frac{2}{L^2} \int_0^L (L-u)\alpha(u;l)\,du = \frac{2}{L^2} \int_0^L u\chi(u;l)\,du. \qquad (II.82)$$

Note the similarity between these formulae and those of (II.78) corresponding to one dimension. In fact, the two-dimensional function $F(L;l)$ is nothing other than a one-dimensional function $F(L)$ with the mean value over l having been taken beforehand.

The two-variable auxiliary function $H(L;l)$ is defined as the mean value of $\gamma(h)$ when one extremity of the vector h is fixed at any one of the corners of the rectangle and the other extremity independently describes the entire rectangle ABCD:

$$H(L;l) = \bar{\gamma}(A, ABCD). \tag{II.83}$$

It can be shown that $H(L;l)$ is also equal to the mean value of $\gamma(h)$ when one extremity of the vector h describes one side of length l and the other extremity independently describes the adjacent side of length L:

$$H(L;l) = \bar{\gamma}(AC, AB). \tag{II.84}$$

The auxiliary function $H(L;l)$ is symmetric in $(L;l)$ and represents a double integral of the isotropic function $\gamma(r)$. It can be shown that $H(L;l)$ can be deduced from the auxiliary functions $\chi(L;l)$ and $F(L;l)$ by simple and double differentiation, respectively:

$$H(L;l) = \frac{1}{2l}\frac{\partial}{\partial l}l^2\chi(L;l) = \frac{1}{4lL}\frac{\partial^2}{\partial l\,\partial L}l^2L^2F(L;l). \tag{II.85}$$

Remarks   The two-variable auxiliary functions can be reduced to the corresponding one-variable auxiliary function by contracting one of the sides of the rectangle ABCD to zero, e.g., $l \to 0$, as in Fig. II.48. The two-variable auxiliary functions then become

$$\begin{aligned}
\alpha(L;0) &= \bar{\gamma}(A,B) = \gamma(L) & \text{but} & \quad \alpha(0;l) = \bar{\gamma}(AC, AC) = F(l); \\
\chi(L;0) &= \bar{\gamma}(A, AB) = \chi(L) & \text{but} & \quad \chi(0;l) = \bar{\gamma}(AC, AC) = F(l); \\
F(L;0) &= \bar{\gamma}(AB, AB) = F(L) & \text{and} & \quad F(0;l) = \bar{\gamma}(AC, AC) = F(l); \\
H(L;0) &= \bar{\gamma}(A, AB) = \chi(L) & \text{and} & \quad H(0;l) = \bar{\gamma}(A, AC) = \chi(l).
\end{aligned} \right\} \tag{II.86}$$

FIG. II.48.   Rectangle ABCD compressed $(l = 0)$.

The auxiliary function $F(L;l)$ corresponds to the highest order (4) multiple integral of $\gamma(h)$, all other auxiliary functions can be deduced from $F(L;l)$ by differentiation or by setting a parameter to zero:

$$\begin{aligned}
\alpha(L;l) &= \frac{1}{2}\frac{\partial^2}{\partial L^2}L^2F(L;l) = \frac{\partial}{\partial L}L\chi(L;l); \\[4pt]
\chi(L;l) &= \frac{1}{2L}\frac{\partial}{\partial L}L^2F(L;l) \quad \text{and} \quad \chi(L) = \chi(L;0); \\[4pt]
H(L;l) &= \frac{1}{4lL}\frac{\partial^2}{\partial l\,\partial L}l^2L^2F(L;l); \\[4pt]
F(L) &= F(L;0).
\end{aligned} \right\} \tag{II.87}$$

In three dimensions, the three-variable auxiliary functions α, χ, F and H can be defined for a rectangular parallelepiped (l_1, l_2, l_3). In practice, these three-variable functions are too awkward to use and the three-dimensional case is restricted to the two-variable functions defined on a parallelepiped $P = ABCDEFGH$ with a *square* base $(L \times l^2)$ as on Fig. II.49. The following auxiliary functions can then be defined:

(i) the mean value of $\gamma(r)$ between the two square faces (l^2) of P,

$$\alpha(L; l^2) = \bar{\gamma}(ACEG, BDFH); \qquad (II.88a)$$

(ii) the mean value of $\gamma(r)$ between one square face and P,

$$\chi(L; l^2) = \bar{\gamma}(ACEG, P); \qquad (II.88b)$$

(iii) the mean value of $\gamma(r)$ within P,

$$F(L; l^2) = \bar{\gamma}(P, P); \qquad (II.88c)$$

(iv) the mean value of $\gamma(r)$ between two adjacent faces of P which is equal to the mean value between a side l and P,

$$H(L; l^2) = \bar{\gamma}(ACEG, ABEF) = \bar{\gamma}(AE, P). \qquad (II.88d)$$

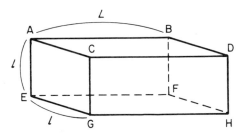

FIG. II.49. Parallelepiped with a square base.

Remark   By contracting one of the sides of the parallelepiped P to zero $(L \to 0)$ these three-dimensional functions can be reduced to the two-dimensional auxiliary functions defined on the square l^2.

Similarly, by contracting two of the three sides of P to zero $(l \to 0)$, these three-dimensional functions can be reduced to the corresponding one-dimensional auxiliary functions defined on the segment L.

The following relations are obtained:

$$\left.\begin{aligned}
\alpha(L; l^2 = 0) &= \gamma(L) \quad \text{and} \quad \alpha(0; l^2) = F(l; l); \\
\chi(L; l^2 = 0) &= \chi(L) \quad \text{and} \quad \chi(0; l^2) = F(l; l); \\
F(L; l^2 = 0) &= F(L) \quad \text{and} \quad F(0; l^2) = F(l; l); \\
H(L; l^2 = 0) &= \chi(L) \quad \text{and} \quad H(0; l^2) = \chi(l; l).
\end{aligned}\right\} \qquad (II.89)$$

Mathematical expressions and charts

Expressions or charts of these auxiliary functions are given for each of the four-point *isotropic* models $\gamma(r)$ which are in current use.

1. *The linear model*

$$\gamma(h)=r \quad \text{with } r=|h|.$$

In one dimension,

$$\chi(L)=L/2 \quad \text{and} \quad F(L)=L/3. \tag{II.90}$$

In two dimensions,

$$\alpha(L;l)=\frac{1}{3}\sqrt{(L^2+l^2)}+\frac{2}{3}\frac{L^2}{l^2}(L-\sqrt{(L^2+l^2)})+\frac{L^2}{l}\log\frac{l+\sqrt{(L^2+l^2)}}{L};$$

$$\chi(L;l)=\frac{1}{6}\frac{L^3}{l^2}+\sqrt{(L^2+l^2)}\left(\frac{1}{4}-\frac{1}{6}\frac{L^2}{l^2}\right)+\frac{1}{3}\frac{L^2}{l}\log\frac{l+\sqrt{(L^2+l^2)}}{L}$$

$$+\frac{1}{12}\frac{l^2}{L}\log\frac{L+\sqrt{(L^2+l^2)}}{l};$$

$$F(L;l)=\sqrt{(L^2+l^2)}\left(\frac{1}{5}-\frac{1}{15}\frac{L^2}{l^2}-\frac{1}{15}\frac{l^2}{L^2}\right)+\frac{1}{15}\left(\frac{L^3}{l^2}+\frac{l^3}{L^2}\right)$$

$$+\frac{1}{6}\frac{L^2}{l}\log\frac{l+\sqrt{(L^2+l^2)}}{L}+\frac{1}{6}\frac{l^2}{L}\log\frac{L+\sqrt{(L^2+l^2)}}{l};$$

$$H(L;l)=\frac{1}{3}\sqrt{(L^2+l^2)}+\frac{l^2}{6L}\log\frac{L+\sqrt{(L^2+l^2)}}{l}+\frac{L^2}{6l}\log\frac{l+\sqrt{(L^2+l^2)}}{L}.$$

$$\tag{II.91}$$

Note the particular case corresponding to the squares, $L=l$:

$$\alpha(l;l)=1\cdot0765l, \qquad \chi(l;l)=0\cdot7351l,$$

$$F(l;l)=0\cdot5213l, \qquad H(l;l)=0\cdot7652l.$$

2. *The logarithmic model*

$$\gamma(h)=\log r \quad \text{with } r=|h|.$$

In one dimension,

$$\chi(L)=\log L-1 \quad \text{and} \quad F(L)=\log L-3/2. \tag{II.92}$$

In two dimensions, with $\theta = \text{arc tan }(l/L)$,

$$\alpha(L;l) = \log l - \log \sin \theta - \frac{3}{2} + \frac{2\theta}{\tan \theta} + \frac{\log \cos \theta}{\tan^2 \theta};$$

$$\chi(L;l) = \log l - \log \sin \theta - \frac{11}{6} + \frac{\theta}{\tan \theta} + \frac{\log \cos \theta}{3.\tan^2 \theta}$$

$$+ \left(\frac{\pi}{2} - \theta\right)\frac{\tan \theta}{3};$$

$$F(L;l) = \log l - \log \sin \theta - \frac{25}{12} + \frac{\tan^2 \theta}{6}\log \sin \theta$$

$$+ \frac{\log \cos \theta}{6 \tan^2 \theta} + \frac{2}{3}\left(\frac{\pi}{2} - \theta\right)\tan \theta. \tag{II.93}$$

These exact formulae are rather tedious to use. In practice, it is better to use their limited expansions as functions of L/l or l/L depending on whether L is less than or greater than l. For $L < l$, we have

$$\alpha(L;l) = \log l - \frac{3}{2} + \pi\frac{L}{l} + \frac{L^2}{l^2}\log\frac{L}{l} - \frac{3}{2}\frac{L^2}{l^2} - \frac{1}{12}\frac{L^4}{l^4} + \ldots;$$

$$\chi(L;l) = \log l - \frac{3}{2} + \frac{\pi}{2}\frac{L}{l} + \frac{1}{3}\frac{L^2}{l^2}\log\frac{L}{l} - \frac{11}{18}\frac{L^2}{l^2} - \frac{1}{60}\frac{L^4}{l^4} + \ldots; \tag{II.94}$$

$$F(L;l) = \log l - \frac{3}{2} + \frac{\pi}{3}\frac{L}{l} + \frac{1}{6}\frac{L^2}{l^2}\log\frac{L}{l} - \frac{25}{72}\frac{L^2}{l^2} - \frac{1}{180}\frac{L^4}{l^4} + \ldots.$$

For $L > l$, we have

$$\alpha(L;l) = \log L + \frac{1}{12}\frac{l^2}{L^2} - \frac{1}{60}\frac{l^4}{L^4} + \ldots;$$

$$\chi(L;l) = \log L - 1 + \frac{\pi}{6}\frac{l}{L} - \frac{1}{12}\frac{l^2}{L^2} + \frac{1}{180}\frac{l^4}{L^4} + \ldots; \tag{II.95}$$

$$F(L;l) = \log L - \frac{3}{2} + \frac{\pi}{3}\frac{l}{L} + \frac{1}{6}\frac{l^2}{L^2}\log\frac{l}{L} - \frac{25}{72}\frac{l^2}{L^2} - \frac{1}{180}\frac{l^4}{L^4} + \ldots.$$

Remark　For values $r = |h| < 1$ the function $\log r$ has negative values. Consequently, for small values of L and l, the auxiliary functions α, χ, F can take negative values. Every estimation or dispersion variance defined on non-point supports and calculated with these auxiliary functions is, nevertheless, always positive, since every regularization of the logarithmic model is a conditional positive definite model.

Linear equivalents

From formula (II.92), the one-dimensional auxiliary function $F(l)$ is written as

$$F(l) = \log l - 3/2.$$

The auxiliary function $F(a; b)$ for a rectangle with sides (a, b) and $a > b$, is given by the first three terms of the limited expansion (II.95):

$$F(a; b) \simeq \log a - \frac{3}{2} + \frac{\pi}{3}\frac{b}{a} \simeq \log a - \frac{3}{2} + \frac{b}{a}$$

(by considering that $\pi/3 \simeq 1$).

These three terms are also the terms resulting from a limited expansion of the function $\log(a + b)$ with $a > b$, i.e.,

$$F(a; b) \simeq \log(a+b) - \frac{3}{2} = \log\left[a\left(1 + \frac{b}{a}\right)\right] - \frac{3}{2} = \log a - \frac{3}{2} + \frac{b}{a} + \dots$$

Thus, the two-dimensional auxiliary function $F(a; b)$, for the rectangle with sides a and b, is approximately equal to the one-dimensional auxiliary function $F(l)$ for the segment $l = a + b$:

$$F(a; b) \simeq F(l) = \log l - 3/2 \quad \text{with } l = a + b. \tag{II.96}$$

The segment of length $l = a + b$ is said to be the "*linear equivalent*" of the rectangles with sides a and b.

More generally, when v is any domain in one, two or three dimensions and not necessarily rectangular, the linear equivalent of v is the segment of length l such that

$$\bar{\gamma}(v, v) = F(l) = \log l - 3/2. \tag{II.97}$$

For the logarithmic model, the approximate linear equivalent of a rectangular parallelepiped of sides $(a \geqslant b \geqslant c)$ is $l = a + b + 0 \cdot 7c$.

Application Let v and $V (v \subset V)$ be two domains with linear equivalents l and L, respectively. The variance of dispersion of v in V is given, according to formula (II.36), by

$$D^2(v/V) = \bar{\gamma}(V, V) - \bar{\gamma}(v, v) = (\log L - 3/2) - (\log l - 3/2) = \log(L/l).$$

If the two domains v and V are geometrically similar, e.g., if they are homothetic parallelepipeds, then

$$V/v = (L/l)^3 \quad \text{and} \quad D^2(v/V) = \tfrac{1}{3}\log(V/v).$$

3. *The spherical model*

$$\gamma(h) = \begin{cases} \dfrac{3}{2}\dfrac{r}{a} - \dfrac{1}{2}\dfrac{r^3}{a^3} & \text{for } r = |h| \in [0, a], \\ 1 = \text{sill} & \text{for } r \geqslant a \text{ (range)} \end{cases}$$

In one dimension, the auxiliary functions become

$$\chi(L) = \begin{cases} \dfrac{3}{4}\dfrac{L}{a} - \dfrac{1}{8}\dfrac{L^3}{a^3}, & \forall L \in [0, a], \\ 1 - \dfrac{3}{8}\dfrac{L}{a}, & \forall L \geqslant a; \end{cases} \qquad (\text{II.98})$$

$$F(L) = \begin{cases} \dfrac{1}{2}\dfrac{L}{a} - \dfrac{1}{20}\dfrac{L^3}{a^3}, & \forall L \in [0, a], \\ 1 - \dfrac{3}{4}\dfrac{a}{L} + \dfrac{1}{5}\dfrac{a^2}{L^2}, & \forall L \geqslant a. \end{cases}$$

In two dimensions, the exact expressions for the auxiliary functions $\chi(L; l)$; $F(L; l)$ and $H(L; l)$ can be found in J. Serra (1967a) and D. Guibal (1973a). In practice, the following charts, which are graduated in terms of L/a and l/a, are used.

Chart no. 1 gives the function $\gamma_G(L) = \alpha(L; l) - \alpha(0; l) = \alpha(L; l) - F(l)$. This function is simply the semi-variogram graded over the constant thickness l, cf. section II.D.3, formula (II.55). As $F(l)$ is given by formula (II.98), this chart also provides the values of the function $\alpha(L; l)$.

Chart no. 2 gives the function $\chi(L; l)$.
Chart no. 3 gives the function $H(L; l)$.
Chart no. 4 gives the function $F(L; l)$.
Chart no. 5 gives the function $F(L; l^2)$.
Chart no. 6 gives the difference $\alpha(L; l^2) - \alpha(0, l^2) = \alpha(L; l^2) - F(l; l)$

4. *The exponential model*

$$\gamma(h) = 1 - \exp(-r/a) \quad \text{with } r = |h|.$$

In one dimension, the auxiliary function becomes

$$\left.\begin{aligned} \chi(L) &= 1 - \frac{a}{L}(1 - e^{-L/a}), \\[2mm] F(L) &= 1 - \frac{2a}{L}\left[1 - \frac{a}{L}(1 - e^{-L/a})\right]. \end{aligned}\right\} \qquad (\text{II.99})$$

In two dimensions, the exact expressions for the auxiliary functions can be found in J. Serra (1967) and D. Guibal (1973a). In practice, the following charts, graduated in terms of L/a and l/a, are used.

Chart no. 11 gives the function $\gamma_G(L) = \alpha(L; l) - \alpha(0; l) = \alpha(L; l) - F(l)$.
Chart no. 12 gives the function $\chi(L; l)$.
Chart no. 13 gives the function $H(L; l)$.
Chart no. 14 gives the function $F(L; l)$.
Chart no. 15 gives the function $F(L; l^2)$.
Chart no. 16 gives the difference $\alpha(L; l^2) - \alpha(0; l^2) = \alpha(L; l^2) - F(l; l)$.

5. The r^θ model

$$\gamma(h) = r^\theta \quad \text{with } r = |h| \quad \text{and} \quad \theta \in \,]0, 2[.$$

The function r^θ is conditional positive definite only when the parameter θ is contained in the interval $]0, 2[$, the limits 0 and 2 being precluded.

Although this model is not currently used in practice, it has some theoretical importance and, therefore, the following one-dimensional auxiliary function can be of interest:

$$\left.\begin{array}{l} \chi(L) = l^\theta/(\theta + 1), \\[2mm] F(L) = 2l^\theta/[(\theta + 1)(\theta + 2)]. \end{array}\right\} \tag{II.100}$$

II.E.3. Using the auxiliary functions

In mining applications, the geometry of the data supports and of the blocks to be estimated is very often rectangular, and, thus, the previously defined and pre-calculated auxiliary functions can be used when calculating the mean values $\bar{\gamma}$.

By way of example, several geostatistical operations will be given, with results that can be read from charts.

More charts for the spherical and exponential models

All developments given hereafter correspond to an *isotropic* point model $\gamma(h)$, i.e., such that $\gamma(h) = \gamma(r)$ with $r = |h|$.

The following elementary estimation variances are given in Charts no. 7 and 17.

(i) The estimation variance of a segment AB of length l by a central sample 0. This variance can be written, according to the general formula (II.27), as

$$\sigma_{E_1}^2 = 2\bar{\gamma}(0, \text{AB}) - \bar{\gamma}(\text{AB}, \text{AB}) - \bar{\gamma}(0, 0)$$

with

$$\bar{\gamma}(0, AB) = \bar{\gamma}(0, 0A) = \bar{\gamma}(0, 0B) = \chi(l/2);$$

because of symmetry,

$$\bar{\gamma}(AB, AB) = F(l) \quad \text{and} \quad \bar{\gamma}(0, 0) = \gamma(0) = 0$$

and the extension variance becomes

$$\sigma_{E_1}^2 = 2\chi(l/2) - F(l)$$

(ii) The estimation variance of a segment AB of length l by two samples located at its extremities A and B. Letting $\mathscr{E} = \{A + B\}$ designate the set of the two sample points, this variance is expressed as

$$\sigma_{E_2}^2 = 2\bar{\gamma}(\mathscr{E}, AB) - \bar{\gamma}(AB, AB) - \bar{\gamma}(\mathscr{E}, \mathscr{E})$$

with

$$\bar{\gamma}(\mathscr{E}, AB) = \bar{\gamma}(A, AB) = \bar{\gamma}(B, AB) = \chi(l),$$

because of symmetry;

$$\bar{\gamma}(AB, AB) = F(l),$$

$$\bar{\gamma}(\mathscr{E}, \mathscr{E}) = \bar{\gamma}(A, \mathscr{E}) = \tfrac{1}{2}[\bar{\gamma}(A, A) + \bar{\gamma}(A, B)] = \tfrac{1}{2}\gamma(l)$$

and, thus,

$$\sigma_{E_2}^2 = 2\chi(l) - F(l) - \gamma(l)/2.$$

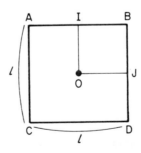

FIG. II.50. Estimation of a square.

(iii) The estimation variance of a square ABCD of side l by a central sample 0, cf. Fig. II.50, is written as

$$\sigma_{E_3}^2 = 2\bar{\gamma}(0, ABCD) - \bar{\gamma}(ABCD, ABCD) - \bar{\gamma}(0, 0)$$

with

$$\bar{\gamma}(0, ABCD) = \bar{\gamma}(0, 0IBJ) = H(l/2; l/2),$$

because of symmetry;

$$\bar{\gamma}(ABCD, ABCD) = F(l; l) \quad \text{and} \quad \bar{\gamma}(0, 0) = \gamma(0) = 0$$

and, thus,

$$\sigma_{E_3}^2 = 2H(l/2; l/2) - F(l; l).$$

(iv) The estimation variance of the square ABCD of side l by the set $\mathscr{E}$ of its four corners, cf. Fig. II.50, is written as

$$\sigma_{E_4}^2 = 2\bar{\gamma}(\mathscr{E}, ABCD) - \bar{\gamma}(ABCD, ABCD) - \bar{\gamma}(\mathscr{E}, \mathscr{E})$$

with

$$\bar{\gamma}(\mathscr{E}, ABCD) = \bar{\gamma}(A, ABCD) = H(l; l),$$

because of symmetry;

$$\bar{\gamma}(\mathscr{E}, \mathscr{E}) = \bar{\gamma}(A, \mathscr{E}) = \tfrac{1}{4}[\bar{\gamma}(A, A) + \bar{\gamma}(A, B) + \bar{\gamma}(A, C) + \bar{\gamma}(A, D)]$$
$$= \tfrac{1}{2}\gamma(l) + \tfrac{1}{4}\gamma(l\sqrt{2})$$

and, thus,

$$\sigma_{E_4}^2 = 2H(l; l) - F(l; l) - \tfrac{1}{2}\gamma(l) - \tfrac{1}{4}\gamma(l\sqrt{2}).$$

Charts no. 8 and 18 give the estimation variance of a rectangle ABCD of sides l and L by its median IJ of length l, cf. Fig. II.51:

$$\sigma_{E_5}^2 = 2\bar{\gamma}(IJ, ABCD) - \bar{\gamma}(ABCD, ABCD) - \bar{\gamma}(IJ, IJ)$$
$$= 2\chi(L/2; l) - F(L; l) - F(l).$$

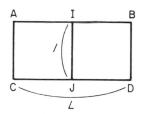

FIG. II.51. Estimation of a rectangle by its median.

Charts no. 9 and 19 give the estimation variance of a rectangular parallelepiped P with a square base $(L \times l^2)$ by its median square, cf. Fig. II.52:

$$\sigma_{E_6}^2 = 2\bar{\gamma}(IJEF, P) - \bar{\gamma}(P, P) - \bar{\gamma}(IJEF, IJEF)$$
$$= 2\chi(L/2; l^2) - F(L; l^2) - F(l; l).$$

As the functions $F(L; l^2)$ and $F(l; l)$ are given elsewhere (Charts no. 4, 5 and 14, 15), these charts no. 9 and 19 can be used to calculate the three-dimensional auxiliary functions $\chi(L; l^2)$.

Charts no. 10 and 20 give the variance of estimation of a parallelepiped P with a square base $(L \times l^2)$ by its axial drill core IJ of length L, cf. Fig. II.53:

$$\sigma_{E_7}^2 = 2\bar{\gamma}(IJ, P) - \bar{\gamma}(P, P) - \bar{\gamma}(IJ, IJ) = 2H(L; (l/2)^2) - F(L; l^2) - F(l).$$

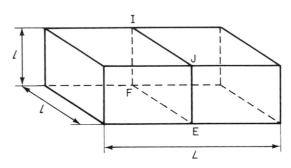

FIG. II.52. Estimation of a parallelepiped by its median square.

As the functions $F(L; l^2)$ and $F(L)$ are given elsewhere (Charts no. 5 and 15 and formulae (II.98) and (II.99)), these Charts no. 10 and 20 can be used to calculate the three-dimensional auxiliary function $H(L; l^2)$.

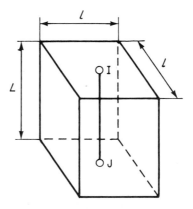

FIG. II.53. Estimation of a parallelepiped by its axial drill core.

Configurations for the estimation of a rectangle

This exercise is intended to familiarize the reader with the use of the various auxiliary functions.

Let P be a rectangular panel with sides a and b, and located within an isotropic regionalization represented by the point model $\gamma(r)$, cf. Fig. II.54.

A practical analogy would be a parallelepipedic panel P with surface $a \times b$ and height c, located within an isotropic horizontal regionalization represented by a model graded over the constant height c.

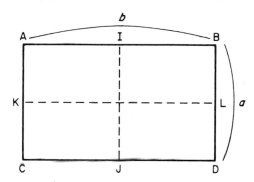

FIG. II.54. Rectangular panel P.

Consider the estimation of the unknown mean grade of panel P by each of the following four estimators:

(i) mean grade Z_1 of the perimeter ABCD of panel P (Z_1 could be obtained by continuous horizontal channel sampling along the length of the four side drives of panel P);
(ii) mean grade Z_2 of the two parallel sides AB and CD;
(iii) mean grade Z_3 of the two medians IJ and KL;
(iv) mean grade Z_4 of the single median IJ.

The estimation variances corresponding to these four data configurations will be calculated and a numerical example will be provided with a linear model $\gamma(r) = r$ and a square panel ($a = b$).

Rectangle evaluated by its perimeter   The estimation variance is written as

$$\sigma_{E_1}^2 = 2\bar{\gamma}(\text{ABCD}, \text{P}) - \bar{\gamma}(\text{P}, \text{P}) - \bar{\gamma}(\text{ABCD}, \text{ABCD})$$

with $\bar{\gamma}(\text{P}, \text{P}) = F(a; b)$.

$$\bar{\gamma}(\text{ABCD}, \text{P}) = \bar{\gamma}(\text{AC} + \text{AB}, \text{P}) = \frac{a\bar{\gamma}(\text{AC}, \text{P}) + b\bar{\gamma}(\text{AB}, \text{P})}{a + b}$$

$$= \frac{a\chi(b; a) + b\chi(a; b)}{a + b}.$$

$$\bar{\gamma}(\text{ABCD}, \text{ABCD}) = \bar{\gamma}(\text{AC} + \text{AB}, \text{ABCD})$$

$$= \frac{a\bar{\gamma}(\text{AC}, \text{ABCD}) + b\bar{\gamma}(\text{AB}, \text{ABCD})}{a + b}.$$

Since

$$\bar{\gamma}(AC, ABCD) = \frac{a\bar{\gamma}(AC, AC) + 2b\bar{\gamma}(AC, AB) + a\bar{\gamma}(AC, BD)}{2(a+b)}$$

$$= \frac{aF(a) + 2bH(a;b) + a\alpha(b;a)}{2(a+b)},$$

we finally obtain

$$\sigma_{E_1}^2 = \frac{2a\chi(b;a) + 2b\chi(a;b)}{a+b} - F(a;b)$$

$$- \frac{a^2F(a) + 4abH(a;b) + b^2F(b) + a^2\alpha(b;a) + b^2\alpha(a;b)}{2(a+b)^2}.$$

Considering formula (II.91) of the linear model and a square panel ($a = b$), we get the following values for the auxiliary functions:

$$F(a) = a/3, \qquad \alpha(a;a) = 1 \cdot 077a, \qquad \chi(a;a) = 0 \cdot 652a,$$

$$F(a;a) = 0 \cdot 521a \quad \text{and} \quad H(a;a) = 0 \cdot 765a.$$

Thus, $\sigma_{E_1}^2 = 0 \cdot 0479a$ for a total sampled length $4a$.

Rectangle evaluated by the two parallel sides AB and CD

$$\sigma_{E_2}^2 = 2\bar{\gamma}(AB+CD, P) - \bar{\gamma}(P, P) - \bar{\gamma}(AB+CD, AB+CD)$$

with $\bar{\gamma}(AB+CD, P) = \bar{\gamma}(AB, P) = \chi(a;b)$.

$$\bar{\gamma}(AB+CD, AB+CD) = \bar{\gamma}(AB, AB+CD)$$

$$= \tfrac{1}{2}[\bar{\gamma}(AB, AB) + \bar{\gamma}(AB, CD)]$$

$$= \tfrac{1}{2}[F(b) + \alpha(a;b)]$$

and, thus, we have

$$\sigma_{E_2}^2 = 2\chi(a;b) - F(a;b) - \tfrac{1}{2}[F(b) + \alpha(a;b)].$$

The numerical application gives $\sigma_{E_2}^2 = 0 \cdot 0778a$, for a total sampled length $2a$.

Rectangle evaluated by its two medians IJ and KL   Similar calculations lead to the following result:

$$\sigma_{E_3}^2 = \frac{2a\chi(b/2; a) + 2b\chi(a/2; b)}{a+b} - F(a;b)$$

$$- \frac{a^2F(a) + 2abH(a/2; b/2) + b^2F(b)}{(a+b)^2}.$$

Considering the values $\chi(a/2; a) = 0 \cdot 454a$ and $H(a/2; a/2) = 0 \cdot 383a$, the numerical application gives $\sigma_{E_3}^2 = 0 \cdot 0288a$, for a total sampled length $2a$.

Rectangle evaluated by a single median IJ   Similarly, the following results can be found:

$$\sigma_{E_4}^2 = 2\chi(b/2; a) - F(a; b) - F(a)$$

and, for the numerical application, $\sigma_{E_4}^2 = 0 \cdot 056a$ for a total sampled length a.

Thus, to sum up, the following inequalities hold:

$$\sigma_{E_3}^2 = 0 \cdot 0288a < \sigma_{E_1}^2 = 0 \cdot 0479a < \sigma_{E_4}^2 = 0 \cdot 056a < \sigma_{E_2}^2 = 0 \cdot 0778a,$$

with the respective total sampled lengths $2a$, $4a$, a, $2a$.

The centred configurations (medians) are obviously better than the closed configurations (sides).

This case study shows how geostatistics, by means of the structural characteristic $\gamma(h)$, is able to balance the quality (the median information is better located) and the quantity of information (the perimeter has the largest total sampled length but does not correspond to the smallest estimation variance).

II.E.4. Charts

All the charts given hereafter correspond to the two following spherical and exponential *isotropic* models with a unit sill value. All the distances considered by the charts are relative distances r/a (divided by the parameter a of the model).

Spherical model:

$$\gamma(r) = \begin{cases} \dfrac{3}{2}\dfrac{r}{a} - \dfrac{1}{2}\dfrac{r^3}{a^3}, & \forall r \in [0, a], \\[2mm] 1 = \text{sill}, & \forall r \geqslant a \text{ (actual range)}. \end{cases}$$

Exponential model:

$$\gamma(r) = 1 - \exp(-r/a), \qquad \forall r \geqslant 0,$$

with a sill equal to 1 and a practical range $a' = 3a$.

	Chart no.	
Object of the chart	Spherical model	Exponential model
Grading $\gamma_G(L) = \alpha(L; l) - \alpha(0; l)$	1	11
Auxiliary functions		
$\chi(L; l)$	2	12
$H(L; l)$	3	13
$F(L; l)$	4	14
$F(L; l^2)$	5	15
$\alpha(L; l^2) - \alpha(0; l^2)$	6	16
Estimation variances		
$\sigma_{E1}^2, \sigma_{E2}^2, \sigma_{E3}^2, \sigma_{E4}^2$	7	17
σ_{E5}^2	8	18
σ_{E6}^2	9	19
σ_{E7}^2	10	20

Regularized semi-variogram over cores of length l	
First point $\gamma_l(l)$ and sill $\gamma_l(\infty)$	21
Semi-variogram $\gamma_l(h)$	22(a), (b)

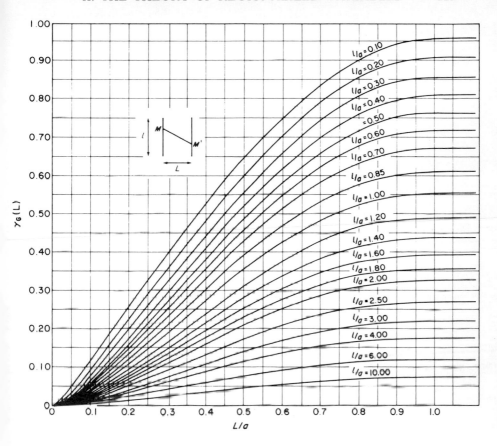

CHART NO. 1. Spherical model. Grading over l: $\gamma_G(L)$.

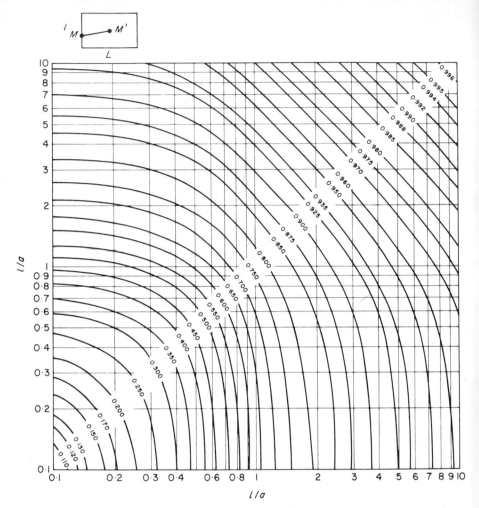

CHART NO. 2. Spherical model. Function $\chi(L;l)$.

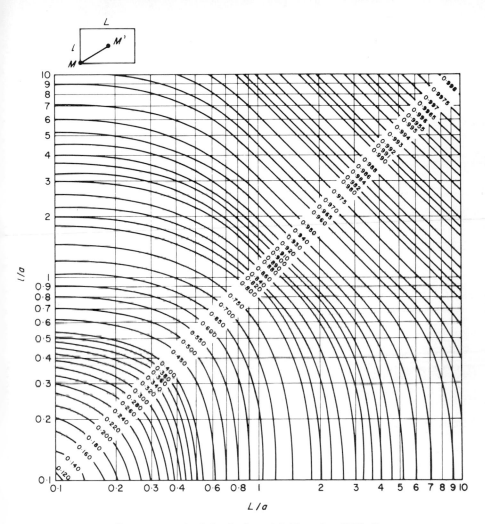

CHART NO. 3. Spherical model. Function $H(L; l)$.

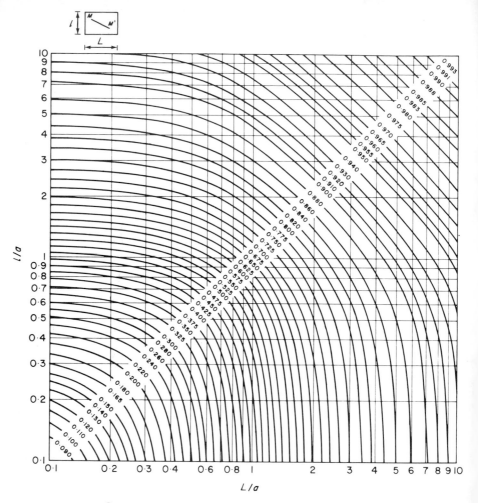

CHART NO. 4. Spherical model. Function $F(L; l)$.

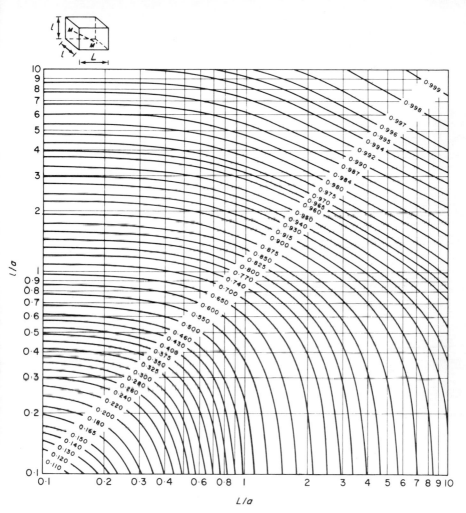

CHART NO. 5. Spherical model. Function $F(L; l^2)$.

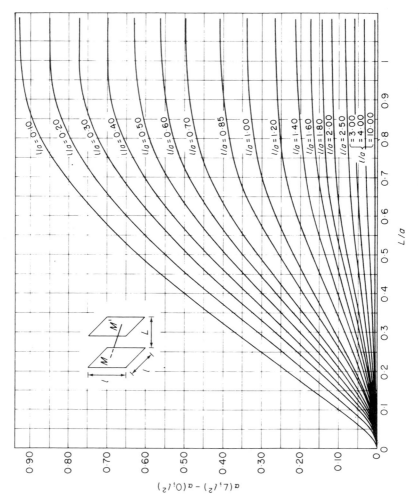

CHART NO. 6. Spherical model. Grading over l^2: $\alpha(L; l^2) - \alpha(0; l^2)$.

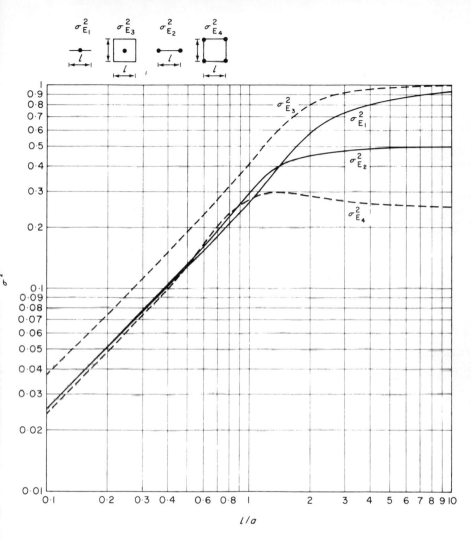

CHART NO. 7. Spherical model. Various extension variances.

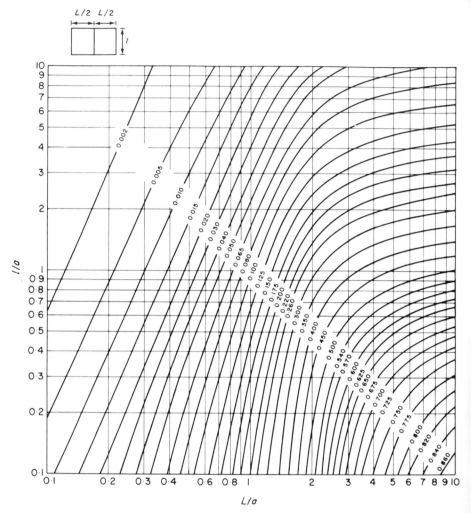

CHART NO. 8. Spherical model. Extension variance σ_{E5}^2.

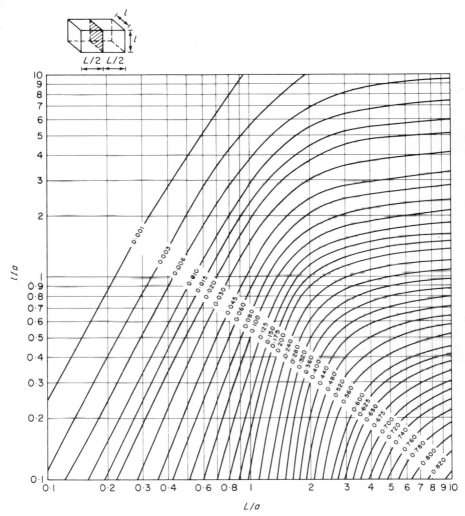

CHART NO. 9. Spherical model. Extension variance $\sigma^2_{E_6}$.

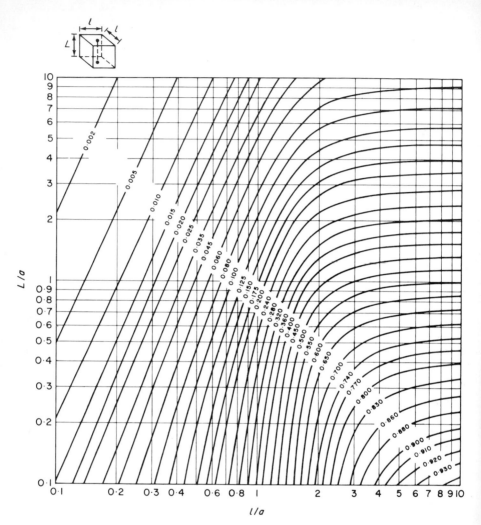

CHART NO. 10. Spherical model. Extension variance σ_{E7}^2.

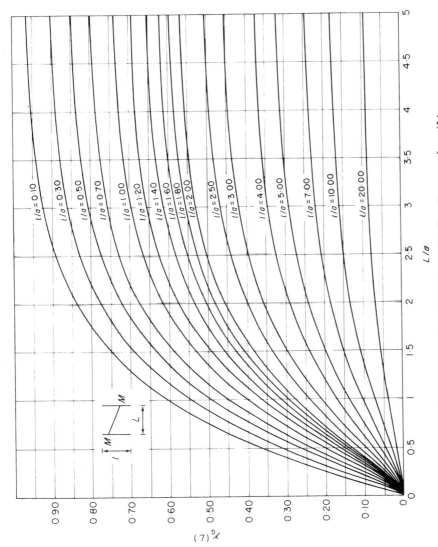

CHART NO. 11. Exponential model. Grading over l: $\gamma_G(L)$.

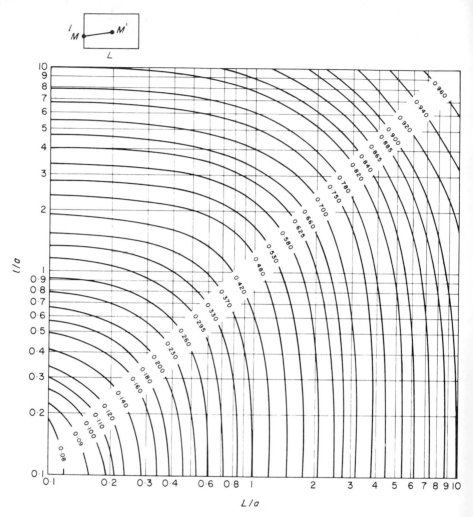

CHART NO. 12. Exponential model. Function $\chi(L; l)$.

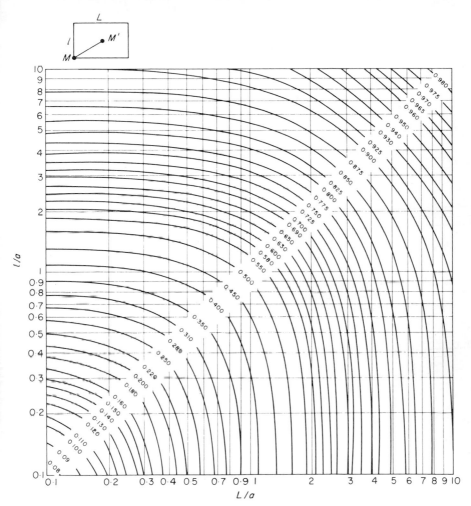

CHART NO. 13. Exponential model. Function $H(L; l)$.

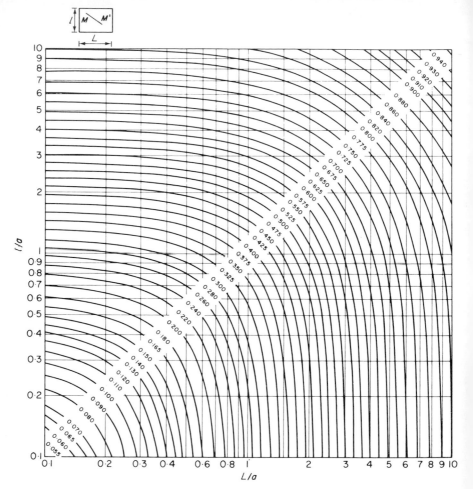

CHART NO. 14. Exponential model. Function $F(L; l)$.

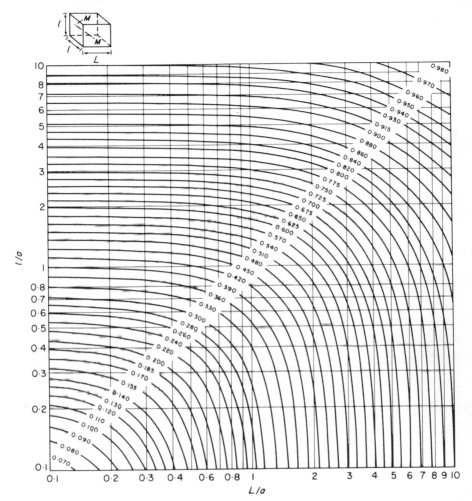

CHART NO. 15. Exponential model. Function $F(L; l^2)$.

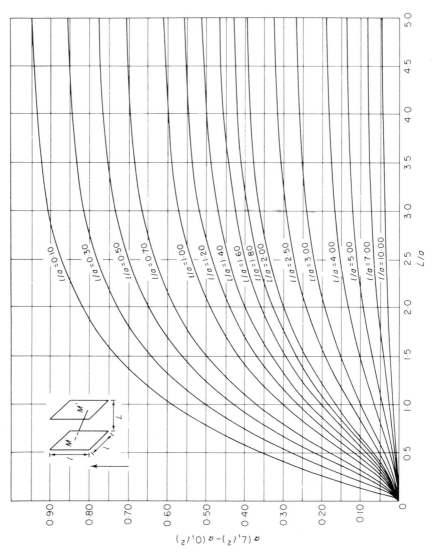

CHART NO. 16. Exponential model. Grading over l^2: $\alpha(L; l^2) - \alpha(0; l^2)$.

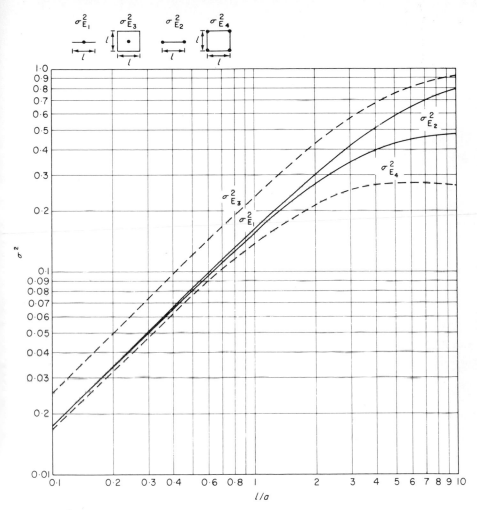

CHART NO. 17. Exponential model. Various extension variances.

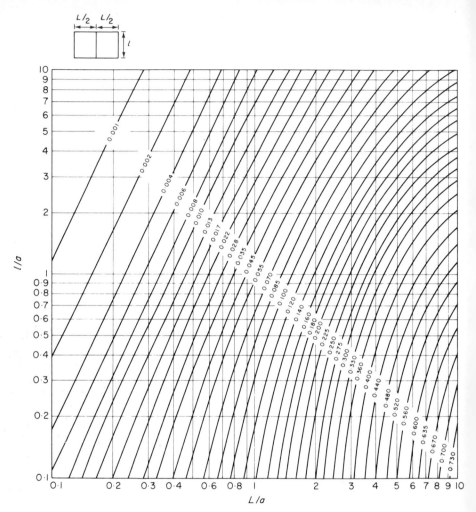

CHART NO. 18. Exponential model. Extension variance $\sigma^2_{E_5}$.

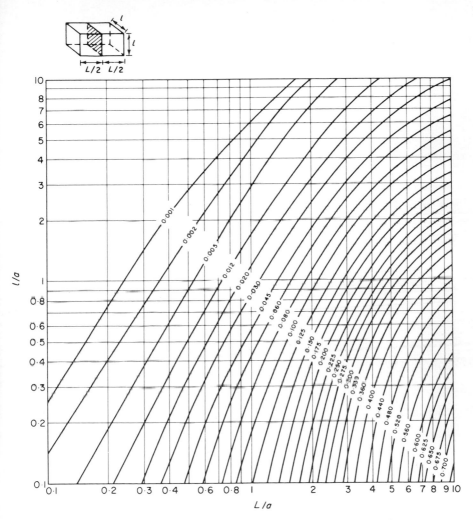

CHART NO. 19. Exponential model. Extension variance $\sigma^2_{E_6}$.

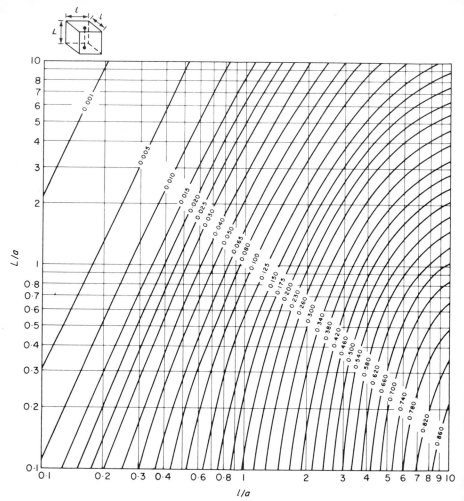

CHART NO. 20. Exponential model. Extension variance σ_{E7}^2.

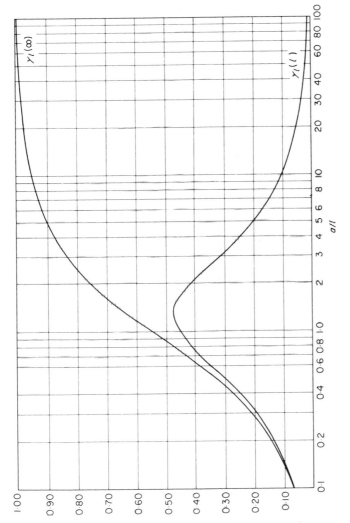

CHART NO. 21. Spherical model. Regularized semi-variogram. First point, $\gamma_l(l)$; sill $\gamma_l(\infty)$.

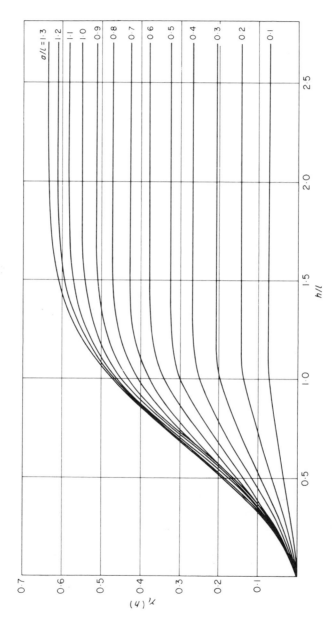

CHART NO. 22(a). Spherical model. Regularized semi-variogram $\gamma_l(h)$.

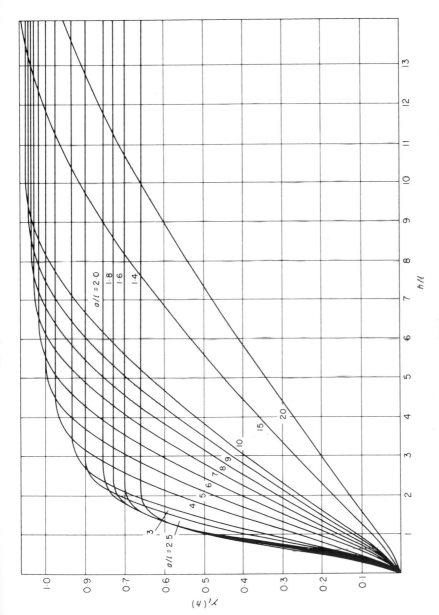

CHART NO 22(b). Spherical model. Regularized semi-variogram $\gamma_l(h)$.

III

Structural Analysis

SUMMARY

The structural analysis of a regionalized phenomenon consists in constructing a model of a variogram which characterizes, in an operational way, the main features of the regionalization. This modelling requires good physical knowledge of the phenomenon under study as well as good "craft" in the practice of fitting geostatistical models.

Section III.A introduces the concept of nested structures to model the succession of various scales of variabilities of a natural phenomenon; e.g., from the petrographic scale to the macro-scale of the distribution of deposits within a metalliferous province. The micro-variabilities, i.e., the variabilities of which scale is overpassed by the available grid of information, are represented on the variogram by an apparent discontinuity at the origin called the "nugget effect".

Section III.B gives the various models of regionalization and coregionalization in current use and shows how they should be combined to provide an operational representation of structural features, such as nested structures, anisotropies, quasi-stationarity. Then, the study of the statistical inference of a moment of order two gives an order of magnitude of the reliability of an experimental variogram or covariance.

Section III.C shows how a structural analysis should be guided in practice. A critical review of the available data, along with elementary statistics of these data, should be made before proceeding to the effective geostatistical analysis. Three main steps can be distinguished in this analysis, successively: the construction of experimental variograms, their interpretation and the construction of a synthetic structural model.

III.A. NESTED STRUCTURES AND THE NUGGET EFFECT

III.A.1. Nested structures

Consider two point or quasi-point (very small support) grades $Z(x)$ and $Z(x+h)$ in a deposit separated by a distance vector h. The variability between $Z(x)$ and $Z(x+h)$, which is characterized by the variogram $E\{[Z(x+h)-Z(x)]^2\}$, is due to many causes, which appear over a range of different scales, for example:

(i) at the level of the support $(h \simeq 0)$, there is a variability due to measurements, i.e., fluctuation in the rate of recovery of the core sample, sampling errors;

(ii) at the petrographic level (e.g., $|h| < 1$ cm) a second variability appears due to the transition from one mineralogical element to another;

(iii) at the level of strata or mineralized lenses (e.g., $|h| < 100$ m), a third variability may be due to the alternation of strata or of lenses with waste material;

(iv) at the level of a metalliferous province (e.g., $|h| < 100$ km), a fourth variability may appear, due to the distribution of the deposits related to the orogenesis of the province;

etc.

All these sources or structures of variability, and possibly many more, come into play *simultaneously* and for all distances h. They are called "*nested structures*".

Scale and support of observation

In practice, all these variabilities are never observed simultaneously, since this would require an enormous amount of data of quasi-point support covering the entire field of variability from 1 μm to 100 km.

At the scale of observation, when counting elements on a thin section (where the information has a quasi-point support), it is not possible to distinguish between the first variability due to errors of measure (e.g., counting errors) and the second due to petrographical variability.

At the scale of evaluation of a mining face by channel samples, the support of the channel samples integrates the first two variabilities into one single undifferentiated variability that will be defined later as a "nugget effect". However, it is possible at this scale to distinguish the third variability, due to lenticular beds in the same stratum.

At the scale of the evaluation of the entire deposit by drill cores spaced at 50- to 100-m intervals, the first three variabilities are indistinguishable. However, it is possible to distinguish the fourth variability due, for example, to the alternation of mineralized strata, or trends in grade values at the borders of the deposit or at lower depths.

Just as in a set of *nested* tables, each table covers all the smaller tables, so each scale of observation integrates the variabilities of all the smaller scales.

Representation of the nested structures

As far as the second-order moments of the RF $Z(x)$ are concerned, these nested structures can be conveniently represented as the sum of a number of variograms (or covariances), each one characterizing the variability at a particular scale:

$$\tfrac{1}{2}E\{[Z(x+h)-Z(x)]^2\} = \gamma(h) = \gamma_0(h)+\gamma_1(h)+\gamma_2(h)+\ldots+\gamma_i(h). \qquad (\text{III.1})$$

For example, $\gamma_0(h)$ may be a transition model (spherical or exponential) which very rapidly reaches its sill value C_0 for distances h that are only slightly larger than the data support. This model thus combines all the micro-variabilities (e.g., measurement errors and petrographic differentiations). $\gamma_1(h)$ may be another transition model with a larger range (e.g., $a_1 = 10$ m) characterizing the lenticular beds and $\gamma_2(h)$ may be a third transition model with a range ($a_2 = 200$ m) representing the alternation of strata or the extent of homogeneous mineralized zones.

At smaller distances ($h < 30$ m), the observed total variability depends on $\gamma_0(h)+\gamma_1(h)$, cf. Fig. III.1, while for large distances it will depend on all the $\gamma_i(h)$.

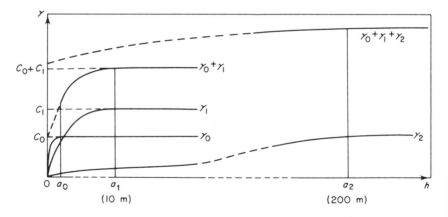

FIG. III.1. Nested structures and the nugget effect.

The linearity of geostatistical operators

The previous representation (III.1) of nested structures is not unique. It is currently used in mining geostatistics because it is convenient in order to fit several experimental variograms with different levels (sills) of variability and, above all, because the expression (III.1) is linear in γ.

As a matter of fact, each of the main geostatistical operators is linear in γ (semi-variogram) or in C (covariance when it exists), cf. formulae (II.27) for the estimation variance, (II.36) for the dispersion variance, (II.41) for regularization. Thus, using the previous linear representation (III.1) for the point semi-variogram $\gamma(h)$, the regularized semi-variogram on a support v is the sum of the regularization on v of each of the component structures:

$$\gamma_v(h) = \sum_i \gamma_{i,v}(h) = \sum_i [\bar{\gamma}_i(v, v_h) - \bar{\gamma}_i(v, v)].$$

More generally, if $\mathscr{E}$ is any one of the previous operators, then

$$\mathscr{E}(\gamma) = \sum_i \mathscr{E}(\gamma_i) \quad \text{if } \gamma(h) = \sum_i \gamma_i(h). \tag{III.2}$$

The operator $\mathscr{E}$ acts identically on each of the components γ_i of the nested model γ.

The same holds true for the covariance; the operator $\mathscr{E}$ operates identically on each of the components C_i of the nested model C:

$$\mathscr{E}(C) = \sum_i \mathscr{E}(C_i) \quad \text{if } C(h) = \sum_i C_i(h). \tag{III.3}$$

III.A.2. Genesis of a nugget effect

If we study the spatial variability of the gold grades of core samples, two core samples that are very close together may have grades $Z(x)$ and $Z(x+h)$ which differ considerably when one of them contains a nugget and the other one does not.

This nugget effect appears on the semi-variogram as a component $\gamma_0(h)$ which rapidly increases to reach a sill C_0 almost as soon as h is greater than zero (more precisely, when h is greater than the range a_0 equal to the dimensions of the gold nugget), cf. Fig. III.1.

To this micro-structure γ_0, of very small range, a structure $\gamma_1(h)$ of larger range a_1 can be added, characterizing, for example, the regionalization of mineralized veins or of a concentration of nuggets. The total variability can then be characterized by the nested model $\gamma(h) = \gamma_0(h) + \gamma_1(h)$.

At the level of sampling a gold deposit by drill cores, the distances of observation $h > 0$ are generally larger than the first range a_0 and, thus,

$$\gamma_0(h) = C_0 \quad \text{and} \quad \gamma(h) = C_0 + \gamma_1(h), \qquad \forall |h| > a_0,$$

with $\gamma(0) = 0$, by definition.

Thus, at the level of observation (e.g., $h \simeq 20 \text{ m} \gg a_0$), the residual micro-structure $\gamma_0(h)$ appears on the total variability as an apparent discontinuity at the origin of amplitude C_0, which is called the "*nugget constant*", cf. Fig. III.1.

Definition

The term "nugget effect" can be generalized from its specific meaning in gold-bearing deposits. It is used to characterize the residual influence of all variabilities which have ranges (a_0) much smaller than the available distances of observation $(h \gg a_0)$. A nugget effect will appear on the variogram (or covariance) as an apparent discontinuity at the origin.

Scale effect   It must be stressed that the definition of a nugget effect is closely related to the scale of observation. The same structure $\gamma_1(h)$ with range $a_1 = 10$ m and sill C_1 would be quite evident on a sampling grid of 3 to 5 m, but would appear as no more than a nugget effect on a grid of 30 to 50 m, cf. Fig. III.1. Experimentally, on such a large grid the two components C_0 and C_1 of the nugget constant would be indistinguishable unless data are available to put into evidence the range a_1.

   For this reason, it is often advisable when a sampling programme has been carried out on a systematic grid (of, for example, sides $b = 100$ m) to take a few other samples on a smaller grid (for example, $b/10 = 10$ m). In this way, data are available at two different scales of observation, which makes it possible to study the nested structures and interpret any possible nugget effect.

Pure nugget effect

When the observed semi-variogram appears as a single nugget effect (completely flat semi-variogram), it is said to represent a pure nugget effect, cf. Fig. III.2 and Case Study 1 in section III.A.5.

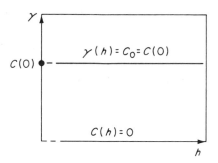

FIG. III.2. Pure nugget effect.

The corresponding model $\gamma_0(h)$ is a transition-type model (spherical, for instance) with a range ε which is very small compared with the smallest distance $|h|$ of observation:

$$\gamma_0(h) = \begin{cases} 0 & \text{for } h = 0, \\ C_0 & \forall |h| > \varepsilon. \end{cases} \qquad\qquad \text{(III.4)}$$

The corresponding covariance is written according to formula (II.4):

$$C_0(h) = C_0(0) - \gamma_0(h).$$

In the course of this book, a pure nugget effect will often be represented by the notation with square brackets: $\gamma_0(h) = [C_0(0) - C_0(h)]$, so as to distinguish it from a structure $\gamma_1(h)$ with a greater range.

Absence of spatial correlation

The corresponding covariance model $C_0(h)$ for a pure nugget is, thus, of transition type, its value at the origin $C_0(0)$ being equal to the sill C_0 of the semi-variogram model, according to formula (II.13), cf. Fig. III.2,

$$C_0(h) = \begin{cases} C_0(0) = C_0 & \text{for } h = 0, \\ C_0 - \gamma(h) = 0, & \forall |h| > \varepsilon. \end{cases} \qquad\qquad \text{(III.5)}$$

The pure nugget effect thus corresponds to a total absence of spatial correlation between two variables $Z(x)$ and $Z(x+h)$, at least for all available distances $(|h| > \varepsilon)$, and is equivalent to the well-known phenomenon of "white noise" in physics. In so far as geostatistics is only concerned with second-order moments, a pure nugget effect will be interpreted as spatial independence.

In order to establish the standard statistical results for independent variables, it is sufficient to apply the various geostatistical formulae to a pure nugget-effect model.

In *practice*, a pure nugget effect at all scales with a function $\gamma(h)$ amounting to a single discontinuity at the origin is exceptional.

In fact, if the pure nugget effect at all scales is associated with stationarity, it represents an incredible degree of homogeneity of the mineralization. It means, for instance, that no local differentiation is possible between mining panels and, thus, no selection is possible. There are no rich zones nor any poor zones (a zone being defined on any scale). At every point in the deposit the best estimator is the mean grade of the deposit.

III.A.3. The nugget effect and geostatistical operations

Let $C(h)$ be the covariance, defined on a strict point support, characterizing the variability of a sequence of nested structures:

$$C(h) = C_0(h) + \sum_{i=1}^{k} C_i(h).$$

The nugget effect can be distinguished as a transition covariance $C_0(h)$ with a very small range $a_0 = \varepsilon$:

$$C_0(h) = 0 \quad \text{when } |h| > a_0.$$

It was shown previously that the main geostatistical operators are linear functions of the covariance C. If $\mathscr{E}$ is one of these operators, then, from (III.3), we have

$$\mathscr{E}(C) = \mathscr{E}(C_0) + \sum_{i=1}^{k} \mathscr{E}(C_i).$$

In this section we shall be concerned with the effect of these operators on a pure nugget effect, i.e., with the term $\mathscr{E}(C_0)$. In terms of the semi-variogram, with $\gamma_0(h) = C_0(0) - C_0(h)$, we shall be concerned with the operation $\mathscr{E}(\gamma_0)$.

Mean values $\bar{C}_0$ and $\bar{\gamma}_0$ of a pure nugget effect

Mean values of the type $\bar{C}_0(v, v)$ or $\bar{C}_0(v, V)$ are calculated for the four possible relative configurations of the two supports v and V. The following results are obtained.

(i) $v \equiv V$, the two supports are identical:

$$\bar{C}_0(v, v) = \frac{A}{v} \quad \text{and} \quad \bar{\gamma}_0(v, v) = C_0(0) - \frac{A}{v}. \qquad \text{(III.6a)}$$

(ii) $v \subset V$, V contains v:

$$\bar{C}_0(v, V) = \frac{A}{V} \quad \text{and} \quad \bar{\gamma}_0(v, V) = C_0(0) - \frac{A}{V}. \qquad \text{(III.6b)}$$

(iii) $v \cap V = \varnothing$, v and V are disjoint and the distance between them is much greater than the range a_0:

$$\bar{C}_0(v, V) = 0 \quad \text{and} \quad \bar{\gamma}_0(v, V) = C_0(0). \qquad \text{(III.6c)}$$

(iv) $v \cap V \neq \varnothing$, v and V have a non-zero intersection, the measure of which has dimensions much greater than the range a_0:

$$C_0(v, V) = A \frac{\text{Meas } v \cap V}{vV} \quad \text{and} \quad \bar{\gamma}_0(v, V) = C_0(0) - A \frac{\text{Meas } v \cap V}{vV}.$$

$$(\text{III.6d})$$

In all four cases, we have v, $V \gg a_0$, i.e., the range a_0 of the pure nugget-effect model is very small compared with the dimensions of the supports v and V; also, $A = \int C_0(h)\,dh$ (recalling that, in the three-dimensional space, h is a vector and this integral in fact represents a triple integral over the whole space).

Proof   By way of example, we shall prove the case $v \subset V$.

$$\bar{C}_0(v, V) = \frac{1}{vV} \int_v dx \int_V C_0(x - y)\,dy.$$

If the distance between every point $x \in v$ and the border of V is greater than the range a_0, then, strictly,

$$\int_V C_0(x - y)\,dy = \int_{-\infty}^{+\infty} C_0(h)\,dh = A, \qquad \forall x \in v,$$

and, thus,

$$\bar{C}_0(v, v) = \frac{1}{v} \int_v \frac{A}{V}\,dx = \frac{A}{V}.$$

If not, i.e., if there exist points $x_0 \in v$ such that the sphere $B_{a_0}(x_0)$, centred on x_0 and with radius a_0, is not entirely contained within V, then, for these points x_0, $\int_V C_0(x - y)\,dy \neq A$. But, as regards the second integration, $\int_v dx$, the influence of these points x_0 is negligible (since a_0 is very small compared with the dimensions of v). Thus, the previous expression holds true: $\bar{C}_0(v, V) = A/V$.

Having established the mean values $\bar{C}_0$ of a pure nugget effect, we can now proceed to the study of its effect on the three main geostatistical operators.

(i) *Regularization on a support v with $v \gg a_0$*   If $\gamma_0(h)$ is the point semi-variogram, then the semi-variogram regularized on a support v is written, according to (II.41), as

$$\gamma_{0v}(h) = \bar{\gamma}_0(v, v_h) - \bar{\gamma}_0(v, v),$$

where v_h is the translation of the support v by the vector h.

As $\gamma_0(h)$ represents a pure nugget effect, the regularized semi-variogram is expressed as

$$\gamma_{0v}(h) = \begin{cases} 0 & \text{if } h = 0, \\[1.5ex] \dfrac{A}{V}\left[1 - \dfrac{\text{Meas } v \cap V}{v}\right] & \text{if } |h| < v, \text{ i.e., if } v \cap v_h \neq \varnothing, \\[1.5ex] \dfrac{A}{v} & \text{if } |h| > v, \end{cases} \qquad \text{(III.7)}$$

with all the dimensions of observation much greater than the range a_0, i.e.,

$$v, \text{Meas } v \cap V, \quad \text{and:} \quad h \gg a_0.$$

Generally, the supports v and v_h do not overlap, so that the only discontinuity at the origin is of amplitude A/v, and the *a priori* variance of the regularized RF $Z_v(x)$ is given by expression (II.45) as

$$D^2(v/\infty) = \text{Var}\,\{Z_v(x)\} = \gamma_{0v}(\infty) = \frac{A}{v}, \qquad \forall v \gg a_0.$$

In *practice*, point data are seldom available and the nugget effect will be revealed by the regularized nugget constant $C_{0v} = A/v$, when v is the smallest support of information (e.g., piece of core). For all other supports V, the regularized nugget constant C_{0v} is given by the inverse proportionality relation:

$$C_{0V} = C_{0v}\frac{v}{V}, \quad \text{with } v, V \gg a_0. \qquad \text{(III.8)}$$

(ii) *Estimation variance of the mean grade Z_V on the support V (mining block) by the mean grade Z_v on the support v* According to formula (II.27) and taking the expressions (III.6) of $\bar{\gamma}_0$ into account, this variance is written

$$\sigma_E^2(v, V) = \begin{cases} A\left(\dfrac{1}{v} - \dfrac{1}{V}\right) = C_{0v}\left(1 - \dfrac{v}{V}\right) & \text{if } V \text{ contains } v, \\[1.5ex] A\left(\dfrac{1}{v} + \dfrac{1}{V}\right) = C_{0v}\left(1 + \dfrac{v}{V}\right), & \text{if } v \text{ and } V \text{ are disjoint,} \\[1.5ex] A\left(\dfrac{1}{v} + \dfrac{1}{V} - 2\dfrac{\text{Meas } v \cap V}{vV}\right) & \begin{array}{l}\text{if } v \text{ and } V \text{ have a} \\ \text{non-zero intersection,}\end{array} \end{cases} \qquad \text{(III.9)}$$

where C_{0v} is the nugget constant regularized on the support v and shown on the regularized semi-variogram $\gamma_{0v}(h)$.

In the estimation of a block V, the measure v of the data support is often negligible when compared with the measure V of the block. In such cases, all terms in A/V in formulae (III.9) are negligible and we have

$$\sigma_E^2(v, V) = \frac{A}{v} = C_{0v}, \qquad \forall V \gg v \gg a_0, \qquad \text{(III.10)}$$

which is the standard relationship found in the statistics of independent variables† (or, more precisely, variables without auto-correlation, the structures of which are, thus, a pure nugget effect). The variance of estimation of the mean grade Z_V of a field V by the arithmetic mean

$$Z_{V'} = \frac{1}{n} \sum_{i=1}^{n} Z_v(x_i)$$

of the grades of n samples $v(x_i)$ located at the n points $\{x_i, i = 1 \text{ to } n\}$ is written, according to (III.10), as

$$\sigma_E^2(V', V) = \frac{A}{V'} = \frac{C_{0v}}{n}, \qquad \forall V \gg nv, \qquad v \gg a_0,$$

since the measure V' of the data support is $V' = nv$.

This estimation variance is directly proportional to the *a priori* variance C_{0v} of the grades $Z_v(x)$ of the samples, and inversely proportional to the number n of samples.

(iii) *Dispersion variance* of the mean grade of a volume v in a field V. According to formula (II.36), and taking the expressions (III.6) of $\bar{\gamma}_0$ into account, this variance is written

$$D^2(v/V) = A\left(\frac{1}{v} - \frac{1}{V}\right) = C_{0v}\left(1 - \frac{v}{V}\right). \qquad \text{(III.11)}$$

In the presence of a pure nugget effect, the dispersion variance $D^2(v/V)$ and also the *a priori* variance $D^2(v/\infty)$ are inversely proportional to the support v.

---

† Strictly speaking, the standard formula σ^2/n of the statistics of independent variables corresponds to the estimation variance of the expectation $E\{Z(x)\}$ of the RF $Z(x)$. However, under the hypothesis of stationarity and ergodicity, the mean grade $Z_V = (1/V)\int_V Z(x)\,dx$ of a fixed domain V, with $V \gg a_0$, tends towards expectation $E\{Z(x)\}$.

III.A.4. Case studies

1. *Pure nugget effect at Adelaïda*

The scheelite (WO_4Ca) of the Adelaïda (Spain) deposit is concentrated in veinlets and small nodules which are distributed more or less homogeneously in a network of quartz veins.

Drilling was carried out in a direction perpendicular to the plane of the greatest density of veins. The variable measured was the cumulative thickness of the quartz veins intersected by a drill core, the support of the variable being core samples of constant length $l = 5$ m. Thus, the variable measured, $Z_l(x)$, is the regularization of the point variable "density of quartz" over the length of a core sample. The core samples can also be grouped in pairs to provide the variable $Z_{2l}(x)$ regularized on core samples of length $2l = 10$ m.

Figure III.3 shows the two regularized semi-variograms $\gamma_l(h)$ and $\gamma_{2l}(h)$, as well as the means and experimental dispersion variances of the data used in the calculations. Both these semi-variograms are flat up to around $h = 50$ m (pure nugget effect), after which ($h > 50$ m) a slow increase indicates the presence of a macro-structure with a range that cannot be determined from the available data.

The respective nugget constants (sills of the flat semi-variograms) adopted are 60 and 30 (cm m$^{-1})^2$, which verify the rule (III.8) of inverse proportionality to the support of the regularization.

The experimental dispersion variances, $s_l^2 = 66\cdot8$ and $s_{2l}^2 = 35\cdot3$ (cm m$^{-1})^2$, verify well enough the rule of inverse proportionality.

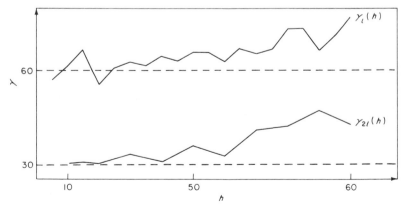

FIG. III.3. Pure nugget-effect semi-variograms. $\gamma_l(h)$: $l = 5$ *m*, $m = 5$ cm m^{-1}, $\sigma_l^2 = 66\cdot8$ (cm m$^{-1})^2$. $\gamma_{2l}(h)$: $2l = 10$ m, $m = 5$ cm m^{-1}, $\sigma_{2l}^2 = 35\cdot3$ (cm m$^{-1})^2$.

These experimental variances are greater than the above nugget constants, the difference corresponding to the additional variability and nesting effect of the macro-structure observed at large distances ($h > 50$ m).

On the scale of observation, $h \in [5, 50$ m$]$, which is also the scale on which production is carried out, this pure nugget effect indicates that selective mining is not possible in this deposit, unless the size of the selection unit is of the order of the vein (5 to 40 cm). In fact, the absence of spatial correlation makes impossible any local differentiation by estimation: at any location in the deposit, the best local estimator of the cumulative quartz thickness is the global mean, $m = 5$ cm m^{-1} in the direction studied (approximately vertical).

This example is one of the rare examples of deposits in which the main variable exhibits no spatial auto-correlation at the scale on which the study is made. There is, however, a macro-structure which appears at greater distances.

2. Nested structure at Mazaugues, after A. Maréchal and J. P. Roullier (1970)

The bauxite deposit at Mazaugues aval (France) consists of subhorizontal stratiform lenses aligned principally in the North–South direction, cf. Fig. III.4(a). These lenses are characterized by a regular hanging wall and a very irregular foot wall which consists of a dolomite karst. The mean dimensions of the lenses are between 200 m and 300 m in the N–S direction and 100 m and 200 m in the E–W direction.

The deposit was sampled by a network of vertical bore-holes with a regular density; on average, each square 100 m × 100 m contains one bore-hole.

Figure III.4(b) shows the semi-variograms of the thickness of bauxite in the two main directions N–S and E–W. These semi-variograms reveal transition models, the ranges of which correspond to the observed mean dimensions of the lenses. Thus, the structure of the variability of the thickness observed on a hectometric scale is that of the lenticular beds.

If a spherical model of range $a_{NS} = 250$ m is fitted to the experimental semi-variogram N–S in Fig. III.4(b), the nugget constant would be $C_0 \simeq 0{\cdot}25$. On the hectometric scale of this observation, it is not possible to specify the structures between 0 and 100 m to any greater degree of precision.

During the course of mining, the bauxite thickness at small sampling distances (2 and 10 m) was measured within one particular lens. Figure III.4(c) shows the semi-variogram calculated from these measurements. A transition with a range of approximately 20 m can be noticed, which is characteristic of the karstic structure of the foot wall, cf. Fig. III.4(a). Beyond this transition, the semi-variogram increases continuously in a way remarkably similar to the linear behaviour at the origin of the previous transition model with range $a_{NS} = 250$ m, which characterizes the macro-structure of the lenticular beds.

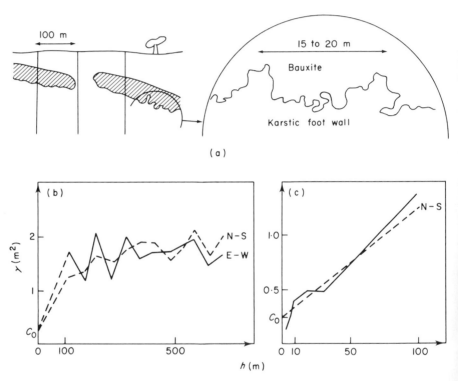

FIG. III.4. (a) Sampling a bauxite thickness. (b) Nested structures at Mazaugues on a hectometric scale. (c) Nested structures at Mazaugues on a decametric scale.

Thus, what is seen on the hectometric scale as a nugget effect is, in fact, the micro-structure of the jagged outline of the karstic foot wall. The total variability of the bauxite thickness, up to $h = 500$ m in the N–S direction, can, thus, be represented as a nested model which is the sum of two transition models (spherical, for example) with ranges $a_0 = 20$ m and $a_1 = 250$ m.

This nested succession of variabilities at different scales is a very frequent phenomenon in mining geostatistics. The best example of it is the nested structures of the ferriferous basin of Lorraine (France) studied by J.

Serra (1967*a*, *b*, 1968). Serra observed and decomposed an entire series of structures from the petrographic scale (100 μm, the regionalization of oolites and limonite) up to the hypermining scale (5 km, the regionalization of deposits related to the stream currents at the time of the deposition of the Lorraine basin).

III.B. MODELS OF VARIOGRAMS

The four main operations of linear geostatistics (variances of estimation and dispersion, regularization and kriging) involve only the structural function of the RF (covariance or variogram). Thus, every geostatistical study begins with the construction of a model designed to characterize the spatial structure of the regionalized variable studied.

III.B.1. Positive definite conditions

It will be recalled (cf. section II.A.3) that the covariances $C(h)$ of a second-order stationary RF $Z(x)$ must be a positive definite function. If the RF is only intrinsic with a semi-variogram $\gamma(h)$, then $-\gamma(h)$ must be a conditional positive definite function.

Positive definiteness and dimensions of the space

It can be shown that the positive definite property of a function $g(h)$ is related to the dimensions of the space on which the distance h is defined. Thus, a function $g(h_u)$ which is positive definite in a one-dimensional space may no longer be so in an n-dimensional space $(n > 1)$; e.g., the function $g(\sqrt{(h_u^2 + h_v^2)})$ of the two parameters (h_u, h_v) may no longer be positive definite in the two-dimensional space.

On the other hand, every function which is positive definite in an n-dimensional space is also positive definite in every space of smaller dimension, $n' < n$.

These properties also hold true for conditional positive definite functions.

In mining practice, the only spaces involved will be those of dimension less than or equal to three. Thus, all the theoretical models presented in the following sections III.B.2 and III.B.3 are positive definite or conditional positive definite in three dimensions and can, therefore, also be used in two or one dimensions.

Two properties

By means of the two following properties of positive definite functions, a vast family of theoretical covariance or covariogram models can be defined in terms of basic positive definite schemes.

(i) Every linear combination of covariances with positive coefficients is a covariance. The same holds true for variograms:

$$\left.\begin{aligned} C(h) &= \sum_{i=1}^{n} \lambda_i^2 C_i(h), \\ \gamma(h) &= \sum_{i=1}^{n} \lambda_i^2 \gamma_i(h), \end{aligned}\right\} \forall \text{ the weights } \lambda_i. \qquad \text{(III.12)}$$

For the proof, it is sufficient to consider the RF $Z(x)$ as the weighted sum of n independent RF's $Y_i(x)$ with covariances $C_i(h)$. The covariance of

$$Z(x) = \sum_{i=1}^{n} \lambda_i Y_i(x)$$

is then given by formula (III.12).

(ii) Any covariance product is a covariance:

$$C(h) = \prod_{i=1}^{n} C_i(h) = C_1(h) C_2(h) \ldots C_n(h) \qquad \text{(III.13)}$$

and the corresponding variogram is written as

$$\gamma(h) = C(0) - C(h) = \prod_{i=1}^{n} C_i(0) - \prod_{i=1}^{n} C_i(h). \qquad \text{(III.14)}$$

Again it is enough for the proof of consider the RF $Z(x)$ as the product of the independent RF's $Y_i(x)$ with covariances $C_i(h)$. The covariance of

$$Z(x) = \prod_{i=1}^{n} Y_i(x)$$

is then given by formula (III.13).

In practice, we can take advantage of the linearity of the geostatistical operators, cf. formula (III.2), and use only the first property. The covariance and variogram models given by formula (III.12) can then be considered as representations of nested structures according to formula (III.1).

III.B.2. Theoretical models of regionalization

Let $Z(x)$ be an intrinsic RF with semi-variogram $\gamma(h)$. The two main characteristics of a stationary variogram are:

(i) its behaviour at the origin, the three types of which are shown on Fig. II.6 (parabolic, linear and nugget effect);

(ii) the presence or absence of a sill in the increase of $\gamma(h)$, i.e., $\gamma(h) =$ constant when $|h| > a$.

Thus, the currently used theoretical models can be classified as:

Models with a sill (or transition models)
  and a linear behaviour at the origin
    (a) spherical model
    (b) exponential model
  and a parabolic behaviour at the origin
    (c) Gaussian model
Models without a sill (the corresponding RF is then only intrinsic and
  has neither covariance nor finite *a priori* variance)
    (a) models in $|h|^{\theta}$, with $\theta \in]0, 2[$
    (b) *logarithmic model*
Nugget effect It has been seen in section III.A.2 that an apparent discontinuity at the origin of the semi-variogram, i.e., a nugget constant C_0, can be interpreted as a transition structure reaching its sill value C_0 at a very small range compared with the available distances of observation.

Remark For the moment, only *isotropic* models will be considered, i.e., RF's which have the same spatial variability in all directions of space. Thus, in the three-dimensional space, the isotropic notation $\gamma(h) = \gamma(r)$ denotes

$$\gamma(\sqrt{(h_u^2 + h_v^2 + h_w^2)}),$$

where (h_u, h_v, h_w) are the three coordinates of the vector h.

1. *Models with a sill, or transition models*

It will be recalled from formula (II.13) and Fig. II.4 that the sill value of a transition structure is the *a priori* variance of the RF $Z(x)$ which is second-order stationary and has a covariance $C(h) = \gamma(\infty) - \gamma(h)$.

The models presented below have been normed to 1, i.e., they correspond to RF's with unit *a priori* variance:

$$\text{Var} \{Z(x)\} = C(0) = \gamma(\infty) = 1.$$

To obtain a model with a sill $C(0) = C \neq 1$, it is enough to multiply the given expressions for $\gamma(h)$ or $C(h)$ by the constant value C.

Linear behaviour at the origin   This is the most frequent type of behaviour encountered in mining practice (variogram of grades and accumulations) and it is most often accompanied by a nugget effect.

(a) *Spherical model*:

$$\gamma(r) = \begin{cases} \dfrac{3}{2}\dfrac{r}{a} - \dfrac{1}{2}\dfrac{r^3}{a^3}, & \forall r \in [0, a], \\ 1 = \text{sill}, & \forall r \geqslant a. \end{cases} \qquad \text{(III.15)}$$

(b) *Exponential model*:

$$\gamma(r) = 1 - \exp(-r/a). \qquad \text{(III.16)}$$

Note that the spherical model effectively reaches its sill for a finite distance $r = a = \text{range}$, while the exponential model reaches its sill only asymptotically, cf. Fig. III.5. However, because of experimental fluctuations of the variogram, no distinction will be made in practice between an effective and an asymptotic sill. For the exponential model, the *practical range a'* can be used with

$$a' = 3a \quad \text{for which } \gamma(a') = 1 - e^{-3} = 0.95 \approx 1.$$

The difference between the spherical and exponential models is the distance (abscissa) at which their tangents at the origin intersect the sill, cf. Fig. III.5:

$r = 2a/3$, two-third of the range for the spherical model;

$r = a = a'/3$, one third of the practical range for the exponential model.

Thus, the spherical model reaches its sill faster than the exponential model.

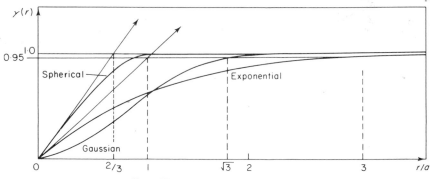

FIG. III.5. Models with a sill.

Parabolic behaviour at the origin This very regular behaviour at the origin is seldom found in mining practice. In the absence of nugget effect, it corresponds to a RF $Z(x)$, which is mean-square differentiable, and, thus, to a ReV $z(x)$, which is very continuous in its spatial variability. Thickness variograms of continuous sedimentary beds sometimes display such a parabolic behaviour at the origin, but are usually accompanied by a slight nugget effect due to errors of measure.

(c) *Gaussian model*:

$$\gamma(r) = 1 - \exp(-r^2/a^2).\qquad\text{(III.17)}$$

The sill is reached asymptotically and a practical range can be considered with $a' = a\sqrt{3}$, for which $\gamma(a') = 0.95 \simeq 1$, cf. Fig. III.5.

A parabolic drift effect (cf. section II.A.3, "Behaviour at infinity" and section II.A.5, Fig. II.8) does not stabilize around a sill for large distances and, thus, cannot be confused with the parabolic behaviour at the origin of a Gaussian model.

However, for small distances $(r < 2a/3)$, it makes little difference whether a very regular local variation is interpreted as a drift effect or as a stationary Gaussian structure.

2. *Models without a sill*

These models correspond to RF's $Z(x)$ with an unlimited capacity for spatial dispersion; neither their *a priori* variances nor their covariance can be defined. The RF's $Z(x)$ are only intrinsic, cf. section II.A.5, Case Study 1.

(a) *Models in r^θ*:

$$\gamma(r) = r^\theta \quad \text{with } \theta \in]0, 2[,\qquad\text{(III.18)}$$

the limits 0 and 2 being precluded.

These models have a particular theoretical and pedagogical importance (they show a whole range of behaviours at the origin when the parameter θ varies and they are easy to integrate).

In practice, only the *linear model* is currently used, cf. Fig. III.6:

$$\gamma(r) = \omega r,$$

with ω the slope at the origin.

Remark 1 For small distances $(h \to 0)$, the linear model can be fitted to any model that has a linear behaviour at the origin (e.g., spherical and exponential models).

Remark 2 As θ increases, the behaviour of $\gamma(r) = r^\theta$ at the origin becomes more regular and corresponds to a RF $Z(x)$ of more and more regular spatial variability.

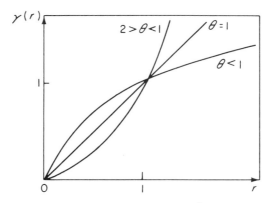

FIG. III.6. Models in r^θ.

Experimentally, models in r^θ for $\theta \in\ ]1, 2[$ are often indistinguishable from a parabolic drift effect, cf. Fig. III.6. The choice of interpretation, as a drift (non-stationarity) or a stationary model in r^θ with θ close to 2, depends on whether or not it is desired to make a drift function $m(x) = E\{Z(x)\}$ obvious.

Remark 3 For $\theta \geq 2$, the function $(-r^\theta)$ is no longer a conditional positive definite function and, in particular, the increase of r^θ at infinity is no longer slower than that of r^2, cf. section II.A.3, formula (II.14). This strict limitation $\theta < 2$ prohibits, in particular, any blind fitting of a variogram by least-square polynomials, even though some polynomials of degree higher than two do satisfy the conditional positive definite property, cf. the expression (III.15) of the spherical model.

(b) *Logarithmic model*:

$$\gamma(r) = \log r. \tag{III.20}$$

Note that $\log r \to -\infty$ as $h \to 0$, so that the logarithmic model cannot be used to describe regionalizations on a strict point support. On the other hand, this model, once regularized on a non-zero support v, cf. section II.D, formulae (II.49) and (II.59), can be used as the model of a regularized semi-variogram $\gamma_v(h)$. In particular, the regularization of a point logarithmic model is a function which is null at the origin: $\gamma_v(0) = 0$.

Logarithmic model or nested transition models? The logarithmic model has been studied exhaustively by the initial workers in geostatistics (D. G.

Krige, 1951; G. Matheron, 1955, 1962; Ph. Formery and G. Matheron, 1963; A. Carlier, 1964). Indeed, up until around 1964–66 this model was practically the only theoretical model used; this is explained by certain analytical properties which make this model a very convenient one (cf. section II.E.2, "Linear equivalents"), and also by the fact that the first geostatistical applications were carried out on deposits (gold on the Rand, uranium) for which the variograms of the characteristic variables had no sill. Since that time, however, geostatistics has been applied to a great number of other ore bodies for which the variograms show more or less clear sills. J. Serra, using a very detailed study of the different scales of variability of the mineralization of the ferriferous basin of Lorraine (France), showed that a logarithmic model is the limit model of a nested succession of transition models, the ranges of which increase geometrically, cf. J. Serra (1976b, 1968).

As the nested model has more parameters, it is more flexible and lends itself more readily to a geological interpretation. Moreover, the particular analytical properties of the logarithmic model are less attractive now that computers are readily available and it is a relatively simple task to program the evaluation of any integral $\bar{\gamma}$ of any nested structure $\gamma(h)$.

This explains why the various sill models have increased in use since around 1966 at the expense of the logarithmic model (except perhaps in South Africa). In our opinion, the use of a logarithmic model or of various nested spherical models is more or less a question of habit and does not alter the result of geostatistical calculations. What is important is that the chosen theoretical model $\gamma(h)$ should be a good fit to the experimental semi-variogram within its limits of reliability ($|h| < b$).

In other words, the results of the geostatistical calculations prove to be *robust* in relation to the choice of the model – provided that the parameters of this model are correctly estimated (see the following example and, in section III.C.6, the exercise on robustness).

Example Consider the regularization of the logarithmic model $\gamma(r) = \log r$ by core samples of length $l = 1$. According to formula (II.50) the regularized model can be written as

$$\gamma_l(r) = \log (r/l) + 3/2 = \log r + 3/2, \qquad \forall r \geqslant l = 1.$$

On Fig. III.7, a nested model has been fitted to this regularized logarithmic model. The nested model consists of a nugget effect and two spherical models γ_1 and γ_2:

$$\gamma_l^*(r) = C_0 + \gamma_1(r) + \gamma_2(r), \qquad \forall r \in [l, 13l]$$

with $C_0 = 0 \cdot 5$; $C_1 = 1 \cdot 5$, $a_1 = 3$; $C_2 = 2$, $a_2 = 13$.

Within the interval $r \in [l, 13l]$ the relative error involved in fitting the nested model $\gamma_l^*(r)$ to the regularized model γ_l is less than 3%, $|\gamma_l - \gamma_l^*|/\gamma_l < 0.03$. In practice, this can be considered as a very satisfactory fit.

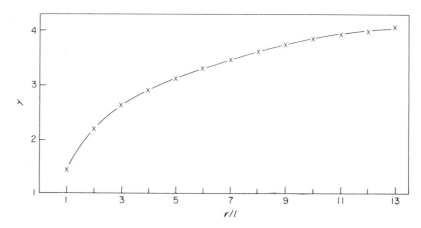

FIG. III.7. Regularized logarithmic model fitted by two spherical nested models ($\times$).

Of course, the sill value $\gamma_l^*(\infty) = C_0 + C_1 + C_2 = 5$, which is attained for distances $r \geq a_2 = 13$, is a pure artefact of the nested model. Similarly, the finite character of the *a priori* variances of the type $D^2(v/\infty)$, calculated using the nested model γ_l^*, is a pure artefact. However, when the use of the regularized model γ_l^* is strictly limited to the interval $r \in [l, 13l]$, there is no difference between calculations using γ_l and those made using γ_l^*.

Now, if the logarithmic growth of an experimental semi-variogram continues well beyond the interval $[l, 13l]$, it would be better to use the regularized logarithmic model γ_l rather than a nested structure γ_l^* consisting of too many transition models.

3. *Hole-effect models*

A semi-variogram $\gamma(h)$ is said to display a hole effect when its growth is not monotonic. The hole effect can appear on models with or without sills, cf. Fig. III.8(a) and (b).

The following hole-effect model, positive definite in the three-dimensional space, is given:

$$C(r) = \sin r/r \quad \text{and} \quad \gamma(r) = 1 - \sin r/r. \qquad (\text{III.21})$$

This model has a sill and parabolic behaviour at the origin, cf. Fig. III.8(a): $\gamma(r) \simeq r^2/6$ when $r \to 0$.

Amplitude of the hole effect   Let α be the relative amplitude of the hole effect, i.e., the minimum value of the covariance divided by the sill value $C(0)$:

$$\alpha = |\min C(h)|/C(0).$$

For the model given in (III.21), this relative amplitude is $\alpha = 0 \cdot 217$, obtained for $r_\alpha \simeq 4 \cdot 4934 \simeq 3\pi/2$.

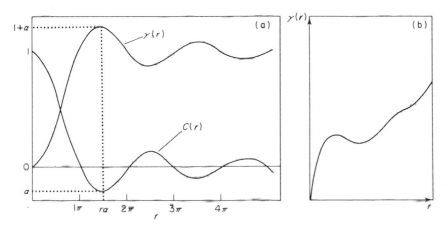

FIG. III.8. Hole-effect model. (a) With sill; (b) without sill.

It can be shown that this value $\alpha = 0 \cdot 217$ is the maximum amplitude of an isotropic hole effect in the three-dimensional space. If an experimental hole effect with an amplitude greater than $0 \cdot 217$ is observed in practice, then either the hole effect is not significant and is nothing more than a fluctuation of the experimental semi-variogram, or the hole effect is directional, i.e., it is not present in every direction of the three-dimensional space.

If we desire to specify a particularly strong directional hole effect, then a positive definite hole-effect model in only one dimension must be used, e.g.,

$$\gamma(h) = 1 - \cos h \tag{III.22}$$

for one-dimensional h.

In this case, the relative amplitude α is equal to 1 and superior to $0 \cdot 217$. The cosine model $C(h) = \cos h$ is not positive definite in three dimensions and, thus, its use must be restricted to one particular direction.

Note that the model (III.22) is periodic without any dampening of the oscillations. In practice, however, such a hole-effect model is always associated, in a nested manner, with other models with or without sills and the considerable oscillations of the cosine will, thus, be dampened, cf. Case Study 6 in section IV.B.

Interpretation   A pseudo-periodic component of the regionalization may appear on the experimental semi-variograms as a hole effect; for example, the stationary sequence of two well-differentiated types of mineralization in an ore body. If this sequence is not isotropic (and, in general, it is not), then the hole effect will be observed only in certain directions, for example, the vertical direction if the phenomenon is a pseudo-periodic sequence of horizontal stratification, cf. Fig. III.9(a).

If the grades of Fig. III.9 are point grades, then the vertical semi-variogram will exhibit a hole effect with an amplitude possibly greater than 0·217 and the abscissae b_1 and b_2 of the oscillations can be related to the mean vertical dimensions of the waste and rich ore strata. The horizontal semi-variograms will differ according to whether they are located within a rich or waste stratum, cf. Fig. III.9(b).

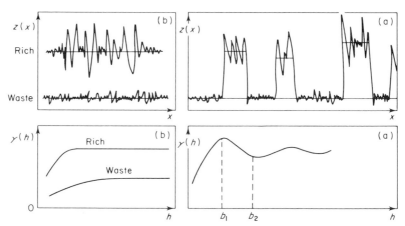

FIG. III.9. Vertical hole effect. (a) Vertical direction; (b) horizontal direction.

A hole effect may also be due to an artificial pseudo-periodicity of the available data. Thus, when a first sampling programme on a square grid b is completed by a second one on a grid $b/2$, the heterogeneity of the drilling material used for the two programmes would appear as a more or less dampened serration on the experimental variograms.

In practice, the natural variability observed in an ore body is, in general, nested, i.e., a pseudo-periodic structure is most often associated with other structures, the consequence of which is a dampening of the observed hole effect. For estimation purposes, an experimental hole effect that is open to a doubtful interpretation, or is not very marked, can simply be ignored, cf. section IV.B, Case Study 6.

III.B.3 Theoretical models of coregionalization

Warning Even though the study of coregionalizations is in every way analogous to that of regionalizations, there is an additional difficulty due to the number of indices involved. Thus, the reader who is not familiar with the index notation of matrices is advised to go directly to the next section, III.B.4. A thorough understanding of the study of regionalizations will make the study of coregionalizations more understandable on a second reading of this chapter.

Consider a set of K second-order stationary RF's $\{Z_k(x), k = 1 \text{ to } K\}$. The object of this section is to establish cross-covariance models $C_{kk'}(h)$, cf. section II.A.4.

The positive definite condition

Just as any function $g(h)$ cannot be considered as the covariance or the variogram of a second-order stationary RF, any set of functions $\{g_{kk'}(h); k, k' = 1 \text{ to } K\}$ cannot be considered as the matrix of covariances or variograms of a set of coregionalized RF's. The matrix of covariances $\{C_{kk'}(h)\}$ must be positive definite so as to ensure that the variances of all finite linear combinations of the basic RF's $Z_k(x)$ are positive.

In practice, to construct a matrix model of coregionalizations, we shall first build a model of the spatial intercorrelation of the RF's $Z_k(x)$. The matrix model of this coregionalization will then be, by definition, positive definite.

The linear model (for coregionalizations)

Consider n second-order stationary RF's $\{Y_i(x), i = 1 \text{ to } n\}$ with the functions $K_i(h)$ as direct covariances. These n RF's are orthogonal, i.e., their cross-covariances are such that $K_{ij}(h) = 0$, $\forall h$ and $i \neq j$.

Consider now the K second-order stationary RF's $\{Z_k(x), k = 1 \text{ to } K\}$ defined by the K linear combinations of the type

$$Z_k(x) = \sum_{i=1}^{n} a_{ki} Y_i(x), \tag{III.23}$$

whatever the parameters a_{ki}: negative, zero or positive.

The set of K intercorrelated RF's thus generated has a positive definite cross-covariance matrix $\{C_{kk'}(h)\}$:

$$C_{kk'}(h) = \sum_{i=1}^{n} \sum_{j=1}^{n} a_{ki} a_{k'j} K_{ij}(h) = \sum_{i=1}^{n} a_{ki} a_{k'i} K_i(h).$$

In the definition expression (III.23) of the RF $Z_k(x)$, some basic RF's, $Y_i(x)$ and $Y_j(x)$, with $i \neq j$, can have the same direct covariance, $K_i(h) = K_j(h)$, while remaining orthogonal, $K_{ij}(h) = 0$. An expression similar to (III.23) can be written by grouping the RF's which have the same direct covariances:

$$Z_k(x) = \sum_{i=1}^{n} \sum_{l} a_{ki}^l Y_i^l(x), \qquad (III.24)$$

with Y_i^l and $Y_i^{l'}$ having the same direct covariance $K_i(h)$; Y_i^l and $Y_j^{l'}$ being orthogonal except when $i = j$ and $l = l'$ simultaneously. The cross-covariance of $Z_k(x)$ with $Z_{k'}(x)$ is then written as

$$C_{kk'}(h) = \sum_{i=1}^{n} [\sum_l a_{ki}^l a_{k'i}^l] K_i(h) = \sum_{i=1}^{n} b_{kk'}^i K_i(h),$$

with

$$b_{kk'}^i = \sum_l a_{ki}^l a_{k'i}^l, \qquad \forall i = 1 \text{ to } n.$$

This last expression gives the matrix of coefficients $[b_{kk'}^i]$ and is nothing more than the general expression of a positive definite matrix.

The linear model for coregionalization is then a model where

(i) all the direct and cross-covariances are derived from linear combinations of n basic direct covariances $\{K_i(h), i = 1 \text{ to } n\}$, i.e.,

$$C_{kk'}(h) = \sum_{i=1}^{n} b_{kk'}^i \cdot K_i(h) \quad \text{with} \quad b_{kk'}^i = b_{k'k}^i, \qquad \forall i; \qquad (III.25)$$

(ii) for a fixed index i, the coefficient matrix $[b_{kk'}^i]$ is positive definite.

Remark 1   The condition enforcing the coefficient matrices $[b_{kk'}^i]$ to be positive definite is a necessary and sufficient condition for a set of functions $\{C_{kk'}(h)\}$, defined as linear combinations of n basic covariances $K_i(h)$, to constitute a *linear* model of coregionalization. The condition is only sufficient for this set of functions $\{C_{kk'}(h)\}$ to constitute a model of coregionalization.

Remark 2   Remember that a matrix $[b_{kk'}]$ is positive definite if all its eigenvalues are real and positive.

In practice, there is no need to look for the eigenvalues of the matrix $[b_{kk'}]$. As this matrix is symmetric, $b_{kk'} = b_{k'k}$, it is positive definite when its determinant as well as its diagonal minors are positive or zero. Thus, for a matrix of dimension K, we have the following K conditions:

$$b_{11} \geq 0, \quad \begin{vmatrix} b_{11} & b_{12} \\ b_{21} & b_{22} \end{vmatrix} \geq 0, \ldots, \quad \begin{vmatrix} b_{11} & b_{12} & \cdots & b_{1K} \\ \vdots & \vdots & & \vdots \\ b_{K1} & b_{K2} & \cdots & b_{KK} \end{vmatrix} \geq 0. \qquad (III.26)$$

In mining practice, coregionalizations more than three variables $(K > 3)$ are seldom considered, mainly for difficulties of statistical inference of the covariances matrix.

For $K = 2$, verification of the positive definiteness of the matrix $[b^i_{kk'}]$ amounts to checking that, $\forall i$, the two following inequalities hold true:

$$b^i_{11} \geq 0; \quad |b^i_{12}| = |b^i_{21}| \leq \sqrt{(b^i_{11} b^i_{22})} \quad \text{(Schwarz's inequality)}.$$

For $K = 3$, in addition to the two preceding inequalities, one must verify that the determinant $|b^i_{kk'}|$ is positive or zero, $\forall i$.

The linear model written in terms of semi-variograms

By replacing the covariances $C_{kk'}(h)$ and $K_i(h)$ by the corresponding semi-variograms $\gamma_{kk'}(h)$ and $\gamma_i(h)$, the definition formula (III.25) can be rewritten as

$$\gamma_{kk'}(h) = \sum_{i=1}^{n} b^i_{kk'} \cdot \gamma_i(h) \quad \text{with } b^i_{kk'} = b^i_{k'k}, \qquad \forall i, \qquad (III.27)$$

where, $\forall i$ fixed, the coefficient matrix $[b^i_{kk'}]$ is positive definite.

Restrictions of the linear model

Restriction 1   As $b^i_{kk'} = b^i_{k'k}$, $\forall i$, the linear model only provides symmetric cross-covariances:

$$C_{kk'}(h) = C_{k'k}(h) \quad \text{and} \quad C_{kk'}(h) = C_{kk'}(-h).$$

Now, there are non-symmetric cross-covariances, cf. formula (II.18).

However, non-symmetric cross-covariances can be obtained from formulae (III.23) and (III.24) by shifting some of the RF's $Z_k(x)$ by a vectorial lag h_k:

$$\left. \begin{aligned} Z_k(x + h_k) &= \sum_i a_{ki} Y_i(x), \\ Z_{k'}(x) &= \sum_i a_{k'i} Y_i(x). \end{aligned} \right\} \qquad (III.28)$$

The cross-covariance $C_{kk'}(h)$ is then rewritten as

$$C_{kk'}(h_k + h') = \sum_i a_{ki} a_{k'i} K_i(h'),$$

or, with the change of variable, $h = h_k + h'$,

$$C_{kk'}(h) = \sum_i a_{ki} a_{k'i} K_i(h - h_k).$$

This last cross-covariance $C_{kk'}(h)$ is no longer symmetric in $(h, -h)$ as, $\forall i$, $K_i(h - h_k) \neq K_i(-h - h_k)$ unless the lag h_k is zero.

Restriction 2 The coefficient matrices $[b_{kk'}^i]$, $\forall i$, must be positive definite. This condition entails Schwarz's inequality:

$$|b_{kk'}^i| \leq \sqrt{(b_{kk}^i \cdot b_{k'k'}^i)}, \qquad \forall i, \qquad \forall k \neq k'. \qquad \text{(III.29)}$$

Consequently, $b_{kk'}^i > 0$ entails $b_{kk}^i \cdot b_{k'k'}^i \geq 0$.

Coming back to the definition (III.25) of the linear model, Schwarz's inequality implies that every basic structure $K_i(h)$ appearing on the cross-covariance $C_{kk'}(h)$ must also appear on the two direct covariances $C_{kk}(h)$ and $C_{k'k'}(h)$.

Thus, for example, if $C_{kk'}(h)$ reveals two nested ranges a_1 and a_2, these two ranges must also appear on the direct covariances $C_{kk}(h)$ and $C_{k'k'}(h)$. On the other hand, a structure may appear on the direct covariances without being present on the cross-covariances. For this, it is sufficient that $b_{kk'}^i = 0$ when $b_{kk}^i > 0$.

Intrinsic coregionalization

This particular example of the linear model corresponds to the case in which *all* the basic RF's $Y_i(x)$ of formulae (III.23) and (III.24) have the same direct covariance $K_0(h)$ (or the same direct semi-variogram $\gamma_0(h)$):

$$K_i(h) = K_0(h), \qquad \forall i = 1 \text{ to } n.$$

The model of intrinsic coregionalization is then

$$\left. \begin{array}{l} C_{kk'}(h) = b_{kk'} \cdot K_0(h), \\[2mm] \gamma_{kk'}(h) = b_{kk'} \cdot \gamma_0(h), \end{array} \right\} \qquad \text{(III.30)}$$

the matrix of coefficients $[b_{kk'}]$ being positive definite.

In practical terms, the intrinsic coregionalization model means that all the direct or cross-covariances (or variograms) can be derived by affinity (multiplication by the constant $b_{kk'}$) from a single basic model $K_0(h)$ or

$\gamma_0(h)$. This model thus corresponds to variables $Z_k(x)$, $Z_{k'}(x)$, the variabilities of which are proportionally related to a unique principal cause characterized by the model $\gamma_0(h)$. This occurs in mining practice in the study of the two proportional variables, thickness $t(x)$ and accumulation $a(x) = t(x) . z(x)$; $z(x)$ being the mean grade of the drill core of length $t(x)$.

Restriction of the intrinsic coregionalization model   The intrinsic coregionalization model is particularly convenient for calculations, since studying the coregionalization amounts to the study of the single regionalization with model $\gamma_0(h)$. On the other hand, this model is a very particular one and has an important limitation in applications: it can be shown, cf. section V.A.4 "Cokriging", that if two variables Z_k and $Z_{k'}$ are in intrinsic coregionalization *and* if these two variables have been sampled in an identical way (e.g., each sample is analysed systematically for Z_k and $Z_{k'}$), then there is no gain in using the $Z_{k'}$ information in the estimation of the Z_k. However, if the variable Z_k has been undersampled with respect to $Z_{k'}$, then the estimation of Z_k can be improved by using $Z_{k'}$ values.

III.B.4. Models of anisotropy

As x represents a point with coordinates (x_u, x_v, x_w) in the three-dimensional space, the notation $\gamma(h)$ represents a function of the vector h of rectangular coordinates (h_u, h_v, h_w) or spherical coordinates $(|h|, \alpha, \varphi)$, where $|h|$ is the modulus of the vector h and α and φ its longitude and latitude.

Isotropy   When the function $\gamma(|h|, \alpha, \varphi)$ depends only on the modulus $|h|$ of the vector h, the phenomenon is said to be "isotropic". The variability of the three-dimensional RF $Z(x_u, x_v, x_w)$, characterized by the semi-variogram $\gamma(|h|)$, is identical in all directions of the space and, thus,

(i) in spherical coordinates,

$$\gamma(|h|, \alpha, \varphi) = \gamma(|h|), \qquad \forall \alpha \text{ and } \varphi;$$

(ii) in rectangular coordinates,

$$\gamma(h_u, h_v, h_w) = \gamma(\sqrt{(h_u^2 + h_v^2 + h_w^2)}).$$

Anisotropy   The phenomenon is said to be "anisotropic" when its variability is not the same in every direction. The structural function $\gamma(|h|), \alpha, \varphi)$ characterizing this spatial variability in *mean square* thus depends on the direction parameters α and φ.

Interpretation   The notion of mean variability in the above definition is important. In a massive stockwerk-type ore body, rich, small, local veinlets can be found in specific directions, but, on average, these veinlets will occur over the entire mass in every possible direction of space and the phenomenon will be isotropic on average.

In practice, a real anisotropy of the structural function $\gamma(h)$ corresponds to the existence of preferential directions at the time of the genesis of the studied phenomenon. These preferential directions are, in general, known *a priori* and the variogram quantifies their respective variabilities. For example,

   (i) the vertical direction in a deposit formed by deltaic deposition;
  (ii) the horizontal directions of the deposition currents in an alluvial deposit;
 (iii) the radial directions around a volcanic pipe in an intrusive deposit.

Anisotropies and nested structures

We have already seen that the observed variability of a phenomenon is most often due to many causes ranging over various scales. Consider a nested model $\gamma(h)$ consisting of the sum of a micro-structure $\gamma_1(h)$ and a macro-structure $\gamma_2(h)$:

$$\gamma(h) = \gamma_1(h) + \gamma_2(h)$$

There is no reason for these two-component structures to have the same directions of anisotropy. Thus, the micro-structure γ_1, characterizing, for example, the measure errors and the phenomena of diffusion and concretion at very small distances, may be isotropic, while the macro-structure γ_2, characterizing, for example, the lenticular deposits of the mineralization, may reveal directions of preferential alignment of these lenses, cf. Fig. II.5. In such an example, the structure γ_1 depends only on the modulus $|h|$ of the vector h and it can be represented by any one of the isotropic models presented in the previous section, III.B.2. On the other hand, the macro-structure γ_2 will require a model $\gamma_2(|h|, \alpha, \varphi)$ which depends not only on the modulus $|h|$ but also on the direction of the vector h.

Representation

All the theoretical models of regionalization presented in section III.B.2 are isotropic, i.e., $\gamma(h)$ depends only on the modulus $r = |h|$ of the vector h.

Anisotropies will be represented by the method of reducing them to the isotropic case either by a linear transformation of the rectangular coordinates (h_u, h_v, h_w) of the vector h in the case of geometric anisotropy, or by representing separately each of the directional variabilities concerned in the case of zonal anisotropy.

1. *Geometric anisotropy*

A semi-variogram $\gamma(h_u, h_v, h_w)$ or a covariance $C(h_u, h_v, h_w)$ has a geometric anisotropy when the anisotropy can be reduced to isotropy by a mere *linear* transformation of the coordinates:

$$\gamma(h_u, h_v, h_w) = \gamma'(\sqrt{(h_u'^2 + h_v'^2 + h_w'^2)}) \qquad \text{(III.31)}$$
$$\underset{\text{anisotropic}}{} \qquad \underset{\text{isotropic}}{\phantom{\gamma'(\sqrt{(h_u'^2 + h_v'^2)})}}$$

with

$$h_u' = a_{11}h_u + a_{12}h_v + a_{13}h_w,$$
$$h_v' = a_{21}h_u + a_{22}h_v + a_{23}h_w,$$
$$h_w' = a_{31}h_u + a_{32}h_v + a_{33}h_w,$$

or, in matrix form,

$$[h'] = [A] \cdot [h],$$

where $[A] = [a_{ij}]$ represents the matrix of transformation of the coordinates, and $[h]$ and $[h']$ are the two column-matrices of the coordinates.

As a first simple example, suppose that we are only interested in two particular directions, α_1 and α_2, in space. Let h_1 and h_2 be the distances measured along these two directions, cf. Fig. III.10. The two semi-variograms $\gamma_{\alpha_1}(h_1)$ and $\gamma_{\alpha_2}(h_2)$ calculated in these two directions have been

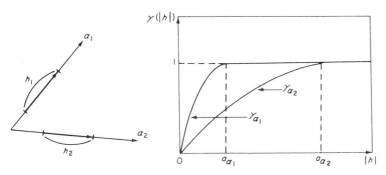

FIG. III.10. Geometric anisotropy.

represented by two spherical models with the same sill 1 and different ranges a_{α_1} and a_{α_2}. Thus, since $\gamma_{\alpha_1}(h) \neq \gamma_{\alpha_2}(h)$, the phenomenon represented is anisotropic: the variability depends on the directions α_1 and α_2.

In order to make the two curves γ_{α_1} and γ_{α_2} in Fig. III.10 coincide, it is sufficient to multiply h_1 by the ratio of affinity $a_{\alpha_2}/a_{\alpha_1}$ or, conversely, to multiply the distances h_2 by the ratio $a_{\alpha_1}/a_{\alpha_2}$. Indeed, consider the two spherical models

$$\gamma_{\alpha_1}(h_1) = \frac{3}{2a_{\alpha_1}} h_1 - \frac{1}{2a_{\alpha_1}^3} h_1^3, \qquad \forall h_1 \leq a_{\alpha_1},$$

and

$$\gamma_{\alpha_2}(h_2) = \frac{3}{2a_{\alpha_2}} h_2 - \frac{1}{2a_{\alpha_2}^3} h_2^3, \qquad \forall h_2 \leq a_{\alpha_2}.$$

The following relation is then derived:

$$\gamma_{\alpha_1}(h_1) = \frac{3}{2a_{\alpha_2}}\left(h_1 \frac{a_{\alpha_2}}{a_{\alpha_1}}\right) - \frac{1}{2a_{\alpha_2}^3}\left(h_1 \frac{a_{\alpha_2}}{a_{\alpha_1}}\right)^3, \qquad \forall h_1 \frac{a_{\alpha_2}}{a_{\alpha_1}} \leq a_{\alpha_2},$$

i.e.,

$$\gamma_{\alpha_1}(h_1) = \gamma_{\alpha_2}(h_1') \quad \text{with} \quad h_1' = h_1 \frac{a_{\alpha_2}}{a_{\alpha_1}}.$$

Subject to the change of coordinates $h_1' = h_1 a_{\alpha_2}/a_{\alpha_1}$, the variability in the two directions α_1 and α_2 is characterized by a single spherical model with a range a_{α_2}.

More generally, if n directional semi-variograms $\{\gamma_{\alpha_i}(h_i), i = 1$ to $n\}$ can be represented by n transition models of the same type (spherical, for example) with the *same sill*, and the n ranges of which form an elliptical-shaped directional graph (in two dimensions), or ellipsoidal (in three dimensions), geometric anisotropy is present.

An example is given on Fig. III.11, which shows four semi-variograms for four horizontal directions $\alpha_1, \alpha_2, \alpha_3, \alpha_4$. Spherical models with identical sills and ranges of $a_{\alpha_1}, a_{\alpha_2}, a_{\alpha_3}, a_{\alpha_4}$ have been fitted to these semi-variograms. The directional graph of the ranges, i.e., the variation of the ranges a_{α_i} as a function of the direction α_i, is also shown on Fig. III.11. There are three possible cases.

(i) The graph can be approximated to a circle of radius a, i.e., $a_{\alpha_i} \simeq a$, for all horizontal directions α_i, and the phenomenon can thus be considered as isotropic and characterized by a spherical model of range a.

(ii) The graph can be approximated by an ellipse, i.e., by a shape which is a linear transform of a circle. By applying this linear transformation to the coordinates of vector h, the isotropic case is produced (circular graph). The phenomenon is a geometric anisotropy.

(iii) The graph cannot be fitted to a second-degree curve and the second type of anisotropy must be considered, i.e., zonal anisotropy in certain directions, α_4, for example, on Fig. III.11.

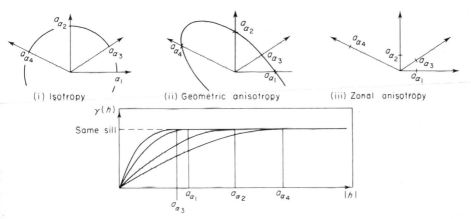

FIG. III.11. Directional graphs of the ranges.

If, instead of transition structures of range a_{α_i} as in Fig. III.11, the directional semi-variograms are of the linear type, $\gamma_{\alpha_i}(h_i) = \omega_{\alpha_i} h_i$, then the directional graphs of the inverses of the slopes at the origin ω_{α_i} will be considered. An hypothesis of isotropy, geometric anisotropy or zonal anisotropy will then be adopted according to whether or not this directional graph can be fitted to a circle or an ellipse.

Correction of geometric anisotropy (in two dimensions)

Consider a geometric anisotropy in two dimensions (e.g., the horizontal space). The directional graph of the ranges a_α is elliptical, cf. Fig. III.12. Let (x_u, x_v) be the *initial* rectangular coordinates of a point, φ the angle made by the major axis of the ellipse with the coordinate axis Ox_u, and $\lambda > 1$ the ratio of anisotropy of the ellipse. The three following steps will transform this ellipse into a circle and, thus, reduce the anisotropy to isotropy.

(i) The first step is to rotate the coordinate axes by the angle φ so that they become parallel to the main axes of the ellipse. The new coordinates (y_1, y_2) resulting from the rotation can be written in matrix notation as

$$\begin{bmatrix} y_1 \\ y_2 \end{bmatrix} = [R_\varphi] \begin{bmatrix} x_u \\ x_v \end{bmatrix} \quad \text{with} \quad [R_\varphi] = \begin{bmatrix} \cos\varphi & \sin\varphi \\ -\sin\varphi & \cos\varphi \end{bmatrix},$$

where $[R_\varphi]$ is the matrix of rotation through the angle φ.

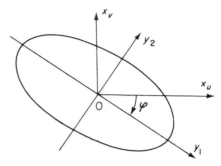

FIG. III.12. Correction of geometric anisotropy.

(ii) The second step is to transform the ellipse into a circle with a radius equal to the major range of the ellipse. This is achieved by multiplying the coordinate y_2 by the ratio of anisotropy $\lambda > 1$. The new coordinates (y_1', y_2') are then deduced from the coordinates (y_1, y_2) by the matrix expression

$$\begin{bmatrix} y_1' \\ y_2' \end{bmatrix} = [\lambda] \begin{bmatrix} y_1 \\ y_2 \end{bmatrix} \quad \text{with} \quad [\lambda] = \begin{bmatrix} 1 & 0 \\ 0 & \lambda \end{bmatrix}.$$

(iii) The initial orientation of the coordinate system is then restored by rotation through the angle $-\varphi$, and the final transformed coordinates (x_u', x_v') are given by

$$\begin{bmatrix} x_u' \\ x_v' \end{bmatrix} = [R_{-\varphi}] \begin{bmatrix} y_1' \\ y_2' \end{bmatrix}.$$

The final transformed coordinates (x_u', x_v') can then be derived from the initial coordinates (x_u, x_v) by the transformation matrix $[A]$ which is the product of the three matrices $[R_{-\varphi}][\lambda][R_\varphi]$, i.e.,

$$\begin{bmatrix} x_u' \\ x_v' \end{bmatrix} = [R_{-\varphi}][\lambda][R_\varphi] \begin{bmatrix} x_u \\ x_v \end{bmatrix} = [A] \begin{bmatrix} x_u \\ x_v \end{bmatrix} \qquad \text{(III.32)}$$

with

$$[A] = \begin{vmatrix} a & c \\ c & b \end{vmatrix}$$

and

$$a = \cos^2 \varphi + \lambda \sin^2 \varphi,$$
$$b = \sin^2 \varphi + \lambda \cos^2 \varphi,$$
$$c = (1 - \lambda) \sin \varphi . \cos \varphi.$$

If h is any vector in the two-dimensional space with initial co-ordinates (h_u, h_v), then, to obtain the value of the anisotropic semi-variogram $\gamma(h) = \gamma(h_u, h_v)$, we first calculate the transformed coordinates (h'_u, h'_v) from

$$\begin{bmatrix} h'_u \\ h'_v \end{bmatrix} = [A] \begin{bmatrix} h_u \\ h_v \end{bmatrix}$$

and then we substitute these coordinates in the isotropic model $\gamma'(|h'|)$, which has a range equal to the major range of the directional ellipse, i.e., according to formula (III.31),

$$\underset{\text{anisotropic}}{\gamma(h_u, h_v)} = \underset{\text{isotropic}}{\gamma'(\sqrt{(h'^2_u + h'^2_v)})}.$$

Remark 1   In practice, in two dimensions, three or four directions α_i must be studied in order to establish the directional graph of ranges and subsequently decide whether a geometric anisotropy is present.

Remark 2   *Optimum sampling grid.* The minor axis of the ellipse (Oy_2 on Fig. III.12) is the direction in which the phenomenon varies fastest, and the major axis the direction in which variability is slowest. Thus, the optimum sampling grid for this phenomenon is a rectangular one, aligned in the directions of these two main axes (Oy_1 and Oy_2) with sides in the ratio λ, the smallest side corresponding to the direction Oy_2 of fastest variability.

Remark 3   It is possible to determine the matrix $[A]$ for the correction of a geometric anisotropy in three dimensions. However, reality seldom conforms to such a particular model as an ellipsoid of geometric anisotropy. In practice, a zonal anisotropy will most often be used to distinguish certain directions (e.g., the vertical direction from the horizontal directions).

2. *Zonal anisotropy*

The model of zonal anisotropy is the one most currently used in practice, since any observed anisotropy which cannot be reduced by a simple linear transformation of coordinates will call for this model.

Let $\gamma(h)$ be a nested model characterizing a variability in the three-dimensional space $\gamma(h) = \sum_i \gamma_i(h)$, where h is a vector with coordinates $(h_u,$

h_v, h_w). Each of the components $\gamma_i(h)$ of this nested model can be aniso-tropic in h, i.e., $\gamma_i(h)$ is a function of the three coordinates $(h_u,\ h_v,\ h_w)$ rather than of just the modulus $|h|$. Moreover, the anisotropy of $\gamma_i(h)$ may be completely different to that of $\gamma_j(h)$. Thus, the structure $\gamma_1(h)$ may have a geometric anisotropy, while $\gamma_2(h)$ is a function of the vertical distance h_w only:

$$\gamma_2(h) = \gamma_2(h_w), \qquad \forall h_u,\ h_v.$$

A third structure may be isotropic in three dimensions: $\gamma_3(h) = \gamma_3(|h|)$.

The model of *zonal anisotropy* can thus be defined as a nested structure in which each component structure may have its own anisotropy.

An obvious anisotropy of the structural function $\gamma(h)$ will most often correspond to a genetic anisotropy known beforehand, so that any pref-erential direction is well known and, thus, can be differentiated when modelling the three-dimensional structural function $\gamma(h)$.

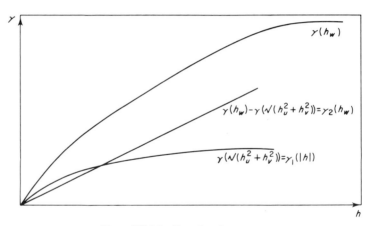

FIG. III.13. Zonal anisotropy.

On Fig. III.13, the semi-variogram $\gamma(h_w)$ constructed in the vertical direction reveals a strong variability due, for instance, to surface weather-ing or enrichment with depth. The mean semi-variogram, isotropic for the horizontal directions $\gamma(\sqrt{(h_u^2 + h_v^2)})$ reveals a weaker variability due, for example, to lenticular beds in the horizontal strata.

On Fig. III.13, the difference $\gamma(h_w) - \gamma(\sqrt{(h_u^2 + h_v^2)})$ is approximately linear, at least over the smaller distances, so that a three-dimensional nested model $\gamma(h)$ can be adopted, $\gamma(h)$ being the sum of an isotropic transition model $\gamma_1(|h|)$ and a linear model $\gamma_2(h_w)$:

$$\gamma(h_u,\ h_v,\ h_w) = \gamma_1(|h|) + \gamma_2(h_w).$$

The model $\gamma_1(|h|)$ is fitted to the mean horizontal semi-variogram and the model $\gamma_2(h_w)$ to the difference $\gamma(h_w) - \gamma(\sqrt{(h_u^2 + h_v^2)})$.

Remark. The model of zonal anisotropy, built from the nest of structures having their own anisotropy, is flexible enough to be used as a model for almost any type of experimental anisotropy.

However, the flexibility of this model must not be misused. When a specific direction is differentiated (e.g., the vertical direction h_w of the previous example), this direction must correspond to a preferential direction of the mineralization. The known geology should always be used as a guide when fitting a mathematical model $\gamma(h)$ to a natural variability.

III.B.5. General structural model

Now that the different models of anisotropy have been studied, the following general expression of a structural model $\gamma(h)$ can be proposed, $\gamma(h)$ being made up of the nested sum of N *isotropic* structures $\{\gamma_i(|h_i|), i = 1$ to $N\}$:

$$\gamma(h) = \sum_{i=1}^{N} \gamma_i(|h_i|). \tag{III.33}$$

The particular anisotropy (geometric or zonal) of each component structure γ_i is characterized by the linear transformation matrix $[A_i]$ which changes the vector h into the vector h_i. More precisely, let (h_u, h_v, h_w) be the initial coordinates of the vector h; the coordinates $(h_{i,u}, h_{i,v}, h_{i,w})$ of the transformed vector h_i are deduced from the initial coordinates by the linear relations

$$[h_i] = [A_i] \cdot [h].$$

Example Consider the following three-dimensional model:

$$\gamma(h_u, h_v, h_w) = \gamma_0(|h|) + \gamma_1(|h|) + \gamma_2(|h_2|) + \gamma_3(h_w) \tag{III.34}$$

with $\gamma_0(|h|) = C_0$ when $|h| > \varepsilon$ characterizing an *isotropic* nugget effect; $\gamma_1(|h|)$ characterizing a first *isotropic* structure; $\gamma_2(|h_2|) = \gamma_2(\sqrt{(h_{2,u}^2 + h_{2,v}^2 + h_{2,w}^2)})$, characterizing a second structure *isotropic* with regard to the coordinates of the transformed vector $[h_2] = [A_2] \cdot [h]$. For example, this second structure may depend only on the horizontal directions (h_u, h_v) with a geometric anisotropy. The previous linear transformation is then written

$$h_{2,u} = a_{11}h_u + a_{12}h_v + 0 \cdot h_w,$$

$$h_{2,v} = a_{21}h_u + a_{22}h_v + 0 \cdot h_w,$$

$$h_{3,w} = 0 \cdot h_u + 0 \cdot h_v + 0 \cdot h_w = 0.$$

The last component structure, $\gamma_3(h_w)$, depends only on the vertical direction (h_w). This structure, $\gamma_3(h_w)$, can be considered as *isotropic* with regard to the coordinates $(h_{3,u}, h_{3,v}, h_{3,w})$ of the transformed vector $[h_3] = [A_3] \cdot [h]$:

$$\gamma_3(h_w) = \gamma_3(|h_3|) \quad \text{with} \quad \begin{cases} h_{3,u} = h_{3,v} = 0, \\ h_{3,w} = h_w. \end{cases}$$

The general structural model of formula (III.33) is particularly convenient since it is made up of the sum of N isotropic structures $\gamma_i(|h_i|)$ which, in the course of a particular geostatistical operation, will be treated separately one by one, cf. section III.A.1, "The linearity of geostatistical operators".

All the case studies and practical exercises given in this book make use of this general structural model.

Computer subroutine

The subroutine FUNCTION GAM(HX, HY, HZ) gives the values of a three-dimensional model $\gamma(h)$ of the general type (III.33). The component structures γ_i are transition models, either spherical or exponential. A nugget effect is represented by a transition model with a zero range.

Example   The parameters of the previous model (III.34) given as an example would be: (HX, HY, HZ) = initial coordinates of the vector h; NST = 4 = component structures, with respective sills C(NST) and ranges A(NST). For the nugget effect, $\gamma_0(|h|)$, we have A(1) = 0.

The matrices of linear transformations of the coordinates are

$$\text{CAX}(4, 3) = \begin{bmatrix} 1 & 1 & 1 \\ 1 & 1 & 1 \\ a_{11} & a_{12} & 0 \\ 0 & 0 & 0 \end{bmatrix} \begin{array}{l} \rightarrow \text{ for the first coordinate of the first component } \gamma_o \\ \\ \rightarrow \text{ for the first coordinate of the third component } \gamma_2 \end{array}$$

$$CAY(4,3) = \begin{bmatrix} 1 & 1 & 1 \\ 1 & 1 & 1 \\ a_{21} & a_{22} & 0 \\ 0 & 0 & 0 \end{bmatrix} \quad \text{and} \quad CAZ(4,3) = \begin{bmatrix} 1 & 1 & 1 \\ 1 & 1 & 1 \\ 0 & 0 & 0 \\ 0 & 0 & 1 \end{bmatrix}.$$

```
      FUNCTION GAM(HX,HY,HZ)
C
C ................................................
C                  SEMI-VARIOGRAM FUNCTION
C DEFINED AS THE SUM OF EXPONENTIAL OR SPHERICAL
C NESTED STRUCTURES..THESE STRUCTURES ARE ISOTROPIC
C AFTER LINEAR TRANSFORMATION OF INITIAL COORDINATES
C HX,HY,HZ..NEW COORDINATES ARE CX,DY,DZ..
C
C ....COMMON  /GAMMA/
C   NST          NUMBER OF ELEMENTARY STRUCTURES
C   C(NST)       SILLS
C   A(NST)       RANGES <0. EXPONENTIAL
C                       =0. NUGGET EFFECT
C                       >0. SPHERICAL
C   CAX(NST,3) |    LINEAR TRANSFORMATION
C   CAY(NST,3) |      MATRICES OF
C   CAZ(NST,3) |    3 COORDINATES
C ................................................
C
      DIMENSION A(1),C(1),CAX(1),CAY(1),CAZ(1)
      COMMON/GAMMA/NST,C,A,CAX,CAY,CAZ
      GAM=0.
      H=SQRT(HX*HX+HY*HY+HZ*HZ)
      IF(H.LT.1.E-03)RETURN
      DO 1 IS=1,NST
      IJS=IS
      IJS1=IJS+NST
      IJS2=IJS1+NST
      CX=HX*CAX(IJS)+HY*CAX(IJS1)+HZ*CAX(IJS2)
      DY=HX*CAY(IJS)+HY*CAY(IJS1)+HZ*CAY(IJS2)
      CZ=HX*CAZ(IJS)+HY*CAZ(IJS1)+HZ*CAZ(IJS2)
      H=SQRT(CX*DX+CY*DY+DZ*CZ)
      IF(A(IS))10,12,11
   10 GAM=GAM+C(IS)*(1.-EXP(H/A(IS)))
      GO TO 1
   11 IF(H.GE.A(IS))GO TO 12
      GAM=GAM+C(IS)*(1.5*H/A(IS)-
     10.5*H*H*H/(A(IS)*A(IS)*A(IS)))
      GO TO 1
   12 GAM=GAM+C(IS)
    1 CONTINUE
      RETURN
      END
```

III.B.6. Proportional effect and quasi-stationarity

Let x and x' be two points with coordinates (x_u, x_v, x_w) and (x'_u, x'_v, x'_w). Without any hypothesis of stationarity,

(i) the expectation of the RF $Z(x)$ depends on the position x, i.e., on the three coordinates (x_u, x_v, x_w),

$$E\{Z(x)\} = m(x_u, x_v, x_w);$$

(ii) the semi-variogram $\gamma(x, x')$ or the covariance $C(x, x')$ depends on the two locations x and x' of the RV's $Z(x)$ and $Z(x')$, i.e., on the six coordinates (x_u, x_v, x_w) and (x'_u, x'_v, x'_w),

$$\tfrac{1}{2}E\{[Z(x) - z(x')]^2\} = \gamma(x, x').$$

It was indicated in section II.A.2 that an hypothesis of quasi-stationarity is generally sufficient for geostatistical applications. This hypothesis amounts to assuming that:

(i) the expectation is quasi-constant over limited neighbourhoods and, thus, $m(x) \simeq m(x') \simeq m(x_0)$ when the two points x and x' are inside the neighbourhood $V(x_0)$ centred on the point x_0, cf. Fig. III.14;
(ii) inside such a neighborhood $V(x_0)$, the structural functions γ or C depend only on the vector of the separating distance $h = x - x'$ and not on the two locations x and x'; however, this structural function depends on the particular neighbourhood $V(x_0)$, i.e., on the point x_0,

$$\gamma(x, x') = \gamma(x - x', x_0) = \gamma(h, x_0), \qquad \forall x, x' \in V(x_0).$$

The construction of a model of quasi-stationarity thus amounts to building a model of the structural function $\gamma(h, x_0)$ which depends on the argument x_0.

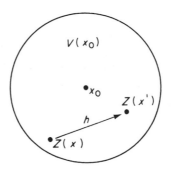

FIG. III.14. Neighbourhood of quasi-stationarity.

Proportional effect

Consider two neighbourhoods of quasi-stationarity $V(x_0)$ and $V(x_0')$ centred on the two different points x_0 and x_0'. Let $\gamma(h, x_0)$ and $\gamma(h, x_0')$ be the semi-variograms defined on these two neighbourhoods.

Quite often, in practice, in mining applications, these two semi-variograms can be made to coincide after multiplication by a factor which is a function of the ratio of the experimental means, $m^*(x_0)$ and $m^*(x_0')$, of the available data in $V(x_0)$ and $V(x_0')$, cf. Fig III.15. This amounts to assuming that there exists a stationary model $\gamma_0(h)$ independent of the neighbourhood $V(x_0)$ and such that

$$\gamma(h, x_0) = f[m^*(x_0)] \cdot \gamma_0(h); \qquad (III.35)$$

thus,

$$\gamma(h, x_0)/f[m^*(x_0)] = \gamma(h, x_0')/f[m^*(x_0')].$$

The experimental mean $m^*(x_0)$ is an estimator of the expectation $E\{Z(x)\} \simeq m(x_0)$, which is constant over the neighbourhood of quasi-stationarity $V(x_0)$. The two quasi-stationarity models $\gamma(h, x_0)$ and $\gamma(h, x_0')$ are said to differ from each other by a *proportional effect*,

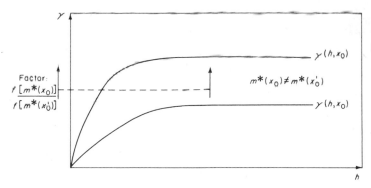

FIG. III.15. Proportional effect.

When a proportional effect is present, the function f will be evaluated by studying the various proportional relationships between the experimental semi-variograms $\gamma(h, x_0)$, $\gamma(h, x_0')$, ..., calculated on their respective neighbourhoods $V(x_0)$, $V(x_0)$, ...

Using the model $f[m^*(x_0)]$ of the proportional effect, the quasi-stationary model of (III.35) provides a structural function corresponding to every neighbourhood $V(x_0)$. Of course, there must be enough data available on $V(x_0)$ to form a correct estimator $m^*(x_0)$ of the expectation

$E\{Z(x)\}$ on $V(x_0)$. Thus, neighbourhoods that are too small compared to the data grid cannot be considered. Examples of the fitting of a proportional effect are given in the various case studies of Chapter IV.

Remark 1   Direct and indirect proportional effect. The proportional effect is said to be direct if the experimental semi-variogram increases with the corresponding experimental mean, i.e., when the function $f[m^*(x_0)]$ in formula (III.35) increases with $m^*(x_0)$.

Such a direct proportional effect can be expected when the studied ReV $z(x)$ has a lognormal-type histogram, the mode M being less than the expectation m. As an example for the regionalizations of Cu, U, Au grades

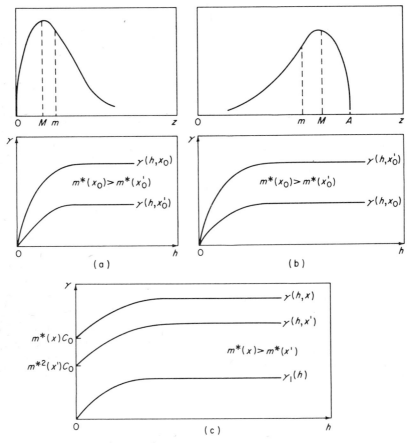

FIG. III.16.   (a) Direct proportional effect. (b) Inverse proportional effect. (c) Proportional effect on the nugget constant only.

or, more generally, for the grades of mineralizations of weak concentration, cf. Fig. III.16(a).

The proportional effect is said to be inverse when the sills of the experimental semi-variograms decrease as the experimental mean increases.

Such an inverse proportional effect can be expected when the studied ReV $z(x)$ has an inverse lognormal-type histogram, the mode M being greater than the expectation m. For example, the grades of Fe-hematite or tricalcite phosphate or, more generally, the grades of mineralizations of strong concentration, cf. Fig. III.16(b).

Remark 2 Proportional effect and geostatistical operations. As the main geostatistical operations are linear in γ (or in C-covariance), cf. formula (III.2), a proportional effect present on the structural function γ or C will appear identically in each of these operations.

More precisely, if $\mathscr{E}$ is any one of the three main geostatistical operations (i.e., estimation and dispersion variances and regularization), then

$$\mathscr{E}(\lambda\gamma) = \lambda\mathscr{E}(\gamma),$$

whatever the factor λ. If we take the proportional effect model of (III.35) with $\lambda = f[m^*(x_0)]$, then,

$$\mathscr{E}[\gamma(h, x_0)] = f[m^*(x_0)] \cdot \mathscr{E}[\gamma_0(h)] \qquad \text{(III.36)}$$

and, thus,

$$\mathscr{E}[\gamma(h, x_0)] = \frac{f[m^*(x_0)]}{f[m^*(x_0')]} \cdot \mathscr{E}[\gamma(h, x_0')].$$

Remark 3 Proportional effect in m^{*2}. Very often, in mining applications, the proportional effect observed is a function of m^{*2}, i.e., the function $f[m^*(x_0)]$ is proportional to the square of the experimental mean; thus,

(i) for a direct proportional effect,

$$f[m^*(x_0)] = [m^*(x_0)]^2;$$

(ii) for an inverse proportional effect,

$$f[m^*(x_0)] = [A - m^*(x_0)]^2,$$

where A is the maximum value that $m^*(x)$ can take.

With a direct effect in m^{*2}, the relative semi-variogram, i.e., the local semi-variogram divided by the square of the local mean, appears as a stationary function independent of the localization x_0:

$$\gamma(h, x_0)/m^{*2}(x_0) = \gamma(h, x_0')/m^{*2}(x_0'), \qquad \forall x_0 \text{ and } x_0'.$$

Consequently, the various relative variances of estimation and dispersion also appear as stationary, i.e., independent of the location x_0 of the neighbourhood $V(x_0)$ within which they have been calculated. Thus, the relative estimation variance of a block $v(x_0)$ by its central core sample $z(x_0)$ is written as

$$\sigma_E^2(z(x_0), v(x_0))/m^{*2}(x_0) = \sigma_E^2(z(x_0'), v(x_0'))/m^{*2}(x_0'), \qquad \forall x_0, x_0'.$$

This last expression explains why, in the presence of a ReV that has a lognormal-type distribution and is thus susceptible to a direct proportional effect in m^{*2}, it is better to use the *relative* standard deviation σ_E/m^* to establish confidence intervals, cf. section II.B.1, formula (II.21).

On the other hand, for a ReV with an inverse lognormal-type distribution, it is better to use the standard deviation divided by the difference $(A - m^*)$; i.e., $\sigma_E/(A - m^*)$.

Remark 4 Proportional effect and nested structures. When nested structures are present, $\gamma = \sum_i \gamma_i$, the proportional effect may be different for each of the component structures, i.e., the function $f_i[m^*(x_0)]$ may be different for each of the models γ_i.

Thus, on Fig. III.16(c), the direct proportional effect affects only the nugget constant C_0:

$$\gamma(h, x) = m^{*2}(x) \cdot C_0 + \gamma_1(h), \qquad \forall x \text{ and } h > 0,$$

the structure $\gamma_1(h)$ being stationary (independent of the location x).

A good practical way to reveal a proportional effect is to draw the relative semi-variogram (when a direct effect is suspected) on transparencies: the visual comparison thus shows whether or not the proportional effect affects the micro-structures (e.g., nugget effect) and the macro-structures equally.

Remark 5 Genesis of a proportional effect. The proportional effect is an experimental observation and can be interpreted as follows.

(i) As an effect of the *non-stationarity* of the RF $Z(x)$ which is only locally stationary or quasi-stationary. This is the interpretation given up to this point.

(ii) As a *conditioning effect* to the local experimental means $m^*(x)$, the RF $Z(x)$ being strictly second-order stationary.

In fact, the local semi-variogram $\gamma(h, x_0)$ calculated over the neighbourhood $V(x_0)$ is relative to the RF $Z(x)$ conditioned by the set of available experimental data on $V(x_0)$ and not to the RF itself. This local conditioning effect on $V(x_0)$ can be represented by a new RF $\{Z(x)/m^*(x_0)\}$ conditioned by the estimator $m^*(x_0)$ of the mean grade

$$Z_v(x_0) = \frac{1}{V} \int_{V(x_0)} Z(x)\,dx$$

on $V(x_0)$. If $\gamma(h)$ is the stationary model of $Z(x)$, the semi-variogram $\gamma(h, x_0)$ of the conditioned RF is no longer stationary (it depends on x_0). Knowing the spatial law of the RF $Z(x)$ and the type of estimator $m^*(x_0)$ considered, it is possible to calculate the theoretical expression of the conditioned model $\gamma(h, x_0)$; the calculations can be found in G. Matheron (1974). Here, we shall limit ourselves to the following principal results.

(i) When the RF $Z(x)$ has a Gaussian spatial law and when it has been conditioned by the best possible linear estimator $m^*(x_0)$ (i.e., by the BLUE kriging) of the mean grade $Z_V(x_0)$ on $V(x_0)$, then the conditioning does not alter the stationary character of the variogram and, thus,

$$\gamma(h, x_0) = \gamma(h), \qquad \forall x_0.$$

This does not hold true for any other definition of the conditioning variable $m^*(x_0)$.

(ii) When the RF $Z(x)$ has a lognormal spatial law (i.e., $Y(x) = \log Z(x)$ has a Gaussian spatial law) and if it has been conditioned by the kriging BLUE estimator $m^*_Y(x_0)$ of the mean grade $Y_V(x_0)$ of $Y(x)$ on $V(x_0)$, then the conditioning is expressed on the local semi-variogram $\gamma_Z(h, x_0)$ of the conditioned RF $\{Z(x)/m^*_Y(x_0)\}$ as a proportional effect in $m^{*2}_V(x_0)$:

$$\gamma_Z(h, x_0) = [m_V(x_0)]^2 \cdot \Gamma(h),$$

where $m_V(x_0) = E\{Z(x)/m^*_Y(x_0)\}$ is the expectation of the conditioned RF. In practice, this expectation will be estimated by the experimental mean $m^*_Z(x_0)$ of the data $Z(x)$ available on $V(x_0)$.

$\Gamma(h)$ is a stationary reference semi-variogram (independent of x_0) and is related to the stationary semi-variogram $\gamma(h)$ of the non-conditioned RF $Z(x)$.

From this brief survey of the possible genesis of a proportional effect, it is enough to remember that the experimental observation of such an effect does not necessarily indicate a non-stationary phenomenon.

III.B.7. Inference of the moments of order 2

In practice, all that is known of the structural function $\gamma(h)$, or $C(h)$, is an estimator $\gamma^*(h)$ derived from the data $z(x_i)$. Since the expressions for the various geostatistical operations have been constructed using the function $\gamma(h)$, it is useful to study the possibility of estimating this function $\gamma(h)$ from a finite network of experimental data $\{z(x_i)\}$.

Let $Z(x)$ be a RF assumed to be stationary on domains of limited extent V. On one such domain V the experimental data $\{z(x_i)\}$ are available, from which the following *experimental* variogram can be calculated:

$$2\gamma^*(h) = \frac{1}{N'} \sum_{i=1}^{N'} [z(x_i+h) - z(x_i)]^2, \qquad (\text{III.39})$$

where N' is the number of pairs of data values on V separated by the distance vector h.

If the domain V was perfectly known, i.e., if values of the realization $z(x)$ were known for every possible location x in V, the following variogram, called the *"local variogram"* on V, could be calculated:

$$2\gamma_{(V)}(h) = \frac{1}{V} \int_{V'} [z(x+h) - z(x)]^2 \, dx, \qquad (\text{III.40})$$

where V' represents the intersection of the domain V and its translation by the vector h; in other words, $x \in V' = V \cap V_h$ is equivalent to $x, x+h \in V$.

The experimental variogram $2\gamma^*(h)$ and the local variogram $2\gamma_{(V)}(h)$ are both relative to the particular realization $z(x)$ of the RF $Z(x)$; these two variograms thus appear as random variables, the expectation of which is precisely the *theoretical variogram* of the RF $Z(x)$:

$$2\gamma(h) = E\{[Z(x+h) - Z(x)]^2\} = E\{2\gamma^*(h)\} = E\{2\gamma_{(V)}(h)\}. \qquad (\text{III.41})$$

In practice, only the experimental variogram $2\gamma^*(h)$ is available, this gives rise to the first deviation $[\gamma_{(V)}(h) - \gamma^*(h)]$, which is the difference between the experimental value and the local value, due to the fact that only a finite number N' of data pairs, $[z(x_i+h), z(x_i)]$ separated by a distance h, is available. This deviation is characterized by the variance $E\{[\gamma_{(V)}(h) - \gamma^*(h)]^2$ called the *"variance of estimation"* of the local semi-variogram.

However, even if the domain V, and, thus, the local semi-variogram $\gamma_{(V)}(h)$ were perfectly known, there would still be a second deviation between this local expression and the theoretical semi-variogram $\gamma(h)$. This deviation is called a "fluctuation" and is characterized by the variance

$E\{[\gamma(h)-\gamma_{(V)}(h)]^2\}$ called the *"fluctuation variance"* of the local semi-variogram.

In fact, since the ReV $z(x)$ is the only available reality (the RF $Z(x)$ being a mere probabilistic model), only the local semi-variogram $\gamma_{(V)}(h)$ and its estimator $\gamma^*(h)$ have any physical meaning; consequently, only the estimation variance of the local semi-variogram is of practical interest. The fluctuation variance merely gives an order of magnitude of the deviation which can be tolerated between the local semi-variogram $\gamma_{(V)}(h)$ and the fitted model $\gamma(h)$ which is chosen to be as simple as possible.

Order of magnitude of the variances of estimation and fluctuation

The estimation variance $E\{[\gamma_{(V)}(h)-\gamma^*(h)]^2\}$ depends on the number N' of experimental data pairs available on V, while the fluctuation variance $E\{[\gamma(h)-\gamma_{(V)}(h)]^2\}$ depends on the dimensions of the domain V.

These two variances involve moments of order 4 of the RF $Z(x)$. Therefore, it is necessary (at least) to calculate moments of order 4 to characterize the quality of the inference of a moment of order 2; this apparently paradoxical fact is a characteristic of all statistical tests.

The calculations involved in these moments of order 4 are tedious and of no real interest in a book devoted to practical mining geostatistics. The study can be found in G. Matheron, (1965, Chap. XIII), and a summary of the principal results follows.

These results are for an intrinsic RF $Z(x)$ with a Gaussian spatial law and a theoretical semi-variogram $\gamma(h)=|h|^\theta$. The domain V is, in the one-dimensional space, a segment of length L (e.g., the length of a drill core). The following results are obtained.

(i) For the estimation variance,

$$E\{[\gamma_{(L)}(h)-\gamma^*(h)]^2\} = 4\gamma(h)\cdot\frac{D^2(0/L)}{N'},$$

where $D^2(0/L)$ is the dispersion variance of the point-support variable $Z(x)$ in L. N' is the number of experimental data pairs $[z(x_i+h), z(x_i)]$ used in calculating the experimental semi-variogram $\gamma^*(h)$.

(ii) For the fluctuation variance,

$$E\{[\gamma(h)-\gamma_{(L)}(h)]^2\} = \begin{cases} A\left(\dfrac{h}{L}\right)[\gamma(h)]^2 & \text{for } h < \dfrac{L}{3} \\[2ex] \text{prohibitive value} & \text{for } h > \dfrac{L}{2} \end{cases}$$

with $A = 4/3$ for the linear model $\gamma(h)=|h|$.

For the linear model and for $h > L/2$, the relative fluctuation variance $E\{[\gamma(h) - \gamma_{(L)}(h)]^2\}/[\gamma(h)]^2$ is about 2 (i.e., a prohibitive value of 200%).

Note that even if the estimation variance vanishes (very large number N' of pairs), the fluctuation variance allows the choice of any possible behaviour of the theoretical model $\gamma(h)$ for distance $|h| > L/2$ greater than the *distance of reliability* which is half the dimension of the field L considered.

Note again that this considerable value of the fluctuation variance prohibits the construction of goodness-of-fit tests for the fitting of a theoretical model $\gamma(h)$ to an experimental semi-variogram $\gamma^*(h)$: such tests would almost never invalidate the fit.

Practical rule

Although there is no reason for the RF $Z(x)$ representing the ReV $z(x)$ to follow a Gaussian spatial law, the preceding expressions for the estimation and fluctuation variances provide the following orders of magnitude:

$$\boxed{N' > 30, 50 \text{ pairs} \quad \text{and} \quad |h| < L/2} \tag{III.42}$$

An experimental semi-variogram should only be considered for small distances ($|h| < L/2$) in relation to the dimension (L) of the field on which it has been computed, and provided that the numbers N' of available data pairs are not too small.

This practical rule holds true for estimators of:

(i) cross-variograms or non-centred cross-covariances;
(ii) *non-centred* covariance, the stationary expectation $m = E\{Z(x)\}$ being supposed known, i.e., for the estimator

$$C^*(h) + m^2 = \frac{1}{N'} \sum_{i=1}^{N'} [z(x_i + h) \cdot z(x_i)].$$

Bias of the covariance estimator

The preceding estimator $[C^*(h) + m^2]$ of the non-centred covariance $[C(h) + m^2]$ is without bias, since

$$E\{C^*(h) + m^2\} = E\{Z(x + h) \cdot Z(x)\} = C(h) + m^2.$$

On the other hand, to get an estimator of the centred covariance $C(h)$, the expectation m must also be estimated from the same set of data $\{z(x_i)\}$, and a bias becomes obvious. Indeed, if N data are available on the quasi-stationary field V, the unbiased estimator of this expectation is

$$m^* = \frac{1}{N} \sum_{j=1}^{N} z(x_j) \quad \text{with} \quad E\{m^*\} = m.$$

A possible estimator of the centred covariance is then written as

$$C^*(h) = \frac{1}{N'} \sum_{i=1}^{N'} [z(x_i + h) . z(x_i)] - \frac{1}{N^2} \sum_j \sum_{j'} z(x_j) z(x_{j'}) \quad \text{(III.43)}$$

and, thus,

$$E\{C^*(h)\} = [C(h) + m^2] - \frac{1}{N^2} \sum_j \sum_{j'} [C(x_j - x_{j'}) + m^2]$$

$$= C(h) - \frac{1}{N^2} \sum_j \sum_{j'} C(x_j - x_{j'}).$$

The estimator $C^*(h)$ of formula (III.43) then shows a bias which is nothing more than the mean value of the theoretical covariance $C(h)$ when the two extremities of the vector h independently describe the set of the N available data:

$$E\{C(h) - C^*(h)\} = \frac{1}{N^2} \sum_{j=1}^{N} \sum_{j'=1}^{N} C(x_j - x_{j'}). \quad \text{(III.44)}$$

This bias can be assimilated to the mean value $\bar{C}(V, V)$ when the N data are uniformly distributed on the field V. When the covariance model $C(h)$ is of transition type, i.e., $C(h) = 0$ when the distance $|h|$ is greater than a range a, then the bias $\bar{C}(V, V)$ vanishes if the dimensions of the field V are much greater than the assumed range a of the covariance model.

This bias is linked to the simultaneous estimation, from the same set of data, of the expectation m and the non-centred covariance $E\{Z(x + h) . Z(x)\}$. *This difficulty does not appear in the estimation of the variogram*, since the expression

$$E\{[Z(x + h) - Z(x)]^2\}$$

does not depend on the unknown expectation m.

III.C. THE PRACTICE OF STRUCTURAL ANALYSIS

Structural analysis, of mining phenomena in particular, is undoubtedly the field of geostatistics which lends itself the least to totally computerized methods. A structural analysis draws continuously on all the practical knowledge gained from the studied phenomena, and also involves a certain amount of "craft" in the choice and use of the various geostatistical tools. While skill can only be gained by experience, structural analysis has certain general attributes. In particular, it should:

(i) be adapted to the proposed aim of the study. Structural analysis is most often a prelude to an intended estimation and, as such, should be limited to the purpose of the estimation. Thus, it is no use to specify structures on a petrographic scale if it is intended to estimate blocks on a hectometric scale.

(ii) take critical account of all qualitative information. The experimental knowledge of a geologist may lead him to give all importance to a data file; conversely, and more rarely, an erroneous geological hypothesis may bias the use of such a file.

Before beginning the geostatistical treatment of a data set, it is advisable to become familiar with both the physical nature of the phenomenon under study (geology, mining technology) and the available data. This preliminary phase is essential in any structural analysis, and also for a correct formulation of the study; very often, it even defines the framework within which geostatistical methods will be applied.

The next step is the "variography" phase, i.e., the construction, analysis and fitting of the variogram models. The purpose of the basic geostatistical tool – the variogram – is to condense the main structural features† of the regionalized phenomenon into an operational form. Variography makes as much use of the previously introduced geostatistical concepts as it does of the gained (qualitative or not) knowledge of the physical nature of the phenomenon (metallogeny, structural geology, tectonics, mineralogy, mining and treatment of ore, etc.).

III.C.1. Review of the data

Before making any geostatistical calculations, it is advisable to carry out a critical assessment of the available data. Errors at all levels can be corrected and, very often, such an assessment can provide simple explanations for results which may seem, at first sight, rather odd.

---

† Most readers have seen feasibility studies in which excellent studies of structural geology or geomorphology are presented, only to be completely ignored in the estimation phase when methods such as polygons of influence are used. If estimation is divorced from geology, one might rightly wonder about the purpose of the geological study. Obviously, geological information must play a part in the estimation and the variogram is used in geostatistics to interpret the structural information to be used in the estimation procedures.

A typical list of questions to be asked when reviewing data is given hereafter.

Study of the reconnaissance plan

What are the criteria upon which the plan is based? Has it been biased in favour of *a priori* geological hypotheses or financial or technical constraints? Is it suited for the purpose at hand? A reconnaissance plan designed for a tectonic survey may prove to be unsuitable for grade estimation. The sampling material should also be suitable, e.g., type of drilling equipment, diameter of cores, channel samples or core samples?

What type of sampling grid is used? e.g., regular grid, pseudo-regular grid with constant sample density, preferential grid. Have the different facies encountered and the variables of interest been sampled in a systematic and homogeneous manner?

To what extent can we ask for additional sampling?

Causes of systematic error

The various geostatistical tools are constructed from the available data. A systematic error in these data will be reflected at every stage of the geostatistical analysis. Geostatistics does not create data, it is a method for treating information; in other words, geostatistics complies with the *in situ* reality only as long as the data are an accurate reflection of this reality. Thus, it is very important, especially at the data review stage, to detect all possible causes of systematic error.

Adherence to the sampling plan

Does the actual layout of data correspond to that of the plan? Geometric errors may have been made, or the drilling platform may have been displaced for accessibility reasons. Such displacements, even though small, may cause considerable biases in estimation, e.g., if preferentially rich or poor zones are the ones accessible to the drilling platform.

Has the deviation of bore-holes been controlled? Quite often, some types of samples will tend to follow preferentially rich pockets of soft ore (e.g., bauxite).

What are the causes of breaks in drill cores? More or less random accidental causes are not so important as definite physical causes, e.g., when a boulder of hard waste is encountered.

Have the different facies been sampled in an homogeneous manner? Samplers do not always take the same care in sampling waste as they do ore, but mining may recover part of this waste.

Representativeness of the data

What is the recovery rate of the core samples and how much does it vary?

Could the various methods of sampling, quartering and analysing have introduced any systematic error? Is it possible to evaluate the variance of measurement errors (account can be taken of this variance in the process of cokriging, cf. section V.A.5, Case study 3)? A fairly large variance of the measurement errors can be tolerated, provided that these errors have a zero mean (non-bias). However, this measurement error variance must not be such as to mask a natural variability of interest.

Homogeneity of the data

Has the representativeness of the data been constant through time and space?

Through time   An original sampling campaign may have been carried out by different companies with different objectives and different equipment.

In space   The same ore body may have been sampled in different places by very different methods, dragline samples on the surface, drill cores in depth, samples taken along a drive, etc. Topography, rock type and equipment used may vary from place to place and can lead to different recovery rates and to biases (loss of fines in blasting holes for example).

Has the sampling of the various variables (thicknesses, grades, densities) been homogeneous? In polymetallic deposits, certain trace metals, although economically important, are not always as well sampled as other metals (e.g., Ag and Au in Pb deposits). Some metal grades are measured from other variables by regression, deconvolution (e.g., U), construction of a metal "balance sheet", etc.

Compiling the data

How and by whom has the data – from drill log sheet to magnetic tape – been compiled? Aberrant grades can sometimes be traced to breaks in the sequence of data compilation, e.g., changes in groups collecting samples or transferring data from one type of computing device to another.

What type of error tests are made when the data file is being completed? Is it possible to retrieve rejected data at a later stage?

On what basis have the geological maps and cross-sections been drawn? how much of the interpretation is ascribable to the geologist or to the draughtsman?

III.C.2. Choice of the regionalized variables

In practice, the following must be specified to define a regionalized variable.

(i) *Significance*, e.g., thickness (in metres) of a given geological horizon or the volumetric grade of uranium obtained by deconvolution of well logs.

(ii) *Support*, i.e., the volume on which the variable is defined, as well as the method used to obtain the samples; e.g., the support may be a piece of drill core of diameter d and constant length l obtained by a specific type of drilling, or the support may be assimilated to a point at which the thickness of a bed is measured, or again it may be a grab sample of variable volume and approximately 1 kg of material.

(iii) *Field of extension*, i.e., the domain in which the spatial distribution of the variable is to be studied. This extension field may cover an entire mining concession or only a zone of primary mineralization, the zone of secondary mineralization being studied separately.

A geostatistician should always participate in the definition of the ReV that he intends to study, i.e., he should at least be consulted at the time that a systematic sampling campaign is being planned. In fact, all potential users of the information to be extracted from the sampling campaign should be consulted to ensure that the information is relevant to the intended studies. Such users would include the mining engineer and metallurgist as well as the geologist. For example, a detailed lithological study in which only the rich facies are analysed is not suitable for estimation in the frame of non-selective mining.

In choosing and defining the ReV for the purpose of estimating a deposit, the following rules should be observed.

1. *Additivity of the variables* The ReV must be such that all linear combinations of its values retain the same meaning. If $f(x)$ is a grade value, $\sum_i \lambda_i f(x_i)$ and especially the arithmetic mean $(1/n)\sum_i f(x_i)$ must have the same meaning as a grade.

Thus, a volumetric grade (in kg m^{-3}, for example) defined on a constant support is an additive regionalized variable. The same volumetric grade defined on variable supports is no longer an additive ReV and, in fact, the mean grade of two different supports is not the arithmetic mean of their grades. In such a case, the additive ReV to be considered should be the accumulation (or quantity of metal in kg), i.e., the product of the volumetric grade (kg m^{-3}) and the support volume (m^3). However, when studying the spatial variability of this accumulation, it would not be possible to distinguish between the variability due to the grade and that due to the variable supports.

If the preceding volumetric grade $z(x)$ is associated with a granulometric fraction $r(x)$ and both analyses are defined on variable supports $v(x)$ centred on point x, then the additive ReV accumulation or quantity of metal is defined as

$$a(x) = r(x) . z(x) . v(x).$$

Other examples of non-additive variables are permeability, direction of a vector $\alpha(x)$ with origin at point x, indices of all sorts (hardness, colour), ratios of grades, certain metallurgical recoveries, etc. The particular case of the non-additive variable $Y(x) = \log Z(x)$, logarithm of a grade, for example, is discussed later.

2. *Adequateness of the variables chosen for the purpose of the study*   The chosen ReV's should be such as to encompass the main characteristics of the problem at hand. They should also lead to operational solutions without any superfluous precision or calculation.

Consider the example of a thin subhorizontal sedimentary bed with obvious foot and hanging walls. Rich and poor zones will be selected by mining on a horizontal plane, but no vertical selection is to be considered (the entire bed thickness will be recovered). Since there is no vertical selection, it is of no use to study the vertical regionalization of the grades, at least for the strict purpose of ore reserve estimations. The chosen additive ReV's for this estimation are:

   (i) the thickness $t(x)$ of the bed measured along the bore-hole located at the point x of the horizontal two-dimensional space;
  (ii) the vertical accumulation $a(x) = t(x) \times z(x)$, defined as the product of the thickness and the corresponding mean grade $z(x)$ sampled on the mineralized section (of length $t(x)$) of the borehole.

Actually, the thickness variable is used in the estimation of the volumes of recoverable ore, while the accumulation variable is used in the estimation of the quantity of recoverable metal. The recovered mean grade within a zone is given by the ratio of the quantity of metal to the recovered tonnage in this zone.

In this example, it can be seen that the following are true.

   (i) It is of no use to consider the three-dimensional regionalization of the grades.
  (ii) It is of no use to cut the cores into small lengths to represent the various facies met along a vertical. Moreover, to do so would risk

biasing the estimation if some facies assumed to be waste were not analysed (in fact, all facies within the bed will be mined and their grades, however small, should be considered).

(iii) The information will be focused on the recognition of the horizontal variabilities since there will be a horizontal selection.

3. *Homogeneity of the variable within its extension field*   The significance and support of a ReV must be constant, or vary very little in its field of definition. In particular, this homogeneity allows the observed variability between two data $z(x)$ and $z(x+h)$ to be ascribed to the single regionalization of $z(x)$.

In a porphyry-copper deposit, for instance, grades from the oxidized zone should not be mixed with those from the sulphide zone. Similarly, data from two drilling campaigns carried out with different drilling equipment should be considered separately, at least in the first step.

It is useful to recall here that the concept of homogeneity and its probabilistic counterpart, the concept of stationarity, are relative to the scale of observation: the same ore considered extremely homogeneous during mining and industrial treatment may appear on a petrographic scale as consisting of a number of very different minerals.

Working on logarithms of grades

In some cases, the spatial variability of an additive ReV $Z(x)$ is so great that it is convenient to study it through a transformed variable; for example, the logarithm $Y(x) = \log Z(x)$. A gold grade in a nugget deposit can vary from 1 to 1000 or more from one sample to another; it is then more convenient to study the regionalization of the grade logarithm. This new ReV is less variable, but is no longer additive.

The structural function $\gamma_Y(h) = E\{[Y(x+h) - Y(x)]^2\}$ of the transformed variable $Y(x)$ is calculated and a linear estimator $Y^*(x_0) = \sum_\alpha Y(x_\alpha) = \sum_\alpha \log Z(x_\alpha)$ is built. It would then be erroneous to think that a mere exponentiation of $Y^*(x_0)$ provides a good estimator of

$$Z(x_0) = e^{Y(x_0)}.$$

First, the non-bias condition may not be met, as $E\{Y^*(x_0)\} = E\{Y(x)\}$ does not entail that

$$E\{e^{Y^*(x_0)}\} = E\{e^{Y(x)}\};$$

then, the required estimation variance $E\{[Z(x_0) + e^{Y^*(x_0)}]^2\}$ is not readily derived from the variance

$$E\{Y(x_0) - Y^*(x_0)]^2\}$$

calculated from the structural function of the transformed variable $Y(x)$.

As a matter of fact, the estimator $e^{Y^*(x_0)}$ is no longer linear with respect to the data $Z(x_\alpha)$; therefore, its study requires not only the second-order moment covariance or variogram of $Z(x)$, but also the knowledge of its entire distribution function. When $Z(x)$ is lognormal, $Y(x) = \log Z(x)$ being normally distributed, it will be shown that a correct estimator $Z^*(x_0)$ can be derived from the linear combination $Y^*(x_0) = \sum_\alpha \lambda_\alpha Y(x_\alpha)$. This technique, called "lognormal kriging" will be presented in Chapter VIII, "Introduction to Non-linear Geostatistics". Until then we shall limit ourselves to the study of additive regionalized variables.

The previous rules of additivity, adequateness and homogeneity must always be applied with flexibility. Obviously, recovery rates from drill cores cannot be perfectly constant. Similarly, a drill core cut into segments of very variable length (non-constant support), as unsuitable as it may be, represents a considerable investment and it is advisable to make the best of what is available, e.g., to create an homogeneous set from this heterogeneous information that will allow a geostatistical treatment of the data.

Reconstitution of core samples of constant length

On a zone considered as homogeneous for the purpose of the study, one of the first requirements of the geostatistical approach is that the available data be defined on a *constant support*. It is recalled from section II.D, "Regularization", that the same spatial variability shows up differently according to the size and geometry of the support on which it has been sampled.

If the cutting and analysis of the core samples on this zone has been carried out heterogeneously, then it is advisable to restore the homogeneity, which, in practice, means defining the data on core sections of constant length. Two cases can be distinguished.

1. *The actual core samples are not of constant length but have all been analysed, cf. Fig. III.17*

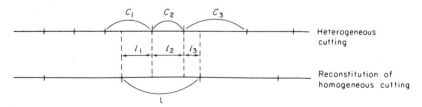

FIG. III.17. Reconstitution of core samples of constant length.

The grades of the core samples of constant length l are obtained by weighted averages of the grades of the actual core samples. Thus, on the drill core of Fig. III.17, the grade z_l of the core sample l is made up of the three actual core sample grades z_1, z_2, z_3, weighted by the respective lengths l_1, l_2, l_3 of their intersections with the reconstituted sample l, i.e.,

$$z_l = \frac{l_1 z_1 + l_2 z_2 + l_3 z_3}{l = l_1 + l_2 + l_3}. \qquad \text{(III.45)}$$

Smoothing error   In such a reconstitution, it is implicitly assumed that the grade is constant within each actual core sample. This causes a smoothing effect as the reconstituted grades give a smoothed idea of the actual variability of the phenomenon. This smoothing becomes even more important as:

(i) the grades become more variable within each actual core sample length, i.e., as the micro-variabilities (nugget effect) become stronger;

(ii) the reconstituted constant length l becomes smaller with respect to the mean length of the actual core samples;

(iii) the distance h considered for the semi-variogram becomes smaller. In fact, two contiguous supports l may be constructed from parts of the same actual core samples. This smoothing at small distance h may mask a micro-structure, bias the evaluation of a nugget effect and sometimes lead to the adoption of excessively regular behaviour at the origin of the variogram.

The theoretical calculation of this smoothing effect would require the knowledge of the distribution of the two variables, length of the actual core sample and its grade, and quickly becomes tedious. However, the following practical rules can be stated.

(i) The dimension l of the reconstituted constant support must not be smaller than the mean dimension of the actual supports; if it is, then the reconstitution amounts in fact to an artificial creation of information. Very often, technical considerations largely determine the choice of the support, i.e., the length l is equal to a submultiple of the bench height in an open-pit operation.

(ii) Use caution when interpreting the behaviour at the origin of an experimental variogram derived from reconstituted data; an excessively regular behaviour may simply be an artefact due to smoothing.

2. *The actual sample lengths are not constant and are not all analysed*

This case is more often associated with a *preferential* analysis (e.g., of rich facies only) rather than with a *random* one (e.g., one sample on two analysed).

Random analysis is more favourable, and the non-analysed samples can be considered as missing data. The missing datum is sometimes in rich facies and sometimes in poor facies and, thus, does not cause any bias in the structural analysis.

Preferential analysis is much more critical, especially when the non-analysed facies (generally the poor ones) may actually be recovered by mining. To reconstruct an homogeneous and unbiased data set, it is possible to act in one or more of the following ways.

(i) To simulate the missing analyses. If the missing analyses correspond to intersections with pure waste, a zero grade is given to the intersection; if the waste is, in fact, poor ore which will be recovered (all or in part) during mining, it is advisable to simulate the missing analyses from the knowledge of the characteristics of this poor ore, deduced, for example, from systematically analysed drill cores.

(ii) To limit the calculation of the variograms to within the rich analysed facies, thus characterizing the variability of these rich facies only. This supposes that, during the mining stage, it will be possible to select perfectly those facies with a negligible amount of contamination from other facies.

(iii) To calculate an inter-facies variogram by assigning a global grade $+1$ or 0 to the various facies, according to whether they are rich or poor. The drill core is then represented by a series of values of the "all or nothing" type with constant support and the sequences of facies may be characterized by the ranges of the variogram.

III.C.3. Statistics of data

Before starting variography (i.e., the study of the spatial auto-correlations), it is advisable to carry out an elementary statistical analysis of the available data and to compare the results with the geological hypotheses.

It is not within the scope of this book to present any of the various statistics or other techniques of data analysis; it is enough to say that such techniques can, and in some cases must, complete the geostatistical approach. Among the many works dealing with the statistical analysis of geological data, some of the most recent are: G. S. Koch and R. F. Link (1970–71), J. C. Davis (1973) and F. P. Agterberg (1974).

The statistical techniques of histogram and correlation plots are sufficiently simple to be carried out systematically before any variography is attempted.

The histogram, or the experimental curve of the frequency of occurence of the different values of the random variable, is a remarkably simple tool by means of which the homogeneity of the spatial distribution can be verified, extreme or suspect data can be identified and the type of proportional effect to be expected can be predicted (cf. section III.B.6, Fig. III.15). In addition, the histogram is an estimator of the stationary distribution function of the RF under study, as well as a fundamental tool for the geostatistical techniques of ore reserves calculation (cf. Chapter VI) and conditional simulations of deposits (cf. Chapter VII).

Correlation diagrams are a means by which the homogeneity of a spatial coregionalization can be verified. Mineralizations or a set of homogeneous data can sometimes be distinguished by apparent groupings in correlation diagrams, cf. the following case study. In addition, the correlation diagrams are essential tools for simulations to the extent that they reveal inequalities of the type

$$\alpha \times \text{grade in metal A} + \beta \times \text{grade in metal B} \leqslant \text{fixed limit.}$$

Such inequalities often characterize the mineral of interest and it is advisable to take them into account when simulating the deposit.

Case study

The mineralization of the precambrian iron deposits of West Africa (Mauritania, Liberia, Guinea) is made up of itabirites concentrated in iron oxides by departure of the silica and is essentially composed of two types of ore:

(i) type α, rich shallow hematite ore corresponding to a maximum departure of the silica (67% Fe, and $SiO_2 < 1\%$);
(ii) type β, mixed poor ore (55% Fe and $SiO_2 > 5\%$) located at the fringes of the deposits or deep at the contact with the clean itabirites.

Of course, when mining within each of these two main types of ore, several subdivisions will be considered according to the granulometry and the degree of alteration (i.e., presence of goethite).

These two types α and β can be clearly distinguished by

(i) their statistical distributions (histograms and correlation diagrams of the three main components, Fe, SiO_2 and Al_2O_3 of the ore);
(ii) the structure of their spatial variability (variograms);
(iii) their spatial localization within each deposit.

Figure III.18 shows the Fe–SiO$_2$ correlation diagram derived from the core samples (of constant support) available on one of the Nimba range deposits (Guinea). This diagram clearly reveals two sets corresponding to

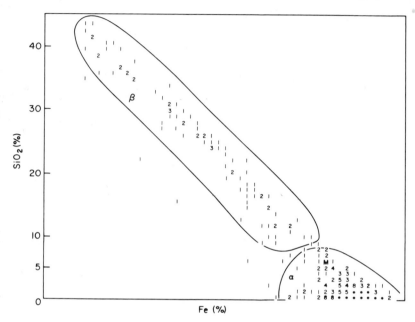

FIG. III.18. Fe–SiO$_2$ correlation diagram (Nimba range).

the two previously defined types of ore. This statistical discrimination is confirmed by the following.

(i) The geographic localization of each core sample. The samples classified on the diagram as type α correspond to the shallow ore while the β-type samples correspond to the deep ore.

(ii) The study of the Fe and SiO$_2$ histograms. The global histograms derived from all the core samples available are multimodal, cf. Fig. II.26 for the Fe grades. The main mode (67% Fe) corresponds to the α-type hematite ore which, when considered separately, shows the unimodal lognormal-type histograms of Fig. II.25(a) and (b). The histograms of the β-type ore remain multimodal with strong dispersion variances, cf. Fig. II.27 for the Fe grade, which confirms the mixed character of the β-type ore.

(iii) The study of the Fe and SiO$_2$ variograms calculated on each of the two distinguished types α and β, cf. Fig. III.19(a) and (b). For the α-type SiO$_2$, there is a quasi-absence of structure (pure nugget effect); the silica appears as mere non-structured traces in the homogeneous hematite ore, whereas the β-type SiO_2 semi-variogram shows a structure quite similar to that of β-type Fe; the variability of iron grades is closely linked to that of the silica grades (since the ore is enriched by departure of the silica).

It is interesting to compare the diagram of Fig. III.18 with those of Fig. III.20(a), (b) and (c) corresponding to the Fe–SiO_2–Al_2O_3 grades of the deposit of F'derick (Mauritania). In his study of the F'derick deposit, G. Matheron (1962) also distinguished two main types of ore, called I and II, the characteristics of which are very similar to those of the two types of ore (α and β) distinguished in the Nimba range deposit.

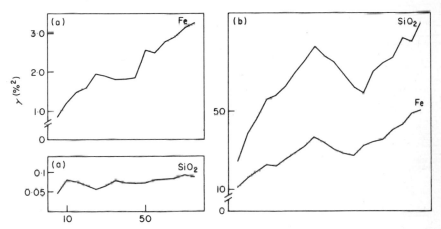

FIG. III.19. Experimental semi-variograms (Nimba range). (a) α-type ore; (b) β-type ore.

III.C.4. Variogram calculation

Let h be a vector of modulus $r = |h|$ and direction α. If there are N' pairs of data separated by the vector h, then the experimental semi-variogram in the direction α and for the distance h is expressed as

$$\gamma^*(r, \alpha) = \frac{1}{2N'} \sum_{i=1}^{N'} [z(x_i + h) - z(x_i)]^2 \qquad (\text{III.46})$$

for a regionalization, and

$$\gamma^*_{kk'}(r, \alpha) = \frac{1}{2N'} \sum_{i=1}^{N'} [z_k(x_i + h) - z_k(x_i)][z_{k'}(x_i + h) - z_{k'}(x_i)]$$

$$(\text{III.47})$$

for a coregionalization.

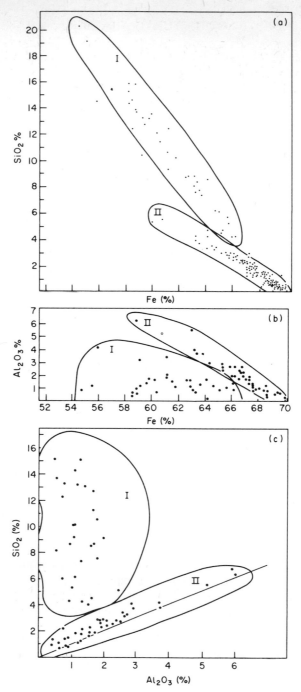

FIG. III.20. (a) Fe–SiO$_2$ correlation diagram (F'derick). (b) Fe–Al$_2$O$_3$ correlation diagram (F'derick). (c) Al$_2$O$_3$–SiO$_2$ (F'derick).

Although these expressions are unique, the methods used in constructing variograms depend on the spatial configuration of the available data. Various cases can be distinguished according to whether or not the data are aligned and to whether or not they are regularly spaced along these alignments.

1. *Data aligned and regularly spaced*

This category covers most configurations resulting from the systematic reconnaissance of a deposit. The preceding experimental expressions can be applied for each direction of alignment α.

Thus, the configuration represented by a rectilinear drill core in the direction α, which has been cut into constant lengths l and analysed, provides an estimator γ_l^* (kl, α) of the semi-variogram regularized by core samples of length l in the direction α and for distances which are multiple of the basic step size l, cf. Fig. III.21(a).

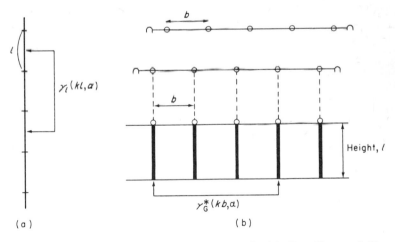

FIG. III.21. Data aligned and regularly spaced. (a) Rectilinear drill core; (b) channel samples along a gallery.

Vertical channel samples of the same height l and spaced at regular intervals b in the same direction α along a horizontal drift also fall into this category and provide an estimator $\gamma_G^*(kb, \alpha)$ of the semi-variogram, graded over the constant thickness l, in the direction α and for distances which are multiples of the basic step size b, cf. Fig. III.21(b).

2. *Aligned but irregularly spaced data*

To construct the experimental semi-variogram in the direction of alignment α, the data are grouped into *distance classes*. Every data pair which is separated by a distance $[r \pm \varepsilon(r)]$ is used to estimate the value $\gamma(r)$.

In practice, it is easier to consider a constant tolerance $\varepsilon(r)$ whatever the distance r, i.e., $\varepsilon(r) = $ constant, $\forall r$. Otherwise, $\varepsilon(r)$ should be smaller for smaller distances and larger for greater distances.

This grouping of data pairs into distance classes causes a smoothing of the experimental semi-variogram $\gamma^*(r)$ relative to the underlying theoretical semi-variogram $\gamma(r)$, cf. Fig. III.22. If the N' available data pairs separated by a distance $r_i \in [r \pm \varepsilon(r)]$ are used instead of those separated by the strict distance r, then it is not $\gamma(r)$ that is being estimated but rather a linear combination of the $\gamma(r_i)$; more precisely, the theoretical mean value

$$\frac{1}{N'} \sum_{i=1}^{N'} \gamma(r_i).$$

The significance of the effect of this smoothing over the interval $[r \pm \varepsilon(r)]$ decreases as the tolerance $\varepsilon(r)$ becomes smaller with respect to the range of the theoretical model $\gamma(r)$ to be estimated.

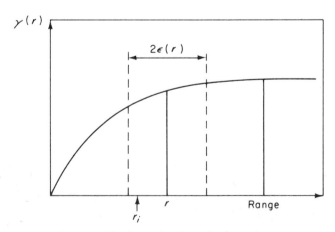

FIG. III.22. Grouping into distance classes.

In practice, the following points should be heeded.

(i) Take the pseudo-periodicities of the data locations into account when defining the tolerances $\varepsilon(r)$ of the classes.

(ii) Ensure that the section of interest of the variogram (e.g., the increase to the sill value) has been calculated using at least three or four classes; thus, the tolerance $\varepsilon(r)$ should not be too large.

(iii) Ensure that each distance class $[r \pm \varepsilon(r)]$ contains enough pairs so that the corresponding estimator of the variogram is reliable; thus, the tolerance $\varepsilon(r)$ should not be too small.

(iv) Detect any risk of bias due to preferential location of data.

3. Non-aligned data

This category can be reduced to one of the former two.

(a) By defining approximately rectilinear pathways passing through the available data locations. Each of these approximate alignments is then treated separately with a possible grouping into distance classes, cf. Fig. III.23(a). The disadvantage of this method is that all the available data are not used, and it is difficult to program.

(b) By grouping the data into *angle classes* followed by distance classes, cf. Fig. III.23(b). To construct the variogram in the direction α, each data value $z(x_0)$ is associated with every other value located within the arc defined by $[\alpha \pm d\alpha]$. Within this angle class, the data can be grouped into distance classes $[r \pm \varepsilon(r)]$.

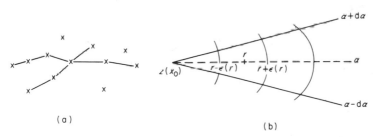

(a)   (b)

FIG. III.23. Non-aligned data. (a) pathways; (b) grouping into angle classes.

In the three-dimensional space, the arc (or angle tolerance) is defined by a solid angle $[\alpha \pm d\alpha, \varphi \pm d\varphi]$. In practice, however, it is seldom necessary to consider the data field in three dimensions, as it is usually possible to discern directions of alignments (bore-holes) or planes of alignment (transverse cross-sectional planes or horizontal levels). Such a three-dimensional data network can be regarded as the sum of several data sets in one or two dimensions which can later be grouped together to provide the information required to construct a three-dimensional variogram model.

This grouping into angle classes provides an experimental variogram without distinction between the directional variabilities within the arc $[\alpha \pm d\alpha]$ and, thus, a smoothing effect will be produced when the theoretical semi-variogram $\gamma(x, \alpha)$ varies in the interval $[\alpha \pm d\alpha]$.

In practice, when the existence of a privileged direction of anisotropy α_0 is suspected (and such a direction is generally preferentially sampled), then an arc $[\alpha_0 \pm d\alpha_0]$ with a small tolerance is defined so as to estimate accurately the variogram in this direction. Alternatively, when such a direction is not suspected, it is then enough to consider four directions in the two-dimensional space, each with a tolerance angle of $\pm \pi/4$, at least as a preliminary step. If an anisotropy becomes obvious, the next step would be to reduce the tolerance of the arcs.

Information associated with each experimental variogram

The following information must be associated with each of the elementary experimental semi-variograms $\gamma^*(r, \alpha)$ given by formula (III.46).

   (i) The number of pairs $N'(r, \alpha)$ of data values used in the construction of the value $y^*(r, \alpha)$. This number of pairs acts as a weight when grouping various elementary experimental semi-variograms into a mean semi-variogram. Moreover, it is an indicator of the possible estimation error, cf. section III.B.7.

  (ii) The dimension of the zone V covering the data used in the calculation of the semi-variogram $\gamma^*(r, \alpha)$. When these data are all aligned in the direction α (e.g., the data from a rectilinear borehole), then the maximum extension L of this alignment defines the significant section of the experimental semi-variogram as $r \leq L/2$, cf. section III.B.7, "Practical rule", formula (III.42).

 (iii) The arithmetic mean $\bar{z}$ and the experimental dispersion variance s^2 of the N data $\{z(x_i), i = 1 \text{ to } N\}$ used in the semi-variogram calculation, i.e.,

$$\bar{z} = \frac{1}{N} \sum_i z(x_i) \quad \text{and} \quad s^2 = \frac{1}{N} \sum_i [z(x_i) - \bar{z}]^2.$$

The arithmetic mean $\bar{z}$ is an estimator of the mean value $m(x_0)$ of the variable $z(x)$ studied in the zone $V(x_0)$ centred on x_0, and can be used to fit a proportional effect between two experimental variograms calculated in two zones $V(x_0)$ and $V(x_0')$ with different mean characteristics, cf. section III.B.6, formula (III.35). When the N data of constant support v are approximately uniformly distributed over the zone V, then the experimental dispersion variance s^2 is an estimator of the theoretical dispersion variance $D^2(v/V)$, cf. section II.C.1.

In the same way, an experimental cross-semi-variogram $\gamma^*_{kk'}(r, \alpha)$ given by formula (III.47) is associated with

(i) The number of pairs $N'(r, \alpha)$ of available data;
(ii) The dimensions and features of the zone V containing the data;
(iii) The means, $\bar{z}_k$ and $\bar{z}_{k'}$, the dispersion variances s^2_k and $s^2_{k'}$ of the data used and the correlation coefficients $\rho_{kk'}$.

Grouping into mean experimental variograms

Consider the two following elementary experimental variograms:

$$2\gamma^*_A(r) = \frac{1}{N'_A(r)} \sum_i [z(x_i+r)-z(x_i)]^2,$$

$$2\gamma^*_B(r) = \frac{1}{N'_B(r)} \sum_j [z(x_j+r)-z(x_j)]^2.$$

γ^*_A and γ^*_B are calculated from two different zones A and B, or from core samples for one and channel samples with the same support for the other, or, again, γ^*_A and γ^*_B correspond to two different directions α_A and α_B.

Suppose that, after examining the two experimental curves, it is concluded that they are not significantly different, and it is required to group them together to form a single mean experimental variogram $2\gamma^*_{A+B}(r)$. This mean variogram is calculated from all pairs of data values separated by a distance r, whether they come from A or B, which amounts to weighting each of the elementary variograms $2\gamma^*_A(r)$ and $2\gamma^*_B(r)$ with the respective number of pairs used in their construction:

$$2\gamma^*_{A+B}(r) = \frac{1}{N'_A(r)+N'_B(r)}\left[\sum_i[z(x_i+r)-z(x_i)]^2 + \sum_j[z(x_j+r)-z(x_j)]^2\right].$$

More generally, the grouping of K elementary semi-variograms $\{\gamma^*_k, k = 1$ to $K\}$ into a mean semi-variogram is expressed as

$$\gamma^*(r) = \frac{\sum_k N'_k(r)\cdot\gamma^*_k(r)}{\sum_k N'_k(r)}. \tag{III.48}$$

The procedure for grouping elementary covariances or cross-variograms is anologous: a mean weighted with the number of data pairs.

Information associated with each mean variogram

Each mean variogram obtained by combining elementary experimental variograms is associated with the following information.

(i) The total number of data pairs, $N'(r) = \sum_k N'_k(r)$.
(ii) The dimensions and features of the zone V containing the data. For example, the mean variogram may be calculated over the zone of primary mineralization from one set of essentially vertical drill cores and another smaller set inclined at 45°.
(iii) The arithmetic mean $\bar{z}$ and the experimental dispersion variance s^2, calculated from the set of the N data used. Care should be taken to consider these N data only, since in practice, there may be data within the zone that are not used in the calculation of any elementary variogram, and therefore play no part in the construction of the mean variogram.

Note that the experimental dispersion variance s^2, calculated from all the N data used in the field V, will only be significant as an estimator of the theoretical dispersion variance $D^2(v/V)$ of the support v in V if these N data are distributed *uniformly* over the field V, and it is not always so in practice. For example, let V be a three-dimensional zone proven by vertical drill-holes. The N data from the vertical core samples of constant length l are thus *not* uniformly distributed through the volume V, since the spatial distribution of these data will favour the vertical interdistances. On the other hand, the data are uniformly distributed in the one-dimensional field represented by the extension of a vertical drill core of mean length $\bar{L}$, and, thus, a significant experimental dispersion variance can be defined on $\bar{L}$ (but not on V).

More precisely, consider K such vertical drill-holes of slightly variable length $\{L_k, k = 1 \text{ to } K\}$. An experimental dispersion variance s_k^2 can be calculated along each elementary drill core k. This dispersion variance is an estimator of the theoretical dispersion variance $D^2(l/L_k)$ of the support l of the core sample in the one-dimensional field of length L_k. As the lengths L_k of the drill cores are variable, a mean length $\bar{L} = (1/K) \sum_k L_k$ can be defined and the experimental dispersion variance on $\bar{L}$ can be calculated from the weighting formula

$$s^2(l/\bar{L}) = \frac{\sum\limits_k N_k s_k^2}{\sum\limits_k N_k}, \qquad \text{estimator of } D^2(l/L), \qquad \text{(III.49)}$$

where s_k^2 is the experimental dispersion variance of the N_k data aligned on drill core no. k of length $L_k = N_k \cdot l$.

In the same way, if the three-dimensional zone V is proven by regular sampling of selected horizontal levels, an experimental dispersion variance can be defined over a mean horizontal surface $\bar{S}$.

A rule for combining the variograms

The rule is to group *homogeneous* information only.

The definition of the ReV (meaning and support) must be reasonably constant from one elementary variogram to another. Thus a variogram calculated from the grades of diamond drill core samples should not be grouped, without caution, with a variogram calculated from the grades of blast-hole cuttings; a variogram for sulphurous copper should not be grouped with a variogram for oxidized copper.

The representativeness of the elementary variograms must be approximately constant. Thus, for large distances r, it is sometimes advisable to distinguish between variograms calculated from long drill-holes and those calculated from short drill-holes, as the fluctuation error for the latter samples can be very high, cf. the practical rule (III.42) for assessing the reliability of an experimental variogram.

When grouping variograms together, care should be taken not to favour a zone of the deposit which is better sampled and perhaps preferentially rich or with a regular variability. Such a zone will contain an increased number of samples that will cause the characteristics of this zone to be predominant in a mean variogram supposed to be representative of the whole deposit.

Exercise. Construction of a variogram

The data set used in this exercise is sufficiently reduced to allow the various directional variograms to be calculated by hand or with the help of a pocket calculator. The example is very simple and is designed to prepare the way for the programming of variogram calculations. If only such a small amount of data were available in practice, the experimental fluctuations on each directional variogram would be so great as to render these variogram curves useless.

The data are located at the corners of a square grid a, cf. Fig. III.24. The directions to be studied are the two main directions α_1 and α_2 and the two diagonal directions α_3 and α_4. Note that the basic step size in the diagonal directions is $a\sqrt{2}$, while it is a in the main directions. Table III.1 gives the number of pairs of data used and the corresponding values of the experimental semi-variogram for each of the four directions and for the first three multiples of the basic step sizes. Isotropy is verified and the mean isotropic semi-variogram is calculated by combining the four directional semi-variograms, cf. Table III.2 and Fig. III.25. A linear model with no nugget effect can be fitted to the mean semi-variogram:

$$\gamma(r) = 4.1\, r/a, \qquad \forall r/a \in [a, 3\sqrt{2}].$$

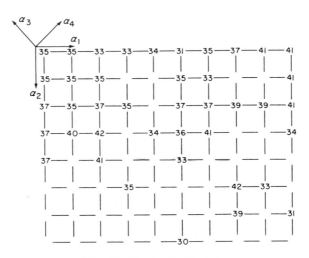

FIG. III.24. Available data.

TABLE III.1. *Directional semi-variograms*

Direction	Step 1		Step 2		Step 3	
	$N'(1)$	$\gamma(1)$	$N'(2)$	$\lambda(2)$	$N'(3)$	$\gamma(3)$
α_1	24	4·1	20	8·4	18	12·1
α_2	22	4·25	18	8·2	15	10·9
α_3	19	5	16	11·9	10	17·3
α_4	18	6·5	14	11·3	8	15·4

TABLE III.2. *Isotropic mean semi-variogram*

	h					
	a	$a\sqrt{2}$	$2a$	$2a\sqrt{2}$	$3a$	$3a\sqrt{2}$
N'	46	37	38	30	33	18
$\gamma(h)$	4·1	5·7	8·3	11·6	11·6	16·3

Computer subroutines

Practical programs for the construction of experimental variograms are, above all, programs for the creation of data files and the classification of data. This section will be limited to the presentation of three basic subroutines which can be adapted to any particular data file.

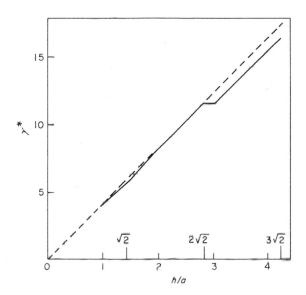

FIG. III.25. Fitting a linear model to the mean isotropic semi-variogram.

These three subroutines correspond to the three types of data configuration most often encountered in practice in mining applications.

1. One-dimensional configuration of aligned and regularly spaced data, cf. Fig. III. 21; e.g., a rectilinear drill core cut into samples of constant length l.
2. Two-dimensional configuration of data at the corners of a regular and rectangular grid, cf. Fig. III.26; e.g., the sampling of a subhorizontal sedimentary deposit by vertical drills analysed over the entire mineralized thickness, the drills being placed horizontally on a regular, rectangular grid.
3. Two-dimensional configuration of data irregularly distributed over a plane, cf. Fig. III.27; e.g., the sampling of the previous sedimentary deposit by vertical drills placed anywhere on the horizontal plane (but not preferentially on rich or poor zones).

Remark The very general configuration corresponding to a three-dimensional mass sampled by *n* drills in all directions can be regarded as either *n* one-dimensional configurations along the length of the *n* drills or *K* two-dimensional configurations on each of *K* horizontal benches, the data on each bench consisting of the intersections of the drill cores with the bench.

1. *Aligned and regularly spaced data* (*subroutine* GAMA1)

The alignment of the ND data is input to subroutine GAMA1 in the form of a vector VR(ND); some of these data may be missing or are to be eliminated. The subroutine calculates the experimental semi-variogram along with the experimental mean and dispersion variance. By mean of another program, this semi-variogram may then be combined with others according to the weighting formula (III.48).

```
        SUBROUTINE GAMA1(VR,ND,KMAX,PAS,NC,G,U,V,N,IS)
C
C ....................................................
C                      SEMI-VARIOGRAM ALONG A LINE
C   REGULAR GRID OF POINTS..MAY HAVE MISSING DATA..
C
C ....PARAMETERS
C  VR(ND)      DATA ARRAY
C  ND          NUMBER OF POINTS OF THE LINE
C  KMAX        MAXIMUM NUMBER OF COMPUTATION LAGS
C  PAS         LENGTH OF LAG
C  NC(KMAX)    NUMBER OF COUPLES/LAG
C  G(KMAX)     SEMI-VARIOGRAM VALUES/LAG
C  U           AVERAGE    |
C  V           VARIANCE   | OF DATA >TEST
C  N           NUMBER     |
C
C ....OPTIONS
C  IS.NE.1     RESULTS ARE PRINTED BY SUBROUTINE
C
C ....COMMONS
C  IOUT        LINE PRINTER UNIT NUMBER
C  TEST        INFERIOR BOUNDARY OF EXISTING DATA
C        IF VR.LE.TEST MISSING OR ELIMINATED DATUM
C ....................................................
C
      DIMENSION VR(1),NC(1),G(1)
      COMMON IOUT,TEST
C
C                    INITIALIZE
C
```

```
      DO 1  IK=1,KMAX
      NC(IK)=0
    1 G(IK)=0.
      IF(VR(NC)-TEST)12,12,13
   12 N=0
      U=0.
      V=0.
      GO TO 11
   13 U=VR(NC)
      V=VR(NC)*VR(NC)
      N=1
   11 CONTINUE
C
C         COMPUTE SEMI-VARIOGRAM AND STATISTICS
C
      NC1=NC-1
      DO 2  I=1,NC1
      VR1=VR(I)
      IF(VR1.LE.TEST)GO TO 2
      N=N+1
      U=U+VR1
      V=V+VR1*VR1
      I1=I+1
      JM=MINO(I+KMAX,NC)
      DO 21 J=I1,JM
      K=J-I
      IF(VR(J).LE.TEST)GO TO 21
      NC(K)=NC(K)+1
      VRR=VR1-VR(J)
      G(K)=G(K)+0.5*VRR*VRR
   21 CONTINUE
    2 CONTINUE
C
C                    RESULTS
C
      IF(N.EQ.0)GO TO 3
      V=(V-U*U/N)/N
      U=U/N
      DO 30 IK=1,KMAX
   30 G(IK)=G(IK)/MAXO(1,NC(IK))
    3 CONTINUE
      IF(IS.EQ.1)GO TO 5
C
C      PRINT RESULTS IF IS.NE.1
C
      WRITE(IOUT,2000)
      WRITE(IOUT,2001)U,V,N
      DO 41 K=1,KMAX
      D=K*PAS
   41 WRITE(IOUT,2002)K,D,NC,G
    4 CONTINUE
```

```
C
 2000 FORMAT(1H1,53X,'SEMI-VARICGRAM '/1H ,
    153X,'************** ',10X,
    2'(REGLLAR GRIC 1 CIMENSICN)')
 2001 FORMAT(1H ,' AVERAGE = ',F1C.5,6X,'VARIANCE = '
    1,E11.5,' NUMBER OF DATA = ',I5/1H ,' LAG  CIST'
    3,'ANCE | NC      VARIOGRAM |')
 2002 FORMAT(1H ,1X,I2,2X,F8.3,2X,'|',I4,2X,E11.5,
    12X,'|')
C
    5 RETURN
      END
```

2. Data on a regular rectangular grid on a plane (subroutine GAMA2)

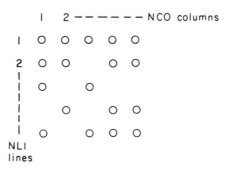

FIG. III.26. Data configuration considered by GAMA2.

The data for subroutine GAMA2 (NLI lines and NCO columns) is input in the form of a column vector VR(NLI*NCO). The subroutine calculates the four experimental semi-variograms corresponding to the four main directions (the two directions of the grid and the two diagonal directions). The subroutine also supplies the experimental mean and dispersion variance corresponding to all the data available on the plane.

```
        SUBROUTINE GAMA2(VR,NLI,NCC,KMAX,NC,G,L,V,N,IS)
C
C .................................................
C             SEMI-VARICGRAM IN TWC CIMENSICNS
C  REGULAR GRIC CF POINTS..MAY HAVE MISSING CATA..
C THIS IS A SAMPLE SUBROUTINE SHCWING CCMPUTATICN
C FOR FOUR CIRECTIONS..HOWEVER VECTORS IC ANC JC MAY
C BE DEFINEC FCR CTHER CIRECTICNS
C
C ....PARAMETERS
C  VR(NLI*NCO) CATA ARRAY (STCREC COLUMNWISE)
```

```
C   NLI          NUMBER OF LINES CF DATA GRID
C   NCO          NUMBER CF COLUMNS
C   KMAX         MAXIMUM NUMBER CF COMPUTATICN LAGS
C   ID(4)        DIRECTION INDICATOR ALONG LINES
C   JD(4)        DIRECTION INDICATOR ALONG COLUMNS
C   NC(KMAX*4)   NUMBER OF COUPLES/LAG/DIRECTION
C   G(KMAX*4)    SEMI-VARIOGRAM VALUES/LAG/DIRECTION
C   U            AVERAGE   |
C   V            VARIANCE  | OF DATA >TEST
C   N            NUMBER    |
C
C   ....OPTIONS
C   IS.NE.1      RESULTS ARE PRINTED
C
C   ....COMMONS
C   ICUT         LINE PRINTER UNIT NUMBER
C   TEST         INFERIOR BOUNDARY CF EXISTING DATA
C        IF VR.LE.TEST MISSING CR ELIMINATED DATUM
C
C   ....COMMENTS:
C   INDICATORS ID,JD SHOW THE RELATIVE LINE AND
C   COLUMN PCSITICNS QF CLOSEST NEIGHBOUR TC PCINT
C   I,J IN THE CONSICEREC DIRECTICN,EXAMPLE:
C     DIRECTICN 1  N-S    (//COLUMNS)     ID=1   JD=0
C               2  W-E    ( //LINES )     ID=C   JD=1
C               3  NW-SE  (DIAGCNAL 1)    ID=1   JD=1
C               3  NE-SW  (DIAGCNAL 2)    ID=-1  JD=1
C   CTHER DIRECTIONS MAY BE DEFINED
C   ...............................................
C
      DIMENSION VR(1),NC(1),G(1),ID(4),JD(4)
      COMMCN ICUT,TEST
      DATA IA/' '/
C
C                   INITIALIZE
C
      KMM=KMAX*4
      U=0.
      V=0.
      N=0
      DC 1 IK=1,KMM
      NC(IK)=0
    1 G(IK)=0.
      DC 10 I=1,4
      ID(I)=1
   10 JD(I)=1
      ID(2)=C
      ID(4)=-1
      JD(1)=0
C
C     COMPUTE SEMI-VARICGRAM,NEW PCINT
C
```

```
      DO 2 I=1,NLI
      DO 2 J=1,NCO
      IJ=I+NLI*(J-1)
C
C                    STATISTICS
C
      VR1=VR(IJ)
      IF(VR1.LE.TEST)GO TO 2
      N=N+1
      V=V+VR1*VR1
      U=U+VR1
    3 CONTINUE
C
C                                NEW DIRECTION
C
      DO 30 KD=1,4
      J3=JD(KD)
      I3=ID(KD)
      I1=I
      J1=J
C
C                                NEW LAG
C
      DO 31 K=1,KMAX
      I1=I1+I3
      IF(I1.LT.1.OR.I1.GT.NLI)GO TO 30
      I1=J1+J3
      IF(J1.LT.1.OR.J1.GT.NCO)GO TO 30
      IJ1=I1+NLI*(J1-1)
      IK=K+KMAX*(KD-1)
      IF(VR(IJ1).LE.TEST)GO TO 31
      NC(IK)=NC(IK)+1
      VRR=VR(IJ1)-VR1
      G(IK)=G(IK)+0.5*VRR*VRR
   31 CONTINUE
   30 CONTINUE
    2 CONTINUE
C
C              RESULTS
C
      IF(N.EQ.0)GO TO 4
      V=(V-U*U/N)/N
      U=U/N
      DO 40 IK=1,KMM
   40 G(IK)=G(IK)/MAXO(1,NC(IK))
    4 CONTINUE
C
C                    PRINT IF IS.NE.1
C
      IF(IS.EQ.1)GO TO 5
      WRITE(IOUT,2000)
      WRITE(IOUT,2001)U,V,N
```

```
      WRITE(IOUT,2002)(IA,IO,IC=1,4)
      IKM=KMAX*3
      WRITE(IOUT,2003)(IA,IO=1,4)
      DO 45 K=1,KMAX
      IK2=K+IKM
   45 WRITE(IOUT,2004)K,(NC(IK),E(IK),IK=K,IK2,KMAX)
C
 2000 FORMAT(1H1,53X,'SEMI-VARICGRAM '/1H ,
     153X,'*************** ',10X,
     2'(REGULAR GRID 2 DIMENSIONS)')
 2001 FORMAT(1H ,' AVERAGE = ',F10.5,6X,'VARIANCE = '
     1,E11.5,' NUMBER OF DATA = ',I5)
 2002 FORMAT(1H ,6X,'|',6(A1,2X,'DIRECTICN ',I2,3X,'|'))
 2003 FORMAT(1H ,' LAG | ',6(A1,'NC   1/2 VARIOGRAM|'))
 2004 FORMAT(1H ,1X,I3,2X,'|',6(I4,2X,E11.5,1X,'|'))
C
    5 RETURN
      END
```

3. Data irregularly spaced over the plane (subroutine GAMA3)

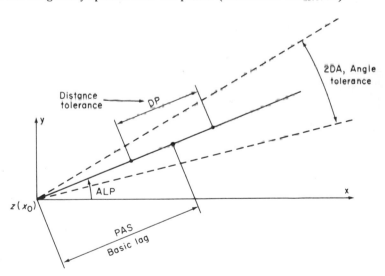

FIG. III.27. Grouping by angle and distance classes in GAMA3.

The ND data values are input to subroutine GAMA3 in the form of a vector VR(ND). The location of each data value is defined by its rectangular coordinates X(ND) and Y(ND). The NDI directions considered in the subroutine are defined by their angles ALP(NDI). The angle tolerance (2DA) and distance tolerance (DP) are unique for each of the NDI directions.

The subroutine supplies the NDI experimental directional semi-variograms as well as the experimental mean and dispersion variance corresponding to all the ND data available on the plane.

```
      SUBROUTINE GAMA3(VR,X,Y,NC,KMAX,PAS,CP,NDI,
     1ALP,DA,NC,G,D,U,V,N,IS)
C
C .................................................
C          SEMI-VARIOGRAM IN TWO CIMENSIONS
C IRREGULAR GRID.THERE MAY BE MISSING DATA..
C CALCULATION BY CLASS OF ANGLE AND DISTANCE..
C
C ....PARAMETERS
C  VR(ND)      DATA ARRAY
C  X(ND),Y(ND)   X AND Y COORDINATES CF FCINTS
C  ND          NUMBER OF POINTS
C  KMAX        MAXIMUM NUMBER CF CCMPUTATICN LAGS
C  PAS         LENGTH OF BASIC LAG
C  DP          WIDTH CF DISTANCE CLASS.IF CP=0.
C                   DP=PAS/2. IS TAKEN
C  NDI         NUMBER OF DIRECTIGNS
C  ALP(NDI )   ANGLES DEFINING DIRECTIONS (WITH
C                   RESPECT TO X AXIS IN DEGREES)
C  DA          WIDTH OF ANGLE CLASS.IF DA=C.
C                   DA=45. DEGREES IS TAKEN
C  NC(KMAX*NDI) NUMBER OF COUPLES/LAG/DIRECTION
C  G(KMAX*NDI)  VARIOGRAM VALUES/LAG/DIRECTION
C  D(KMAX*NDI)  AVERAGE DISTANCE/LAG/DIRECTION
C  U           AVERAGE |
C  V           VARIANCE | OF DATA >TEST
C  N           NUMBER
C
C ....OPTIONS
C  IS.NE.1 RESULTS ARE PRINTED:ONE PAGE CF PRINTCUT
C                 PER EACH 5 DIRECTICNS
C
C ....COMMONS
C  IOUT       LINE PRINTER UNIT NLMBER
C  TEST        INFERICR BCUNCARY CF EXISTING DATA
C       IF VR.LE.TEST MISSING OR ELIMINATED CATUM
C
C ....CAPACITY:
C  10 DIRECTIONS:SAN(10),CAN(1C)
C .................................................
C
      DIMENSION VR(1),X(1),Y(1)
      DIMENSION ALP(1),CAN(1C),SAN(10)
      DIMENSION G(1),D(1),NC(1)
      COMMCN ICUT,TEST
      DATA IA/' '/
```

```
C
C                    INITIALIZE
C
      PI=3.14159265
      TOL=DP
      IF(TOL.LE.0)TOL=PAS/2.
      DALPHA=DA
      IF(DA.LE.0)DALPHA=45.
C
      DO 1 KD=1,NDI
      ALPHA=PI*ALP(KD)/180.
      CAN(KD)=COS(ALPHA)
    1 SAN(KD)=SIN(ALPHA)
      THETA=PI*DALPHA/180.
      CDA=COS(THETA)
      KMM=KMAX
C
      DO 10 IK=1,KMM
      NC(IK)=0
      D(IK)=0.
   10 G(IK)=0.
      IF(VR(ND)-TEST)12,12,13
   12 N=0
      U=0.
      V=0.
      GO TO 11
   13 N=1
      U=VR(ND)
      V=VR(ND)*VR(ND)
   11 CONTINUE
C
C          COMPUTE SEMI-VARIOGRAM,NEW POINT
C
      ND1=NC-1
      DO 2 I=1,ND1
      VR1=VR(I)
      IF(VR1.LE.TEST)GO TO 2
      N=N+1
      U=U+VR1
      V=V+VR1*VR1
      I1=I+1
C
C                                  NEW LAG
C
      DO 21 J=I1,ND
      IF(VR(J).LE.TEST)GO TO 21
      DX=X(J)-X(I)
      DY=Y(J)-Y(I)
      H=SQRT(DX*DX+DY*DY)
      IF(H.LT.1.E-03)GO TO 25
      K=INT(H/PAS+0.5)+1
C
```

```
      H1=ABS(H-(K-1)*PAS)
      IF(K.GT.KMAX.OR.H1.GT.TOL)GO TO 21
C
C                                        NEW DIRECTION
C
      DO 22 KD=1,NDI
      COSD=(DX*CAN(KD)+DY*SAN(KD))/H
      IF(ABS(COSD).GE.CDA)GO TO 23
   22 CONTINUE
      GO TO 21
   23 IK=K+KMAX*(KD-1)
      NC(IK)=NC(IK)+1
      D(IK)=D(IK)+H
      VRR=VR(J)-VR1
      G(IK)=G(IK)+0.5*VRR*VRR
      GO TO 21
   25 WRITE(IOUT,2000)I,X(I),Y(I),J,X(J),Y(J)
   21 CONTINUE
    2 CONTINUE
C
C              RESULTS
C
      IF(N.EQ.0)GO TO 3
      V=(V-U*U/N)/N
      U=U/N
      DO 30 IK=1,KMM
      D(IK)=D(IK)/MAXO(1,NC(IK))
   30 G(IK)=G(IK)/MAXO(1,NC(IK))
    3 CONTINUE
C
C              PRINT RESULTS IF IS.NE.1
C
      IF(IS.EQ.1)GO TO 5
      IMP=(NDI-1)/5+1
      IDM=FLOAT(NDI)/FLOAT(IMP)+0.9999
      DO 42 IM=1,IMP
      WRITE(IOUT,2001)IM
      WRITE(IOUT,2002)U,V,N
      ID1=1+IDM*(IM-1)
      ID2=MINO(NDI,IDM*IM)
      WRITE(IOUT,2003)DALPHA,PAS,TOL,
     1(IA,ID,ID=ID1,ID2)
      WRITE(IOUT,2004)(ALP(ID),ID=ID1,ID2)
      WRITE(IOUT,2005)(IA,ID=ID1,ID2)
      IKO=KMAX*(ID1-1)
      IKM=IKO+KMAX*(ID2-ID1)
      DO 43 K=1,KMAX
      IK1=K+IKO
      IK2=K+IKM
   43 WRITE(IOUT,2006)K,(NC(IK),D(IK),G(IK),
     1IK=IK1,IK2,KMAX)
   42 CONTINUE
```

```
   41 CONTINUE
C
   2000 FORMAT(1H ,'**DOUBLY DEFINED POINT**DATUM',
      1I4,' X=',F9.4,' Y=',F9.4,' DATUM',I4,' X=',
      2F9.4,' Y=',F9.4)
   2001 FORMAT(1H1,53X,'SEMI-VARIOGRAM ',3EX,'***PAGE:'
      1,I2/54X,'*************',1CX,'(IRREGULAR GRID',
      2' 2 DIMENSIONS)'/)
   2002 FORMAT(1H ,' AVERAGE = ',F1C.5,6X,'VARIANCE = '
      1,E11.5,' NUMBER OF DATA = ',I5)
   2003 FORMAT(1H ,'DIRECTION TOLERANCE = ',F4.0,
      1' DEGREES ,LAG = ',F7.2,' ,DISTANCE TOLERANCE'
      2,' = ',F7.2/1H ,5X,'|',5(A1,5X,' DIRECTION',I2
      3,6X,'|'))
   2004 FORMAT(1H ,5X,'|',5(3X,F7.1,'DEGREES',6X,'|'))
   2005 FORMAT(1H ,' LAG |',5(A1, 'NC DISTANCE ',
      1'1/2 VARIO | '))
   2006 FORMAT(1H ,1X,I3,' |',5(I3,1X,F7.2,1X,
      1E11.5,'|'))
C
    5 RETURN
      END
```

III.C.5. Variogram analysis

The object of variogram analysis is to detect the major structural characteristics of the regionalized phenomenon under study by analyzing the
various experimental variograms. The structural information gathered
from such an analysis must be continually compared with the known
physical characteristics of the phenomenon (geology, tectonics,
mineralogy, sampling procedures, etc.). Under no circumstances should a
variographical study take the place of a good geological survey of the
phenomenon; on the contrary, the variography should always be guided by
geology. The variography completes and enriches the geological knowledge of the phenomenon, since the variogram quantifies the structural
information for use in estimation procedures.

For instance, geological studies may indicate a direction of preferred
variability of the mineralization due, for example, to alteration or diffusion.
A study of the anisotropies of the various directional variograms will
permit the verification of such an hypothesis and, moreover, will allow a
quantitative evaluation of the amount of total variability due to the
particular phenomenon of diffusion or alteration (e.g., in section IV.F,
Case Study 13 – the case study of the radial anisotropy of El Teniente
deposit). Conversely, a significant anisotropy displayed by the variograms

must be accounted for, either by the geology of the phenomenon or as a structural artefact due to measurement (preferential data locations, biased analyses, etc.).

Again, through geological studies the various scales of variability of the phenomenon can be qualitatively evaluated, e.g., the petrographic scale of the transitions from one element to another, the lithological scale of the various facies, the tectonic scale of the transitions delimited by the main faults, etc. Through the concept of nested structures, variography quantifies the respective part played by each of these scales of variability and evaluates their importance in relation to the various problems of estimation, optimization of production, quality control, etc. For example, what is the residual macroscopic effect of the micro-variabilities on the grade fluctuations of a production size unit? The geostatistical answer to this problem is provided by a study of the effect of regularization of these micro-variabilities when changing from the data support (core sample of several kilograms) to the support of a production size unit (several hundred tonne block) cf. section IV.G, Cast Study 17 – the case study of the grade fluctuations in a nickel deposit.

Each of the case studies given in this book, and particularly those in Chapter IV, demonstrates the complementary nature of the quantitative approach provided by variography and the more qualitative geological approach.

Because of this essential interaction between the geostatistical approach and a qualitative experimental knowledge of the studied phenomenon, it is not possible to define a universal model or plan for the analysis of variograms which would apply to all types of deposits and to all methods of proving these deposits. At the most, several general remarks can be made on the various data artefacts which may bias the structural picture given by the variograms.

Causes of heterogeneity between variograms

Before ascribing the observed difference between two variograms to a physical heterogeneity of the phenomenon under study, it should be ensured that this difference is not due to unrepresentative variograms or, more generally, to an artefact of measurement or calculation. Causes of heterogeneity due to measurement are numerous.

First, each of the two variograms may be calculated from an insufficient number of pairs or a neighbourhood which is too small (cf. practical rule (III.42)) to be representative. For example, if the available data consists of vertical drill cores and horizontal channel samples, and it is intended to detect anisotropy, then it is advisable, if possible, to compare the two mean experimental variograms calculated from the two sets of data, rather than to compare a single vertical drill core with a single horizontal channel sampling profile.

Second, if the two experimental variograms are defined on different supports (channel sample and core sample in the preceding example) it is advisable to determine whether the difference in support is sufficient to explain the observed difference between the variograms.

Third, if the means of the data used to calculate the variograms are different, then it is advisable to look for a possible proportional effect; for example, by plotting and comparing the relative semi-variograms γ^*/m^{*2} if a direct proportional effect is suspected.

Fourth, the preliminary review of the data may immediately reveal the causes of an observed difference between two variograms: a single aberrant data value, when squared, is enough to completely change the behaviour of the variogram; the two sets of data may not have been collected with the same equipment, etc.

Variography does not create data, it simply presents it in the synthetic and quantified form of the variogram. A significant heterogeneity displayed by a variogram usually has an obvious physical explanation; when this is not the case, an artefact of measurement or calculation should be suspected. If only a slight difference is observed between two variograms and there is no obvious physical explanation, then it is best to ignore it – by assuming isotropy, for instance – rather than developing a complicated model for anisotropy that may turn out to be illusory. The rule in such a case is to adopt the simplest model.

Pure nugget effect or experimental fluctuations?

Let $\gamma_0(h)$ be a point transition model with a range a_0 which is very small when compared with the dimensions of the data support: $a_0 \ll v$.

In practice, when $v > a_0$, the regularization on the support v covers the entire variability γ_0 so that it no longer appears on the regularized semi-variogram except for a nugget constant, cf. formula (II.7):

$$\gamma_{0v}(h) = C_{0v} = \text{constant}, \qquad \forall |h| > v,$$

with the rule of inverse proportionality to the support: $C_{0v} = (v'/v)C_{0v'}$

When the variability consists only of micro-structures of the type $\gamma_0(h)$, i.e., structures with ranges much smaller than the supports of the data, the experimental semi-variograms will appear flat or, more exactly, fluctuating slightly around their sill values, as in Fig. III.28(a).

Before assuming the hypothesis of a pure nugget effect, it should be ensured that nugget-like behaviours (i.e., without apparent increase) of the

experimental variograms are not due to a smoothing effect in the calculation of the variograms (e.g., due to reconstruction of core samples of
constant length, cf. formula (III.45)) or a fluctuation effect (e.g., too few
data used for the construction of the variogram). Contrary to an apparent
nugget effect due to experimental fluctuations, a pure and stationary
nugget effect only shows slight fluctuations around the sill value, cf. Fig.
III.28(a) and (b). A good way to verify a pure nugget effect is to test the
rule of inverse proportionality of the sills to the supports, cf. in section
III.A.4, the case study of the Adelaïda mine.

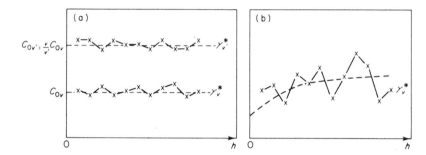

FIG. III.28. Nugget-like variograms. (a) Actual pure nugget effect; (b) apparent
nugget effect.

It should be stressed that under the hypothesis of stationarity a pure
nugget effect corresponds to an hypothesis of extreme homogeneity of the
underlying mineralization, cf. section III.A.2. In mining practice, an actual
pure nugget effect is extremely rare; on the other hand, insufficient or
inadequate data will quite often result in experimental variograms having
the appearance of a pure nugget effect.

III.C.6. Fitting a model

As was the case for the analysis of the variograms, there is no unique
technique or method for fitting a theoretical model to a regionalization
identified by a set of experimental variograms. At the most, the following
remarks can be made.

Operational representation

The representation of a natural variability by a point or regularized variogram can never be unique. In practice, it is enough that this representation

reflects the various experimental features which have been found (experimental variograms, anisotropies, *a priori* geological knowledge, etc.). This representation must also be operational, i.e., simple and adapted to the purpose of the study. In particular, certain minor features revealed by structural analysis may be omitted from the adopted representation. Similarly, when two geological or mineralogical causes contribute in the same way to the observed variability at some scale, it is useless to complicate the model by distinguishing between these two causes.

Support and regularization

The dimensions of the support v of the data are often very small with respect to the distances involved in the intended estimations, and also with respect to the dimensions of the units V to be estimated. In such cases, the support v of the data can be approximated to a quasi-point support ($v \simeq 0$) and the experimental regularized semi-variogram $\gamma_v^*(h)$ can be treated as if it were an estimator of the point model $\gamma(h)$.

The techniques of regularization (passing from γ to γ_v) or, conversely, of deconvolution (passing from γ_v to γ) are required only when comparing variograms defined on different supports (γ_v and $\gamma_{v'}$ with $v \neq v'$), cf. in section II.D.4, remark 3.

These techniques should not be used to deduce a point model which is illusively detailed, e.g., a sophisticated point model with a large number of parameters. To be rigorous, it is not possible to reach a greater degree of precision than that of the smallest support v of the data without introducing supplementary and unverifiable hypotheses. It is always better to avoid such hypotheses and stick to the available data; for example, by assuming a quasi-point support.

Sill and dispersion variance

The choice of a theoretical model with or without sill is always made by examining the experimental variograms, taking account of the fact that these variograms are subject to significant fluctuations at large distances.

In the same way, a sill is chosen by fitting a line through the experimental fluctuations. The experimental dispersion variance

$$s^2 = \frac{1}{N} \sum_i [z(x_i) - \bar{z}]^2$$

can sometimes be used in fitting the sill, provided that it is in fact a good estimator of this sill (i.e., of the *a priori* variance $D^2(v/\infty)$, when v is the support on which the semi-variogram is defined), cf. section II.C.1 and section III.C.4 (the note about the experimental dispersion variance s^2

associated with each mean variogram). In particular, it should be stressed that the existence of a sill is not related to the existence of an experimental dispersion variance s^2; the experimental variance of N data can always be calculated even when a sill does not exist (the corresponding RF being only intrinsic).

Behaviour at the origin

Most variables encountered in mining applications yield a semi-variogram with a linear behaviour at the origin and a nugget effect, cf. Fig. III.29. The nugget constant C_0 is obtained by extrapolating the average linear behaviour of the first experimental points to the ordinate axis.

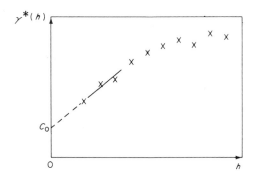

FIG. III.29. Linear behaviour at the origin.

Nested models

It will be recalled from III.A.1 that the nested structure $\gamma(h) = \sum_i \gamma_i(h)$ is a model particularly well suited to applications because of the linearity of geostatistical operators. Thus, it is not strictly necessary that the different components γ_i have a distinct physical interpretation. However, two components γ_i and γ_j with markedly different sills and ranges will, in practice, have a clear physical meaning.

Automatic fitting

As far as mining geostatistics is concerned in this book, any attempt at blind automatic fitting of parameters to experimental variograms – such as least squares methods – should be avoided.

It should be recalled that each estimator point $\gamma^*(h)$ of an experimental semi-variogram is subject to an error of estimation and fluctuation which is particular to that point; this error varies according to the separation distance h, the zone or data set from which the variogram is calculated, etc. The best method for weighting each estimated point $\gamma^*(h)$ with its "coefficient of reliability" comes from a critical review of the data and from practical experience.

Among other things, physical knowledge of the phenomenon (geology, for example), common sense involved in approximations and adapting available information to the ultimate purpose of the study are all as essential to a structural analysis as the calculated values of the experimental variograms.

Robustness of the geostatistical results with respect to the chosen type of model

Consider the two transition-type models in current use which have a linear behaviour at the origin, i.e., the exponential and the spherical models.:

$$\gamma(r) = 1 - \exp(-r/a);$$

and

$$\gamma(r) = \begin{cases} \dfrac{3}{2}\dfrac{r}{a} - \dfrac{1}{2}\dfrac{r^3}{a^3}, & \forall r \leq a, \\ 1, & \forall r \geq a. \end{cases}$$

The object of the study is to demonstrate the robustness of the geostatistical results with respect to the choice of either one of these two models, provided that a correct fit to the experimental semi-variogram has been made.

Consider a spatial regionalization characterized by a point isotropic exponential model with a unit sill and a parameter $a = 1$:

$$\gamma(r) = 1 - e^{-r}.$$

Suppose now that this theoretical model is not known, and that a nested model $\gamma^*(r)$ consisting of two spherical models is fitted to the experimental semi-variogram $(1 - e^{-r})$ in the interval $r \in [0, 3]$, cf. Fig. III.30:

$$\gamma^*(r) = C_1 \gamma_1(r) + C_2 \gamma_2(r),$$

the parameters of the two spherical component structures $C_1 \gamma_1$ and $C_2 \gamma_2$ being

$$\text{sills} \begin{cases} C_1 = 0.45, \\ C_2 = 0.53, \end{cases} \qquad \text{ranges} \begin{cases} a_1 = 1.2 \\ a_2 = 3.6. \end{cases}$$

For these two models, exponential $\gamma(r)$ and its fit $\gamma^*(r)$, the values of the following geostatistical operations in the one, two and three-dimensional spaces are given in Table III.3:

(i) The dispersion variance of a point within a segment, square or cube of side $l = 3a = 3$, cf. formulae (II.98) and (II.99) and Charts no. 4, 14 and 5, 15;

(ii) The estimation variance of this segment and this square by a central point, and of the cube by an axial core of length l, cf. Charts no. 7, 17 and 10, 20.

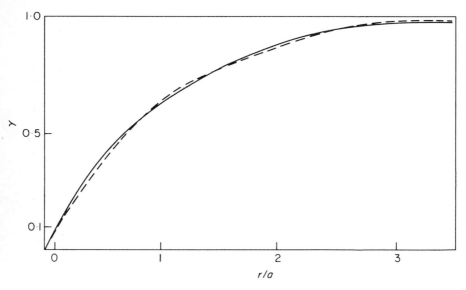

FIG. III.30. Exponential model (———) approximated by two nested spherical models (– – – –).

TABLE III.3. *Robustness of the geostatistical variance with respect to the type of model chosen*

	Dispersion variance			Estimation variance		
	R^1	R^2	R^3	R^1	R^2	R^3
Exponential γ	0·544	0·73	0·82	0·42	0·57	0·18
Fit γ^*	0·535	0·72	0·81	0·419	0·571	0·179
Relative deviation	0·017	0·014	0·012	0·002	0·002	0·006

Note that the relative deviation between two results derived from the two models is of the order of 2% for the dispersion variances and of several per thousand for the estimation variances. The excellent concordance of the results is to be expected since the fitting of $\gamma^*(r)$ to $\gamma(r)$ on Fig. III.30 is remarkable, at least within the working interval $r \in [0, 3]$.

IV

Case Studies of Structural Analysis

SUMMARY

The 15 case studies given in this chapter were selected to cover the main structural models of variograms in current use as well as the most critical problems which may be encountered in the course of an actual structural analysis.

The first five examples of section IV.A show how to fit the various variogram models currently used and introduced in section III.B.2.

Section IV.B gives a typical example of model fitting to a regionalization which shows a hole effect and a proportional effect.

Section IV.C gives two case studies of coregionalization and their fit to the linear model introduced in section III.B.3.

The two examples in section IV.D present the two models of anisotropy introduced in section III.B.4, i.e., the geometric and the zonal anisotropies.

While the presentation of each of the previous ten case studies is focused on a limited number of theoretical concepts, the three case studies of section IV.E give three examples of structural analyses in three dimensions, where most of the previous concepts are involved simultaneously.

The two examples of section IV.F are particularly important in mining practice, as they show how the two geostatistical tools – estimation and dispersion variances – can be used in a simple but operational way to control grades during production.

IV.A. FITTING MODELS

Case Study 1. Regionalization of oil grades, after P. A. Dowd (1977)

This study concerns a section of the Athabasca tar sands (Canada) that were proven by 40 vertical diamond drills, the cores of which were cut into approximately constant lengths, $l \in (1 \cdot 8 \text{ to } 2 \cdot 2 \text{ ft})$. The drill-holes themselves had an approximately constant length $L = 350$ ft, thus providing a set of some 7000 samples analysed for oil expressed in per cent saturated weight.

As there was little variation in the data support, it was not deemed necessary to reorganize the data into strictly constant lengths. However, an effective grouping into distance classes was made by approximating the data to a point grade $z(x)$ centred on the mid-point of the corresponding core sample, the class interval being equal to the mean length, $l = 2$ ft, of a core sample. It will be recalled from section III.C.4 that the grouping into distance classes causes a smoothing effect particularly significant for very small distances, and, for this reason, the first point of the experimental variogram, i.e., $\gamma^*(2)$, has been rejected.

The mean vertical semi-variogram obtained from the weighted average of all the elementary vertical semi-variograms for each drill core is shown on Fig. IV.1. The following information is relevant to this variogram.

(i) The first experimental point $\gamma^*(4)$ was estimated with 3363 pairs of data values, and all points up to $h = 150$ ft were estimated with at least 2000 pairs each. Toward the limit of reliability $(h = L/2 \simeq 175$ ft, one half of the mean length L of the drill cores), the number of pairs was still greater than 700. Thus, the reliability of this mean vertical semi-variogram can be considered as excellent.

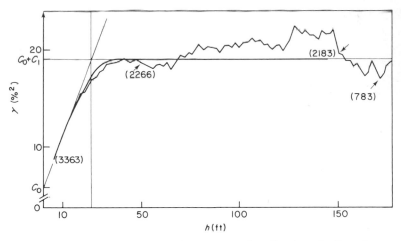

FIG. IV.1. Spherical model fitted to a regionalization of oil grades.

(ii) The arithmetic mean of the 7000 grade values used is $m^* = 8 \cdot 8\%$ saturated weight.

(iii) The mean experimental dispersion variance of a core sample of length l within the total core length L is, $s^2(l/L) = 17 \cdot 3\%^2$. This value was obtained by weighting the elementary dispersion variances for each drill core according to formula (III.49). This mean experimental variance is an estimator of the theoretical dispersion variance $D^2(l/L)$ of the grade of a sample, $l \simeq 2$ ft, in the mean length of a vertical drill-hole $L \simeq 350$ ft.

A spherical model with a nugget effect was fitted to the experimental semi-variogram $\gamma(h) = C_0 + \gamma_1(h)$, $\forall h$ vertical $\in [2, 175 \text{ ft}]$, with $C_0 = 6\%^2$ for the nugget constant, and $\gamma_1(h)$ a spherical model with a sill $C_1 = 13\%^2$ and a range $a_1 = 36$ ft, cf. formula (III.15).

The fit was achieved by considering the following features of the experimental semi-variogram.

(i) The tangent to the curve at the origin was approximated by the average linear behaviour of the first three experimental points. This average linear behaviour gives the value $C_0 = 6\%^2$ when extrapolated to $h = 0$.

(ii) The total sill value, $C_0 + C_1 = 19\%^2$ was estimated as the value around which the semi-variogram becomes stable.

(iii) The value of h at which the preceding tangent intersects the sill was taken as two-thirds of the value of the range a_1 of the spherical model, i.e., $2\,a_1/3 \simeq 24$ ft, and, thus, $a_1 = 36$ ft.

Verification   One way to verify the chosen model is to compare the theoretical dispersion variance $D^2(l/L)$, derived from the fitted model, to the experimental dispersion variance $s^2(l/L) = 17 \cdot 3\%^2$.

This dispersion variance is a linear operator of the semi-variogram γ, cf. formula (III.2), and is the sum of the variances $D_0^2(l/L)$ contributed by the nugget effect and $D_1^2(l/L)$ contributed by the structure γ_1:

$$D^2(l/L) = D_0^2(l/L) + D_1^2(l/L).$$

According to formula (III.11), the nugget-effect contribution is written as

$$D_0^2(l/L) = \frac{A}{l} - \frac{A}{L} = \frac{A}{l}\left(1 - \frac{l}{L}\right) = C_0\left(1 - \frac{2}{350}\right) = 5 \cdot 96$$

and, from formula (II.36), the variance contributed by the structure γ_1 is given by

$$D_1^2(l/L) = \bar{\gamma}_1'(L, L) - \bar{\gamma}_1'(l, l),$$

where $\gamma_1'(h)$ is the underlying point model of the model $\gamma_1(h)$ regularized by core samples of length l. The range a'_1 of the point model is thus equal to $a'_1 = a_1 - l = 34$ ft, and the point sill value is given by formula (II.51) as

$$C_1 = C'_1\left[1 - \frac{l}{2a'_1} - \frac{l^3}{20a'^2_1}\right] = 0 \cdot 97 C'_1,$$

giving $C'_1 = 13 \cdot 39$. Approximating the point model $\gamma'_1(h)$ to a new spherical model with range $a'_1 = 34$ ft and sill value $C'_2 = 13 \cdot 39\%^2$ and using formulae (II.51) and (II.98), the following mean values $\bar{\gamma}'_1$ are obtained:

$$\bar{\gamma}'_1(l, l) = C'_1 - C_1 = 0 \cdot 39,$$

$$\bar{\gamma}'_1(L, L) = C'_1\left[1 - \frac{3}{4}\frac{a'_1}{L} + \frac{1}{5}\frac{a'^2_1}{L^2}\right] = 0 \cdot 947 C'_1 = 12 \cdot 67.$$

Thus, the variance contributed by the structure γ_1 is

$$D^2_1(l/L) = 12 \cdot 67 - 0 \cdot 39 = 12 \cdot 28$$

and the theoretical dispersion variance is

$$D^2(l/L) = 5 \cdot 96 + 12 \cdot 28 = 18 \cdot 24,$$

which is close enough to the experimental value $s^2(l/L) = 17 \cdot 3$, and this can be taken as a verification of the chosen model.

Remark 1   As the support $l = 2$ ft of the data is small with respect to the experimental range $a_1 = 36$ ft, it could have been approximated by a point. The fitted model, $\gamma(h) = C_0 + \gamma_1(h)$, would then be considered as a point model and the experimental dispersion variance could be accepted as an estimator of the theoretical dispersion variance $D^2(0/L)$ of a point in L. This variance $D^2(0/L)$ can then be calculated directly from the model $\gamma(h)$, i.e.,

$$D^2(0/L) = \bar{\gamma}(L, L) = C_0 + C_1\left[1 - \frac{3}{4}\frac{a_1}{L} + \frac{1}{5}\frac{a^2_1}{L^2}\right] = 18 \cdot 02.$$

This value of $D^2(0/L)$ is very close to the value $18 \cdot 24$ calculated assuming a support of length $l = 2$ ft. Note that the quasipoint support approximation greatly simplifies the calculations (see the remark on support and regularization in section III.C.6).

Remark 2   Exponential models, like spherical models, also have a linear behaviour at the origin and a sill. However, if such a model is adopted for $\gamma_1(h)$, the practical range would be three times the abscissa ($h = 24$ ft) of intersection of the tangent at the origin with the sill, cf. section III.B.2, remark on the definition formulae (III.15) and (III.16), i.e., $a_1 = 3 \times 24 = 72$ ft. Now, this value (72 ft) is too large to correctly represent the rapid increase of the experimental semi-variogram towards its sill.

Case Study 2. Regionalization of copper grades, after J. Damay and M. Durocher (1970)

The copper deposit of Los Bronces (Chile) is a breccia mass impregnated with chalcopyrite and can be seen as a series of fault planes delimiting zones of homogeneous grades, cf. Fig. IV.2(a).

The deposit has been sampled by vertical drill-holes and horizontal drills from the galleries. The drill cores have been cut into constant 2-m lengths and analysed for copper.

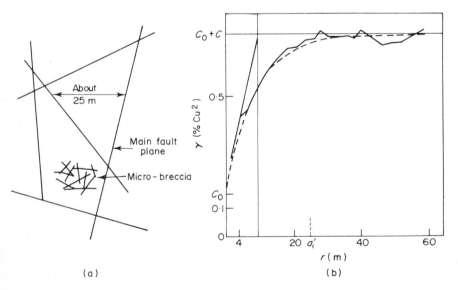

FIG. IV.2. Regionalization of Cu grades (Los Bronces). (a) Breccia structure; (b) fitted exponential model.

The experimental semi-variograms for each direction verify the hypothesis of isotropy of the mineralization, and they can thus be grouped together to provide a single isotropic semi-variogram, $\gamma(h) = \gamma(r)$, with $r = |h|$.

This experimental isotropic semi-variogram constructed from 8000 m of drill core or approximately 4000 samples is shown on Fig. IV.2(b). For a distance up to $r = 60$ m, equal to the limit of reliability, an exponential model with a nugget effect can be fitted remarkably well:

$$\gamma_0(r) = C_0 + \gamma_1(r), \qquad \forall r \in [2, 60 \text{ m}]$$

with nugget constant $C_0 = 0 \cdot 16$ (% $\text{Cu})^2$ and $\gamma_1(h)$ an exponential model with sill $C_1 = 0 \cdot 57\%^2$ and practical range $a_1' = 3a_1 = 25 \cdot 5$ m, cf. formula (III.16).

This model was fitted in a similar manner to that of the previous example, except that an exponential model was chosen instead of a spherical model, to take account of the slower increase of the experimental curve towards its sill.

The practical range $a_1' \approx 25$ m of the model γ_1 corresponds to the mean distance, observed during mining, between the main fault planes defining the zones of extreme homogeneity of the mineralization. The nugget effect C_0 is due both the the measurement errors and to the micro-variability introduced by a network of micro-faults through which the chalcopyrite has been able to impregnate. In fact, during mining, it was seen that the mineralized breccia consisted principally of two levels of heterogeneity, the first one corresponding to the main fault planes with range $a_1' \approx 25$ m and the second one corresponding to a micro-breccia with a 20 to 50 cm range. For a survey on a support equal to the core sample length (2 m) this micro-structure appears as a nugget effect.

Case Study 3. Regionalization of geopotential, after P. Delfiner (1973b)

In meteorology, geopotential is the name given to the height $Z(x)$ at a point x of a given isobar surface (in the present case study this isobar surface is 500 mbar). Taking the earth as a sphere, the distance h between two points x and $x + h$ is a geodesic distance, i.e., a distance measured on the great circle passing through these two points.

The experimental semi-variograms of the 500-mbar geopotential for the two main directions N–S (longitude) and E–W (latitude) are shown on Fig. IV.3. These two directions are perpendicular and parallel to that of the general circulation of winds, which is from West to East.

The N–S semi-variogram shows a marked parabolic-type increase which is characteristic of a drift effect, cf. section II.A.3, formula (II.14). The E–W semi-variogram, however, shows a very regular behaviour at the origin and attains a sill at a distance of about 3000 km. A Gaussian model with a nugget effect has been fitted to this E–W semi-variogram:

$$\gamma(h) = C_0 + \gamma_1(h), \qquad \forall h \in]0, 4000 \text{ km}],$$

where h is the geodesic E–W distance, $C_0 = 300 \text{ m}^2$ is the nugget constant, and $\gamma_1(h)$ is a Gaussian model with sill $C_1 = 25\ 000 \text{ m}^2$ and practical range $a'_1 = a_1\sqrt{3} = 2500$ km, cf. formula (III.17).

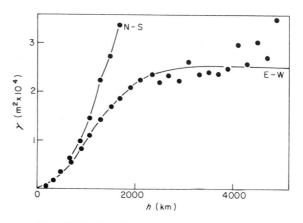

FIG. IV.3. Semi-variogram of geopotential.

The N–S drift can be seen directly on the data themselves; the N–S profile of the mean values of the geopotential within zones of 5° of latitude is shown on Fig. IV.4. By contrast, the E–W profile of the mean values of the geopotential within vertical bands of 10°, shown on Fig. IV.5, appears stationary.

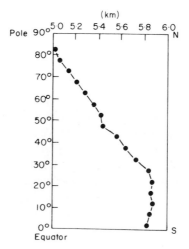

FIG. IV.4. N–S profile of geopotential.

The non-stationary RF geopotential $Z(x)$ can then be modelled as the sum of a drift function $m(x)$ and a stationary RF $Y(x)$:

$$Z(x) = m(x) + Y(x),$$

where $m(x) = E\{Z(x)\}$ is the meridian drift, i.e., $m(x) = m(x+h)$, whatever h in the E–W direction may be.

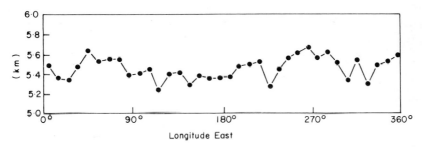

FIG. IV.5. E–W profile of geopotential.

$Y(x)$ is a stationary RF with zero expectation, $E\{Y(x)\} = 0$, and with an isotropic Gaussian model with nugget effect fitted to the experimental semi-variogram in the E–W direction.

It is advisable to verify that this model is consistent with the two experimental semi-variograms of Fig. IV.3. As these semi-variograms were calculated from the quadratic deviations of the data $z(x_i)$, they are, in fact, estimates of the moment $\frac{1}{2}E\{[Z(x+h) - Z(x)]^2\}$ which, according to the above model, is

$$\tfrac{1}{2}E\{[Z(x+h) - Z(x)]^2\} = \tfrac{1}{2}E\{[Y(x+h) - Y(x)]^2\} + \tfrac{1}{2}[m(x+h) - m(x)]^2$$

$$= \gamma(h) + \tfrac{1}{2}[m(x+h) - m(x)]^2.$$

It will be seen that if h is a vector in the E–W direction, the difference $[m(x+h) - m(x)]$ will be zero and the experimental E–W semi-variogram is an estimator of the Gaussian model $\gamma(h)$ with nugget effect. Also, if h is a vector in any direction except E–W (N–S, for example), the experimental semi-variogram will consist essentially of a marked parabolic increase due to the drift term $\tfrac{1}{2}[m(x+h) - m(x)]^2$.

Case Study 4. Topographic regionalization, after J. P. Chiles and P. Delfiner (1975).

The first step in the mapping of the Noirétable region of France by kriging was to construct the semi-variogram of the variable $Z(x)$, defined as the altitude at point x.

The directional experimental semi-variograms showed no clear anisot-ropy over small distances ($h < 300$ m), and so they were grouped together to provide a mean isotropic semi-variogram, as shown on Fig. IV.6.

A model without sill was fitted to this semi-variogram: $\gamma(r) = Cr^{\theta}$, with $C = 0.7$, and $\theta = 1.4 < 2$, cf. formula (III.18).

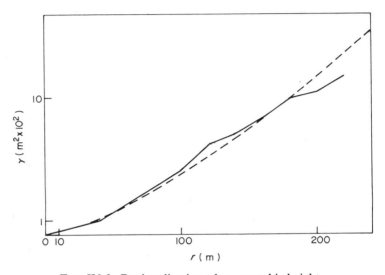

FIG. IV.6. Regionalization of topographic heights.

Case Study 5. Regionalization of bauxite accumulations, after G. Matheron (1962)

The bauxite deposit at Mehengui (Guyana) consists of a horizontal bed with a vertical thickness varying from 5 to 10 m. The deposit was sampled by vertical drills on a 50-m-square grid and the drill cores were analysed for Al_2O_3 over all the mineralized thickness, cf. Fig. IV.7. On two 50 m × 50 m squares, samples were taken on a 5-m grid in order to study small range structures. Horizontally, the mineralized section of the deposit can be approximated to a rectangle 1125 m × 670 m. Vertically, the mean thickness was found to be $\bar{t} = 7.25$ m.

The support on which the Al_2O_3 grade analyses were made is the mineralized thickness $t(x)$ at point x. As this support is not constant, the ReV grade $z(x)$ defined over $t(x)$ is not an additive variable, cf. section II.C.2. For this reason, the additive variable $a(x)$ – accumulation at point x – was defined as the weighting of the grade $z(x)$ by the vertical dimension

$t(x)$ of the support, i.e., $a(x) = t(x) . z(x)$, and is expressed in per cent Al_2O_3-metres.

Using the data on the 50-m grid, the semi-variograms for the N–S and E–W directions were calculated for distances h in multiples of 50 m.

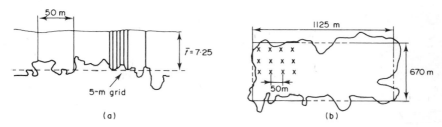

(a)   (b)

FIG. IV.7. Survey of Mehengui deposit. (a) Vertical section; (b) horizontal section.

Semi-variograms were also calculated in the diagonal directions NW–SE and NE–SW for distances h in multiples of $50\sqrt{2} = 70{\cdot}7$ m. Finally, these same four directional semi-variograms were calculated from the data on the 5-m grid for distances $h \in [5, 50$ m$]$. No significant directional anisotropy was evident on the semi-variograms and, thus, they were grouped together to form a single mean semi-variogram representing the horizontal regionalization of the accumulation $a(x)$, cf. Fig. IV.8(a) and (b) on a logarithmic and arithmetic scale, respectively.

A logarithmic model was fitted to the semi-variogram drawn on the logarithmic scale, and, on that scale, this model appears as a straight line. Matheron's preliminary fit (broken lines on Fig. IV-8(a) and (b)) corresponds to the model

$$\gamma(r) = 2{\cdot}88 \log r - 0{\cdot}69, \qquad \forall r \geqslant 5 \text{ m.}$$

The semi-variogram drawn on the arithmetic scale gives a greater visual importance to the larger distances $(r > 350$ m$)$ and a sill model seems more appropriate in this case. A nested model was fitted, consisting of a nugget effect and two spherical models:

$$\gamma(r) = C_0 + C_1\gamma_1(r) + C_2\gamma_2(r), \qquad \forall r > 0,$$

with

$$C_0 = 3{\cdot}2, \qquad C_1 = 7{\cdot}3, \qquad C_2 = 5{\cdot}5 \quad (\% \ Al_2O_3 \text{ m})^2$$

and γ_1 and γ_2 being two spherical models with unit sill values and respective ranges $a_1 = 50$ m and $a_2 = 350$ m.

Although the two chosen models are very different, the first one having no sill while the second one has a sill equal to $C_0 + C_1 + C_2 = 16$, it is not possible to determine visually which is the best fit, at least within the interval of reliability of the experimental semi-variogram ($r < 500$ m).

The theoretical calculations using both models lead to completely similar results, cf. the following evaluation of the theoretical dispersion variance.

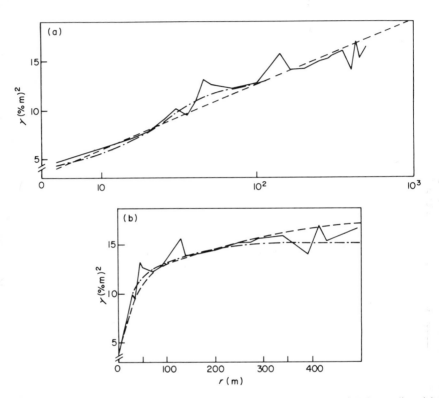

FIG. IV.8. Horizontal regionalization of accumulations (Mehengui). (a) Logarithmic scale. ———, Experimental variogram; —·—, two nested spherical models; – – – –, logarithmic model. (b) Arithmetic scale. —·—, Two nested spherical models, $\gamma(h) = C_0 + C_1\gamma_1 + C_2\gamma_2$, $C_0 = 3\cdot2$, $C_1 = 7\cdot3$, $C_2 = 5\cdot5$, $a_1 = 50$, $a_2 = 350$; – – – –, logarithmic model $\gamma(r) = 2\cdot88 \log r - 0\cdot69$.

This case study shows that the important thing in fitting variograms is not the type of models, but rather the agreement of the model with the various experimental curves available, cf. section III.C.6 (the paragraph on robustness). Note the importance of the graph scale in plotting the variograms.

Calculation of the dispersion variance

The experimental dispersion variance of the accumulations $a(x)$ within the deposit is $s^2(0/D) = 15\cdot3$ (% m)^2. This variance was calculated from the data on the regular square 50-m grid (uniform distribution within D) and can thus be considered as an estimator of the theoretical dispersion $D^2(0/D)$ within the deposit approximated by a rectangle $D = 1125$ m $\times 670$ m. The support of the horizontally regionalized variable $a(x)$ is the horizontal section of a core and is assumed to be a point.

The experimental value $s^2(0/D) = 15\cdot3$ will now be compared with the two theoretical values derived from the two previous models.

From the general formula (II.36), the theoretical dispersion variance is written as

$$D^2(0/D) = \bar{\gamma}(D/D) - \gamma(0) = \bar{\gamma}(D, D).$$

For the logarithmic model, the mean value $\bar{\gamma}(D, D)$ in the rectangle D is given by formula (II.97), which considers the linear equivalent ($l = 1125 + 670 = 1791$ m) of the rectangle D;

$$\bar{\gamma}(D, D) \simeq 2\cdot88[\log l - 3/2] - 0\cdot69 \simeq 16\cdot6.$$

For the nested sill model, $\gamma(r) = C_0 + C_1\gamma_1(r) + C_2\gamma_2(r)$, the mean values $\bar{\gamma}_1(D, D)$ and $\bar{\gamma}_2(D, D)$ of the spherical models γ_1 and γ_2 are found from chart no. 4 in section II.E.4:

$$\bar{\gamma}(D, D) = C_0 + C_1 \times 0\cdot999 + C_2 \times 0\cdot918 = 15\cdot5.$$

The latter theoretical value of $15\cdot5$ agrees with the experimental value $15\cdot3$. The theoretical value $16\cdot6$ derived from the logarithmic model seems a little large, due to the fact that this model assumes a continuous increase of the theoretical semi-variogram for large distances $r \in [500, 1125$ m] which are beyond the limit of reliability of the experimental semi-variogram. In this case, a slight preference should be given to the second model.

IV.B. HOLE EFFECT AND PROPORTIONAL EFFECT

Case Study 6. Regionalization of uranium grades

This study concerns the regionalization of uranium grades along a vertical cross-section of a sedimentary uranium deposit in Niger (Central Africa). Twenty-five vertical drill-holes were available, each of which was cut into core samples of 20 cm length and chemically analysed for U. The mineralized length L_i is defined for each drill-hole i. The experimental mean m_i^* of uranium grades and the corresponding dispersion variance s_i^2 have been calculated† along each of these lengths L_i. This experimental variance s_i^2 is an estimator of the dispersion variance $D^2(l/L_i)$ of a core sample l in the vertical segment of length L_i.

Figures IV.9 and IV.10 show the experimental vertical semi-variograms for four holes S_4, S_{14}, S_{10} and S_{16}. The semi-variograms are clearly different from each other both in shape and variance: semi-variograms S_{14} and S_{10} show a marked hole effect between 1 m and 3 m; the increase of S_{14} is very rapid while that of S_{16} stabilizes quickly. A proportional effect is also evident: the variability increases with the experimental mean m_i^*.

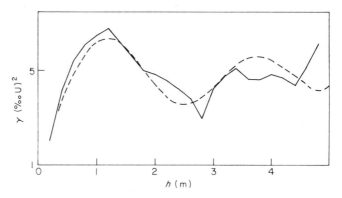

FIG. IV.9. Semi-variogram S_4 and its fit. ———, Experimental; – – – –, model.

The experimental mean semi-variogram obtained by grouping all 25 elementary semi-variograms is shown on Fig. IV.11. This mean semi-variogram corresponds to a mean mineralized length $\bar{L} = \frac{1}{25}\sum_i L_i = 13$ m and is calculated from more than 1500 samples with a mean grade $m^* = 2\cdot2\%$ U. The hole effect is no longer evident. Two nested spherical models with a nugget effect have been fitted to it, i.e.,

$$\gamma(h) = C_0 + C_1\gamma_1(h) + C_2\gamma_2(h), \qquad \forall h \text{ vertical} \in [0\cdot2 \text{ to } 4 \text{ m}]$$

with $C_0 = 3\cdot5$ $(\%\ \text{U})^2$, as nugget constant, and $C_1\gamma_1$ and $C_2\gamma_2$ being two spherical models with sills $C_1 = 3\cdot7$, $C_2 = 1\cdot1$ $(\%\ \text{U})^2$ and ranges $a_1 = 1$ m and $a_2 = 3$ m.

The two ranges correspond roughly to the abscissae of the maximum and minimum of the hole effect observed on the semi-variograms S_4 and S_{10}. The two ranges can be interpreted as the mean thickness of the rich seams $(a_1 \simeq 1$ m$)$ and the mean distance between two such seams $(a_2 \simeq 3$ m$)$ which

† For proprietary reasons, all the mean grades quoted in this example have been multiplied by a factor.

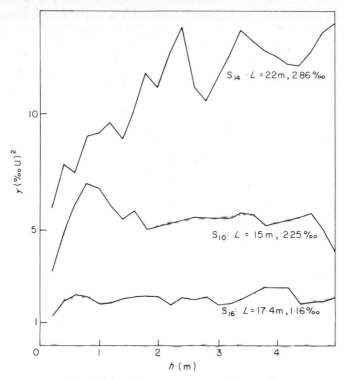

FIG. IV.10. Semi-variograms S_{14}, S_{10}, S_{16}.

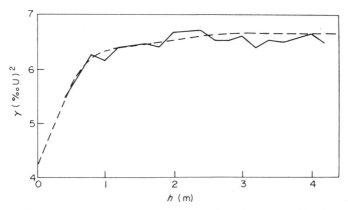

FIG. IV.11. Mean experimental semi-variogram and its fit.

can also be seen on the vertical profile of the grades of drill-hole S_4 on Fig. IV.12.

Fitting a proportional effect

The preceding nested spherical model is only representative of the mean vertical structure over the zone sampled by the 25 drill-holes with mean grade $m^* = 2\cdot2\%$ U. In order to represent local values on a zone with a mean grade $m_i^* \neq 2\cdot2$, a proportional effect should be added to the model.

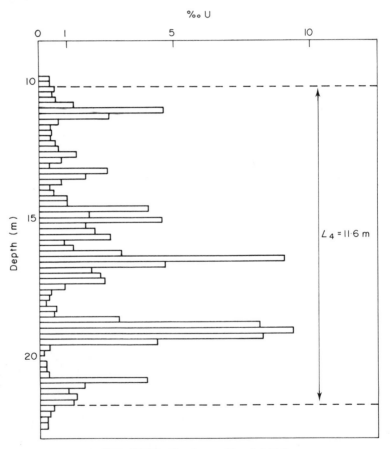

FIG. IV.12. Grade profile of drill S_4.

To do this, the experimental dispersion variance s_i^2 of the grades in each mineralized length L_i were considered, and, in order to make a significant comparison, the 12 drill-holes with the longest mineralized lengths were selected, $L_i \in [11 \text{ to } 33 \text{ m}]$. As these mineralized lengths are larger, the variance s_i^2 is calculated with a larger number of samples (five per metre) and can be considered as an estimator of the *a priori* variance $D^2(l/\infty)$ of the grades of core samples of support $l = 20$ cm. A scatter diagram of experimental variance s_i^2 is shown on Fig. IV.13. A parabola has been fitted to the scatter diagram:

$$s^2 = 2 \cdot 2 m^{*2} - 2 \cdot 3, \qquad \forall m^* \in [1 \cdot 3 \text{ to } 3‰ \text{ U}]$$

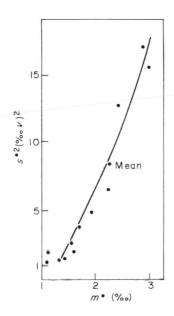

FIG. IV.13. Fitting the proportional effect (experimental variance s_i^2 versus experimental mean m_i).

The mean experimental variance $s^2 = 8 \cdot 3$ calculated by weighting the 25 elementary variances s_i^2 and the vertical model $\gamma(h) = C_0 + C_1\gamma_1(h) + C_2\gamma_2(h)$ corresponds to the overall mean value $m^* = 2 \cdot 2$. So, for a local vertical semi-variogram in a zone with constant mean grade m, the following quasi-stationary model applies:

$$\gamma(h; m) = \frac{2 \cdot 2 m^2 - 2 \cdot 3}{8 \cdot 3} [C_0 + C_1\gamma_1(h) + C_2\gamma_2(h)]$$

with h vertical $\in [0 \cdot 20$ to 4 m$]$ and $m \in [1 \cdot 3$ to $3‰$ U$]$. However, note that this model may not be adapted for a precise local estimation around drill S_4.

Fitting a hole-effect model

By way of example, an attempt has been made to fit the best possible model with a hole effect to the experimental semi-variogram of core S_4. There is no practical purpose in fitting this model, as this hole effect has only a local significance and disappears on the mean structure of Fig. IV.11, which is the only one used in the geostatistical calculations. It does, however, serve as a good exercise in fitting such a hole-effect model.

The first step is to estimate the amplitude of the hole effect (cf. section III.B.2, subsection 3, "Hole effect models"),

$$\alpha = |\min\{C(h)\}|/C(0) \simeq \frac{6\cdot8 - 4\cdot8}{4\cdot8} = 0\cdot417.$$

The maximum value of the experimental semi-variogram on Fig. IV.9 is $6\cdot8$ and, thus, $(6\cdot8 - s_4^2)$ is the minimum absolute value of the corresponding covariance. The experimental variance $s_4^2 = 4\cdot8$ is an estimator of the *a priori* variance $C(0)$.

As the value of $0\cdot417$ is larger than the maximum amplitude of an isotropic hole effect in three dimensions, $\alpha_{max} = 0\cdot217$, the hole effect observed on the semi-variogram S_4 will not be reproduced in the non-vertical directions of the deposit. And, indeed, in such a subhorizontal sedimentary deposit, the pseudo-periodicity of the alternance of rich mineralized seams and waste has no reason to be isotropic.

As this hole effect of amplitude $\alpha = 0\cdot417$ is limited to the vertical direction, it can be modelled by positive-definite-type models in only one dimension, in particular the cosine model, $\gamma(h) = 1 - \cos h$, for which the hole-effect amplitude is $\alpha = 1$, cf. formula (III.22). The following nested model has been fitted and is shown on Fig. IV.9:

$$\gamma(h) = \gamma_1(h) + \gamma_2(h), \qquad \forall h \text{ vertical} \in [0\cdot2 \text{ to } 4 \text{ m}]$$

with

$$\gamma_1(h) = A[1 - \exp(-\mu h/\lambda).\cos \mu h],$$

$$A = 2\cdot26(\text{‰ U})^2, \qquad \lambda = 10 \text{ m}, \qquad \mu = 2\cdot5;$$

$$\gamma_2(h) = \begin{cases} Bh, & \forall h \in [0, a], \\ Ba, & \forall h \geqslant a, \end{cases}$$

with

$$B = 9(\text{‰ U})^2 \text{ m}^{-1}, \qquad a = 0\cdot28 \text{ m}.$$

The model $\gamma_1(h)$ is constructed from formula (III.13) for the covariance function $\exp(-\mu h/\lambda).\cos \mu h$, which is the product of an exponential covariance and the cosine covariance.

The model $\gamma_2(h)$ is a transition model with range a and is conditional positive definite only in the one-dimensional space.

Note that the sill of this model is $A + Ba = 4\cdot78 (\text{‰ U})^2$ and the amplitude of its hole effect is

$$\alpha \simeq \frac{6\cdot45 - 4\cdot78}{4\cdot78} = 0\cdot349.$$

IV.C. COREGIONALIZATIONS

Case Study 7. Coregionalization of uranium radiometry, after M. Guarascio (1975)

The pitchblende mineralization of the Novazza deposit (Italy) is contained within a tuff bed dipping 20 to 30°, see Fig. IV.14.

The deposit has been sampled by bore-holes of various directions drilled from the galleries. The samples have been analysed, either chemically in

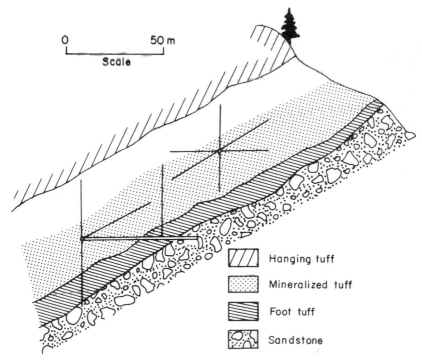

0   50 m
Scale

| Hanging tuff |
| Mineralized tuff |
| Foot tuff |
| Sandstone |

FIG. IV.14. Vertical cross-section of the Novazza deposit.

U_3O_8 grade on cores of variable length, or by radiometric measurements. The purpose of the study is to provide a local and global estimation of the deposit through "cokriging", i.e., considering simultaneously both types of information available, the chemical analyses and the radiometric measurements. For this, it is necessary to study the U grade-radiometry coregionalization.

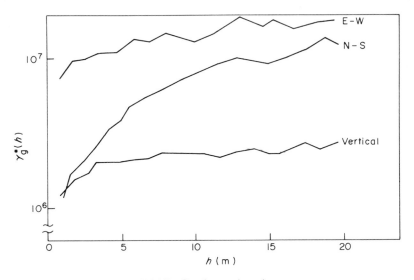

FIG. IV.15. Grade semi-variograms.

The analysed grades of core samples of constant length $l = 1$ m were reconstituted according to formula (III.45), and experimental semi-variograms along each of the reconstituted drill-holes were calculated. These elementary semi-variograms were then grouped into three mean semi-variograms which are characteristic of the three directions N-S, E-W and vertical (the angle tolerance around each direction being $\pm d\alpha = \pm 15°$). The chemical-grade semi-variograms are shown on Fig. IV.15. They indicate very different levels of variability (the ordinate axis of Fig. IV.15 is graduated in logarithmic scale). However, a direct proportional effect can be observed. Such an effect is classical in low-grade uranium deposits and is corroborated by the scatter diagram (s_i^2, m_i^*) between the experimental mean m_i^* and the experimental dispersion variance s_i^2 calculated on each chemically analysed drill core: a parabolic relation (s_i^2, m_i^*) is observed. Thus, a direct proportional effect in m_i^*, cf. formula (III.37) has been considered, and the three semi-variograms of Fig. IV.15 have been divided

by the squares of their respective mean. These three relative semi-variograms are shown on Fig. IV.16, the coordinate axis being graduated now in
arithmetic scale. A remarkable isotropy between the three directions
studied is observed and a last grouping provides the mean isotropic relative
semi-variogram of chemical grades, $\gamma_g^2(h)/m_g^{*2}$, cf. Fig. IV.17.

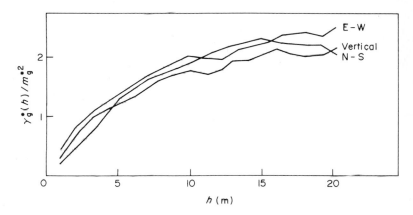

FIG. IV.16. Relative grade semi-variograms.

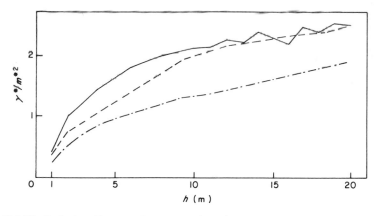

FIG. IV.17. Relative direct and cross semi-variograms. ——, γ_{g-R}; – – – –, γ_g;
—·—, γ_R.

The radiometric measures reconstituted on the same support (cores of
constant length $l = 1$ m) provide experimental semi-variograms with structural features quite similar to those of the grade semi-variograms. A direct
proportional effect in m_R^2 has been considered and the various directional

semi-variograms have been grouped into a unique mean isotropic relative semi-variogram of radiometric measurements, $\gamma_R^*(h)/m_R^{*2}$, cf. Fig. IV.17.

This same figure, Fig. IV.17, shows the mean cross semi-variogram, $\gamma_{g-R}^*/m_g^* \cdot m_R^*$, isotropic and relative to the product of the two experimental means of the grades (m_g^*) and radiometries (m_R^*) which have been used to calculate it.

On Fig. IV.17 it should be noted that the cross-variability $\gamma_{g-R}^*/m_g^* \cdot m_R^*$ is always greater than the direct variabilities γ_g^*/m_g^{*2} and γ_R^*/m_R^{*2}, which would entail that the experimental correlation coefficient $\rho^*(h)$ between the two increment variables $[Z_g(x+h)-Z_g(x)]$ and $[Z_R(x+h)-Z_R(x)]$ is superior to 1. Indeed, the theoretical correlation coefficient is written

$$\rho(h)=\frac{E\{[Z_g(x+h)-Z_g(x)][Z_R(x+h)-Z_R(x)]\}}{\sqrt{(E\{[Z_g(x+h)-Z_g(x)]^2\} \cdot E\{[Z_R(x+h)-Z_R(x)]^2\})}}$$

$$=\frac{\gamma_{g-R}(h)}{\sqrt{(\gamma_g(h) \cdot \gamma_R(h))}}.$$

Thus, the three experimental curves of Fig. IV.17 are not consistent. One possible explanation is that the experimental cross-semi-variogram γ_{g-R}^* has not been calculated from the same data as the direct semi-variograms (in practice it is difficult to do otherwise, unless we consider only the few points where the two measurements – grade and radiometry – exist simultaneously). Inconsistent and excessively small values of the means m_g^* and m_R^* can then explain the excessive variability of the experimental relative cross-semi-variogram $\gamma_{g-R}^*/m_g^* \cdot m_R^*$ as observed on Fig. IV.17.

Case Study 8. Coregionalization Pb–Zn–Ag, after P. A. Dowd (1971)

The Broken Hill lead–zinc ore body in Australia is "currently" considered as an exhalative sedimentary deposit with a secondary metamorphism. Three types of ore can be clearly distinguished in the studied zone, according to their gangue minerals:

(i) rich ore in a quartz gangue;
(ii) medium grade ore in a calcite gangue;
(iii) poor ore in a rhodonite gangue.

A schematic vertical cross-section of the mineralized horizon with its three gangues is shown on Fig. IV.18(a). The horizontal cross-section on a particular mining level exhibiting the various mineralized surfaces is shown on Fig. IV.18(b). Note, at this level, that the principal ore is the one associated with the calcite gangue.

Mining is carried out using a bore and pillar method in an E–W direction; pillars are 19 ft wide and stopes are 33 ft wide, cf. Fig. IV.19. On each mining level, both faces of each pillar were sampled continuously by 10-ft chip samples, providing a set of data on a regular grid measuring 10 ft in the E–W direction and alternately 33 ft and 19 ft in the N–S direction. Each chip sample is analysed for Pb, Zn and Ag.

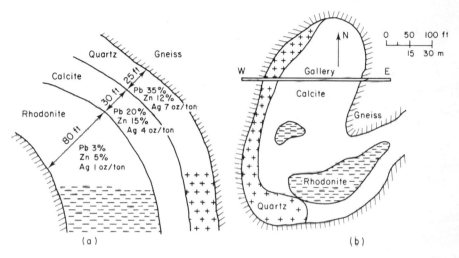

FIG. IV.18. Cross-sections showing the three types of ore. (a) Vertical; (b) horizontal.

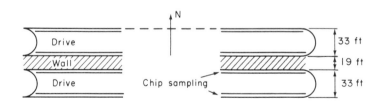

FIG. IV.19. Sampling of a mining level.

For the E–W and N–S directions, the experimental direct semi-variograms for lead, zinc and silver grades are shown on Fig. IV.20(a), (b), (c), as well as the experimental means and dispersion variances of the data used in the calculation. The experimental variances calculated from formula (III.49) are estimators of the dispersion variances $D^2(l/\bar{L})$ of the grades of a chip sample of length l within the mean length $\bar{L}$ of the drives ($\bar{L} \approx 400$ ft in the E–W direction).

The following points can be stressed about these experimental variograms.

(i) There is a particularly clear structural isotropy for small distances ($h < 150$ ft), confirmed by the extreme homogeneity of the mineralization within a given gangue.

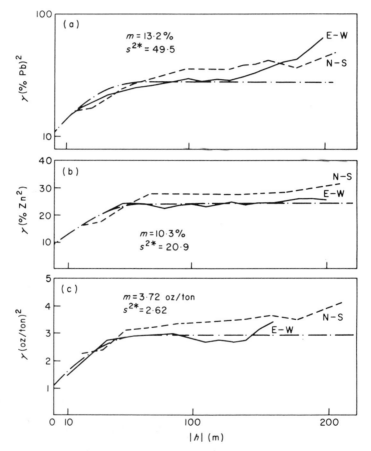

FIG. IV.20. Direct semi-variograms. (a) Pb grades regionalization; (b) Zn grades regionalization; (c) Ag grades regionalization.

(ii) There are obvious sills which are well approximated by the experimental dispersion variances s^2 and which, in all cases, are reached at distances of about $|h| = a_1 = 60$ ft.

(iii) There are obvious nugget effects.

(iv) The variabilities begin to increase above the sills for distances $|h|$ greater than 150 ft. Although located in a zone close to the limit of reliability of the experimental semi-variograms ($L/2 \simeq 200$ ft), this increase may be due to the non-stationarity at large distances caused by the presence of three different gangues: for distances $|h|$ greater than 150 ft, when a point x is in one gangue, the point $x + h$ will probably be in a different gangue causing the square deviation $[z(x + h) - z(x)]$ to become particularly large.

The experimental cross-semi-variograms for lead–zinc, silver–lead and silver–zinc are shown on Fig. IV.21(a), (b) and (c). According to the preceding hypothesis of isotropy, only the E–W direction has been considered. There are no nugget effects on these cross-semi-variograms, and there is a transition structure with a range approximating that found for the direct semi-variograms ($a_1 \simeq 60$ ft).

Fitting a model

The simultaneous presence of the same transition structure with range $a_1 \simeq 60$ ft on all direct and cross-semi-variograms suggests that the linear model of coregionalization (III.27) would provide a good fit. This model is given by

$$\gamma_{kk'}(h) = \sum_i b^i_{kk'} \gamma_i(h), \qquad \forall h \text{ horizontal} \in [0, 150 \text{ ft}].$$

The indices ($k, k' = 1$ to 3) represent, respectively, the grade in lead, zinc and silver. The index ($i = 0$ and 1) represents the two basic structures; a transition model $\gamma_0(h)$ with a very small range or nugget effect, and a transition model $\gamma_1(h)$ with range $a_1 = 60$ ft, respectively.

As the nugget effect is not present on the cross-semi-variograms, the coefficients $b^0_{kk'}, \forall k \neq k'$ are all zero.

A spherical model with a unit sill and a range $a_1 = 60$ ft has already been fitted to the basic structure $\gamma_1(h)$, and Table IV.1 gives the two matrices of coefficients $[b^i_{kk'}]$ for $i = 0$ (nugget effect) and for $i = 1$ (γ_1 structure). The corresponding fits obtained using this model are shown by the dotted lines on Figs IV.20 and IV.21.

TABLE IV.1. *Coefficient matrices of the linear model*

	Pb	Zn	Ag				
Pb	11	0	0		39	14.5	5
Zn	0	9	0		14.5	15	3.8
Ag	0	0	1.1		5	3.8	1.8

$$i = 0 \qquad\qquad\qquad i = 1$$

Remark 1   The coefficients $b^i_{kk'}$ verify the constraints of the linear model, i.e., both màtrices of Table IV.1 are positive definite, cf. conditions (III.26).

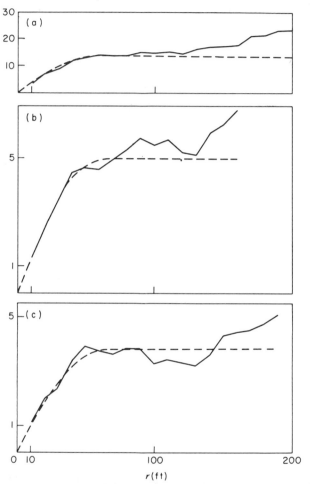

FIG. IV.21. Cross-semi-variograms. (a) Pb–Zn coregionalization; (b) Ag–Pb coregionalization; (c) Ag–Zn coregionalization.

Remark 2   Note the quality of the fit achieved for small distances ($h <$ range $a_1 = 60$ ft). This coregionalization model is thus ideally suited to the geostatistical procedure of co-estimation of one grade (e.g., Ag) by the three data sets (Pb, Zn and Ag), cf. section V.A.5, Case Study 2.

This co-estimation will be carried out within an horizontal neighbourhood of approximately 60 ft around the unit to be estimated.

Remark 3   One possible way of interpreting the chosen linear model of coregionalization is to consider the three RF's $Z_k(x)$ as issuing from two principal causes of variability represented by the basic RF's $Y_{0k}(x)$ and $Y_1(x)$, cf. the genesis of the linear model in formula (III.23):

$$Z_k(x) = Y_{0k}(x) + a_{k1} Y_1(x), \qquad \forall k = 1 \text{ to } 3.$$

The RF's $Y_{0k}(x)$ represent the micro-variabilities of the grades for distances ($h < 10$ ft) which are not accessible from the available data; these micro-variabilities vary from one metal to the other and appear as different nugget constants on the direct semi-variograms. The unique RF $Y_1(x)$ represents a second variability common to all three metals. When the hypothesis of a sedimentary origin of the deposit is accepted, this common variability can be explained in terms of horizontal beds of deposition micro-basins, the mean horizontal dimension of which is around 60 ft, cf. Fig. IV.22.

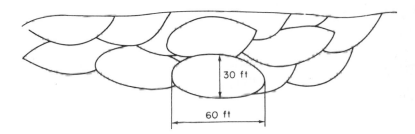

FIG. IV.22. Micro-basins of deposition.

It is interesting to note that the study of the vertical regionalizations of these three grades, from one mining level to the other, produced direct semi-variograms with the following features.

(i) Nugget constants strictly equal to those found for the horizontal structures: the micro-variabilities $Y_{0k}(x)$ would then be isotropic in the three-dimensional space.

(ii) A common transition-type structure with a range $a_1' = 30$ ft representing the mean vertical dimension of the micro-basins of deposition, cf. Fig. IV.22. A proportional effect is added to this geometric anisotropy of ratio $a_1'/a_1 = 0.5$, due to the fact that the three types of gangue are not encountered in the same proportions in each level.

IV.D. ANISOTROPIES

Case Study 9. Geometric anisotropy, after J. Serra (1967a)

As presented here, the study concerns the regionalization of iron grades on a hectometric scale (from 100 m to 2 km) in the ferriferous basin of Lorraine (France). The data come from two mines, Hayange and Bois d'Avril which, although geographically distinct, mine the same sedimentary iron seam known as "red seam".

This red seam is subhorizontal, very regular, and has a fairly constant thickness of about 3 m. The horizontal regionalization of the iron grades of this seam has been sampled along the length of mining galleries by regularly-spaced small drills (90, 100 and 120 m according to direction and mine). Data is thus available from both mines on a regular grid in the E–W and N–S directions of the galleries. These galleries are very long, from 2 to 3 km, and, consequently, the limit of reliability of the experimental variograms extends up to 1 km.

The E–W and N–S experimental semi-variograms for Hayange and Bois d'Avril are shown on Figs IV.23 and IV.24, respectively. The N–S semi-variograms for both mines are compared on Fig. IV.25.

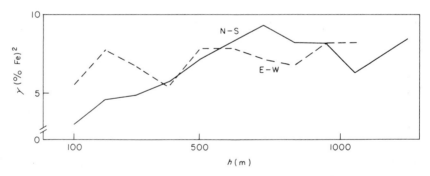

FIG. IV.23. Horizontal regionalization at Hayange (red seam).

Continuity of the red seam

Although both mines are geographically distinct (6 km distant from each other), their variograms reveal exactly the same structures on the hectometric scale. This is quite normal, since both variograms are relative to the same seam, which is found over a large portion of the Lorraine ferriferous basin. In addition, the means of both sets of data used in the calculations are almost identical (34% Fe).

Hole effect and geometric anisotropy

There is a marked hole effect on the E–W semi-variograms, with a maximum and a minimum variability occurring at around $|h| = 200$ m and $|h| = 350$ m, respectively. This hole effect is also present on the N–S semi-variograms with the same mean sill of variability (around $8 \cdot 3$ (% Fe)^2) but with the maximum and minimum variabilities occurring at around $|h| = 700$ m and $|h| = 1000$ m, respectively.

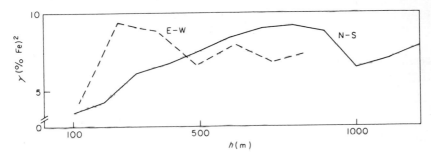

FIG. IV.24. Horizontal regionalization at Bois d'Avril (red seam).

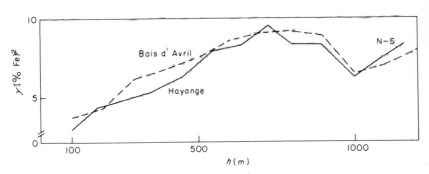

FIG. IV.25. N–S direction variograms at Hayange and Bois d'Avril.

This right shift of the hole effect along the distance axis $|h|$ suggests a model of geometric anisotropy between the E–W and N–S semi-variograms. When the E–W distances are multiplied by the anisotropy ratio $\lambda = 700/200 = 3 \cdot 5$, the E–W and N–S maximum values of the hole effect are made to coincide, cf. section III.B.4, Fig. III.10.

Genetic interpretation

It is generally accepted that the origin of the Lorraine ferriferous basin is sedimentary with marine deltaic deposition of the upper toarcian, cf.

J. Serra, (1968). Iron-rich sediments coming from inland were deposited in the sea between the main current lines of the river and formed thin horizontal pseudo-elliptical sedimentation units, cf. Fig. IV.26. Sandbanks in a river bed provide another example of this type of sedimentation. Thus, the transition structure with hole effect and horizontal geometric anisotropy as shown on the variograms, corresponds to the alternance of these sedimentary units; the N–S direction corresponds to the main direction of the ancient fresh-water currents and the mean N–S and E–W dimensions of these units are given by the abscissae (700 m for N–S, 200 m for E–W) of the maximum variability observed on the variograms.

It should be mentioned that sedimentologists explain the vertical succession of the different mineralized seams in terms of transgressions and regressions of the marine sedimentation: during certain epochs and at certain places of the delta, sedimentation was stopped and then started again at a later period. Thus, the red seam occurring in both mines corresponds to the same sedimentation sequence (and, more precisely, within the granulometric differentiation 80 to 300 μm in the sequence).

It is interesting to note that there are similar öolitic iron deposits presently in the course of being formed in the sea at the estuaries of the Meuse (North Sea) and the Senegal (West African coast) rivers.

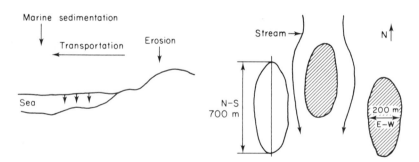

FIG. IV.26. Deltaic marine sedimentation.

The grey seam at Sancy

The Sancy mine is a third geographically distinct mine in which a different seam (the grey seam) is mined. The same structural analysis was carried out on the horizontal regionalization of the iron grades.

The experimental semi-variograms for the four directions E–W, N–S, NW–SE and NE–SW are shown on Fig. IV.27(a) and (b). The length of the galleries sampled at Sancy is much shorter than at Hayange and Bois

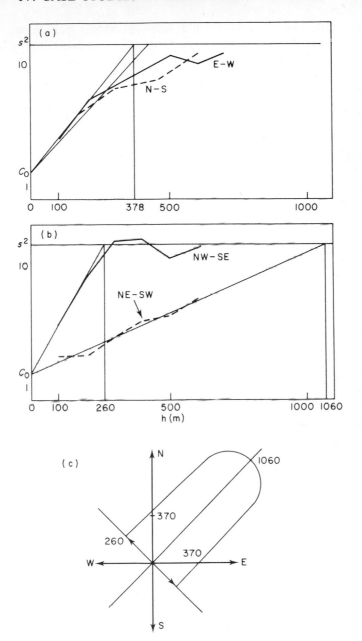

FIG. IV.27. Horizontal regionalization in the grey seam (Sancy). (a) N–S and E–W directions; (b) NW–SE and NE–SW directions; (c) ellipse of geometric anisotropy.

d'Avril, and, thus, the experimental semi-variogram from Sancy will not reveal the hole effect quite so clearly. However; the beginning of the hole effect is quite obvious on the NW–SE semi-variogram which shows the most rapid variability.

The first step in establishing a model of the geometric anisotropy at Sancy was to approximate the sill of the various semi-variograms by the experimental dispersion variance, $s^2 = 11·5$ $(\%\,Fe)^2$ of all the data considered in the calculations. Next, a line representing the average slope of the first few experimental points of each semi-variogram was extrapolated to this sill, providing an abscissa which was transferred to the directional graph on Fig. IV.27(c) (e.g., 260 m for the NW–SE direction). An ellipse of geometric anisotropy can be fitted to this directional graph and shows that, at Sancy and in the grey seam, the sedimentary units seem to be elongated in the NE–SW direction and appreciably longer than those of the other two mines, all facts which are corroborated by various lithological observations.

Case Study 10. Zonal anisotropy, after M. Guarascio and G. Raspa (1974)

The Salafossa zinc deposit is located in the dolomitic Alps (Italy) and consists of an ellipsoidal mass with a main axis in the N–S direction, dipping between 0 and 30° W. The mineralization (blend and galena) consists of dolomitic breccia cement. Three types of ore can be clearly distinguished during mining, each one being related to the dimension of the elements of the host breccia, cf. Fig. IV.28(a) and (b):

(i) type 1 ore in the heart of the deposit, associated with gravelly breccia (elements with dimensions of several centimetres);
(ii) type 2 ore in the upper part of the deposit, associated with fractured dolomitic rock (elements with dimensions of several decimetres);
(iii) type 3 ore at the base of the deposit in karstic cavities consisting of dolomitic sand rich in sulphurous mineralized cement.

Thus, the two upper ores (types 1 and 2) are associated with altered dolomite, within which stratification can still be distinguished, cf. Fig. IV.28(b). The direction D perpendicular to this stratification is vertical in the upper part of the deposit (type 2 ore) and then tends progressively towards a slope of 30° West in the central part of the deposit (type 1 ore). No stratification can be distinguished in the type 3 ore. The following structural analysis by variography will identify these preferential structural directions.

Data available

The data come from fans of drills set in an E–W vertical plane approximately perpendicular to the N–S direction of advance of the main drives, cf. Fig. IV.28(b). The cuttings from the drills were analysed in 1·5-m and 3-m lengths according to the zone. Because of this configuration, no data is available in the N–S direction.

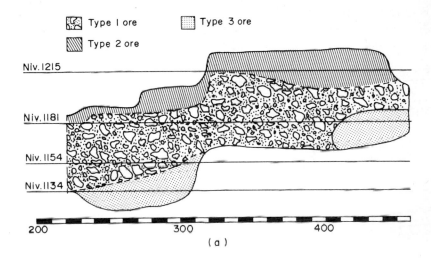

(a)

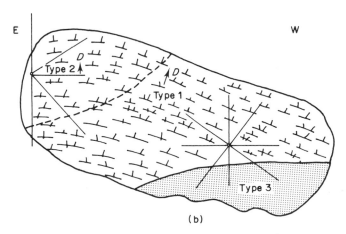

(b)

FIG. IV.28. The three types of ore at Salafossa mine. (a) Vertical N–S cross-section; (b) Vertical E–W cross-section and location of the N–S drives.

Regionalization of the zinc grades

The experimental semi-variograms for the zinc grades were calculated for each ore type and for each of the four main directions on an E–W vertical plane.

The semi-variograms for direction 1 and the two combined directions 3 + 4 are shown on Fig. IV.29. They correspond to the upper type 2 ore for which the data support is 3 m. There is a marked anisotropy: the variability is obviously greater in the vertical direction (1) than in the subhorizontal directions (3 + 4), the horizontal direction corresponding to the mean direction of stratification in the type 2 ore.

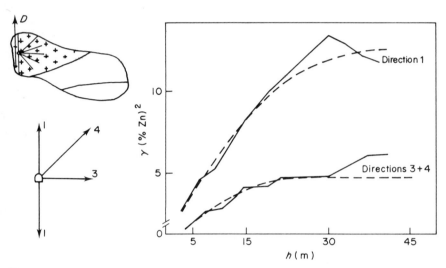

FIG. IV.29. Regionalization in the type 2 ore. ———, Experimental; – – – –, model; support, $l = 3$ m.

The combined semi-variograms for directions 1 + 4 and 2 + 3 are shown on Fig. IV.30. They correspond to the central type 1 ore for which the data support is 1·5 m. In this case, the main variability direction (1 + 4) tends towards directions D (30° W), perpendicular to the mean direction of stratification in type 1 ore.

The semi-variograms for directions 2 and 4 and the combined semi-variogram for directions 4 + 2 + 1 are shown on Fig. IV.31. They correspond to the lower type 3 ore for which the data support is 1·5 m. In this case, no significant directional anisotropy can be distinguished, which confirms the absence of stratification in the type 3 ore (dolomitic sand filling the karstic cavities).

Fitting models to the zonal anisotropy

The three-dimensional point support model adopted for each of the three ore types consists of an isotropic structure with a nugget effect and an anisotropic structure depending only on the direction D of zonality ($D =$ direction 1 for the type 2 ore, and $D = 30°\ W$ for the type 1 ore):

$$\gamma(h) = \{C_0(0) - C_0(r)\} + C_1\gamma_1(r) + C_2\gamma_2(h_D), \qquad \text{(IV.1)}$$

where $r = |h|$ is the modulus of the three-dimensional vector h; h_D is the coordinate of the vector h along the direction D of zonality; $\{C_0(0) - C_0(x)\}$ represents an isotropic nugget effect which appears on the

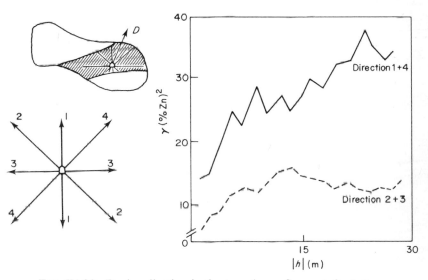

FIG. IV.30. Regionalization in the type 1 ore. Support, $l = 1·5$ m.

regularization of support l as a nugget constant C_{0l}; $C_1\gamma_1(r)$ is an isotropic spherical model with sill value C_1 and range a_1; $C_2\gamma_2(h_D)$ is a spherical model with sill C_2 and range a_2, depending only on the distances h_D in direction D.

The values of the various parameters of this three-dimensional point model for each of the three ore types are given on Fig. IV.32.

Fitting the parameters

The parameters of the point model are derived from the experimental semi-variograms regularized on the support l by the standard procedure of deconvolution and identification, cf. section II.D.4.

For type 2 ore and the combined direction $3+4$ perpendicular to the zonality direction D, Fig. IV.29 gives the following estimators of the parameters, regularized on $l = 3$ m, of the isotropic point spherical model:

$$C_{0l} = 1, \qquad C_{1l} = 3 \cdot 7 \ (\% \ \text{Zn})^2, \qquad a_{1l} = 24 \ \text{m}.$$

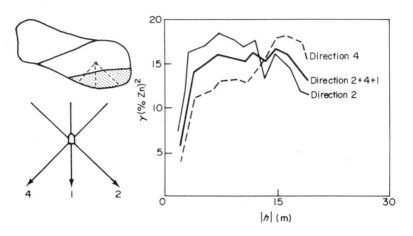

FIG. IV.31. Regionalization in the type 3 ore. Support, $l = 1 \cdot 5$ m.

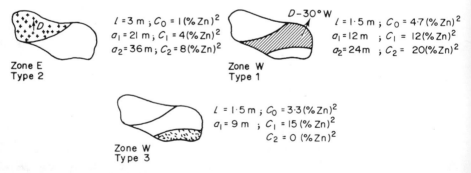

FIG. IV.32. Parameters of the three-dimensional point model.

The parameters for the point model are then

$$a_1 = a_{1l} - 3 = 21 \ \text{m} \quad \text{and} \quad C_1 = C_{1l}/[1 - \bar{\gamma}_1(l, l)] = C_{1l} \times 1 \cdot 075 = 4.$$

The regularized theoretical semi-variogram of this point model for a vector h_0 perpendicular to the direction D of zonality is then written

$$\gamma_l(h_0) = C_{0l} + C_{1l}\gamma_{1l}(h_0), \qquad \forall h_0 \geqslant 3 \ \text{m},$$

$C_{1l}\gamma_{1l}(h_0)$ being a spherical model of sill $C_{1l} = 3\cdot7$ and range $a_{1l} = 24$ m. This regularized model $\gamma_l(h_0)$ is shown on Fig. IV.29 and is a good fit to the experimental semi-variogram for direction $3+4$.

Using the experimental difference $[\gamma_i^*(h_D) - \gamma_l(h_0)]$ between the experimental semi-variogram $\gamma_i^*(h_D)$ in the zonality direction $(D = 1$ on Fig. IV.29) and the preceding isotropic theoretical semi-variogram $\gamma_l(h_0)$, the regularized parameters can be fitted to the anisotropic component of the point model (formula (IV.1)), giving

$$a_{2l} = 39 \text{ m} \quad \text{and} \quad C_{2l} = 7\cdot7 \ (\% \text{ Zn})^2.$$

The point model parameters are then

$$a_2 = a_{2l} - 3 = 36 \text{ m}$$

and

$$C_2 = C_{2l}/[1 - \bar{\gamma}_2(l,l)]$$
$$= C_{2l} \times 1\cdot04 = 8.$$

Finally, the regularized model $\gamma_l(h_D)$ in the direction D of zonality is written as

$$\gamma_l(h_D) = C_{0l} + C_{1l}\gamma_{1l}(h_D) + C_{2l}\gamma_{2l}(h_D), \qquad \forall h_D \geqslant 3 \text{ m},$$

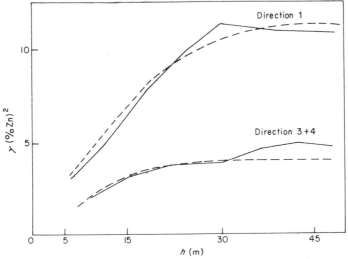

FIG. IV.33. Regularization of support $2l = 6$ m in type 2 ore.

$C_{2l}\gamma_{2l}(h_D)$ being a spherical model with sill C_{2l} and range a_{2l}. This regularized model $\gamma_l(h_D)$ is shown on Fig. IV.29 and is a good fit to the experimental semi-variogram for direction 1.

Verification The expressions for the theoretical semi-variograms regularized on the double support $2l = 6$ m, $\gamma_{2l}(h)$, were calculated from the previous point model (formula (IV.1)). These theoretical curves were then compared with the corresponding experimental semi-variograms (obtained from the data of support $2l$) on Fig. IV.33, for the type 2 ore. The verification is positive.

IV.E. VARIOUS STRUCTURAL ANALYSES

Case Study 11. Structural analysis of El Teniente Cu deposit after A. Journel and R. Segovia (1974)

The El Teniente copper mine in Chile has a daily production of 65 000 tons of crude ore assaying from 1·5 to 1·7% total copper and ranks among the largest copper mines in the world.

Two mineralizations can essentially be distinguished:

  (i) an intensive so-called "secondary" mineralization with high grades of around 2% in the upper levels within a very fractured andesite favourable to the flow of solutions.
  (ii) a so-called "primary" mineralization with lower grades of about 1% in the lower levels within a much clearer andesite.

On a given horizontal level, the mineralization shows a radial drift around a vertical volcanic chimney: the copper grades decrease as the distance from the chimney and its aureole of fracturation increases, cf. Fig. IV.34.

The underground mining is done by standard block-caving. The secondary ore, the degree of fracturation of which allows excellent recovery, is being exhausted, and, thus, the present study concentrates on primary ore only.

Available data

Two sampling campaigns are available:

  (i) systematic horizontal channel sampling along the drives and cross-cuts. The channel samples have been cut into 10-ft lengths and analysed.

(ii) drills are collared in each drive and developed in fans. The directions are variable but the cores have been analysed in constant 10-ft lengths.

The channel sample and core sample supports can be mixed with good approximation and statistical studies have shown that there is no significant bias between the two data sets.

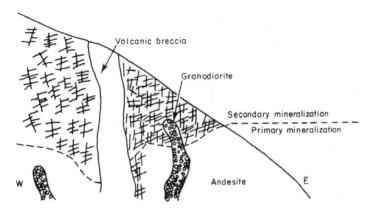

FIG. IV.34. E–W cross-section of El Teniente deposit.

Object of the study

Each mining block (60 m × 60 m × 180 m) is evaluated by the channel samples of the drives above and below the block and by the drill cores through the mass of the block. The object is to define the optimal proportion of each type of sample and the optimal configuration of the drills.

Regionalization of the channel sample grades

Six levels of drives and cross-cuts were considered within the primary mineralization. The N–S direction of the cross-cuts is distinguished from the E–W direction of the drives.

By way of example, the E–W and N–S semi-variograms of the 10-ft channel samples on level T4 are shown on Fig. IV.35. The variability of the grades in the E–W direction is obviously greater than that in the N–S direction; in fact, at the T4 level studied here, the E–W direction is a radial direction with regard to the chimney and therefore takes the previously mentioned drift of the grades into account. A proportional effect is also present in the micro-structures: the experimental nugget constant C_0^*

increases with the mean m^* of the data from which the semi-variogram was calculated.

All the data of the channel samples from six levels were combined to provide the two mean semi-variograms for the N–S and E–W directions

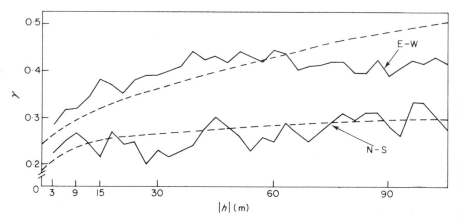

FIG. IV.35. Channel-sample semi-variograms (at level T4). E–W: $N = 1619$, $m = 1\cdot66$, $C_0^* = 0\cdot240$. N–S: $N = 207$, $m = 1\cdot51$, $C_0^* = 0\cdot190$.

shown on Fig. IV.36. A considerable number of data were used: 3360 for the E–W direction and 2730 for the N–S direction. Again, the radial drift effect on the E–W semi-variogram and the proportional effect on the

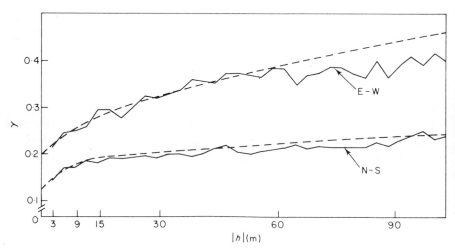

FIG. IV.36. Channel-sample mean-variograms (at six levels). E–W: $N = 3360$, $m = 1\cdot53$, $C_0^* = 0\cdot200$. N–S: 2730, $m = 1\cdot51$, $C_0^* = 0\cdot128$.

nugget constants are observed. On the N–S curve, there is a change of slope around 10 m and the curve stabilizes around 100 m, which suggests the fitting of two nested structures with 10-m and 100-m ranges, respectively.

Regionalization of the drill core grades

As the inclinations of the drill cores are very variable, they have been grouped into angle classes, cf. section III.C.4 Fig. III.23(b). Five classes, corresponding to the following five mean directions, were considered: α_1, vertical direction; α_2, N–S horizontal; α_3, E–W inclined at $-45°$; α_4, E–W inclined at $+45°$; α_5, E–W horizontal.

The mean semi-variograms for each angle class are shown on Fig. IV.37(a) to (e). The following features can be discerned:

(i) a very significant radial drift effect in the horizontal E–W direction, and a less significant one in the E–W directions inclined at $+45°$ and $-45°$;

(ii) a change in slope around 10 m;

(iii) the curves for directions α_1 and α_2, which are unaffected by the radial drift, stabilize around 100 m;

(iv) a proportional effect on the nugget constant values.

Fitting a three-dimensional model

If all the preceding semi-variograms not affected by the E–W drift are superimposed, they will coincide except for the nugget constant. This means that the two nested structures with 10-m and 100-m ranges are isotropic. First of all, these two structures were fitted to the best-known experimental semi-variogram, i.e., that for the channel samples in the N–S direction, as shown on Fig. IV.36. The adopted model is

$$\gamma(h) = C_0 + C_1\gamma_1(r) + C_2\gamma_2(r), \qquad \forall r = |h| > 10 \text{ ft}, \qquad \text{(IV.2)}$$

where $C_0 = 0{\cdot}128$, $C_1\gamma_1(r)$ is a spherical model with sill $C_1 = 0{\cdot}045$ and range $a_1 = 9$ m, $C_2\gamma_2(r)$ is a second spherical model with sill $C_2 = 0{\cdot}065$ (% Cu)^2 and $a_2 = 105$ m.

To determine the experimental nugget constant C_0^* for each of the semi-variograms, a transparency of the preceding model was moved over each of the semi-variograms along the ordinate axis until the best fit was achieved. Table IV.2 shows the increase of this experimental nugget constant C_0^* as a function of the experimental mean m^* of the data used in the corresponding semi-variogram calculation (direct proportional effect).

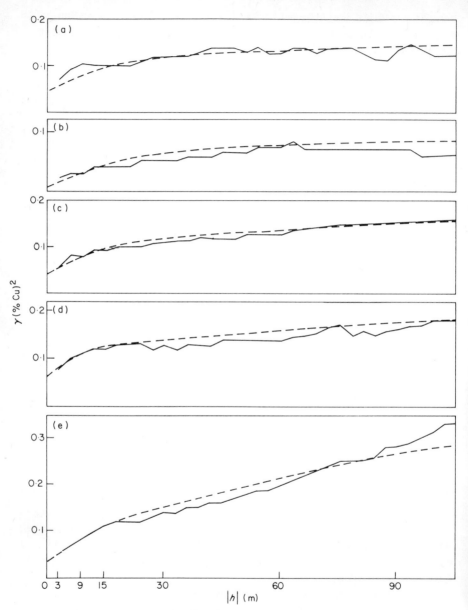

FIG. IV.37. Core sample directional semi-variograms. (a) Vertical: $N = 694$, $m = 0.95$, $C_0^* = 0.043$. (b) N–S horizontal: $N = 543$, $m = 0.65$, $C_0^* = 0.004$. (c) E–W minus 45°: $N = 1588$, $m = 0.94$, $C_0^* = 0.035$. (d) E–W plus 45°: $N = 1331$, $m = 1.12$, $C_0^* = 0.063$. (e) E–W horizontal: $N = 1199$, $m = 0.$ $C_0^* = 0.030$.

TABLE IV.2. *Proportional effect observed on the nugget constant*

	Direction								
	Drills					Channel samples			
	Vertical	Horizontal N-S	-45° E-W	+45° E-W	Horizontal E-W	T4 N-S	T4 E-W	All levels N-S	All levels E-W
m^*	0·95	0·65	0·94	1·12	0·88	1·51	1·66	1·31	1·53
C_0^*	0·043	0·004	0·035	0·063	0·030	0·19	0·24	0·128	0·20
Fitted parabola C_0^*	0·037	0·004	0·035	0·064	0·026	0·184	0·242	0·118	0·19

These experimental points (C_0^*, m^*) are shown graphically on Fig. IV.38; the parabola $C_0 = 0.18 \, [m - 0.5]^2$ is a remarkably good fit.

To take account of the radial drift in the E–W directions, it is necessary to add an additional anisotropic term to the preceding isotropic model (formula (IV.2)). The deviations from the isotropic model of the E–W

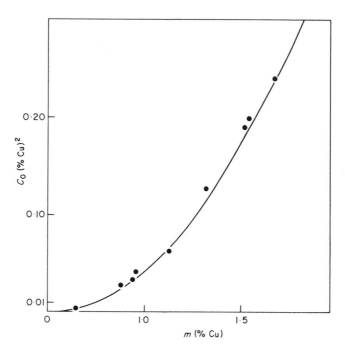

FIG IV.38. Parabola of proportional effect. ●, C_0^*, experimental;. ———, $C_0 = 0.18[m - 0.5]^2$, model.

horizontal semi-variograms are shown on Fig. IV.39. A linear semi-variogram can be fitted to these deviations, at least for small distances ($h_u <$ 60 m):

$$\gamma_3(h_u) = C_3 h_u \quad \text{with} \quad C_3 = a_u = 14 + 10^{-4} \, (\% \text{ Cu})^2 \, \text{m}^{-1}.$$

The drift effect, only slightly evident in the $\pm 45°$ E–W directions, is no longer visible in the N–S and vertical directions. This calls for a three-dimensional model which cancels out the drift in all directions except the E–W. An ellipsoidal model with an anisotropy ratio of $a_u/a_v = 20$ was adopted for the slope C_3, cf. Fig. IV.40. The slope C_3 is equal to the radius vector ρ of the ellipsoid and is given in terms of spherical coordinates (ρ, α, φ) as

$$C_3 = \frac{a_u^2 \cdot a_v^2}{a_v^2 \sin^2 \varphi \cos^2 \varphi + a_u^2 [\sin^2 \varphi \sin^2 \alpha + \cos^2 \varphi]}$$

$$= \frac{196 \times 10^{-8}}{\sin^2 \varphi \cos^2 \alpha + 400 [\sin^2 \varphi \sin^2 \alpha + \cos^2 \varphi]} \quad (\% \text{ Cu})^2 \text{ m}^{-1},$$

from which it follows that

for the E–W horizontal direction, $C_3 = 14 \times 10^{-4} = a_u$;
for the $\pm 45°$ E–W directions, $\varphi = \pi/4$, $\alpha = 0$ and $C_3 = 0 \cdot 07 a_u$;
for the N–S and vertical direction, $C_3 = a_v = 0 \cdot 05 a_u \approx 0$.

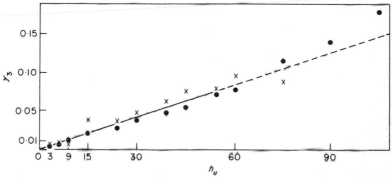

FIG. IV.39. E–W horizontal deviations from the isotropic model. ●, Core samples;
×, channel samples.

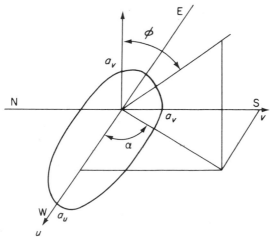

FIG. IV.40. Ellipsoidal model for the radial E–W drift.

To sum up, the three-dimensional model adopted to represent the regionalization of the Cu grades of 10-ft samples is:

$$\gamma(h) = 0 \cdot 18[m - 0 \cdot 5]^2 + 0 \cdot 045 \gamma_1(r) + 0 \cdot 065 \gamma_2(r) + C_3(\alpha, \varphi)r,$$

where $m > 0 \cdot 5\%$ Cu is the mean grade of the zone considered.

γ_1 and γ_2 are spherical models with ranges $a_1 = 9$ m and $a_2 = 105$ m, the three-dimensional vector h being defined by its spherical coordinates (r, α, φ). This model is valid for $r = |h| < 150$ m and for horizontal E–W distances less than 60 m.

This model proves to be an excellent fit to all the available experimental figures, cf. the dashed curves on Figs. IV.35 to IV.37.

Case Study 12. Structural analysis of Exotica Cu deposit, after Ch. Huijbregts and R. Segovia (1973)

The mineralization of the Exotica copper deposit in Chile is generally supposed to be of detritic sedimentary origin, consisting of material coming from the neighbouring massive Chuquicamata porphyry-copper deposit. The deposit consists of a large stratiform slab with an average plunge of 10°. The total surface area is approximately 1 km × 2 km, with a thickness varying between 50 and 100 m.

Available data

The results from three types of sampling campaigns are available.

 (i) The entire deposit was sampled by vertical diamond drilling (DDH) on a pseudo-regular 100 m × 100 m horizontal grid. The diamond drill-holes were cut into constant lengths $l = 13$ m and analysed for copper. The 13-m core sample length corresponds to the height of mining benches.

 (ii) The primary production zones were sampled by vertical diamond drilling (RDH) on a regular 50 m × 50 m horizontal grid. These drill-holes were also cut into 13-m lengths.

 (iii) The production benches were sampled by the same rotary diamond drills used for blast-holes. The drill cuts coming from these blast-holes (BH) were approximately 14 m long. The blast-holes are located on an irregular horizontal grid with constant density and a mean 8-m distance between two BH. At the time this study was carried out, the ore production had just begun and the information from the BH was very limited.

All samples were analysed for total copper (Cu T) and soluble copper (Cu S).

The three types of samples (DDH, RDH and BH) were studied separately. The vertical regionalization can only be studied through the DDH, which are the only sampled long enough. On the horizontal plane (by grading over the constant 13-m thickness of a horizontal bench), these DDH provide data over large horizontal distances (h ⩾ 100 m.) The RDH provide a comparable horizontal information, i.e., graded over 13 m, for horizontal distances (h ⩾ 50 m). The BH provide horizontal information graded over 14 m (14 ≃ 13), but at smaller distances ($h \geqslant 8$ m).

Horizontal variograms graded over 13 m

DDH samples. The horizontal semi-variograms of the Cu T and Cu S were calculated bench by bench over 14 benches and in three directions (the two directions of the grid and one diagonal direction). The number of data pairs used in the calculations were too few for the bench variograms to be significant. However, a strong proportional effect can be seen on the nugget effect: the nugget constant increases with the mean grade of the data from the bench considered.

For each of the three directions considered, the 14 bench semi-variograms were combined into a mean directional semi-variogram, cf. Fig. IV.41. As the three directional semi-variograms proved to be isotropic, they were combined into a single mean isotropic semi-variogram, cf. Fig. IV.41. For both Cu T and Cu S, the mean isotropic semi-variogram shows a macro-structure with a range $a_2 \simeq 200$ m and a very strong nugget effect engulfing all the structure with ranges less than the horizontal scale of observation of the DDH (100 m).

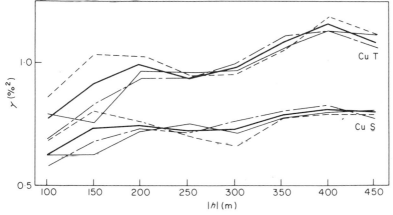

FIG. IV.41. Horizontal regionalization of DDH samples (graded over 13 m).
————, — – —, – – – – –, Directional; ————, Mean isotropic.

The two other groups of samples – RDH and BH – were used to evaluate these small-range structures. Care was taken to choose benches and zones for which the Cu T and Cu S grades were similar to the corresponding DDH grades (1·1% Cu T and 0·8% Cu S). This precaution cancelled the influence of the proportional effect.

RDH and BH samples. The mean isotropic semi-variograms obtained from the RDH samples (on a regular 50-m grid) and the BH samples (grouped into classes because of the irregular grid) are shown in Fig. VI.42. There is a remarkable agreement between the structures revealed by the three types of samples. The long-distance structures revealed on the RDH and DDH semi-variograms are almost exact extensions of the small-distance ($h < 50$ m) structures revealed by the BH samples. Conversely, the BH semi-variogram explains part of the strong nugget effect observed on the RDH and DDH semi-variograms.

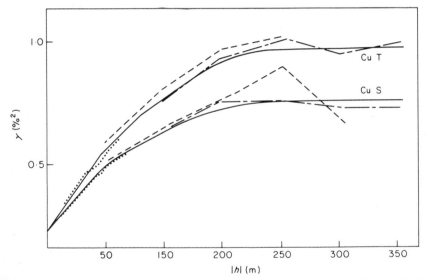

FIG. IV.42. Horizontal regionalization of the three types of samples (DDH, RDH, BH). ————, Theoretical model graded over 13 m; – – – – –, RDH; — – —, DDH; ·········, BH.

Horizontal model fitted

The following model of the horizontal, isotropic, *point* support regionalization has been considered:

$$\gamma(h) = [C_0(0) - C_0(r)] + C_1\gamma_1(r) + C_2\gamma_2(r), \qquad \forall \text{ horizontal vector } h \text{ with}$$
$$\text{modulus } r \in [0, 400 \text{ m}]. \qquad\qquad (IV.3)$$

$\{C_0(0) - C_0(r)\}$ represents the nugget effect which appears on the grading over the $l = 13$-m support as the nugget constant C_{0l}, cf. formula (III.7). $C_1\gamma_1(r)$ and $C_2\gamma_2(r)$ are two spherical models with respective sills C_1 and C_2 and ranges $a_1 = 45$ m $a_2 = 200$ m.

The theoretical model of the semi-variogram graded over the constant 13-m thickness is obtained from the point model by applying the practical rules of section II.D.4, and is given by

$$\gamma_G(h) = C_{0l} + C_{1l}\gamma_1(r) + C_{2l}\gamma_2(r), \qquad \forall \text{ horizontal vector with modulus } r \in$$
$$[13, 400 \text{ m}], \qquad\qquad\qquad (IV.4)$$

where the graded sills are written as

$$C_{1l} = C_1[1 - \bar{\gamma}_1(l, l)] \quad \text{and} \quad C_{2l} = C_2[1 - \bar{\gamma}_2(l, l)].$$

The graded parameters C_{0l}, C_{1l}, C_{2l} are fitted to the experimental graded semi-variograms. By way of example, the fitted parameters for total copper are

$$C_{0l} = 0.22, \qquad C_{1l} = 0.10, \qquad C_{2l} = 0.64 \ (\% \text{ Cu T})^2$$

and, thus, the parameters for the point model (IV.3) are

$$C_1 = C_{1l} \times 1.16 = 0.12 \quad \text{and} \quad C_2 = C_{2l} \times 1.03 = 0.66 \ (\% \text{ Cu T})^2.$$

The graded model (IV.4) thus considered is shown on Fig. IV.42 and is a good fit to the experimental curves.

Vertical variograms regularized over 13 m

Only the DDH have a vertical extension long enough for the construction of vertical variograms regularized over core samples of 13 m length. The mean vertical semi-variogram calculated from the set of DDH samples is shown on Fig. IV.43, along with the preceding 13-m graded model. There is a marked anisotropy between the vertical and the horizontal directions: the vertical variability is clearly greater than the horizontal variability.

Three-dimensional model fitted

A three-dimensional model can be fitted to the 13-m regularized vertical semi-variogram by adding to the 13-m graded model (IV.4) an additional variability $C_3\gamma_3(h_w)$ which depends only on the vertical coordinate h_w of the vector $h = (h_u, h_v, h_w)$. The following point support three-dimensional model is then obtained:

$$\gamma(h) = [C_0(0) - C_0(r)] + C_1\gamma_1(r) + C_2\gamma_2(r) + C_3\gamma_3(h_w), \qquad \forall r = |h|. \quad (IV.5)$$

This three-dimensional model is the sum of an isotropic component and an anisotropic, specifically vertical component $C_3\gamma_3(h_w)$. For every horizontal vector h (such that $h_w = 0$), this three-dimensional model reduces to the preceding horizontal point model (IV.3).

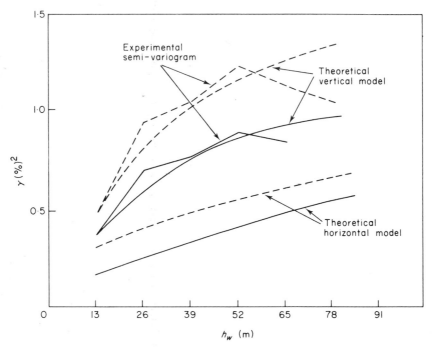

FIG. IV.43. Vertical regionalization of the DDH samples. ————, Cu S; – – – –, Cu T.

For a vertical vector h_w (such that $h_u = h_v = 0$), the following point model is obtained:

$$\gamma(h_w) = [C_0(0) - C_0(h_w)] + C_1\gamma_1(h_w) + C_2\gamma_2(h_w) + C_3\gamma_3(h_w),$$

and the theoretical model obtained from regularization by core samples of length $l = 13$ m is, thus,

$$\gamma_l(h_w) = C_{0l} + C_{1l}\gamma_1(h_w) + C_{2l}\gamma_2(h_w) + C_{3l}\gamma_3(h_w). \qquad (IV.6)$$

Regularization over $l = 13$ m reduces the isotropic component to the preceding theoretical 13-m graded model $\gamma_G(h_w)$. An estimation of the vertical anisotropic component $C_{3l}\gamma_3(h_w)$ regularized on l, is then given by the experimental difference $[\gamma_l^*(h_w) - \gamma_G(h_w)]$ between the experimental vertical semi-variogram and the theoretical 13-m graded semi-variogram, cf. Fig. IV.43.

A spherical model, $C_{3l}\gamma_3(h_w)$, with range $a_{3l} = 65$ m and sill $C_{3l} = 0\cdot61$ (% Cu T)^2, without nugget effect, was fitted to this experimental difference $[\gamma_i^*(h_w) - \gamma_G(h_w)]$. The underlying *point* model $C_3\gamma_3(h_w)$ is, thus, spherical without nugget effect, with range

$$a_3 = a_{3l} - l = 65 - 13 = 52 \text{ m},$$

cf. section II.D.2, and sill

$$C_3 = C_{3l}/[1 - \bar{\gamma}_3(l, l)] = C_{3l} \times 1\cdot14 = 0\cdot70 \text{ (\% Cu T)}^2.$$

The regularized model (IV.6) thus considered is shown on Fig. IV.43 and is a good fit to the experimental vertical semi-variogram.

The three-dimensional point model adopted for the Cu S grades is completely similar, with the same vertical anisotropy, the same values for the nested ranges a_1, a_2, a_3 and the same nugget constant C_{0l}, but with different values for the sills C_1, C_2, C_3.

Remark.   The values of the previously fitted parameters C_{0l}, C_1, C_2, C_3 correspond to the well-determined mean grades (1·1% Cu T and 0·8% Cu S), which are equal to the mean grades of the deposit as determined by the DDH. These parameter values thus define an average model valid for a global study of the deposit. For local estimations on quasi-stationary zones or benches, which have different mean grades, it is advisable to take the proportional effect into account (this will not be presented here).

Case Study 13. Structural analysis of the Mea nickel deposit, after J. Deraisme (1976)

The nickel silicate deposit of Mea (New Caledonia) consists of altered basic rocks. The following three components can be schematically distinguished on the vertical, cf. Fig. IV.44.

(i) A seam of yellow laterites with low, uneconomic nickel grades.

(ii) A seam of silicate ore consisting of various facies of altered basic rocks (dunites, harzburgites). These facies are more or less nickel rich. Unaltered inclusions of variable sizes are found within this silicate ore and consist of hard blocks called pure waste, that can be distinguished both when drilling and mining (selective crushing and selection at the face by the miner himself).

(iii) A karstic bed-rock consisting of peridotites.

The deposit was evaluated on a pseudo-regular, square ($80\,\mathrm{m} \times 80\,\mathrm{m}$) grid of vertical drill-holes. These holes were used to determine the hanging and foot walls of the silicate seam. Kriging was used to map the isobaths of the hanging wall and foot wall as well as the isopacks of the silicate thickness, cf. section II.B.3, Case Study 2.

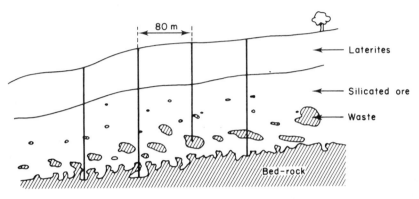

FIG. IV.44. Vertical cross-section of the Mea deposit.

The drill cores were cut into constant 1-m lengths within the silicate seam. The pure waste was removed from the softer, altered material with a geologist's pick and the lengths of the two separated materials were measured and analysed for density and grades (of dry ore) cf. Fig. IV.45.

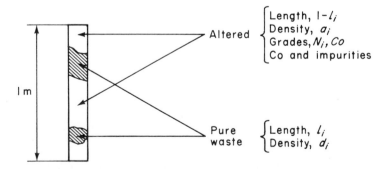

FIG. IV.45. Ore and waste materials separated on a core sample.

Besides mapping of the silicate seam, the study consisted of a local estimation on panels and a global estimation on the deposit of the four quantities

tonnage of pure waste, T_W;

tonnage of altered ore, T_A;

tonnage of nickel metal, Q_A;

mean nickel grade, $g_A = Q_A/T_A$.

The three additive regionalized variables regularized on the 1-m core sample support that were used in the study, are

the accumulation of waste, $z_W(x_i) = l_i \times d_i$;

the accumulation of ore, $z_T(x_i) = (1 - l_i) \times a_i$;

the accumulation of metal, $z_Q(x_i) = (1 - l_i) \times a_i \times g_i$;

where l_i is the length of pure waste measured on each core sample, d_i is the density of this pure waste, a_i is the density of the altered ore (of length $1 - l_i$) and g_i is the corresponding Ni grades of this altered ore, cf. Fig. IV.45.

The preceding tonnages are obtained by multiplying these three regionalized variables by the volume in cubic metres.

The following regionalizations were studied, using the data available from the vertical drill cores cut into $l = 1$-m lengths:

(i) the vertical regionalizations through the semi-variograms $\gamma_l(h)$ regularized on l, cf. Fig. IV.46(a);

(ii) the horizontal regionalizations through the semi-variograms $\gamma_G(h)$ graded over the constant thickness $3l = 3$ m of a horizontal mining bench, cf. Fig. IV.46(b).

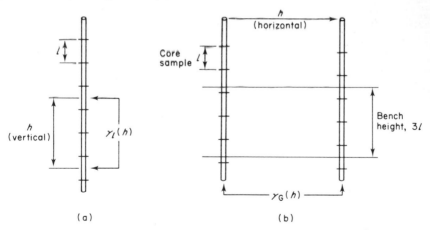

(a)   (b)

FIG. IV.46. Available structural information. (a) Regularization over l; (b) grading over $3l$.

Vertical regionalization

The 1-m and 3-m regularized vertical semi-variograms, $\gamma_l(h)$ and $\gamma_{3l}(h)$ for each of the above three variables, were calculated from the 167 vertical drill-holes and are shown on Fig. IV.47(a), (b), (c).

The semi-variograms of the waste and ore accumulations are remarkably similar, and a transitive model with a range $a_{1l} = 4$ m (regularized on $l = 1$ m) was fitted. The semi-variogram for the metal accumulation shows a net increase after reaching the first range, which suggests a second nested structure with a range around $a_{2l} \simeq 15$ m.

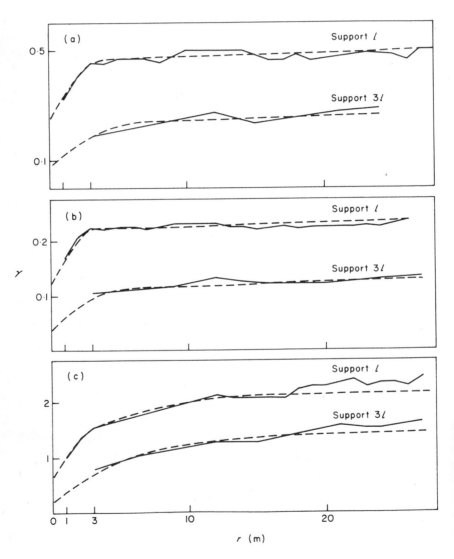

FIG. IV.47. Regularized vertical semi-variograms. (a) Waste accumulation Z_W; (b) Ore accumulation Z_T; (c) metal accumulation Z_Q.

Horizontal regionalization

The three accumulation variables graded over a constant thickness of $3l = 3$ m were defined by the intersections of the vertical drill-holes with each of the 3 m thick horizontal mining benches, cf. Fig. IV.46(b). The graded variables were defined by weighting the analyses of the 1-m core lengths by the intersection of these lengths with the bench, according to formula (III.45).

Using these variables graded over $3l$, the graded experimental semi-variograms were calculated, bench by bench, and for various horizontal directions. The bench variograms are not significant, due to the small number of data pairs used in the calculations, and they were combined into mean directional variograms for all benches. These mean directional variograms show no significant anisotropy and, thus, were combined into a single mean isotropic, horizontal variogram valid for all benches. These three mean isotropic semi-variograms, graded over $3l$, are shown on Fig. IV.48(a), (b), (c).

On each of these experimental curves, there is a horizontal macro-structure with a range $a_3 \simeq 180$ m and a very prominent nugget effect with an amplitude approximately equal to the sills of variability of the vertical $3l$ regularized experimental semi-variograms $\gamma_{3l}(h)$, cf. Fig. IV.47.

Thus, a three-dimensional model can be established by adding a macro-structure with a range $a_3 \simeq 180$ m to the structure revealed by the vertical regionalization.

Fitting a three-dimensional model

The theoretical, point support, three-dimensional model adopted for each of the three accumulation variables considered, is of the type

$$\gamma(h) = [C_0(0) - C_0(r)] + C_1 \gamma_1(r) + C_2 \gamma_2(r) + C_3 \gamma_3(r), \qquad (IV.7)$$

where $r = |h|$, $[C_0(0) - C_0(r)]$ represents the nugget effect that appears on the l and $3l$ support regularizations as nugget constants $C_{0,l}$ and $C_{0,3l}$. $C_1 \gamma_1(r)$, $C_2 \gamma_2(r)$ and $C_3 \gamma_3(r)$ are isotropic spherical models with respective sills C_1, C_2, C_3 and ranges $a_1 = 3$ m, $a_2 = 14$ m, $a_3 = 180$ m.

This model was fitted in two stages.

(i) The parameters of the small-distance structures, i.e., the nugget effect and the first two structures $C_1 \gamma_1$ and $C_2 \gamma_2$ were fitted using the vertical regularized semi-variograms. The influence of the macro-structure $C_3 \gamma_3$ can be neglected for these small distances ($r \leqslant 25$ m).

(ii) The parameters of the macro-structure $C_3 \gamma_3$ are estimated from the horizontal graded semi-variograms. Over large distances ($r \geqslant 80$ m), the short-distance structures (nugget effect $+ C_1 \gamma_1 + C_2 \gamma_2$) show up as a single nugget constant equal to

$$C'_{0,3l} = C_{0,3l} + C_{1,3l} + C_{2,3l}$$

(the grading support being $3l$). The semi-variogram graded over $3l$ can then be written

$$\gamma_G(h) = C'_{0,3l} + C_{3,3l} \gamma_3(r), \qquad \forall r = |h| \geqslant 80 \text{ m.}$$

The nugget constant $C'_{0,3l}$ is obtained from the preceding study of the short-distance structures, while the parameters $C_{3,3l}$ and a_3 are estimated from the experimental semi-variogram graded over $3l$.

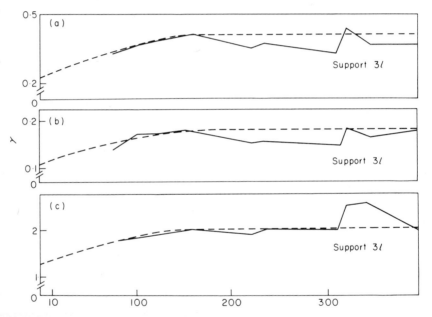

FIG. IV.48. Graded horizontal semi-variograms. (a) Waste accumulation Z_W; (b) ore accumulation Z_T; (c) metal accumulation Z_Q.

Remark   Strictly speaking, the theoretical model $C_{3,3l} \gamma_3(r)$ of the macro-structure, estimated from only the horizontal graded semi-variograms, describes only large-distance ($r \geqslant 80$ m) horizontal regionalizations. This horizontal macro-structure is then extended to three dimensions by adopting an hypothesis of structural isotropy to give the isotropic term $C_3 \gamma_3(r)$ of the model (IV.7).

This hypothesis has little effect on the vertical study as the vertical thickness of the Mea silicate seam never exceeds 35 m and for such small distances the influence of the macro-structure $C_3\gamma_3$ is very weak, if not negligible.

To sum up, nested models of the type (IV.7) have been adopted for the three-dimensional point support regionalizations of the three accumulation variables. The values of the sills, regularized on l and $3l$, are given on Table IV.3. The corresponding theoretical models, regularized by core samples of lengths l and $3l$ and graded over $3l$, agree remarkably well with all the available experimental curves.

TABLE IV.3. *Fitted sill values*
(a) *Support l*

	$C_{0,l}$	$C_{1,l}$	$C_{2,l}$	$C_{3,l}$	$\Sigma\, C_{i,l}$
Waste accumulation	0·24	0·22	0	0·184	0·644
Ore accumulation	0·12	0·105	0	0·07	0·295
Metal accumulation	0·63	0·79	0·61	0·70	2·73

(b) *Support 3l*

	$C_{0,3l}$	$C_{1,3l}$	$C_{2,3l}$	$C_{3,3l}$	$\Sigma\, C_{i,3l}$
Waste accumulation	0·08	0·145	0	0·185	0·41
Ore accumulation	0·04	0·07	0	0·07	0·18
Metal accumulation	0·21	0·52	0·57	0·70	2·0

Logarithmic model or nested spherical model

One drawback of the previous structural models is the large number of estimated parameters, five for the accumulation of waste and ore and seven for the metal accumulation, namely the four sill values C_0, C_1, C_2, C_3 and the three ranges a_1, a_2, a_3.

The roughly geometric progression of the three ranges a_1, a_2, a_3 suggests a logarithmic structural model for the metal accumulation, cf. section III.B.2, the subsection entitled "Logarithmic model or nested models?" The advantage of such a logarithmic model is that it depends on only two parameters rather than the seven parameters of the nested spherical model.

The logarithmic model with nugget effect is expressed as

$$\gamma(h) = [C_0(0) - C_0(r)] + C \log r,$$

for the point support;

$$\gamma_{3l}(h) = C_{0,3l} + C\left[\log\frac{r}{3l} + \frac{3}{2}\right],$$

for the $3l$ support; cf. section II.D.2, regularization of the logarithmic model.

For the two supports l and $3l$ considered here, these models become

$$\gamma_l(h) = [C_{0,l} + 1\cdot5C] + C\log\frac{r}{l}, \qquad \forall r \geqslant l;$$

$$\gamma_{3l}(h) = [C_{0,3l} + C(1\cdot5\text{-}\log 3)] + C\log\frac{r}{l}, \qquad \forall r \geqslant 3l;$$

thus

$$\gamma_l(h) - \gamma_{3l}(h) = C_{0,l} - C_{0,3l} + C\log 3$$

$$= 2C_{0,3l} + C\log 3.$$

Consequently, if the logarithmic model is adopted, the experimental semi-variograms for metal accumulation plotted on logarithmic paper would be approximated by straight lines with slope C. The two lines corresponding to the two supports l and $3l$ should differ by a translation of modulus $2C_{0,3l} + C\log 3$.

On Fig. IV.49, the two parameters fitted to the experimental curve regularized on $3l$ are $C_{0,3l} = 0\cdot21$ and $C = 0\cdot36$.

Over small distances ($r \leqslant 25$ m), there is a very good agreement between this logarithmic model and the experimental curves. On the other hand, the logarithmic model slightly overestimates the variabilities at large distances

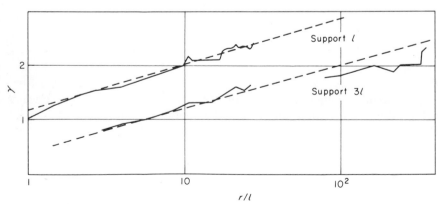

FIG. IV.49 Fitting a logarithmic model to the regionalization of metal accumulation.

$(r \geqslant 80 \text{ m})$. The nested structure of spherical models (IV.7) is preferred in practice, essentially for reasons of homogeneity with the models chosen for the other two variables used, accumulations of waste and ore.

IV.F. GRADE CONTROL DURING PRODUCTION

The two following case studies show how the characterization of a spatial variability through the variogram model allows the prediction and, thus, the control of the production grades. These two case studies can also be considered as examples of practical applications of the two geostatistical variances: the estimation and the dispersion variances.

Case Study 14. Prediction of the grade of planned production from several stopes

Consider a deposit the daily production of which comes from N blocks $\{B_i, i = 1 \text{ to } N\}$ of same size V. The true mean grade θ_i of the block B_i is estimated without bias by the estimator θ_i^* with estimation variance $\sigma_{e_i}^2 = E\{[\theta_i - \theta_i^*]^2\}$.

For a day j, the total production is programmed from N mined blocks. Each block B_i provides p_{ij} tonnes with a true mean grade t_{ij} estimated by the global estimator θ_i^* of block B_i, cf. Fig. IV.50. The elementary

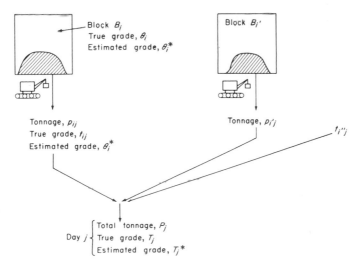

FIG. IV.50. Production planned from several blocks.

tonnages p_{ij} are calculated so that the predicted mean grade T_j^* of the total production P_j of the day j satisfies the sales or milling constraints. Thus, for a day j, the following relations hold:

total tonnage produced, $P_j = \sum_{i=1}^N p_{ij}$;

predicted mean grade, $T_j^* = \sum_i p_{ij}\theta_i^*/P_j$;

true mean grade actually recovered, $T_j = \sum_i p_{ij} \cdot t_{ij}/P_j$.

The object of the study is to control the difference between the predicted grade T_j^* and the true grade actually recovered. This is done by means of the estimation variance:

$$\sigma_{E_j}^2 = E\{[T_j - T_j^*]^2\}.$$

Suppose that the N blocks B_i are within the same stationary regionalization characterized by the grade expectation m. The various grades θ_i^*, θ_i, t_{ij} thus have the same expectation m and the non-bias for the prediction of the daily grade is verified:

$$E\{T_j - T_j^*\} = \sum_i p_{ij}E\{t_{ij} - \theta_i^*\}/P_j = 0.$$

The estimation variance thus becomes

$$\sigma_{E_j}^2 = \sum_i \sum_{i'} p_{ij} \cdot p_{i'j} \cdot E\{(t_{ij} - \theta_i^*)(t_{i'j} - \theta_{i'}^*)\}/P_j^2.$$

If the distance is sufficient between the N simultaneously mined blocks, then the two errors, $(t_{ij} - \theta_i^*)$ and $(t_{i'j} - \theta_{i'}^*)$, for $i \neq i'$, can be considered as independent and, thus,

$$\sigma_{E_j}^2 = \sum_i p_{ij}^2 \cdot E\{[t_{ij} - \theta_i^*]^2\}/P_j^2.$$

The elementary error can be written as

$$t_{ij} - \theta_i^* = (t_{ij} - \theta_i) + (\theta_i - \theta_i^*)$$

and, thus,

$$\sigma_{E_j}^2 = \frac{1}{P_j^2} \sum_i p_{ij}^2 [E\{(t_{ij} - \theta_i)^2\} + E\{(\theta_i - \theta_i^*)^2\} + 2E\{(t_{ij} - \theta_i)(\theta_i - \theta_i^*)\}].$$

The fluctuation $(t_{ij} - \theta_i)$ of the elementary grades t_{ij} within the block B_i can be assumed to be independent of the estimation error $(\theta_i - \theta_i^*)$ of this block, and, thus,

$$E\{(t_{ij} - \theta_i)(\theta_i - \theta_i^*)\} = 0.$$

If the exact location of the tonnage p_{ij} in block B_i were known, the extension variance of the grade t_{ij} of this tonnage p_{ij} to the grade θ_i of block B_i, i.e., the term $E\{(t_{ij} - \theta_i)^2\}$, could be calculated. In practice, it is sufficient to consider the mean value of this term $E\{(t_{ij} - \theta_i)^2\}$ when the tonnage p_{ij} is taken anywhere within block B_i, i.e., the dispersion variance of an average production unit p_i in block B_i:

$$\frac{1}{V} \int_{B_i} E\{(t_{ij} - \theta_i)^2\} \, dx \simeq D^2(p_i/B_i),$$

cf. formula (II.33).

If the block B_i is mined in k days, then the average production unit p_i can be expressed in terms of tonnage as

$$p_i = \frac{1}{k} \sum_{j=1}^{k} p_{ij}.$$

As for the term $E\{(\theta_i - \theta_i^*)^2\}$, it is, by definition, the estimation variance $\sigma_{e_i}^2$ of the mean grade of block B_i. If this block has been estimated by kriging, then $\sigma_{e_i}^2$ is simply the kriging variance of the block given by the kriging system of equations.

Finally, the estimation variance of the mean grade T_j of the total production P_j of day j is written

$$\sigma_{E_j}^2 = \frac{1}{P_j^2} \sum_{i=1}^{N} p_{ij}^2 [D^2(p_i/B_i) + \sigma_{e_i}^2]. \qquad \text{(IV.8)}$$

This variance thus depends not only on the dispersion of the grades of the production units p_i in each block B_i but also on the quality of the estimators of the mean grades of these blocks B_i.

Remark 1   If the programming of the mining for a day j is based on the estimators t_{ij}^* of the grades t_{ij} of each production unit rather than on the estimators θ_i^* of the blocks or stopes B_i, then the previous formula reduces to the sum of N independent elementary estimations:

$$\sigma_{E_j}^2 = \frac{1}{P_j^2} \sum_{i=1}^{N} p_{ij}^2 E\{(t_{ij} - t_{ij}^*)^2\}. \qquad \text{(IV.9)}$$

This case corresponds, for example, to an open-pit deposit for which the mining program is based on blast-hole analyses which provide the estimator t_{ij}^* of each production unit.

Remark 2   However, it sometimes happens that it is not possible to directly estimate the grades t_{ij}, either because the exact location of the corresponding production unit p_{ij} is not known (e.g., extraction by block caving of p_{ij} tonnes from block B_i), or because the necessary data is not available. These two cases will be studied later in the numerical applications.

In such cases, it is obvious that the grade control has an absolute limit imposed by the dispersion of the production grade units in the blocks, and obtained by equating the estimation variances $\sigma_{e_i}^2$ to zero in formula (IV.8) (i.e., assuming that the true grades θ_i of blocks B_i are perfectly known). The minimum value of the estimation variance $\sigma_{E_j}^2$ of the daily grades is, thus,

$$\min \{\sigma_{E_j}^2\} = \frac{1}{P_j^2} \sum_i p_{ij}^2 D^2(p_i/B_i).$$
(IV.10)

If the daily production units p_{ij} extracted from each block B_i are more or less equal, then

$$D^2(p_i/B_i) = D^2(p/B), \qquad \forall i,$$

and, thus,

$$\min \{\sigma_{E_j}^2\} = \frac{1}{N} D^2(p/B).$$
(IV.11)

Remark 3   If it is desired to improve the quality of the control of daily production grades (i.e., to decrease on average the deviations between the predicted grades T_j^* and the actual recovered grades T_j), formula (IV.8) must be used, either to decrease the estimation variances $\sigma_{e_i}^2$ so that, with the grades of the blocks B_i being better estimated, the production planning is on a much surer basis, or, preferably, to reduce the average dispersion variance $(1/N)\sum_i D^2(p_i/B_i)$ by increasing the number N of stopes – the compensation effect of the estimation errors $(t_{ij} - \theta_i^*)$ between the different stopes is then increased.

Numerical application 1   Consider a massive copper deposit mined by block-caving with blocks of constant dimensions 60 m × 60 m × 180 m (e.g., El Teniente deposit in Chile). The estimation variance of such blocks by the available holes is calculated to be $\sigma_e^2 = 0.003$ (% Cu)^2, and, with a mean grade $m = 1\%$ Cu, the mean relative standard deviation is $\sigma_e/m = 0.055 = 5.5\%$.

The daily production of 65 000 tons of ore is obtained from $N = 8$ blocks. This production is assumed to be equally distributed among the eight blocks, providing an average daily mining unit in each block of approximately 8125 tons, corresponding to a mean support p of dimensions $60\text{ m} \times 60\text{ m} \times 0.75\text{ m}$.

The mean dispersion variance $D^2(p/B)$ is calculated from the three-dimensional semi-variogram model of the grades. This results in $D^2(p/B) = 0.026$, i.e., a mean relative standard deviation of dispersion

$$D(p/B)/m = 0.16 = 16\%.$$

The estimation variance of the daily production grades is then given by formula (IV.8):

$$\sigma_E^2 = \frac{1}{N}[D^2(p/B) + \sigma_e^2] = \frac{0.029}{8} = 0.003\,63 \text{ (\% Cu)}^2,$$

i.e., a relative standard deviation $\sigma_E/m = 0.06 = 6\%$ and, thus, when a Gaussian standard is adopted, in 95% of cases the actual grades T_j will differ from the predicted grades T_j^* by less than $\pm 2\sigma_E/m = \pm 12\%$, i.e., $T_j \in T_j^* \ (1 \pm 0.12)$.

If the grades of the blocks B_i were perfectly known, i.e., with an almost unlimited sampling budget, this variance could be brought to its lower limit (IV.11):

$$\min\{\sigma_E^2\} = \frac{1}{N}D^2(p/B) = \frac{0.026}{8} = 0.003\,25 \text{ (\% Cu)}^2,$$

which amounts to a maximum relative decrease of $(363-325)/325 = 0.116$, less than 12%. For a better daily grade control in this deposit, it is more efficient to make mining flexible by increasing the number N of blocks being simultaneously mined, than spending more money on extra sampling.

Numerical application 2   Consider now a subhorizontal deposit mined by open-pit with 5-m benches (e.g., a New Caledonian silicate nickel deposit). The horizontal regionalization of the nickel grades graded over the constant thickness of the bench is characterized by an isotropic model with a range $a = 300$ m, a sill $C = 0.3$ (% Ni)^2 and a nugget constant, defined on the support $l = 5$ m of the grading, $C_{0l} = 0.2$ (% Ni)^2. The mean nickel grade over the deposit is $m = 2\%$ Ni.

A direct proportional effect in m^{*2} is observed, which means that, locally, in a production zone with an estimated mean grade m^*, the relative semi-variogram γ/m^{*2} is a spherical model with the parameters range $a = 300$ m, sill $C/m^2 = 0.3/2^2 = 0.075$, nugget constant $C_{0l}/m^2 = 0.05$.

The deposit has been evaluated by vertical drills on a regular square grid of $60\text{ m} \times 60\text{ m}$. The workings on which each shovel is located can be approximated to this grid unit, i.e., to a panel B of dimensions $60 \text{ m} \times 60 \text{ m} \times 5 \text{ m}$, cf. Fig. IV.51. A preliminary estimation of the mean grade θ_i of each panel B_i is obtained by taking the mean grade θ_i^* of its central drill-hole (classical but not optimal estimation procedure by polygons of influence), the estimation variance σ_e^2 is, thus, given by chart no. 7, and the relative estimation variance is

$$\sigma_{e_i}^2/\theta_i^{*2} = C_0/m^2 + (C/m^2) \times 0 \cdot 074 = 0 \cdot 056, \qquad \forall \text{ the panel } i.$$

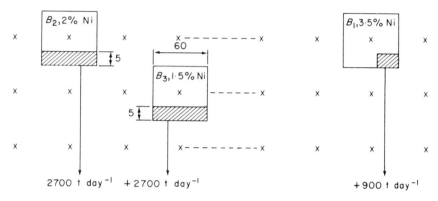

FIG. IV.51. Daily production model.

It is desired to produce 6300 t of ore per day with a grade average of 2% Ni from three shovels located in workings with different mean grades, cf. Fig. IV.51.

The following situation is typical of any given day j.

(i) Shovel no.1, which has a small capacity (900 t day^{-1}) is located in workings B_1 with a rich mean grade estimated to be $\theta_1^* = 3 \cdot 5\%$ Ni. This shovel moves a unit of size $p_1 = 20 \text{ m} \times 5 \text{ m} \times 5 \text{ m}$ each day.

(ii) Shovels no. 2 and 3 have thrice the capacity (2700 t day^{-1}) of shovel no.1. Shovel no.2 is located in a workings of average grade ($\theta_2^* = 2\%$ Ni), while shovel no. 3 is in a poor grade site ($\theta_3^* = 1 \cdot 5\%$ Ni).

Each shovel moves a unit of size $p_2 = p_3 = 60 \text{ m} \times 5 \text{ m} \times 5 \text{ m}$ per day. The predicted mean grade of the 6300 t day^{-1} so produced is

$$T_j^* = \frac{900 \times 3 \cdot 5 + 2700\,(2 + 1 \cdot 5)}{6300} = 2 \cdot 00 \% \text{ Ni}.$$

The dispersion variances $D^2(p_i/B_i)$ of the grades of units p_i in the working sites B_i are given by the standard procedure of the difference between the two auxiliary functions $F(p_i)$ and $F(B_i)$ defined on the two rectangles p_i and B_i, cf. section II.E.2, formula (II.82), e.g.,

$$D^2(p_1/B_1) = \bar{\gamma}(B_1, \beta_1) - \bar{\gamma}(p_1, p_1) = F(60; 60) - F(20; 5).$$

(For the calculations of the values $D^2(p_i/B_i)$, note that as the supports p_i and B_i are considerably larger than the support of the 5 m-long core samples, the influence of the nugget effect disappears and the relative spherical model with range $a = 300$ m and sill $C/m^2 = 0.075$ is sufficient.)

Immediate calculation and chart reading provide the following values:

$$D^2(p_1/B_1) = \overline{3.5}^2 \,(0.0117 - 0.001\,95) = 0.119 \;(\% \text{ Ni})^2,$$

$$D^2(p_2/B_2) = \overline{2}^2 \,(0.0117 - 0.003\,38) = 0.033 \;(\% \text{ Ni})^2,$$

$$D^2(p_3/B_3) = \overline{1.5}^2 \,(0.0117 - 0.003\,38) = 0.019 \;(\% \text{ Ni})^2.$$

Thus, from formula (IV.8), the estimation variance of the daily grade production is

$$\sigma_{E_j}^2 = \frac{1}{6300^2}[\overline{900}^2 \,(0.119 + 0.686) + \overline{2700}^2 \,(0.0033 + 0.224)$$
$$+ \overline{2700}^2 \,(0.019 + 0.126)] = 0.0903 \;(\% \text{ Ni})^2,$$

i.e., a standard deviation $\sigma_{E_j} = 0.30\%$ Ni.

Thus, adopting the Gaussian standard, the mean grade T_j actually recovered each day will, in 95% of the cases, fluctuate in the interval $[2 \pm 0.6] \%$ Ni around the predicted value $T_j^* = 2\%$ Ni.

The mediocre result of this grade control is due to the rather hasty estimation of the panel grade θ_i by the grade of the central drill core (polygon of influence method). Using the same data (drills on a regular 60 m $\times$ 60 m grid), simple kriging would, *without* any additional cost, reduce the estimation variance $\sigma_{e_i}^2$ by half, and daily grades could then be controlled with a variance of $\sigma_{E_j}^2 = 0.0511$ (% Ni)^2, i.e., a standard deviation of $\sigma_{E_j} = 0.23\%$ Ni, which represents a noted improvement.

The lower limit of the variance $\sigma_{E_j}^2$ corresponding to the perfect knowledge of the mean grades θ_i of the working sites is given by formula (IV.10):

$$\min \{\sigma_{E_j}^2\} = 0.0120, \quad \text{i.e., } \min \{\sigma_{E_j}\} = 0.11\% \text{ Ni}.$$

The interval of fluctuation is, thus, $T_j \in [2 \pm 0.22] \%$ Ni.

In practice, if the grade control is deemed insufficient in the interval $T_j \in [2 \pm 0.46]\%$ Ni provided by kriging and the data on a 60 m $\times$ 60 m grid, it is advisable to establish some sort of equilibrium between the cost of reorganizing the daily production plan and the cost of additional sampling.

Case Study 15. Dispersion of production grades

Consider again the New Caledonian type of nickel silicate deposit considered in the latter numerical applications of Case Study 14.

In the present example, only the two large capacity shovels are retained, providing a weekly production of $5400 \times 6 = 32\,400$ t, which can be achieved in either of the two following manners, cf. Fig. IV.52.

> 1. The two shovels can advance in line on the same front. The weekly production unit s can be approximated by a rectangle with horizontal dimensions $360\,\text{m} \times 5\,\text{m}$. The height of the bench is 5 m, and the average density of the dry ore is 1·8.
>
> 2. The two shovels can advance in parallel on two fronts, getting progressively farther away from each other. The weekly production unit s thus consists of two rectangles s_1 and s_2 with equal horizontal dimensions $180\,\text{m} \times 5\,\text{m}$.

In the first case, the annual production (1·62 million tonnes over 50 weeks) can be approximated by a rectangle S with horizontal dimensions $720\,\text{m} \times 125\,\text{m}$; in the second case, it will be approximated by a rectangle S with horizontal dimensions $500\,\text{m} \times 180\,\text{m}$.

Taking the weekly production as the homogenization unit, it is desired to calculate the dispersion variance $D^2(s/S)$ of the weekly mean grades actually obtained in one production year, using both types of shovel advancement.

The horizontal regionalization of the Ni grades *graded* over the constant thickness (5 m) of the benches is stationary with expectation $m = 2\%$ Ni and characterized by an isotropic spherical model with a nugget effect,[†] range $a = 300$ m and sill $C = 0·3$ (% Ni)^2

The dispersion variance is given by the standard formula:

$$D^2(s/S) = \bar{\gamma}(S, S) - \bar{\gamma}(s, s).$$

The first term $\bar{\gamma}(S, S)$ is simply the auxiliary function F defined on the rectangle S. Chart no. 4 gives

$$\bar{\gamma}(S, S) = F(720; 125) = C \times 0·754 = 0·226$$

for method 1, and

$$\bar{\gamma}(S, S) = F(500; 180) = C \times 0·700 = 0·210.$$

for method 2

---

† There is no need to give the value of the nugget constant, as the nugget effect vanishes when regularized on the previous supports s and S of the weekly and annual productions, cf. formula (III.8).

Similarly, for the first method, the term $\bar{\gamma}(s, s)$ is given by

$$\bar{\gamma}(s, s) = F(360; 5) \simeq F(360) = C \times 0 \cdot 514 = 0 \cdot 154.$$

The calculation of the term $\bar{\gamma}(s, s)$ for the second method is a little more involved. As s is the union of the two rectangles s_1, s_2, it follows that

$$\bar{\gamma}(s, s) = \bar{\gamma}(s_1, s) = \tfrac{1}{2}[\bar{\gamma}(s_1, s_1) + \bar{\gamma}(s_1, s_2)].$$

$\bar{\gamma}(s_1, s_1)$ is given immediately as

$$\bar{\gamma}(s_1, s_1) = F(180; 5) \simeq F(180) = C \times 0 \cdot 289 = 0 \cdot 087.$$

The term $\bar{\gamma}(s, s_2)$ depends on the distance $2x$ between the two working sites s_1 and s_2, cf. Fig. IV.52(b). The production year consists of 50 weeks ($i = 1$ to 50) and, during week i, the distance between the two sites is $5(2i - 1)$ metres. Approximating each working site by its median line of 180 m length, the term $\bar{\gamma}(s_1, s_2)$ for week i becomes

$$\bar{\gamma}_{(i)}(s_1, s_2) = \alpha[5(2i - 1); 180],$$

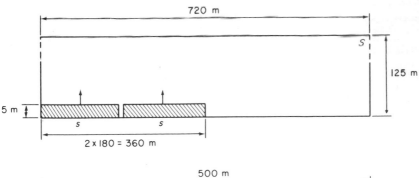

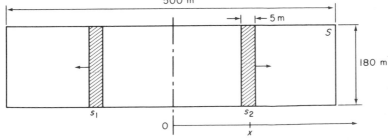

FIG. IV.52. Two methods of shovel advancement. (a) Two shovels in line; (b) two shovels in parallel.

cf. the definition (II.79) of the auxiliary function $\alpha(L, l)$. On average, over the 50 weeks of production, the term $\bar{\gamma}(s_1, s_2)$ can be written as

$$\bar{\gamma}(s_1, s_2) = \frac{1}{50} \sum_i \alpha[5(2i-1); 180] = C \times 0.84 = 0.252$$

and, thus, for the second mining method,

$$\bar{\gamma}(s, s) = \tfrac{1}{2}[0.087 + 0.252] = 0.170.$$

Finally, the desired dispersion variances are

$$D^2(s/S) = 0.072, \quad \text{i.e., } D(s/S)/m = 0.13 = 13\%$$

for method 1, and

$$D^2(s/S) = 0.040, \quad \text{i.e., } D(s/S)/m = 0.10 = 10\%.$$

for method 2

Configuration no. 2, with the two shovels in parallel, thus provides a significant decrease in the dispersion variance.

Remark The knowledge of the dispersion over one year may not be sufficient to control the variablity of grades. For example, it is important to ensure that weekly mill-head grades will be sufficiently regular over a short space of time (e.g., 1 month). For this purpose, the dispersion variance $D^2(s/M)$ of the weekly grades within a month's production can be calculated. Krige's additivity relationship (II.37) gives

$$D^2(s/S) = D^2(s/M) + D^2(M/S).$$

Thus, the distribution of the total annual variability $D^2(s/S)$ between the short-term variability $D^2(s/M)$ and the long-term variability $D^2(M/S)$ can be evaluated.

If it is desired to further increase the control of variabilities, the deposit can be simulated, i.e., the spatial regionalization of the grades can be simulated, cf. Chapter VII. A simulation of the mining method can then be applied to this simulated deposit to give an idea of the fluctuations of grades from one period to another, e.g., day per day, month to month, etc., cf. the case study given in section VII.B.3.

V

Kriging and the Estimation of *in situ* Resources

SUMMARY

The *in situ* resources of a deposit are distinct from the recoverable reserves which can be mined within some given technical and economical context. The object of this chapter is the estimation of the *in situ* resources prior to the application of any cut-off; the estimation of recoverable reserves will be treated in Chapter VI.

Section V.A introduces the technique of kriging, which provides a minimum-variance unbiased linear estimator of the mean characteristic of a block of any given geometry. In the case of non-stationarity, several non-bias conditions are required and this leads to the technique known as "universal kriging", or unbiased kriging of order k. When a group of variables are correlated, the cokriging technique can be used to estimate any variable (e.g., silver grade) from the data available on all the correlated variables (e.g., lead and zinc grades in addition to the silver grades).

Section V.B deals with applying kriging techniques in practice and demonstrates that it is possible to krige thousands of blocks with reasonable speed and cost. Some typical subroutines are given along with numerous examples of kriging plans which can be adapted to the estimation of a large range of deposits of variable morphology.

Section V.C deals with the global estimation of a deposit or a large zone. The global estimator is obtained by combining the various local kriged estimators of the blocks or units which make up the zone. When calculating the global estimation variance, a distinction is made between the geometrical error involved in evaluating the limits of the mineralized surface or volume and the quality error due to the estimation of the mean characteristics (e.g., grade) within a zone of fixed limits. These two errors are then combined to provide the estimation variance attached to global estimators, such as quantities of ore and metal and the corresponding mean grades.

V.A. THEORY OF KRIGING

The problem

The problem of *local* estimation is to find the best estimator of the mean value of a regionalized variable over a limited domain, the dimensions of which are small compared to the dimensions of the quasi-stationary (homogeneous) zones of the deposit, e.g., the mean grade of a block located well within a zone of homogeneous mineralization. Local estimation differs from *global* estimation in that global estimation considers distances larger than the limits of quasi-stationarity and, thus, sometimes engulfs various heterogeneous mineralizations.

The available information used for local estimation within a quasi-stationary zone is generally made up of a set of data (e.g., n core grades) and structural information (e.g., the variogram model characterizing the spatial variability in the studied zone).

Kriging is a local estimation technique which provides the best linear unbiased estimator (abbreviated to BLUE) of the unknown characteristic studied. This limitation to the class of linear estimators is quite natural, since it means that only the second-order moment of the RF (i.e., the covariance or variogram) is required, and, in general, it is possible in practice to infer the moment, cf. section II.B.2.

Of course, when the available structural information includes more than the simple knowledge of the second-order moment, then non-linear estimators – more precise than simple kriging – can be defined, cf. Chapter VIII, "Introduction to Non-linear Geostatistics". The present chapter will be limited to linear estimators only.

V.A.1. Equations of kriging

Let $Z(x)$ be the RF under study. $Z(x)$ is defined on a *point* support and is second-order stationary, with

expectation, $\quad E\{Z(x)\} = m$,

a constant m which is generally unknown;

centred covariance, $\quad E\{Z(x+h)Z(x)\} - m^2 = C(h)$;

variogram, $\quad E\{[Z(x+h) - Z(x)]^2\} = 2\gamma(h)$.

Either of these two second-order moments is supposed known. When only the variogram exists, then the RF $Z(x)$ is intrinsic only.

The estimation of the mean value

$$Z_V(x_0) = \frac{1}{V} \int_{V(x_0)} Z(x)\, dx$$

over a domain $V(x_0)$ is required.

The experimental data to be used consist of a set of discrete grade values $\{Z_\alpha, \alpha = 1 \text{ to } n\}$. These grades are either defined on point or quasi-point supports, or else they are the mean grades $Z_{V_\alpha}(x_\alpha)$ defined on the supports v_α centred on the points x_α; the n supports may differ from each other. Note that under the hypothesis of stationarity the expectation of each of these data is m: $E\{Z_\alpha\} = m, \forall \alpha$.

The linear estimator Z_K^* considered is a linear combination of the n data values $Z_K^* = \sum_{\alpha=1}^n \lambda_\alpha Z_\alpha$. The n weights λ_α are calculated to ensure that the estimator is unbiased and that the estimation variance is minimal (the estimator Z_K^* is then said to be optimal).

Non-bias condition

To obtain a zero error $[Z_V - Z_K^*]$ in expectation (i.e., an average over a great number of similar estimations) it is enough to impose the condition $\sum_{\alpha=1}^n \lambda_\alpha = 1$, and, thus,

$$E\{Z_K^*\} = m \sum_\alpha \lambda_\alpha = m = E\{Z_V\},$$

which entails $E\{Z_V - Z_K^*\} = 0$.

Minimum estimation variance

The estimation variance $E\{[Z_V - Z_K^*]^2\}$ can be expanded as follows:

$$E\{[Z_V - Z_K^*]^2\} = E\{Z_V^2\} - 2E\{Z_V Z_K^*\} + E\{Z_K^{*2}\},$$

with

$$E\{Z_V^2\} = \frac{1}{V^2} \int_V dx \int_V E\{Z(x)Z(x')\} dx' = \bar{C}(V, V) + m^2,$$

$$E\{Z_V Z_K^*\} = \sum_\alpha \lambda_\alpha \frac{1}{V v_\alpha} \int_V dx \int_{v_\alpha} E\{Z(x)Z(x')\} dx' = \sum_\alpha \lambda_\alpha \bar{C}(V, v_\alpha) + m^2,$$

$$E\{Z_K^{*2}\} = \sum_\alpha \sum_\beta \lambda_\alpha \lambda_\beta \frac{1}{v_\alpha v_\beta} \int_{v_\alpha} dx \int_{v_\beta} E\{Z(x)Z(x')\} dx'$$

$$= \sum_\alpha \sum_\beta \lambda_\alpha \lambda_\beta \bar{C}(v_\alpha, v_\beta) + m^2.$$

The terms m^2 are eliminated and, thus, we have

$$E\{[Z_V - Z_K^*]^2\} = \bar{C}(V, V) - 2 \sum_\alpha \lambda_\alpha \bar{C}(V, v_\alpha) + \sum_\alpha \sum_\beta \lambda_\alpha \lambda_\beta \bar{C}(v_\alpha, v_\beta).$$

(The standard notation $\bar{C}(V, v_\alpha)$ designates the mean value of the covariance function $C(h)$ when the two extremities of the vector h independently describe the domains V and v_α, respectively.)

The estimation variance can, thus, be expressed as a quadratic form in λ_α, λ_β, and is to be minimized subject to the non-bias condition $\sum_\alpha \lambda_\alpha = 1$. The optimal weights are obtained from standard Lagrangian techniques by setting each of the n partial derivatives $\partial[E\{[Z_V - Z_K^*]^2\} - 2\mu \sum_\alpha \lambda_\alpha]/\partial\lambda_\alpha$ to zero. This procedure provides a system of $(n+1)$ linear equations in $(n+1)$ unknowns (the n weights λ_α and the Lagrange parameter μ) which is called the "*kriging system*".

$$\left.\begin{array}{l} \displaystyle\sum_{\beta=1}^{n} \lambda_\beta \bar{C}(v_\alpha, v_\beta) - \mu = \bar{C}(v_\alpha, V), \qquad \forall \alpha = 1 \text{ to } n, \\[3ex] \displaystyle\sum_{\beta=1}^{n} \lambda_\beta = 1. \end{array}\right\} \qquad \text{(V.1)}$$

The minimum estimation variance, or "*kriging variance*", can then be written as

$$\sigma_K^2 = E\{[Z_V - Z_K^*]^2\} = \bar{C}(V, V) + \mu - \sum_{\alpha=1}^{n} \lambda_\alpha \bar{C}(v_\alpha, V) \qquad \text{(V.2)}$$

This kriging system can also be expressed† in terms of the semi-variogram function $\gamma(h)$, particularly when the RF $Z(x)$ is intrinsic only and the covariance function $C(h)$ is not defined:

$$\left.\begin{array}{l} \displaystyle\sum_\beta \lambda_\beta \bar{\gamma}(v_\alpha, v_\beta) + \mu = \bar{\gamma}(v_\alpha, V), \qquad \forall \alpha = 1 \text{ to } n, \\[3ex] \displaystyle\sum_\beta \lambda_\beta = 1 \end{array}\right\} \qquad \text{(V.3)}$$

and

$$\sigma_K^2 = \sum_\alpha \lambda_\alpha \bar{\gamma}(v_\alpha, V) + \mu - \bar{\gamma}(V, V).$$

---

† *In practice*, even when the covariance is not defined, the system (V.1) written in terms of covariance is preferred for reasons of programming efficiency. For this purpose, it is enough to define the "pseudo-covariance" $C(h)$ such that $\gamma(h) = A - C(h)$, the constant A being any positive value greater than the greatest mean value $\bar{\gamma}$ used in the kriging system (V.3). The non-bias condition $\sum_\alpha \lambda_\alpha = 1$ allows the elimination of this constant A from the system (V.3), which then becomes a (V.1)-type system written in terms of the pseudo-covariance $C(h)$.

Matrix form

Both kriging systems can be expressed in matrix form as

$$[K] . [\lambda] = [M2] \tag{V.4}$$

from which, for system (V.1),

$$[\lambda] = [K]^{-1} . [M2],$$

and, $[\lambda]^t$ being the transposed matrix of $[\lambda]$,

$$\sigma_K^2 = \bar{C}(V, V) - [\lambda]^t . [M2],$$

where the matrix of unknowns $[\lambda]$ and the second member $[M2]$ can be written as two column matrices:

$$[\lambda] = \begin{bmatrix} \lambda_1 \\ \lambda_2 \\ \vdots \\ \lambda_\alpha \\ \vdots \\ \lambda_n \\ -\mu \end{bmatrix}, \qquad [M2] = \begin{bmatrix} \bar{C}(v_1, V) \\ \vdots \\ \bar{C}(v_\alpha, V) \\ \vdots \\ \bar{C}(v_n, V) \\ 1 \end{bmatrix}$$

The first member $[K]$, or "kriging matrix" is written as:

$$[K] = \begin{bmatrix} \bar{C}(v_1, v_1) & \cdots & \bar{C}(v_1, v_\beta) & \cdots & \bar{C}(v_1, v_n) & 1 \\ \vdots & & \vdots & & \vdots & \vdots \\ \bar{C}(v_\beta, v_1) & \cdots & \bar{C}(v_\beta, v_\beta) & \cdots & \bar{C}(v_\beta, v_n) & 1 \\ \vdots & & \vdots & & \vdots & \vdots \\ \bar{C}(v_n, v_1) & \cdots & \bar{C}(v_n, v_\beta) & \cdots & \bar{C}(v_n, v_n) & 1 \\ 1 & \cdots & 1 & \cdots & 1 & 0 \end{bmatrix}$$

Main diagonal

Note that the kriging matrix is symmetric, i.e.,

$$\bar{C}(v_\alpha, v_\beta) = \bar{C}(v_\beta, v_\alpha), \qquad \forall \alpha, \beta.$$

Seven important practical remarks

Remark 1 *The existence and uniqueness of the solution.* The kriging system (V.1) has a unique solution if and only if the covariance matrix

$[\bar{C}(v_\alpha, v_\beta)]$ is strictly positive definite and, thus, necessarily has a strictly positive determinant. For this purpose, it is enough that the point covariance[†] model $C(h)$ used is positive definite and that no data support v_α coincides completely with another one. In fact, $v_\alpha \equiv v_{\alpha'}$ entails that $\bar{C}(v_\alpha, v_\beta) = \bar{C}(v_{\alpha'}, v_\beta), \forall \beta$, and the determinant $|\bar{C}(v_\alpha, v_\beta)|$ is, thus, zero.

This condition for the existence and the uniqueness of the solution of the kriging system thus entails that the kriging variance (V.2) is non-negative, cf. section II.A.3, relation (II.8).

Remark 2 Kriging, which is an unbiased estimator, is also an *exact* interpolator, i.e., if the support V to be estimated coincides with any of the supports v_α of the available data, then the kriging system provides:

(i) an estimator Z_K^* identical to the known grade Z_α of the support $v_\alpha = V$;
(ii) a zero kriging variance, $\sigma_K^2 = 0$.

In cartography, it is said that "the kriged surface passes through the experimental points". Not every estimation procedure has this property, especially procedures using least square polynomials.

Remark 3 The expressions of the systems and the kriging variances using the notations $\bar{C}$ and $\bar{\gamma}$ are completely general,

(i) *whatever the supports* v_α, v_β of the data and the support V to be estimated, some data supports may partially overlap, $v_\alpha \cap v_\beta \neq \varnothing$, but for $\alpha \neq \beta$ it is imperative that $v_\alpha \neq v_\beta$ – some data supports may be included in the volume V to be estimated, $v_\alpha \subset V$;
(ii) *whatever the underlying structure* characterized by the model $C(h)$ or $\gamma(h)$, the structure may be isotropic or anisotropic, nested or not.

Remark 4 The kriging system and the kriging variance depend only on the structural model $C(h)$ or $\gamma(h)$ and on the relative geometries of the various supports v_α, v_β, V, but not on the particular values of the data Z_α. Consequently, once the data configuration is known and before drilling is undertaken, the kriging system can be solved and the corresponding minimum estimation variance can be forecast. Thus, the kriging variance can be used to balance the cost of a new drilling campaign with its forecast utility (new data would decrease the estimation variance, thus providing narrower confidence intervals) cf. O. Bernuy and A. Journel (1977).

[†] If the semi-variogram is used, then $-\gamma(h)$ must be conditional positive definite.

Remark 5 The kriging matrix [K] depends only on the relative geometries of the data supports v_α, v_β and not at all on the support V of the domain to be estimated.

Thus, two identical data configurations would provide the same kriging matrix [K] and it is then enough to take the inverse matrix $[K]^{-1}$ only once. The two solution column matrices $[\lambda]$ and $[\lambda']$ are then obtained by taking the products of the single inverse matrix $[K]^{-1}$ and the respective second-member matrices:

$$[\lambda] = [K]^{-1}[M2] \quad \text{and} \quad [\lambda'] = [K]^{-1} . [M2'].$$

In practice, this amounts to using a program for the resolution of linear systems with two or more second members (such as subroutine GELG in the IBM package).

If, in addition to identical data configurations, i.e. $[K] = [K']$, the two panels to be estimated have the same geometry, then $[M2] = [M2']$ and the resulting kriging weights are identical: $[\lambda] = [\lambda']$. It is then enough to solve the kriging system once, which results in a considerable saving of computing time. For this reason, when evaluating the same stationary or quasi-stationary phenomenon, it is advisable to ensure, as far as possible, that the data configurations are systematic and regular. In this way, a single kriging plan that can be repeated identically over all or part of the deposit, can be used.

Remark 6 The kriging system and the kriging variance take into account the four essential and intuitive facts, previously mentioned in section II.B.2, remark 4, which condition every estimation. These are as follows.

(i) The geometry of the domain V to be estimated, expressed in the term $\bar{\gamma}(V, V)$ in the expression of the kriging variance σ_K^2.

(ii) The distances between V and the supports v_α of the information, expressed by the terms $\bar{\gamma}(v_\alpha, V)$ of the matrix [M2].

(iii) The geometry of the data configuration as expressed by the terms $\bar{\gamma}(v_\alpha, v_\beta)$ of the kriging matrix [K]. The accuracy of an estimation depends not only on the number of data but also on their configuration in relation with the main features of the regionalization as characterized by the structural function $\gamma(h)$ in the various terms $\bar{\gamma}(v_\alpha, v_\beta)$.

(iv) The main structural features of the variability of the phenomenon under study as characterized by the semi-variogram model $\gamma(h)$.

Consider the estimation in two dimensions of the panel V by the symmetric configuration of the four data A, B, C, D as shown on Fig. V.1. The underlying mineralization shows a preferential direction u of continuity which appears on the anisotropic semi-variogram $\gamma(h_u, h_v)$ as a slower variability in the direction u. The kriging system thus gives a greater weight to the data B and D, although they are at the same distance from V as A and C.

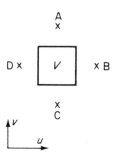

Direction of continuity

FIG. V.1. Influence of the variability structure on kriging.

Remark 7 *The influence of a nugget effect.* In practice, a partial nugget effect is very often found, i.e., a nested structure, sum of a nugget effect, and a macro-structure:

$$C(h) = C_0(h) + C_1(h).$$

The mean values $\bar{C}(v_\alpha, v_\beta)$ appearing in the kriging system are written, according to formulae (III.6), as

$$\bar{C}(v_\alpha, v_\beta) = \bar{C}_1(v_\alpha, v_\beta) + \begin{cases} A/v_\alpha & \text{if } v_\beta \subset v_\alpha, \\ 0 & \text{if } v_\alpha \text{ and } v_\beta \text{ are disjoint,} \end{cases}$$

where A is a constant given by $A = \int C_0(h)\,dh$. (It is recalled – cf. section III.A.3, formula (III.7) – that this constant A appears in the nugget constant value $C_{0v} = A/v$ of a semi-variogram regularized over the support v.)

The kriging system (V.1) can then be written as

$$\left. \begin{aligned} \lambda_\alpha[A/v_\alpha + \bar{C}_1(v_\alpha, v_\beta)] + \sum_{\beta \neq \alpha} \lambda_\beta \bar{C}_1(v_\alpha, v_\beta) - \mu &= \bar{C}_1(v_\alpha, V), \quad \forall \alpha, \\ \sum_\beta \lambda_\beta &= 1, \end{aligned} \right\} \quad \text{(V.5)}$$

if all supports v_α, v_β, V are disjoint, and

$$\sigma_K^2 = \frac{A}{V} + \bar{C}_1(V, V) + \mu - \sum_\alpha \lambda_\alpha \bar{C}_1(v_\alpha, V).$$

If all supports v_α, v_β, V are not disjoint, then each of the terms $\bar{C}$ of the kriging system must be calculated.

A case often found in practice is that of a configuration of data disjoint among themselves, some of which ($v_{\alpha'}$) being located within the support V to be estimated:

$$
\begin{matrix}
v_{\alpha'} \subset V, & & v_{\alpha'} \cap v_\beta = \varnothing, & \\
& \text{but} & & \forall \alpha, \alpha', \beta, \\
v_\alpha \cap V = \varnothing & & v_\alpha \cap v_\beta = \varnothing,
\end{matrix}
$$

$$
\left.
\begin{matrix}
\lambda_{\alpha'}\left[\dfrac{A}{v_{\alpha'}} + \bar{C}_1(v_{\alpha'}, v_{\alpha'})\right] + \sum\limits_{\beta \neq \alpha'} \lambda_\beta \bar{C}_1(v_{\alpha'}, v_\beta) - \mu \\
\\
\qquad\qquad\qquad = \dfrac{A}{V} + \bar{C}_1(v_{\alpha'}, V), \qquad \forall\alpha', \\
\\
\lambda_\alpha\left[\dfrac{A}{v_\alpha} + \bar{C}_1(v_\alpha, v_\beta)\right] + \sum\limits_{\beta \neq \alpha} \lambda_\beta \bar{C}_1(v_\alpha, v_\beta) - \mu = \bar{C}_1(v_\alpha, V), \qquad \forall\alpha, \\
\\
\sum\limits_\beta \lambda_\beta = 1
\end{matrix}
\right\}\text{(V.6)}
$$

and

$$\sigma_K^2 = \frac{A}{V} + \bar{C}_1(V, V) + \mu - \sum_{\alpha'} \lambda_{\alpha'}\left[\frac{A}{V} + \bar{C}_1(v_{\alpha'}, V)\right] - \sum_\alpha \lambda_\alpha \bar{C}_1(v_\alpha, V).$$

In mining applications, the support V is very often much larger than the data supports v_α, $v_{\alpha'}$, and the term A/V is negligible when compared with the terms A/v_α or $A/v_{\alpha'}$. When all the data supports v_α, v_β are disjoint, system (V.5) is obtained.

If the semi-variogram $\gamma(h) = \{C_0(0) - C_0(h)\} + \gamma_1(h)$ is to be used instead of the covariance $C(h)$, it is enough to substitute $(-\gamma_1)$ for C_1 in the preceding systems (V.5) and (V.6).

Note in system (V.5) that the influence of a nugget effect with nugget constant $C_{0v} = A/v$ affects only the elements of the main diagonal of the kriging matrix as an additional term A/v_α.

The influence of a pure nugget effect

If the macro-structure $C_1(h)$ disappears, leaving only the pure nugget effect $C_0(h)$, the kriging system (V.5) becomes

$$\lambda_\alpha A/v_\alpha - \mu = 0, \qquad \forall \alpha,$$

$$\sum_\beta \lambda_\beta = 1$$

and, thus,

$$\left. \begin{array}{c} \lambda_\alpha = v_\alpha \Big/ \sum_\beta v_\beta, \qquad \forall \alpha, \\[2ex] \mu = A \Big/ \sum_\beta v_\beta \end{array} \right\} \qquad \text{(V.7)}$$

when all the supports v_α, v_β, V are disjoint, and

$$\sigma_K^2 = A\left[\frac{1}{\sum_\beta v_\beta} + \frac{1}{V}\right].$$

The weights for the n available data values Z_α are proportional to the volumes v_α of their respective supports.

Moreover, if the n supports v_α are equal (e.g., n core samples of constant length and volume v), then the system (V.7) becomes

$$\left. \begin{array}{c} \lambda_\alpha = 1/n, \qquad \forall \alpha = 1 \text{ to } n, \\[2ex] \sigma_K^2 = \dfrac{A}{v}\left[\dfrac{1}{n} + \dfrac{v}{V}\right] \simeq \dfrac{A}{nv} = \dfrac{C_{0v}}{n}, \qquad \text{if } V \gg v. \end{array} \right\}$$

The n weights λ_α are all equal and the kriging variance $\sigma_K^2 = C_{0v}/n$ is simply the variance of estimation given by the statistics of spatially uncorrelated variables. The nugget constant C_{0v} is the *a priori* variance of the support v data, according to formula (III.11):

$$C_{0v} = A/v = D^2(v/\infty) = \text{Var}\{Z_v(x)\}.$$

Thus, for a pure nugget effect, we have the following.

(i) All data having the same support v (disjoint from V) have the same weight whatever their distances from the volume V to be estimated. The close data no longer screen the influence of the more distant data and the nugget effect is said "*to remove the screening effect*".

(ii) Within a given neighbourhood of quasi-stationarity, it is no longer possible to make different local estimations: all estimators Z_K^* of all panels V within the neighbourhood are equal to the unique mean value $\sum_\alpha Z_\alpha/n$ of the n data available in this neighbourhood. The nugget effect is said *"to smooth the estimation"*.

V.A.2. Taking account of a drift

In all the kriging equations presented heretofore, it is assumed that the RF $Z(x)$ is stationary or at least quasi-stationary over a given estimation neighbourhood.

In practice, it may be known that a drift (non-stationary expectation $E\{Z(x)\}$) exists within certain zones of the deposit and that there is not enough data available in these zones to allow quasi-stationary estimations (by considerably reducing the neighbourhoods of estimation). It then becomes necessary to take the drift into account when estimating these zones by kriging. For example, consider the mineralized seam on Fig. V.2, which thins out towards one end. We want to estimate the mean thickness (or the ore tonnage) of a panel V located on the fringe of the seam, by drill-holes all located further along the seam.

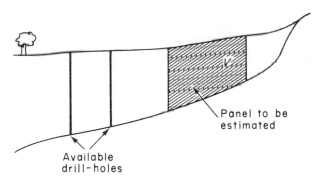

Panel to be
estimated

Available
drill-holes

FIG. V.2. Drift of a seam thickness.

Kriging under a stationary hypothesis, like any other "stationary" estimation procedure such as inverse-squared distances weighting, will inevitably overestimate the real mean thickness of the panel V. For this reason, it is essential to take the drift of the thickness into account in the direction of thinning of the seam.

The following method of *"unbiased kriging of order k"* (also called "universal kriging") provides an unbiased linear estimator which takes the drift into account, provided that both the form of the drift $E\{Z(x)\} = m(x)$ and the covariance or variogram of the non-stationary RF $Z(x)$ are known.

1. *The form of the drift*

Let $Z(x)$ be the non-stationary point RF in the studied zone. By definition, the drift is the non-stationary expectation of this RF:

$$E\{Z(x)\} = m(x).$$

The non-stationary RF $Z(x)$ can then be expressed as the sum of a drift term and a residual term $Y(x) = Z(x) - m(x)$. This residual term $Y(x)$ may or may not be stationary but has a zero expectation:

$$Z(x) = m(x) + Y(x) \quad \text{with} \quad E\{Y(x)\} = 0, \qquad \forall x. \qquad \text{(V.8)}$$

The drift $m(x)$ expresses the regular and continuous variation of the RF $Z(x)$ at the scale of observation of the experimental data. The RF $Y(x)$ accounts for the erratic fluctuations around the drift, cf. Fig. V.3.

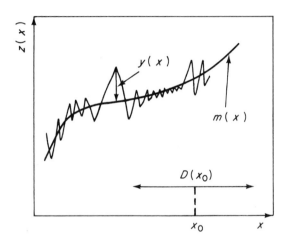

FIG. V.3. Drift and residual fluctuations.

The form of the drift $m(x)$ is supposed known. For example, $m(x)$ may be a linear combination of any known functions $\{f_l(x), \ l = 1 \text{ to } k\}$, the coefficients a_l of which are unknown, so that the drift $m(x)$ remains unknown:

$$m(x) = \sum_{l=1}^{k} a_l f_l(x). \qquad \text{(V.9)}$$

In mining applications, the estimation of a panel $V(x_0)$, centred on point x_0, is made from all the available data within a limited neighbourhood $D(x_0)$ around this panel. It is usually enough to express the drift in this neighbourhood $D(x_0)$ in the form of a polynomial of the first or second degree (the first terms of the Taylor expansion of the function $m(x)$, i.e.,

$$m(x) = a_1 + a_2 x \qquad \text{(V.10a)}$$

for a so-called "linear" drift,

$$m(x) = a_1 + a_2 x + a_3 x^2, \qquad \text{(V.10b)}$$

for a so-called "quadratic" drift.

It should be recalled that x represents a point in the working space. If this space is one-dimensional (e.g., time with coordinate t), then a quadratic drift would take the form

$$m(t) = a_1 + a_2 t + a_3 t^2.$$

If the working space is two-dimensional (e.g., horizontal regionalization of vertical thickness with the two rectangular coordinates u, v), then a quadratic drift would take the form

$$m(u, v) = a_1 + a_2 u + a_3 v + a_4 u^2 + a_5 v^2 + a_6 uv.$$

If the working space is three-dimensional, with the three rectangular coordinates (u, v, w), then a quadratic drift would take the form

$$m(u, v, w) = a_1 + a_2 u + a_3 v + a_4 w + a_5 u^2 + a_6 v^2$$
$$+ a_7 w^2 + a_8 uv + a_9 uw + a_{10} vw.$$

Of course, this drift may only be evident in some directions. For example, the phenomenon may be stationary in the horizontal plane with a drift in the vertical direction, in which case the previous quadratic drift would reduce to

$$m(u, v, w) = a_1 + a_2 w + a_3 w^2.$$

2. The covariance or variogram of the non-stationary RF

As in the stationary case, the kriging of a non-stationary RF requires the knowledge of either one of its structural functions, variogram or covariance:

$$C(x, y) = E\{Z(x)Z(y)\} - m(x) \cdot m(y) = E\{Y(x)Y(y)\},$$
$$2\gamma(x, y) = \text{Var}\{Z(x) - Z(y)\} = E\{[Z(x) - Z(y)]^2\} - [m(x) - m(y)]^2$$
$$= \text{Var}\{Y(x) - Y(y)\} = E\{[Y(x) - Y(y)]^2\}.$$

The semi-variogram $\gamma(x, y)$ is simply the semi-variogram of the residuals $Y(x) = Z(x) - m(x)$. Because of the unknown drift $m(x)$, it is not possible to estimate $\gamma(x, y)$ directly from the initial experimental data $z(x_\alpha)$. This would entail the simultaneous estimation of the drift $m(x)$ and the semi-variogram $\gamma(x, y)$ from a single realization $z(x)$ of the RF $Z(x)$, which is not rigorously possible.[†]

In mining applications, experience has shown that, in most instances, it is possible to take either of the following approximations for $\gamma(x, y)$:

  (i) the quasi-stationary semi-variogram $\gamma(h)$ calculated on neighbouring zones with a possible correction for a proportional effect;
 (ii) more simply, the behaviour at the origin (in general, linear, with or without nugget effect) of the neighbouring quasi-stationary semi-variogram, $\gamma(x, y) = \gamma(h) \simeq \varpi|h|$, where $h = (x - y)$ is the vector joining the two points x and y.

This uncertainty of the structural function $\gamma(x, y)$ or $C(x, y)$ means that the unbiased kriging of order k, as used in practice, will not result in the minimum estimation variance. This is not particularly important because the major interest is in avoiding any risk of bias due to the drift, while the minimization of the estimation variance is a secondary consideration.

Equations of the unbiased kriging of order k

Consider the estimation of the mean grade $Z_V(x_0)$, defined on panel $V(x_0)$, from n mean grades Z_α defined on the supports v_α.

  The corresponding *point support* RF $Z(x)$ is non-stationary and, within the estimation neighbourhood $D(x_0)$, it has an unknown drift of the form

$$E\{Z(x)\} = m(x) = \sum_{l=1}^{k} a_l f^l(x),$$

in which the functions $f_l(x)$ are known.

[†] An approximate iterative solution for the simultaneous inference of the drift $m(x)$ and the semi-variogram $\gamma(x, y)$ is proposed in G. Matheron (1969). Besides this iterative technique, Matheron has proposed a model of intrinsic RF, the so-called "generalized covariances", which can be inferred from a single experimental realization. The advantage of the structural analysis of intrinsic RF of order k is that it can be entirely programmed and is commercially available in packages such as Krigepack or Bluepack. On the other hand, this automatic procedure does not have the same flexibility as the "hand-made" structural analysis presented in Chapters III and IV; in mining applications, all sorts of physical and qualitative information (geology) intervene in a structural analysis.
  The following references are pertinent to these techniques: G. Matheron (1969), Ch. Huijbregts and G. Matheron (1970), G. Matheron (1971, p. 139) and G. Matheron (1973) for the theory of universal kriging and intrinsic RF of order k (these publications are of a rather advanced mathematical level and are not recommended to mining engineers or geologists who have no prior geostatistical background); P. Delfiner (1975), A. Haas and C. Jousselin (1975) for presentations of the packages Bluepack and Krigepack, respectively.

$D(x_0)$ is a neighbourhood centred on x_0 and includes the panel V and the supports of all the data used in the estimation. Within $D(x_0)$, the covariance $C(h)$ or the semi-variogram $\gamma(h)$ are known.

A linear combination of the n data is then formed to provide the estimator:

$$Z_K^* = \sum_{\alpha=1}^{n} \lambda_\alpha Z_\alpha.$$

Non-bias condition   The non-bias condition can be written as

$$E\{Z_V - Z_K^*\} = E\{Z_V\} - E\{Z_K^*\} = 0,$$

where

$$E\{Z_V\} = \frac{1}{V} \int_{V(x_0)} E\{Z(x)\}\, dx = \sum_l a_l \frac{1}{V} \int_{V(x_0)} f_l(x)\, dx = \sum_l a_l b_V^l,$$

$$E\{Z_K^*\} = \sum \lambda_\alpha E\{Z_\alpha\} = \sum_l a_l \frac{1}{v_\alpha} \int_{v_\alpha} f_l(x)\, dx = \sum_l a_l b_{v_\alpha}^l,$$

denoting the mean value on the support v of the known function $f_l(x)$ by b_v^l, i.e.,

$$b_v^l = \frac{1}{v} \int_v f_l(x)\, dx.$$

To obtain a zero-error expectation, it is enough to impose the k following conditions, called the "non-bias conditions":

$$\sum_{\alpha=1}^{n} \lambda_\alpha b_{v_\alpha}^l = b_v^l, \qquad \forall l = 1 \text{ to } k. \tag{V.11}$$

Minimum estimation variance   The preceding non-bias conditions allow all the terms involving the unknown drift $m(x)$ to be eliminated from the expression for the estimation variance, leaving

$$E\{[Z_V - Z_K^*]^2\} = \bar{C}(V, V) - 2 \sum_\alpha \lambda_\alpha \bar{C}(V, v_\alpha) + \sum_\alpha \sum_\beta \lambda_\alpha \lambda_\beta \bar{C}(v_\alpha, v_\beta).$$

This estimation variance then appears as a quadratic form in λ_α, λ_β, just as it did in the case of kriging under a stationary hypothesis. However, in this case the estimation variance must be minimized subject to the k non-bias constraints, instead of the single constraint for stationary kriging. Using the standard Lagrangian technique, a system of $(n + k)$ linear equations in $(n + k)$ unknowns is obtained (the n weights λ_α and the k Lagrange parameters μ_l). This system is called the "*unbiased kriging system of order k*" or the "universal kriging system".

$$\sum_{\beta=1}^{n} \lambda_\beta \bar{C}(v_\alpha, v_\beta) - \sum_{l=1}^{k} \mu_l b_{v_\alpha}^l = \bar{C}(v_\alpha, V), \qquad \forall \alpha = 1 \text{ to } n, \\ \left. \sum_{\beta=1}^{n} \lambda_\beta b_{v_\beta}^l = b_V^l, \qquad \forall l = 1 \text{ to } k. \right\} \qquad (V.12)$$

The minimum estimation variance or "kriging variance" can then be written as

$$\sigma_K^2 = \underset{\text{minimized}}{E\{[Z_V - Z_K^*]^2\}} = \bar{C}(V, V) + \sum_{l=1}^{k} \mu_l b_V^l - \sum_{\alpha=1}^{n} \lambda_\alpha \bar{C}(v_\alpha, V). \qquad (V.13)$$

This system can also be established from the semi-variogram $\gamma(h)$ instead of the covariance $C(h)$. A direct result is obtained by replacing the $\bar{C}$ by $(-\bar{\gamma})$ in system (V.12):

$$\sum_\beta \lambda_\beta \bar{\gamma}(v_\alpha, v_\beta) + \sum_l \mu_l b_{v_\alpha}^l = \bar{\gamma}(v_\alpha, V), \qquad \forall \alpha = 1 \text{ to } n, \\ \left. \sum_\beta \lambda_\beta b_{v_\beta}^l = b_V^l, \qquad \forall l = 1 \text{ to } k, \right\} \qquad (V.14)$$

and

$$\sigma_K^2 = \sum_\alpha \lambda_\alpha \bar{\gamma}(v_\alpha, V) + \sum_l \mu_l b_V^l - \bar{\gamma}(V, V).$$

Remark 1 The kriging system (V.1) in the stationary case can be seen as an unbiased kriging of order $k = 1$, the stationary expectation being reduced to an unknown constant a_1:

$$m(x) = a_1 f_1(x) = a_1 \quad \text{with} \quad f_1(x) = 1, \qquad \forall x.$$

Note that the mean value of the constant function $f_1(x) = 1$ over any volume v is equal to 1: $b_v^1 = 1, \forall v$.

Remark 2   Existence and uniqueness of the solution. As was the case under the stationary hypothesis, cf. section V.A.1, remark 1, the covariance matrix $[\bar{C}(v_\alpha, v_\beta)]$ is assumed to be strictly positive definite. It can then be shown that the system (V.12) of unbiased kriging of order k has a unique solution if and only if the k functions $f_l(x)$ are linearly independent on the set of the n data, i.e., if the n following linear relations

$$\sum_{l=1}^{k} c_l b_{v_\alpha}^l = 0, \qquad \forall \alpha = 1 \text{ to } n, \qquad (V.15)$$

imply that $c_l = 0$, $\forall l = 1$ to k. For example, consider a data set in which all the data are aligned perpendicularly to the direction u, as on Fig. V.4. It is obviously impossible to take a drift in the direction u into account, using just this set of data, even if this drift is only linear: $m(u) = a_1 + a_2 u$.

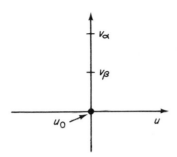

FIG. V.4. Aligned data.

In fact, in such a case, $b_{v_\alpha}^1 = 1$ and $b_{v_\alpha}^2 = u_0$, $\forall \alpha = 1$ to n, and the n relations (V.15) can be written as

$$c_1 b_{v_\alpha}^1 + c_2 b_{v_\alpha}^2 = c_1 + c_2 u_0 = 0, \qquad \forall \alpha = 1 \text{ to } n.$$

The condition $c_1 + c_2 u_0 = 0$ does not imply that c_1 and c_2 must be zero, since $c_1 = -u_0$ and $c_2 = 1$ also satisfy the condition.

Remark 3   Matrix form. The first member matrix $[K_k]$ of the system (V.12) of unbiased kriging of order k can be obtained by adding $(k-1)$ rows and $(k-1)$ columns to the first member matrix $[K]$ of the system (V.1) for kriging under the stationary hypothesis, i.e.,

$$[K_k] = \begin{bmatrix} \bar{C}(v_1, v_1) & \cdots & \bar{C}(v_1, v_n) & 1 & b_{v_1}^2 & \cdots & b_{v_1}^k \\ \vdots & & & & \vdots & & \vdots \\ \bar{C}(v_n, v_1) & \cdots & \bar{C}(v_n, v_n) & 1 & b_{v_n}^2 & \cdots & b_{v_n}^k \\ 1 & \cdots & 1 & 0 & 0 & \cdots & 0 \\ b_{v_1}^2 & \cdots & b_{v_n}^2 & 0 & 0 & \cdots & 0 \\ \vdots & & \vdots & \vdots & \vdots & & \vdots \\ b_{v_1}^k & \cdots & b_{v_n}^k & 0 & 0 & \cdots & 0 \end{bmatrix} \begin{array}{l} \left.\rule{0pt}{40pt}\right\} \begin{array}{l}(n+1)\\ \text{rows}\end{array} \\ \left.\rule{0pt}{40pt}\right\} \begin{array}{l}(k-1)\\ \text{rows}\end{array} \end{array}$$

$$\underbrace{\qquad\qquad}_{(n+1) \text{ columns}} \quad \underbrace{\qquad}_{(k-1) \text{columns}} \text{Main diagonal}$$

Note that the matrix $[K_k]$ is still symmetric.

The column matrices of the unknown $[\lambda]$ and of the second members $[M2_k]$ are obtained likewise:

$$
[\lambda] = \begin{bmatrix} \lambda_1 \\ \vdots \\ \lambda_n \\ -\mu_1 \\ -\mu_2 \\ \vdots \\ -\mu_k \end{bmatrix}, \qquad [M2_k] = \begin{bmatrix} \bar{C}(v_1, V) \\ \vdots \\ \bar{C}(v_n, V) \\ 1 \\ b_V^2 \\ \vdots \\ b_V^k \end{bmatrix}.
$$

The system (V.12) for unbiased kriging of order k can then be expressed in matrix notation as

$$[K_k] \cdot [\lambda] = [M2_k] \qquad\qquad (V.16)$$

$$[\lambda] = [K_k]^{-1} \cdot [M2_k],$$

and, thus,

$$\sigma_K^2 = \bar{C}(V, V) - [\lambda]^t \cdot [M2_k],$$

$[\lambda]^t$ being the row-matrix transpose of the column matrix $[\lambda]$.

V.A.3. Combining kriged estimates

Very often, in practice, we will be required to combine the kriged estimates of local units to form a "global" estimator of a zone or even of the entire deposit. Assuming an hypothesis of stationarity for the entire zone, the theorem for combining kriged estimates shows that the direct kriging of the required global mean yields the same results as combining the local kriged estimates. In practice, the second procedure is preferred for its greater flexibility, and because it does not require the preceding stationarity over the entire zone.

Kriging the global mean

Let $Z(x)$ be a point RF supposed to be either stationary of second order over the entire zone or deposit D, or to have a drift of known form on this extension D. The covariance or the variogram of $Z(x)$ is also supposed to be known.

Instead of estimating the mean value Z_{V_i} of each of the panels V_i which make up the zone D, it is possible to estimate the global mean value

$Z_D = (1/D)\int_D Z(x)\,dx$ directly over the entire zone D. A linear combination of the available data can be formed to provide the estimator $Z_D^* = \sum_\alpha \lambda_\alpha Z_\alpha$ and the standard kriging systems, (V.12) or (V.1), to be considered can be used by replacing the index V by D (depending on whether or not a drift is to be considered).

In mining applications, the global mean of an entire deposit D would very rarely be kriged directly for the two following reasons.

(i) It is not usually possible to assume stationarity or a single drift of known form over the entire deposit D, but only over limited neighbourhoods (local quasi-stationarity).

(ii) Even if stationarity could be verified over the entire deposit D, there are usually too many data in D to construct a kriging matrix and then to solve the kriging system. Moreover, the construction of such a kriging matrix would imply that the structural function, $C(h)$ or $\gamma(h)$, is known for distances $|h|$ of the order of the dimensions of the deposit D and, as was seen in section III.B.7, rule (III.42), the limit of reliability of an experimental semi-variogram is a distance $L/2$, half the dimensions of the field D.

Thus, to estimate the global mean value† over D, it is usually advisable to combine the estimates of the component volumes of D. If these local estimations are done by kriging, then the term "combination of kriged estimates" is used.

Theorem of combining kriged estimates

Consider again the point RF $Z(x)$, which is assumed to be either second-order stationary over the extent of the zone or deposit D, or to have a drift with a known form over the extension D.

The kriging systems (V.1) and (V.12) have the remarkable property that only the second-member matrices [M2] or [M2$_k$] depend on the support V to be estimated. The solution of the kriging system, $[\lambda] = [K]^{-1} \cdot [M2]$, is thus linear in V.

More precisely, let $\lambda_\alpha(x)$ be the solution of the kriging of the point value $Z(x)$ at point x from the data configuration $\{Z_\alpha, \alpha = 1$ to $n\}$. The mean

---

† *Remark* The global mean $Z_D = (1/D)\int_D Z(x)\,dx$, which has a precise physical meaning, should not be confused with the stationary mathematical expectation $m = E\{Z(x)\}$ of the probabilistic model. Z_D is an estimator of the expectation m only when the dimensions of the deposit D are very large with respect to the actual range of the stationary variogram or covariance representing the model (ergodic property).

value $Z_D = (1/D) \int_D Z(x) \, dx$ on the support D is then kriged from the *same* configuration and the solution $\lambda_\alpha(D)$ is written as

$$\lambda_\alpha(D) = \frac{1}{D} \int_D \lambda_\alpha(x) \, dx, \qquad \forall \alpha = 1 \text{ to } n. \qquad (V.17)$$

The kriging of the global mean on D can be expressed as a linear combination of the point krigings.

This relation also applies to the Lagrange parameters μ, which are simply the $(n + l)$th terms of the solution matrix $[\lambda]$:

$$\mu_l(D) = \frac{1}{D} \int_D \mu_l(s) \, dx, \qquad \forall l = 1 \text{ to } k.$$

To be entirely rigorous, this theorem is based on the assumption that all the partial estimations are carried out using the *same* data configuration, which means using all the data values available over D. In practice, partial kriging estimations are usually carried out within sliding neighbourhoods and the data base for each estimation is limited to the values within this neighbourhood. This practice can be justified for two reasons:

   (i) in most cases, these data neighbourhoods effectively screen out the influence of data values further away;
   (ii) the hypotheses of stationarity or form of the drift are usually limited to distances $|h|$ which are of the order of the dimensions of sliding neighbourhoods.

Remark   Forming the global estimator of a zone or a deposit D of any form by combining local elementary kriging estimates is done in the standard way by weighting the estimators by their respective supports.

For example, let $\{V_i, i = 1 \text{ to } N\}$ be N units of unequal supports V_i, and let $\{Z_{V_i}^*, i = 1 \text{ to } N\}$ be their kriged mean values. If the zone $D = \sum_{i=1}^{N} V_i$ consists of the sum of all the N units V_i, then its mean value can be written as

$$Z_D = \frac{1}{D} \sum_i V_i Z_{V_i} \qquad (V.18)$$

and its estimator is

$$Z_D^* = \frac{1}{D} \sum_i V_i Z_{V_i}^*.$$

If the units V_i are the mesh-points x_i of a regular grid, the estimator of a zone D is obtained by taking the mean of the kriged point values $Z_K^*(x_i)$

inside the polygonal contour which best approximates the form of the zone to be estimated, cf. Fig. V.5.

Variance of the global estimator

Kriging variances cannot be combined as simply as the kriging estimators. Let $Z_D^* = (1/D)\sum_i V_i Z_{V_i}^*$ be the combination of the N kriged values $Z_{V_i}^*$ and let $\sigma_{KV_i}^2 = E\{[Z_{V_i} - Z_{V_i}^*]^2\}$ be the N elementary kriging variances of the N units V_i. The global estimation variance of Z_D by the combination Z_D^* can be written as

$$\sigma_{ED}^2 = E\{[Z_D - Z_D^*]^2\} = E\left\{\left[\frac{1}{D}\sum_i V_i(Z_{V_i} - Z_{V_i}^*)\right]^2\right\}$$

and, thus,

$$\sigma_{ED}^2 = \frac{1}{D^2}\left[\sum_i V_i^2 \sigma_{KV_i}^2 + \sum_i \sum_{j \neq i} V_i V_j E\{(Z_{V_i} - Z_{V_i}^*)(Z_{V_j} - Z_{V_j}^*)\}\right]. \qquad \text{(V.19)}$$

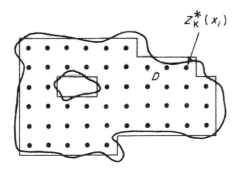

$$z_K^*(x_i)$$

FIG. V.5. Combining point-kriged estimates.

If the N elementary supports V_i are all equal, this expression simplifies to

$$\sigma_{ED}^2 = \frac{1}{N}\sum_i \sigma_{KV_i}^2 + \frac{1}{N^2}\sum_i \sum_{j \neq i} E\{(Z_{V_i} - Z_{V_i}^*)(Z_{V_j} - Z_{V_j}^*)\}.$$

Each time two elementary krigings $Z_{V_i}^*$ and $Z_{V_j}^*$ $(j \neq i)$ have considered common data, the two corresponding errors of estimation $[Z_{V_i} - Z_{V_i}^*]$ and $[Z_{V_j} - Z_{V_j}^*]$ are correlated. Thus, most of the $N(N-1)$ error covariances $E\{[Z_{V_i} - Z_{V_i}^*][Z_{V_j} - Z_{V_j}^*]\}$ are non-zero, and their sum cannot be neglected with regard to the first term $(1/N)\sum_i \sigma_{KV_i}^2$ of the expression of the global variance σ_{ED}^2. Theoretically, the calculation of these error covariances is possible if the semi-variogram $\gamma(h)$ is known for all distances $|h|$ between the data in D.

In mining applications, such a calculation is very awkward and, in addition, the semi-variogram $\gamma(h)$ is usually not defined for the large distances $|h|$ approaching the dimensions of the deposit D. The solution is then to calculate the global estimation variance σ_{ED}^2 with the aid of approximation principles which will be given in section V.C.1.

V.A.4. Cokriging

Sometimes, one variable may not have been sampled sufficiently to provide estimates of acceptable precision. The precision of this estimation may then be improved by considering the spatial correlations between this variable and other better-sampled variables. For example, the silver grade of a panel can be estimated by lead and zinc grades in addition to those of silver.

From a theoretical point of view, cokriging is no different from kriging; difficulties arise essentially from notations, since an extra index k is required to differentiate between the variables.

Consider a regionalization characterized by a set of K spatially intercorrelated random functions $\{Z_k(x), k = 1 \text{ to } K\}$. The first- and second-order moments of these RF, assuming stationarity, are

$$E\{Z_k(x)\} = m_k, \qquad \forall x;$$

$$E\{Z_{k'}(x+h) . Z_k(x)\} - m_{k'} . m_k = C_{k'k}(h), \quad \text{cross-covariance};$$

$$E\{[Z_{k'}(x+h) - Z_{k'}(x)][Z_k(x+h) - Z_k(x)]\} = 2\gamma_{k'k}(h), \quad \text{cross-variogram};$$

cf. section II.A.4.

The estimation of the mean value of the point RF $Z_{k_0}(x)$ over the support $V_{k_0}(x_o)$ is required, i.e.,

$$Z_{V_{k_0}} = \frac{1}{V_{k_0}} \int_{V_{k_0}(x_0)} Z_{k_0}(x)\,dx.$$

The available data $\{Z_{\alpha_k}, \alpha_k = 1 \text{ to } n_k, \text{ for each } k = 1 \text{ to } K\}$ are defined on the supports $\{v_{\alpha_k}\}$, e.g.,

$$Z_{\alpha_k} = \frac{1}{v_{\alpha_k}} \int_{v_{\alpha_k}} Z_k(x)\,dx.$$

The estimator $Z^*_{V_{k_0}}$ is a linear combination of all the available data values of all the K variables in coregionalization:

$$Z^*_{V_{k_0}} = \sum_{k=1}^{K} \sum_{\alpha_k=1}^{n_k} \lambda_{\alpha_k} Z_{\alpha_k}. \qquad (V.20)$$

The expectation of the error involved is

$$E\{Z_{V_{k_0}} - Z^*_{V_{k_0}}\} = E\{Z_{V_{k_0}}\} - \sum_{\alpha_{k_0}} \lambda_{\alpha_{k_0}} E\{Z_{\alpha_{k_0}}\} - \sum_{k \neq k_0} \sum_{\alpha_k} \lambda_{\alpha_k} E\{Z_{\alpha_k}\}$$

$$= m_{k_0}\left[1 - \sum_{\alpha_{k_0}} \lambda_{\alpha_{k_0}}\right] - \sum_{k \neq k_0} m_k \sum_{\alpha_k} \lambda_{\alpha_k}.$$

The non-bias, $E\{Z_{V_{k_0}} - Z^*_{V_{k_0}}\} = 0$, is then expressed by the K conditions†

$$\sum_{\alpha_{k_0}} \lambda_{\alpha_{k_0}} = 1 \quad \text{and} \quad \sum_{\alpha_k} \lambda_{\alpha_k} = 0, \qquad \forall k \neq k_0.$$

Note that the condition $\sum_{\alpha_{k_0}} \lambda_{\alpha_{k_0}} = 1$ cannot hold true if $n_{k_0} = 0$. In other words, cokriging requires at least one experimental value of the main variable Z_{k_0}.

The minimization of the variance $\sigma^2_{V_{k_0}} = E\{[Z_{V_{k_0}} - Z^*_{V_{k_0}}]^2\}$ under the K non-bias constraints results in a system of $(\sum_k n_k + K)$ linear equations (the unknown being the $\sum_k n_k$ weights λ_{α_k} and the K Lagrange parameters μ_k) called the "*cokriging system*":

$$\left.\begin{array}{l} \displaystyle\sum_{k'=1}^{K} \sum_{\beta_{k'}=1}^{n_{k'}} \lambda_{\beta_k'} \bar{C}_{k'k}(v_{\beta_{k'}}, v_{\alpha_k}) - \mu_k = \bar{C}_{k_0 k}(V_{k_0}, v_{\alpha_k}), \\[4mm] \qquad\qquad\qquad \forall \alpha_k = 1 \text{ to } n_k, \qquad \forall k = 1 \text{ to } K, \\[4mm] \displaystyle\sum_{\beta_{k_0}=1}^{n_{k_0}} \lambda_{\beta_{k_0}} = 1, \\[4mm] \displaystyle\sum_{\beta_k=1}^{n_k} \lambda_{\beta_k} = 0, \qquad \forall k \neq k_0. \end{array}\right\} \qquad (V.21)$$

The minimized estimation variance or "cokriging variance" can then be written as

$$\sigma^2_{V_{k_0}} = \bar{C}_{k_0 k_0}(V_{k_0}, V_{k_0}) + \mu_{k_0} - \sum_{k=1}^{K} \sum_{\alpha_k=1}^{n_k} \lambda_{\alpha_k} \bar{C}_{k_0 k}(V_{k_0}, v_{\alpha_k}) \qquad (V.22)$$

---

† It may happen that one of the secondary variables $Z_{k'}$, with $k' \neq k_0$, has the same meaning and the same expectation ($m_{k'} \equiv m_{k_0}$) as Z_{k_0}, e.g., Z_{k_0} and $Z_{k'}$ are two lead grades, but one is obtained from core analyses and the other one from cutting analyses. The two variables Z_{k_0} and $Z_{k'}$ may have the same expectation, but different structures of spatial variability (generally, their nugget effect would be different due to different measurement errors). The K previous non-bias conditions are then reduced to

$$\sum_{\alpha_{k_0}} \lambda_{\alpha_{k_0}} + \sum_{\alpha_{k'}} \lambda_{\alpha_{k'}} = 1 \quad \text{and} \quad \sum_{\alpha_k} \lambda_{\alpha_k} = 0, \qquad \forall k \neq k_0, k'.$$

(the notation $\bar{C}_{k'k}(v_{\beta_{k'}}, v_{\alpha_k})$ is the standard representation of the mean value of the cross-covariance $C_{k'k}(h)$ when one extremity of the vector h describes the support $v_{\beta_{k'}}$ and the other extremity describes independently the support v_{α_k}).

Remark 1 Unlike the kriging system, the cokriging system can be written simply in terms of the cross-semi-variogram $\gamma_{k'k}(h)$ only if $C_{k'k}(h) = C_{kk'}(h)$, which entails that $\gamma_{k'k}(h) = C_{k'k}(0) - C_{k'k}(h)$, i.e., if the cross-covariances are symmetric in $(h, -h)$, cf. section II.A.4. The cokriging system written in terms of cross-semi-variograms is then simply derived from system (V.21) by replacing $\bar{C}_{k'k}$ by $(-\bar{\gamma}_{k'k})$.

Remark 2   The existence and uniqueness of the solution. It can be shown that the cokriging system (V.21) provides a unique solution if the covariance matrix $[\bar{C}_{k'k}(v_{\beta_{k'}}, v_{\alpha_k})]$ is strictly positive definite, for which it is enough that a positive point coregionalization model $[C_{k'k}(h)]$ be adopted, cf. section III.B.3, and that no data value is totally redundant with respect to another, i.e., that $v_{\alpha_k} \neq v_{\beta_{k'}}$, $\forall \alpha_k \neq \beta_{k'}$.

However, as explained above, the non-bias condition is satisfied only if the set of data values of the main variable Z_{k_0} is not empty, i.e., it is necessary that $n_{k_0} \neq 0$.

Remark 3   Intrinsic coregionalization. It will be recalled from section III.B.3, formula (III.30), that the variables $\{Z_k(x), k = 1 \text{ to } K\}$ are said to be in intrinsic coregionalization if their cross-covariances or cross-variograms are all proportional to the same basic model:

$$C_{k'k}(h) = b_{k'k}K_0(h), \qquad \forall k, k'.$$

If the K data configurations of each variable Z_k are *identical*, then it can be shown that the cokriging of a particular variable Z_{k_0} is identical to the kriging of this variable from its corresponding data values $\{Z_{\alpha_{k_0}}, \alpha_{k_0} = 1 \text{ to } n_{k_0}\}$.

In the presence of an intrinsic coregionalization, the complication of cokriging can be justified only if one of the variables (k_0) is under-sampled when compared with the others $(k \neq k_0)$. This practical remark can be extended to all other coregionalization models.

In mining applications, a coregionalization study and subsequent cokriging are carried out only if one of the variables to be estimated is under-sampled with respect to the other variables with which it is spatially correlated, cf. Case Studies 2 and 3 of the following section V.A.5.

V.A.5. Case studies

The following case studies are simple enough to be carried out by hand, without the use of a computer. They are intended as a comparison of the kriging method and the more usual weighting methods, and to demonstrate the influence of the structural function, $\gamma(h)$ or $C(h)$.

Case Study 1. Estimation on a square

Consider a two-dimensional regionalization characterized by a point RF $Z(u, v)$ which, as a first approximation, is supposed to be intrinsic with an isotropic stationary semi-variogram: $\gamma(h) = \gamma(r)$ with $r = |h|$. For example, $Z(u, v)$ may be the vertical thickness of a sedimentary seam.

 The mean value Z_V is to be estimated over a square panel ABCD of side l from the non-symmetric configuration of the four data of support v, located at S_1 (central sample) and S_3, O_4, O_5 (peripheral samples) as shown on Fig. V.6.

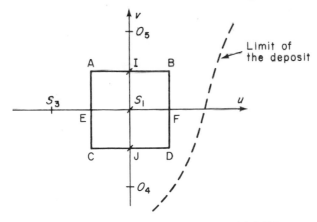

FIG. V.6. Estimation of square panel ABCD.

 For reasons of symmetry and for an isotropic semi-variogram $\gamma(r)$ the two data O_4 and O_5 receive the same weight, and they can thus be grouped together to form the set $S_2 = \{O_4 \cup O_5\}$ of support $2v$. On the other hand, the two data S_1 and S_3 must be treated separately. The linear estimator thus considered contains three weights:

$$Z_V^* = \sum_{\alpha=1}^{3} \lambda_\alpha Z(S_\alpha) \quad \text{with} \quad Z(S_2) = \tfrac{1}{2}[Z(O_4) + Z(O_5)].$$

The kriging system (V.3) is written as

$$\left.\begin{array}{l} \lambda_1\bar{\gamma}(S_1, S_1)+\lambda_2\bar{\gamma}(S_1, S_2)+\lambda_3\bar{\gamma}(S_1, S_3)+\mu = \bar{\gamma}(S_1, V), \\ \lambda_1\bar{\gamma}(S_2, S_1)+\lambda_2\bar{\gamma}(S_2, S_2)+\lambda_3\bar{\gamma}(S_2, S_3)+\mu = \bar{\gamma}(S_2, V), \\ \lambda_1\bar{\gamma}(S_3, S_1)+\lambda_2\bar{\gamma}(S_3, S_2)+\lambda_3\bar{\gamma}(S_3, S_3)+\mu = \bar{\gamma}(S_3, V), \\ \lambda_1+\lambda_2+\lambda_3 = 1 \end{array}\right\} \quad \text{(V.23)}$$

and

$$\sigma_K^2 = \lambda_1\bar{\gamma}(S_1, V)+\lambda_2\bar{\gamma}(S_2, V)+\lambda_3\bar{\gamma}(S_3, V)+\mu - \bar{\gamma}(V, V)$$

with, for the various mean values $\bar{\gamma}$,

$$\bar{\gamma}(S_1, S_1) = \bar{\gamma}(S_3, S_3) = \bar{\gamma}(v, v),$$

mean value over the data support,

$$\bar{\gamma}(S_2, S_2) = \bar{\gamma}(O_4, S_2) = \tfrac{1}{2}[\bar{\gamma}(v, v)+\gamma(2l)]$$

(the dimensions of the support v are negligible with respect to the length l),

$$\bar{\gamma}(S_1, S_3) = \bar{\gamma}(S_1, S_2) = \gamma(l) \quad \text{and} \quad \bar{\gamma}(S_2, S_3) = \gamma(l\sqrt{2}).$$

$$\bar{\gamma}(S_1, V) = \bar{\gamma}(S_1, \text{AIES}_1) = H(l/2; l/2),$$

$$\bar{\gamma}(S_3, V) = \bar{\gamma}(O_4, V) = \bar{\gamma}(O_4, \text{AIJC})$$

$$= \frac{2}{l^2}\left[\frac{3l^2}{4}H(3l/2; l/2)-\frac{l^2}{4}H(l/2; l/2)\right]$$

$$= \tfrac{3}{2}H(3l/2; l/2)-\tfrac{1}{2}H(l/2; l/2),$$

$$\bar{\gamma}(V, V) = F(l; l),$$

cf. in section II.E.2 the definition of the auxiliary functions H and F.

Numerical application A linear model with nugget effect is considered: $\gamma(r) = \{C_0(0)-C_0(r)\}+\varpi r$. The nugget effect $\{C_0(0)-C_0(r)\}$ is characterized on the regularization over the data support v by a nugget constant $A/v = C_{0v}$, cf. formula (III.7). The nugget constant on the regularization of support V is negligible with respect to that of support v:

$$V \gg v \text{ entails} \quad A/V = C_{0v}(v/V) \approx 0.$$

The auxiliary functions F and H for the linear model are given by formulae (II.91). By way of example, we have

$$\bar{\gamma}(V, V) = C_0(0)-A/V +0\cdot5213\varpi l \approx C_0(0)+0\cdot5213\varpi l,$$

$$\bar{\gamma}(v, v) = C_0(0)-A/v +0\cdot5213\varpi\varepsilon \approx C_0(0)-A/v,$$

as the dimension ε of the support v is very small compared to the distance l.

Three cases are considered:

(i) the pure nugget effect with $C_{0v} = 1$, $\varpi = 0$;
(ii) the partial nugget effect with $C_{0v} = 0.5$, $\varpi l = 0.959$;
(iii) the absence of nugget effect with $C_{0v} = 0$, $\varpi l = 1.918$.

The previous values of the parameters C_{0v} and ϖl have been chosen so as to ensure, in three cases, a constant dispersion variance of data of support v in the panel V:

$$D^2(v, V) = \bar{\gamma}(V, V) - \bar{\gamma}(v, v) = C_{0v} + 0.5213 \varpi l = 1.$$

The solution of system (V.23) is then immediate and the results are given in Table V.I and Fig. V.7.

TABLE V.I. *Estimation of a square panel (stationary case).*

Nugget effect	Kriging	Poly	ID	ID2
Pure	$\lambda_1 = 0.25$ $\lambda_2 = 0.50$ $\lambda_3 = 0.25$ $\sigma_K^2 = 0.25$	$\lambda_1 = 1$ $\lambda_2 = 0$ $\lambda_3 = 0$ $\sigma_E^2 = 1$	0.484 0.344 0.172 0.324	0.727 0.182 0.091 0.553
Partial	$\lambda_1 = 0.768$ $\lambda_2 = 0.172$ $\lambda_3 = 0.060$ $\sigma_K^2 = 0.651$	*idem* $\sigma_E^2 = 0.734$	*idem* 0.78	*idem* 0.671
Absent	$\lambda_1 = 1.163$ $\lambda_2 = -0.126$ $\lambda_3 = -0.037$ $\sigma_K^2 = 0.422$	*idem* $\sigma_E^2 = 0.468$	*idem* 1.23	*idem* 0.754

Remark 1 In the case of a pure nugget effect, the conditions of formula (V.7) are met: $\sigma_K^2 = C_{0v}/4 = 0.25$ and the weights are proportional to the supports, $\lambda_1 = \lambda_3 = \lambda_2/2$. There is no advantage linked to the privileged location of the central datum S_1.

Remark 2 As the nugget effect decreases and, thus, the spatial correlation increases, the influence of the central information becomes preponderant (λ_1 increases from 0.25 to 1.16). Note that for $\varpi \neq 0$ the weight of the datum S_2 (of support $2v$) is always greater than twice the weight of

S_3 (of support v); this is due to the fact that S_2 is nearer to the right-hand side of panel V, which is less sampled.

Remark 3   Note the negative (but still optimal) values of the weights λ_2 and λ_3 in the case of absence of nugget effect.

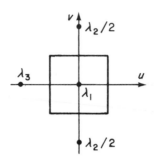

FIG. V.7. Weights used for the estimation of the square panel.

Other estimation procedures   The standard weighting methods (Poly, ID, ID2) do not take into account any of the characteristics peculiar to the regionalization under study and, consequently, provide the same estimator whatever the structural function $\gamma(h)$ may be, cf. Table V.I.

The polygon of influence method gives all the weight to the central datum: $\lambda_1 = 1$ and $\lambda_2 = \lambda_3 = 0$.

Both the inverse and inverse-squared distance weighting methods (ID and ID2) give the same weight to the three peripheral data: $\lambda_3 = \lambda_2/2$. The central datum S_1 is given greater weight by ID2 ($\lambda_1 = 0.727$) than by ID ($\lambda_1 = 0.484$). The mean distance from S_1 to the panel has been taken as a quarter of the diagonal, i.e., $l\sqrt{2}/4$.

In the stationary case, these standard methods ensure the unbiasedness of the estimation, as they satisfy the unique non-bias condition $\sum \lambda_\alpha = 1$, but they do not, by themselves, provide the estimation variances σ_E^2. For this purpose, it is necessary to characterize in some form the spatial variability of the phenomenon under study. The geostatistical method is to use the structural function $\gamma(h)$ with which the estimation variance of any unbiased linear estimator can be calculated from formula (II.28).

The following remarks can be made on the results of Table V.I.

   (i) In all cases, kriging provides the best unbiased linear estimator (BLUE).
   (ii) Depending on the nugget effect (i.e., the degree of spatial cor-
        relation), one or the other of the three standard estimators is close

enough to the kriging optimum, but only a structural analysis, i.e., a geostatistical approach, can tell which one is the closest.

(iii) Generally, the greater the nugget effect (thus, the greater the screen removing effect), the better are procedures (such as ID) which do not give too much weight to data preferentially placed with respect to the panel to be estimated. On the other hand, the smaller the nugget effect, the better are procedures (such as ID2) which favour well-located data (by screening the influence of furthest data).

(iv) An unfortunate choice (ID, for instance) in the absence of a nugget effect would result in an estimation variance almost equal to three times the kriging variance, i.e., a standard Gaussian confidence interval $\pm 2\sigma_E = \pm 2 \cdot 2$, almost *twice* the kriging confidence interval: $\pm 2\sigma_K = \pm 1 \cdot 3$.

Non-stationary case   Let us consider now a non-stationary regionalization characterized by a RF $Z(u, v)$ with an *anisotropic* linear drift which is evident only in the direction of the coordinate u, i.e.,

$$E\{Z(u, v)\} = m(u) = a_1 + a_2 u.$$

For example, $Z(u, v)$ may be the vertical thickness of a subhorizontal sedimentary seam which thins out towards one end in the direction u, or it may be the depth of the continental shelf, the direction u being perpendicular to the coastline, cf. Fig. V.8.

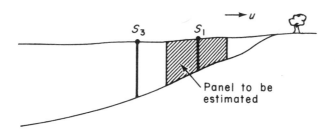

FIG. V.8. Linear drift in the direction u.

As previously, the mean value Z_V is to be estimated over the square panel V from the non-symmetric configuration of the four point data located at S_1, O_4, O_5, S_3, cf. Fig. V.6.

For reasons of symmetry, and since the drift does not occur in the direction v, the two data O_4 and O_5 will receive the same weight and they can thus be grouped into the set $S_2 = \{O_4 \cup O_5\}$ of support $2v$. And, as previously, a linear estimator with three weights can be defined:

$$Z_V^* = \sum_{\alpha=1}^{3} \lambda_\alpha Z(S_\alpha).$$

The drift $m(u) = a_1 + a_2 u = a_1 f_1(u) + a_2 f_2(u)$ consists of the two known functions, $f_1(u) = 1$, $\forall u$, and $f_2(u) = u$. The mean values of these functions over the supports S_α and V can thus be written as

$$b_{S_\alpha}^1 = b_V^1 = 1, \qquad b_{S_\alpha}^2 = u_\alpha \quad \text{(coordinate } u_\alpha \text{ of the support } S_\alpha\text{)},$$

$$b_V^2 = \frac{1}{V} \int_V m(u)\, du = \frac{1}{l} \int_{EF} u\, du = \frac{1}{l} \int_{c-l/2}^{c+l/2} u\, du = c,$$

if the coordinate u_1 of the central data S_1 is equal to c.

The system (V.14) of unbiased kriging of order $k = 2$ can then be written as

$$\left.\begin{array}{l} \displaystyle\sum_{\beta=1}^{3} \lambda_\beta \bar{\gamma}(S_\alpha, S_\beta) + \mu_1 + \mu_2 u_\alpha = \bar{\gamma}(S_\alpha, V), \qquad \forall \alpha = 1 \text{ to } 3, \\[2mm] \displaystyle\sum_\beta \lambda_\beta = 1, \\[2mm] \displaystyle\sum_\beta \lambda_\beta u_\beta = c \end{array}\right\} \qquad \text{(V.24)}$$

and

$$\sigma_K^2 = \sum_\alpha \lambda_\alpha \bar{\gamma}(S_\alpha, V) + \mu_1 + \mu_2 c - \bar{\gamma}(V, V).$$

With the coordinates $u_1 = u_2 = c$ and $u_3 = c - l$, the second non-bias condition can be expressed as

$$(\lambda_1 + \lambda_2)c + \lambda_3(c - l) = (\lambda_1 + \lambda_2 + \lambda_3)c - \lambda_3 l = c - \lambda_3 l = c,$$

which gives $\lambda_3 = 0$, i.e., the non-bias condition introduced by the presence of the linear drift amounts to ignoring the datum S_3, whatever the underlying structure $\gamma(h)$ may be.

By putting $\lambda_3 = 0$ and considering the tensorial invariance property of the kriging systems (cf. the addendum at the end of the present case study), it will be shown that the system (V.24) can be reduced to the system (V.26) of two linear equations in two unknowns.

Review of other weighting methods In the absence of a structural analysis providing evidence of the linear directional drift $m(u)$, the three usual weighting methods (Poly, ID, ID2) would provide the same weights as in the stationary case, i.e., the weights given in Table V.1 and reproduced in Table V.2. Consequently, the ID and ID2 estimators are biased, since they do not fulfil the second non-bias condition ($\lambda_3 = 0$). As for the Poly-estimator, it remains unbiased, since it gives a zero weight to the datum S_3 (but also – and wrongly – to the information S_2).

If, as in the case of the unbiased kriging of order 2, the form of the drift $m(u) = a_1 + a_2u$ is supposed known, this drift can then be estimated by least squares (abbreviated to LS). This estimated drift obviously passes through the value $Z(S_3)$ and also through the mean of the three values $Z(O_4)$, $Z(S_1)$ and $Z(O_5)$, i.e.,

$$m^*(u_1) = \tfrac{1}{3}[Z(S_1) + Z(O_4) + Z(O_5)].$$

The estimator of the mean drift over V can then be written as

$$m_V^* = \frac{1}{V}\int_V m^*(u)\,du = m^*(u_1) = \tfrac{1}{3}[Z(S_1) + Z(O_4) + Z(O_5)].$$

If that least-square estimator, m_V^*, is used to estimate the mean value Z_V over the panel V, then the weights are $\lambda_1 = 1/3$, $\lambda_2 = 2/3$, $\lambda_3 = 0$, and, thus, the non-bias condition is satisfied. The estimation variance, $\sigma_E^2 = E\{[Z_V - m_V^*]^2\}$, is calculated from formula (II.28), using the semi-variogram $\gamma(h) = \tfrac{1}{2}\mathrm{Var}\,\{Z(x+h) - Z(x)\}$ – the knowledge of which requires a geostatistical approach.

It is stressed that the estimation variance σ_E^2 of the LS estimator m_V^* has *nothing to do* with the minimized variance Q of the experimental residues. In this example, Q is equal to the experimental variance of the three data $Z(S_1)$, $Z(O_4)$, $Z(O_5)$:

$$Q = \tfrac{1}{3}[[Z(S_1) - m^*(u_1)]^2 + [Z(O_4) - m^*(u_1)]^2 + [Z(O_5) - m^*(u_1)]^2] \neq \sigma_E^2.$$

Numerical application As in the previous stationary case, an isotropic linear model with nugget effect is considered:

$$\gamma(h) = \tfrac{1}{3}\mathrm{Var}\,\{Z(x+h) - Z(x)\} = \{C_0(0) - C_0(r)\} + \varpi r, \qquad \forall r = |h|.$$

The same three cases (pure, partial and absent nugget effect) are considered with the same values for the parameters C_{0v} (nugget constant) and ϖl.

The results of kriging (unbiased of order 2) are compared to those of the four "non-regionalized" usual methods (i.e., Poly, ID, ID2 and LS) in Table V.2.

Remark 1   When there is a pure nugget effect (total absence of spatial correlation of the true residuals $Z(x) - m(x)$) the results obtained from the least squares method are the same as the optimal results given by kriging. It can be shown that this property remains true whatever data configuration is used (cf. G. Matheron, 1971, p. 162).

TABLE V.2. *Estimation of a square panel in the presence of a drift $m(u)$*

Nugget effect	Kriging	LS	Poly	ID	ID2
Pure	$\lambda_1 = 1/3$ $\lambda_2 = 2/3$ $\lambda_3 = 0$ $\sigma_K^3 = 1/3$	$\lambda_1 = 1/3$ $\lambda_2 = 2/3$ $\lambda_3 = 0$ $\sigma_E^2 = 1/3$	1 0 0 1	0·484 0·344 0·172 σ_E^2 non-defined	0·727 0·182 0·091
Partial	$\lambda_1 = 0·791$ $\lambda_2 = 0·209$ $\lambda_3 = 0$ $\sigma_K^3 = 0·659$	*idem* $\sigma_E^2 = 1·016$	*idem* 0.734	*idem*	*idem*
Absent	$\lambda_1 = \ \ 1·148$ $\lambda_2 = -0·148$ $\lambda_3 = \ \ 0$ $\sigma_K^3 = \ \ 0·426$	*idem* $\sigma_E^2 = 1·699$	*idem* 0.468	*idem*	*idem*
	Unbiased	Unbiased	Unbiased	Biased	Biased

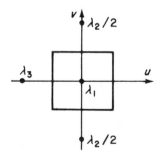

FIG. V.9. Weights used for estimation of the square panel.

Remark 2   As the nugget effect becomes smaller, the LS method deviates more and more from the optimum provided by kriging. In the case of absence of nugget effect, the deviation amounts to 400% in terms of estimation variance, $1·699/0·426 = 3·99$!!

On the other hand, the smaller the nugget effect, the less the Poly results, deviate from the kriging optimum.

Remark 3   In the presence of the directional drift $m(u)$, the ID and ID2 estimators are biased. This bias, $E\{Z_V - Z_V^*\}$, depends on the unknown drift $m(u)$ and is itself unknown. The existence of this unknown bias thus prohibits the definition and calculation of the estimation variance $\sigma_E^2 = \text{Var}\{Z_V - Z_V^*\}$.

Addendum. The tensorial invariance property of the kriging systems

(This subsection may be omitted in a first reading.)

Obviously, neither the kriging system (V.24) nor the corresponding kriging variance σ_K^2 should depend on the arbitrary choice of the origin of the coordinates, i.e., of the parameter c. In particular, this system must be equivalent to the following system (V.25) corresponding to the choice $c = 0$:

$$\left.\begin{array}{l} \sum_\beta \lambda'_\beta \bar\gamma(S_\alpha, S_\beta) + \mu'_1 + \mu'_2(u_\alpha - c) = \bar\gamma(S_\alpha, V), \qquad \forall \alpha = 1 \text{ to } 3, \\[2mm] \sum \lambda_\beta = 1, \\[2mm] \lambda_3 = 0. \end{array}\right\} \qquad \text{(V.25)}$$

In fact, with the change of variables $\mu'_1 = \mu_1 + \mu_2 c$, $\mu'_2 = \mu_2$, the two systems (V.24) and (V.25) are equivalent. Thus, $\lambda'_\beta = \lambda_\beta$, $\forall \beta$, since the solution of the kriging system is unique.

This invariance property of the kriging system with respect to changing the origin of coordinates is related to the polynomial character of the component functions $f_l(x)$ of the drift $m(x)$ and is the reason why only functions of this type are used, in practice, to characterize the form of the drift.

The equation corresponding to $\alpha = 3$ can be eliminated from system (V.24) as it only serves to define the parameter μ_2, which has no purpose when $c = 0$. The system (V.24) then reduces to (V.26) which, in the stationary case, is the same as kriging panel V by the two data S_1 and S_2:

$$\left.\begin{array}{l} \lambda_1 \bar\gamma(S_1, S_1) + (1 - \lambda_1)\bar\gamma(S_1, S_2) + \mu_1 = \bar\gamma(S_1, V), \\[2mm] \lambda_1 \bar\gamma(S_2, S_1) + (1 - \lambda_1)\bar\gamma(S_2, S_2) + \mu_1 = \bar\gamma(S_2, V) \end{array}\right\} \qquad \text{(V.26)}$$

and

$$\sigma_K^2 = \lambda_1 \bar\gamma(S_1, V) + (1 - \lambda_1)\bar\gamma(S_2, V) + \mu_1 - \bar\gamma(V, V).$$

Case Study 2. Cokriging on a square

Consider the coregionalization of the two stationary RF's $Z_1(x)$ and $Z_2(x)$ defined on a horizontal plane and characterized by the matrix of isotropic cross-semi-variograms $[\gamma_{kk'}(r)]$, $k, k' = 1$ to 2.

The mean value Z_{1V} of the main variable $Z_1(x)$ is to be estimated over a square panel V of dimensions $l \times l$, as shown on Fig. V.10. The available data consist of the central point value $Z_1(S_0)$ of the main variable and the five point values $Z_2(S_0)$, $Z_2(A)$, $Z_2(B)$, $Z_2(C)$, $Z_2(D)$ of the secondary variable located at the centre and the four corners of the square. This corresponds to the case of a variable Z_1, undersampled with respect to another variable (Z_2), and the problem is to evaluate the gain in precision provided by cokriging (Z_{1V} is estimated from the two data sets on Z_1 *and* Z_2) over kriging (Z_{1V} is estimated from the single data set on Z_1).

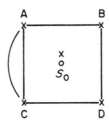

FIG. V.10. Cokriging on the square ABCD. ○, Data set Z_1; ×, data set Z_2.

Kriging amounts to estimating the panel V from its central value $Z_1(S_0)$. The kriging variance is, thus, given by the general formula (II.27):

$$\sigma_K^2 = E\{[Z_{1V} - Z_1(S_0)]^2\} = 2\bar{\gamma}_{11}(S_0, V) - \bar{\gamma}_{11}(V, V) - \bar{\gamma}_{11}(S_0, S_0)$$
$$= 2H_{11}(l/2, l/2) - F_{11}(l, l), \tag{V.27}$$

where H_{11} and F_{11} are the auxiliary functions corresponding to the point model $\gamma_{11}(r)$ which characterizes the regionalization of $Z_1(x)$.

For cokriging, reasons of symmetry (the coregionalization matrix being isotropic) allow the four peripheral samples in Z_2 to be grouped together into one single set S_R with a grade of

$$Z_2(S_R) = [Z_2(A) + Z_2(B) + Z_2(C) + Z_2(D)]/4.$$

The cokriging estimator is then a linear combination of three data:

$$Z_{1V}^* = \lambda_1 Z_1(S_0) + \lambda_2 Z_2(S_0) + \lambda_3 Z_2(S_R).$$

The two non-bias conditions of the cokriging system (V.21) can then be written as

$$\lambda_1 = 1 \quad \text{and} \quad \lambda_2 + \lambda_3 = 0.$$

By putting $\lambda = \lambda_3$, λ being the weight of $Z_2(S_R)$, the cokriging estimator is to be written as

$$Z_{1V}^* = Z_1(S_0) - \lambda Z_2(S_0) + \lambda Z_2(S_R). \tag{V.28}$$

The two non-bias conditions have already been taken into account by the form of the estimator (V.28) itself, and the cokriging system (V.21) reduces to the following three linear equations in three unknowns (λ and the two Lagrange parameters, μ_1 and μ_2):

$$\begin{cases} \bar{\gamma}_{11}(S_0, S_0) - \lambda \bar{\gamma}_{12}(S_0, S_0) + \lambda \bar{\gamma}_{12}(S_0, S_R) + \mu_1 = \bar{\gamma}_{11}(S_0, V) \\ \bar{\gamma}_{12}(S_0, S_0) - \lambda \bar{\gamma}_{22}(S_0, S_0) + \lambda \bar{\gamma}_{22}(S_0, S_R) + \mu_2 = \bar{\gamma}_{12}(S_0, V) \\ \bar{\gamma}_{12}(S_R, S_0) - \lambda \bar{\gamma}_{22}(S_R, S_0) + \lambda \bar{\gamma}_{22}(S_R, S_R) + \mu_2 = \bar{\gamma}_{12}(S_R, V) \end{cases}$$

and

$$\sigma_{CK}^2 = \underset{\text{minimized}}{E\{[Z_{1V} - Z_{1V}^*]^2\}}$$

$$= \bar{\gamma}_{11}(S_0, V) - \lambda \bar{\gamma}_{12}(S_0, V) + \lambda \bar{\gamma}_{12}(S_R, V) + \mu_1 - \bar{\gamma}_{11}(V, V),$$

where the mean values $\bar{\gamma}_{kk'}$ are

$$\bar{\gamma}_{kk'}(S_0, S_0) = \gamma_{kk'}(0) = 0, \qquad \bar{\gamma}_{kk'}(S_0, S_R) = \bar{\gamma}_{kk'}(S_0, A) = \gamma_{kk'}(l\sqrt{2}/2),$$

$$\bar{\gamma}_{kk'}(S_R, S_R) = \bar{\gamma}_{kk'}(S_R, A) = [2\gamma_{kk'}(l) + \gamma_{kk'}(l\sqrt{2})]/4,$$

$$\bar{\gamma}_{kk'}(S_0, V) = H_{kk'}(l/2; l/2),$$

$$\bar{\gamma}_{kk'}(S_R, V) = \bar{\gamma}_{kk'}(A, V) = H_{kk'}(l; l),$$

$$\bar{\gamma}_{kk'}(V, V) = F_{kk'}(l; l).$$

By simplifying, the preceding system gives the following solution:

$$\lambda = \frac{\bar{\gamma}_{12}(S_0, S_R) + \bar{\gamma}_{12}(S_0, V) - \bar{\gamma}_{12}(S_R, V)}{2\bar{\gamma}_{22}(S_0, S_R) - \bar{\gamma}_{22}(S_R, S_R)} = \frac{N \text{ (numerator)}}{D \text{ (denominator)}}.$$

$$\sigma_{CK}^2 = \sigma_K^2 - \lambda N, \tag{V.29}$$

σ_K^2 being given by formula (V.27). The gain in precision of cokriging over kriging is thus equal to

$$g = \sigma_K^2 - \sigma_{CK}^2 = \lambda N = N^2/D.$$

This gain is always positive since D is the extension variance of $Z_2(S_0)$ to $Z_2(S_R)$ and thus always positive or zero:

$$\sigma_E^2(S_0, S_R) = D \geq 0,$$

according to formula (II.27).

Numerical application 1   Consider first the case in which all the semi-variograms consist of a linear isotropic model with nugget effect:

$$\gamma_{kk'}(r) = C_{kk'} + \varpi_{kk'} r, \qquad \forall r > \varepsilon \text{ (very small range)}$$

The auxiliary functions H and F of the linear model are given by formulae (II.91) and, for $l \gg \varepsilon$, give

$$H_{kk'}(l; l) = C_{kk'} + 0.7652 \varpi_{kk'} l$$

and

$$F_{kk'}(l; l) = C_{kk'} + 0.5213 \varpi_{kk'} l.$$

For calculation purposes, $\gamma_{11}(r)$ will be fixed so that $\sigma_K^2 = 1$ and the gain $g = \lambda N$ will thus be relative: $g = g_r = \lambda N / \sigma_K^2$. For example, the parameters $C_{11} = 0.5$ and $\varpi_{11} l = 2.05$ will be taken, which entails $\sigma_K^2 = 1$, according to formula (V.27).

Similarly, $\gamma_{22}(r)$ will be fixed so that the variance D is small, which means that the gain, $g = N^2/D$, will be large. For example, the model $\gamma_{22}(r) = \frac{1}{5}\gamma_{11}(r)$ will be taken with the parameters $C_{22} = 0.1$ and $\varpi_{22} l = 0.41$. According to formula (V.29), D thus becomes

$$D = 2\gamma_{22}(l\sqrt{2}/2) - \tfrac{1}{2}\gamma_{22}(l) - \tfrac{1}{4}\gamma_{22}(l\sqrt{2})$$

$$= \tfrac{5}{4} C_{22} + 0.5607 \varpi_{22} l = 0.355.$$

With the two direct fixed models γ_{11} and γ_{22}, the cross-model $\gamma_{12}(r)$ will be made to vary, taking care that the coregionalization matrix $[-\gamma_{kk'}(r)]$ remains conditionally positive. For this purpose, it is enough that the two parameter matrices $[C_{kk'}]$ and $[\varpi_{kk'}]$ be positive definite, cf. section III.B.3, conditions (III.26). It is thus enough that the two following determinants be positive or zero:

$$\begin{vmatrix} C_{11} & C_{12} \\ C_{12} & C_{22} \end{vmatrix} \geq 0 \Leftrightarrow C_{12} \leq \sqrt{(C_{11} . C_{22})} = 0.2236.$$

$$\begin{vmatrix} \varpi_{11} & \varpi_{12} \\ \varpi_{12} & \varpi_{22} \end{vmatrix} \geq 0 \Leftrightarrow \varpi_{12} l \leq \sqrt{(\varpi_{11} l . \varpi_{22} l)} = 0.9168.$$

Table V.3 gives nine pairs $(C_{12}, \varpi_{12} l)$ which satisfy these conditions, and gives for each one of them the value λ of the weight assigned to the secondary sample $Z_2(S_R)$, as well as the relative gain g_r provided by cokriging.

Remark 1   When the increments $[Z_1(x+h) - Z_1(x)]$ and $[Z_2(x+r) - Z_2(x)]$ of the two variables are independent, the cross-semi-variogram $\gamma_{12}(r)$ is zero, and the secondary data on $Z_2(x)$ adds nothing to the estimation of the main variable $Z_1(x)$.

Remark 2   When the cross-variability (i.e., the parameters C_{12} and ϖ_{12}) increases, the gain of cokriging also increases. The maximum 77% gain is obtained with the maximum cross-variability.

Numerical application 2   Consider now the silver–lead coregionalization model adopted for the Broken Hill deposit (Australia) in section IV.C, Case Study 8. The silver grade is considered here as the main variable $Z_1(x)$, which is undersampled with respect to the lead grade $Z_2(x)$, this being often the case in silver–lead deposits.

TABLE V.3. *Influence of the cross-structure on cokriging*

C_{12}	$\varpi_{12}l$	λ	g_r	
0	0	0	0	Increments of Z_1 and Z_2 independent
0	0·411	0·38	0·05	
0	0·916	0·84	0·25	
0·1	0	0·28	0·03	γ_{12} is a pure nugget effect
0·1	0·411	0·66	0·15	Intrinsic coregionalization
0·1	0·916	1·12	0·45	
0·223	0	0·63	0·14	γ_{12} is a pure nugget effect
0·223	0·411	1·01	0·36	
0·223	0·916	1·47	0·77	γ_{12} has its maximum value

The adopted semi-variograms $\gamma_{kk}(r)$ (with a quasi-point support) consist of an isotropic spherical model γ_1 with nugget effect and range $a = 60$ ft, cf. Table (IV.1):

for silver grades, $\gamma_{11}(r) = 1\cdot1 + 1\cdot8\gamma_1(r)(\text{oz Ag t}^{-1})^2$;

for lead grades, $\gamma_{22}(r) = 11 + 39\gamma_1(r)(\% \text{ Pb})^2$;

for silver–lead $\gamma_{12}(r) = 5\gamma_1(r)(\text{oz Ag t}^{-1})(\% \text{ Pb})$.

The auxiliary functions H_1 and F_1 for the spherical model γ_1 are given by charts no. 3 and 4 in section II.E.4. The following results are thus obtained: for the kriging variance,

$$\sigma_K^2 = 2\cdot38(\text{oz Ag t}^{-1})^2;$$

for cokriging, according to formulae (V.29),

$$N = 4\cdot125(\text{oz Ag t}^{-1})(\% \text{ Pb}); \qquad D = 62.5(\% \text{ Pb})^2;$$

hence, $\lambda = N/D = 0\cdot066(\text{oz. Ag t}^{-1})(\% \text{ Pb})^{-1}$, which corresponds to a relative gain $g_r = \lambda N/\sigma_K^2 = 0\cdot114 \approx 11\%$

Thus, although the silver grade is quite undersampled with respect to the lead grade, the gain of cokriging over kriging is negligible for the configuration studied here. This is due to both the fact that the only available data for silver, $Z_1(S_0)$ is well located, and that the cross-variability γ_{12} is not high.

To sum up this case study on cokriging, the following advice can be given: before undertaking a coregionalization study, first proceed to the calculation of the possible gain of the planned cokriging, considering a simple data configuration.

Case Study 3. Cokriging with systematic errors

In a deposit mined by open-pit, two types of data are available on the grade of interest.

(i) An unbiased data set derived from drill cores. The analyses of the cores are considered as the true grades $Z_1(x)$ of these cores.

(ii) A *biased* (e.g., because of selective loss of fines being blown away) and imprecise data set derived from blast-hole cuttings. The analyses of these cuttings are considered as realizations of the RF $Z_2(x)$, which differs from the true grade $Z_1(x)$ by a random error $\varepsilon(x)$, i.e.,

$$Z_2(x) = Z_1(x) + \varepsilon(x).$$

In general, the poor quality of the cuttings data set is balanced by the large number of blast-holes on any mining bench.

To estimate the true mean grade Z_{1V} of a block with support V, the following estimator is used:

$$Z_{1V}^* = \sum_{\alpha_1 \in S_1} \lambda_{\alpha_1} Z_1(x_{\alpha_1}) + \sum_{\alpha_2 \in S_2} \lambda_{\alpha_2} Z_2(x_{\alpha_2}),$$

S_1 and S_2 being the two sets of core data and cuttings data, respectively. The supports of the core lengths and cuttings are negligible with respect to the support V of the block.

As the matrix of coregionalization of the two variables Z_1 and Z_2 is $[\gamma_{kk'}(h), k, k' = 1, 2]$, the cokriging system (V.21) is written in terms of semi-variograms:

$$
\left.
\begin{aligned}
&\sum_{\beta_1 \in S_1} \lambda_{\beta_1} \bar{\gamma}_{11}(x_{\alpha_1}, x_{\beta_1}) + \sum_{\beta_2 \in S_2} \lambda_{\beta_2} \bar{\gamma}_{12}(x_{\alpha_1}, x_{\beta_2}) + \mu_1 \\
&\qquad\qquad\qquad = \bar{\gamma}_{11}(x_{\alpha_1}, V), \qquad \forall \alpha_1 \in S_1, \\[2mm]
&\sum_{\beta_1 \in S_1} \lambda_{\beta_1} \bar{\gamma}_{12}(x_{\alpha_2}, x_{\beta_1}) + \sum_{\beta_2 \in S_2} \lambda_{\beta_2} \bar{\gamma}_{22}(x_{\alpha_2}, x_{\beta_2}) + \mu_2 \\[1mm]
&\qquad\qquad\qquad = \bar{\gamma}_{12}(x_{\alpha_2}, V), \qquad \forall \alpha_2 \in S_2, \\[2mm]
&\sum_{\beta_1 \in S_1} \lambda_{\beta_1} = 1, \\[2mm]
&\sum_{\beta_2 \in S_2} \lambda_{\beta_2} = 0
\end{aligned}
\right\} \quad \text{(V.30)}
$$

and

$$\sigma_{CK}^2 = \sum_{\alpha_1 \in S_1} \lambda_{\alpha_1} \bar{\gamma}_{11}(x_{\alpha_1}, V) + \sum_{\alpha_2 \in S_2} \lambda_{\alpha_2} \bar{\gamma}_{12}(x_{\alpha_2}, V) + \mu_1 - \bar{\gamma}_{11}(V, V).$$

It should be recalled, from section V.A.4, remark 2, that the set S_1 of the unbiased data $Z_1(x_{\alpha_1})$ must not be empty. This restriction is quite natural, since the unknown bias of the data set S_2 cannot be corrected unless there are actually some unbiased data in S_1. Under the condition $S_1 \neq \varnothing$, the influence of the unknown bias of the data set S_2 is made nil by the relation

$$\sum_{\beta_2 \in S_2} \lambda_{\beta_2} = 0,$$

and there is no need to evaluate this bias $E\{\varepsilon(x)\}$ when cokriging.

However, when this bias is known, i.e., when an unbiased cutting data set S_2 can be defined, the two non-bias conditions of the preceding system (V.30) reduce to the unique standard non-bias condition:

$$\sum_{\beta_1 \in S_1} \lambda_{\beta_1} + \sum_{\beta_2 \in S_2} \lambda_{\beta_2} = 1.$$

But there is still a need for cokriging, since the two spatial structures γ_{11} and γ_{22} of the two types of data may be different.

Numerical application   The three RF's $Z_1(x)$, $Z_2(x)$, $\varepsilon(x)$ have the following characteristics.

First, they are all stationary with unknown expectations:

$$E\{Z_1(x)\} = m_1; \qquad E\{Z_2(x)\} = m_2;$$
$$E\{\varepsilon(x)\} = E\{Z_2(x) - Z_1(x)\} = m_2 - m_1,$$

equal to the unknown bias between the two data sets.

Second, the structure of the true grades $Z_1(x)$ is characterized by the quasi-point model $\gamma_{11}(h)$ derived from drill-core data.

Third, as a first approximation, the error $\varepsilon(x)$ can be assumed to be spatially uncorrelated (the loss of fines not being regionalized). Thus, the semi-variogram of the error $\varepsilon(x)$ reduces to a pure nugget effect:

$$\gamma_2(h) = \tfrac{1}{2} E\{[\varepsilon(x+h) - \varepsilon(x)]^2\} = \mathrm{Var}\{\varepsilon(x)\} = C_\varepsilon, \qquad \forall |h| > a_0,$$

a_0 being a very small range compared to the distance $|h|$ of the experimental observation.

Fourth, the error $\varepsilon(x)$ is supposed to be independent of the true grade $Z_1(x)$ (this is not always true, e.g., in banded hematite quartzite deposits, the quantity of fines is inversely proportional to the iron grade Z_1). Thus,

$$\gamma_{12}(h) = \tfrac{1}{2}E\{[Z_1(x+h)-Z_1(x)][Z_2(x+h)-Z_2(x)]\} = \gamma_{11}(h)$$

as

$$Z_2(x+h)-Z_2(x) = [Z_1(x+h)-Z_1(x)]+[\varepsilon(x+h)-\varepsilon(x)],$$

and the increments of $\varepsilon(x)$ are independent of the increments of Z_1. Similarly,

$$\gamma_{22}(h) = \tfrac{1}{2}E\{[Z_2(x+h)-Z_2(x)]^2\} = \gamma_{11}(h)+\gamma_\varepsilon(h)$$

$$= \gamma_{11}(h)+C_\varepsilon.$$

Thus, the following regionalization model is obtained:

$$\gamma_{11}(h) = \gamma_{12}(h) \quad \text{and} \quad \gamma_{22}(h) = \gamma_{11}(h)+C_\varepsilon, \qquad \forall|h|>a_0.$$

Indeed, in practice, the two direct semi-variograms γ_{11} and γ_{22} are calculated from the corresponding experimental data and they usually differ only by the nugget effect:

$$\gamma_{22}(h)-\gamma_{11}(h) = C_\varepsilon, \qquad \forall|h|>a_0.$$

The cokriging configuration of Case Study 2, above, cf. Fig. V.10, will be taken as an example, i.e., the estimation of the mean grade Z_{1V} over a square block V of side $l = 20$ m by the two data sets $Z_1(S_0)$ with weight 1 for the central core analysis, $Z_2(S_0)$ with weight $-\lambda$ and $Z_2(S_R)$ with weight $+\lambda$ for the cutting analyses at the centre S_0 and the corners S_R of the block.

The quasi-point semi-variogram $\gamma_{11}(h)$ is the model fitted to the experimental semi-variogram of the iron grades of core length at Tazadit (Mauretania), cf. section II.D.5, Case Study 1, i.e., a spherical model without nugget effect, range $a = 40$ m and sill $C = 133$ (% Fe)^2.

For the direct estimation of Z_{1V} by the only central core grade $Z_1(S_0)$, the variance σ_K^2, given by (V.27), is

$$\sigma_K^2 = 0.185C = 24.6(\% \text{ Fe})^2.$$

Cokriging, i.e., taking the cutting data into account, gives the following results from (V.29):

$$N = 0.243C = 32.3 \qquad \text{and} \qquad D = 1.25\, C_\varepsilon + 0.451\, C.$$

By taking the variance of the measurement errors C_ε equal to the first experimental value $\gamma_{11}(3)$, i.e., $C_\varepsilon = 15$, D becomes 78.7, and the weight assigned to the corner cuttings data $Z_2(S_R)$ is $\lambda = N/D = 0.41$. The gain of cokriging over kriging is thus equal to

$$g = \sigma_K^2 - \sigma_{CK}^2 = AN = 13.3 \ (\% \text{ Fe})^2,$$

i.e., a fairly good relative gain: $g_r = g/\sigma_{CK}^2 = 0.54$.

In spite of their bias, taking the cuttings data into account provides a relative gain of more than 50% in terms of estimation variance.

V.B. THE APPLICATION OF KRIGING

V.B.1. A comparison of kriging with other weighting methods

If we are limited to the class of linear estimators (the estimator $Z^* = \sum_{\alpha=1}^{n} \lambda_{\alpha} Z_{\alpha}$ is a linear combination of the n data values) and if the structural analysis is sufficient to infer correctly the underlying structural function (covariance or variogram), then, by definition, kriging provides the best unbiased estimator (in the sense of minimum estimation variance). This property is irrefutable and is not shared by the more classical linear estimation methods such as polygons of influence (Poly), inverse distances (ID), inverse-square distances (ID2) and least-square polynomials (LS), to cite only the methods more commonly used in mining applications.

The two criticisms most often levelled at kriging are:

(i) the gain in terms of estimation variance provided by kriging is negligible in practice when compared with more common estimation procedures;

(ii) the application of kriging, as well as geostatistics in general, is difficult and costly in terms of both sampling and computer time.

The following responses can be made to these criticisms.

First, as kriging is optimal, small deviations in the choice of weights around their optimal values will only result in a second-order increase in the estimation variance. This means that if, for any given data configuration, the weights provided by any of the usual estimation methods differ little from the kriging weights, then the gain due to kriging can be neglected, and the more usual method will be preferred for this particular data configuration because of the simplicity of its application. *However*, it still must be known which method most closely approximates kriging, and this can only be determined by a structural analysis to calculate the variogram to be used in the kriging solution. When a strong screen effect is present, Poly will usually be adopted, whereas ID will usually be better when the screen effect is weak. A modified version of ID2 may prove better when anisotropies are present, and LS should usually be preferred in cases of non-stationarity, cf. Case Study 1, section V.A.5. Before a structural analysis is made, it can never be determined which of these methods could, in practice, replace kriging; it depends not only on the variogram (i.e., on the structure of variability), but also on the data configuration. Since, in mining applications, the configuration of the data used to estimate a panel may vary considerably throughout the deposit, kriging remains the surest solution (and not necessarily the costliest).

Second, the understanding of geostatistics and the application of a structural analysis followed by kriging is, without doubt, more difficult than the arbitrary choice of one of the usual weighting methods.

Is geostatistics really costly in terms of sampling? Geostatistics quantifies the quality of the available data by means of the kriging variance, and this allows cases in which the amount of data is dangerously insufficient to be identified. To reject the determination of variability structure by using a method which does not have an associated estimation variance† amounts to ignoring completely the problems involved in estimation.

Geostatistical studies can also result in the conclusion that there are too many data for a given objective. Used in time, geostatistics can result in considerable savings in sampling, cf. A. Journel (1973b).

Is the computer cost of kriging really prohibitive? Even though computer costs are always negligible when compared to the cost of obtaining a few core samples, the answer to this question is yes, if no precaution is taken when writing and solving the kriging system. Now, if all the symmetries and approximations are taken into account by establishing a "kriging plan", then both the dimension of the kriging system and its solving cost can be reduced.

The theoretical optimum of kriging is of little practical use if the application of kriging is not competitive in terms of computer cost.

V.B.2. Elaborating a kriging plan.

Consider the estimation of the mean grade Z_V of a panel of support V from the n data Z_α defined on supports v_α. The kriging system (V.1) is written as

$$\begin{cases} \sum_\beta \lambda_\beta \bar{C}(v_\alpha, v_\beta) - \mu = \bar{C}(v_\alpha, V), & \forall \alpha = 1 \text{ to } n, \\\\ \sum_\beta \lambda_\beta = 1 \end{cases}$$

and

$$\sigma_K^2 = \bar{C}(V, V) + \mu - \sum_\alpha \lambda_\alpha \bar{C}(v_\alpha, V).$$

(When a drift is present, system (V.12) should be used.)

---

† Once again, it should be recalled here that the residual variance of the least-squares polynomial method (LS) is not the estimation variance, cf. Case Study 1, section V.A.5 and G. Matheron (1967).

The various steps in the practical application of the kriging of each panel V are as follows.

(i) Select the data Z_α to be used in the estimation of Z_V
(ii) Calculate the mean covariances $\bar{C}(v_\alpha, v_\beta)$ between all the selected data (these are the elements of the first member matrix [K] of the kriging system).
(iii) Calculate the mean covariances $\bar{C}(v_\alpha, V)$ between each of the selected data and the panel V to be estimated (these are the elements of the second member matrix [M2]).
(iv) Choose an algorithm for the solution of the linear system.

To carry out a "crude" kriging without any consideration for computer cost would consist of taking all available data without distinction within a neighbourhood of quasi-stationarity centred on each panel V, calculating exactly the various mean values $\bar{C}$ and then applying a precise algorithm which would give an exact solution of the linear system (such as SIMQ-SSP IBM). Such an application would yield very precise results, but would also be very costly in terms of computer time.

To establish a practical alternative, each one of the preceding steps must be critically examined to provide an approximate procedure which will minimize computer time while yielding results that are within acceptable limits of approximation.

The five essential points to be used in reducing the kriging cost are:

1. the reduction of the dimension of the kriging system;
2. the reduction of the number of kriging systems;
3. the rapid calculation of the various mean values $\bar{C}$ (or $\bar{\gamma}$);
4. the preparation of a data file adapted to the kriging plan;
5. the choice of a time-competitive algorithm for the resolution of the kriging system.

1. *Reduction of the dimension of the kriging systems*

The dimension $(n+1)$ of a kriging system is directly related to the number n of data, Z_α, to which a weight λ_α is to be assigned. To reduce this dimension, either the estimation neighbourhood can be reduced or data can be grouped together.

Limiting the estimation neighbourhood   The first limitation of the estimation neighbourhood is that of the neighbourhood of quasi-stationarity defined by the structural analysis: the semi-variogram $\gamma(h)$ is known only for distances $|h| < b$. Thus, to estimate a panel $V(x_0)$ centred on x_0, only the data inside a neighbourhood $b(x_0)$ centred on x_0 are considered, cf. Fig. V.11. This neighbourhood is not necessarily isotropic.

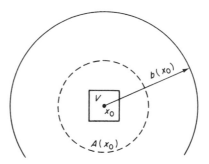

FIG. V.11. Neighbourhood of estimation $A(x_0)$.

It may happen that the data within $b(x_0)$ are either too few, in which case more distant data may be included in considering a non-stationary kriging system, or too many, in which case a reduced neighbourhood $A(x_0)$ may be considered. In fact, this procedure amounts to neglecting the weights of data exterior to $A(x_0)$, cf. Fig. V.11. This approximation is justified when the data within $A(x_0)$ can be considered as sufficiently screening the influence of the data exterior to $A(x_0)$. This screen effect increases as the nugget constant of the underlying structural function becomes smaller (cf. section V.A.1. remark 7, "The influence of the nugget effect") and also as the data within $A(x_0)$ surrounds more completely the panel $V(x_0)$ in all directions, without gaps.

A prime objective in defining a data neighbourhood $A(x_0)$ should be to ensure that it avoids all risks of bias in the estimation. Thus, in two dimensions, for example, when the data configuration is such that one direction β is undersampled, although the structural continuity in this direction is significant, it is advisable to extend the estimation neighbourhood $A(x_0)$ in this direction so as to include more data Z_β, which will take into account the β-direction continuity in the estimation, cf. Fig. V.12. Examples of such a procedure are the automatic mapping programs that partition the space around x_0 into equal parts (e.g., four quadrants in R^2, or eight octants in R^3), and only the two or three data closest to x_0 are considered within each of these parts.

It often happens when estimating panels on the border of a deposit that most of the data lie on one side of the panel, i.e., they are preferentially located in a rich zone, cf. Fig. V.13. In such a case, it is advisable to do one of the following:

(i) limit the extent of the neighbourhood $A(x_0)$ to the favoured side of the panel, so as not to give too much influence to the preferentially rich data;

(ii) introduce, with due care, a supplementary non-bias condition corresponding, for example, to a linear drift $m(x_\alpha) = a_0 + a_1 x_\alpha$ in the direction α of impoverishment;

(iii) not carry out the estimation without any additional data – this would be the proper course if preferential data positioning as well as heterogeneity of the mineralization were suspected.

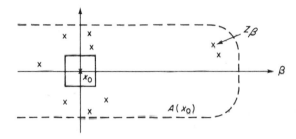

FIG. V.12. Anisotropic neighbourhood of estimation.

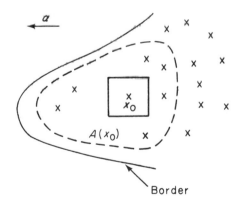

FIG. V.13. Estimation at the border of a deposit.

Grouping of the data   A second method of reducing the dimension of the kriging system consists in grouping the n data $\{Z_i$ of support $v_i, i = 1$ to $n\}$ available within the estimation neighbourhood into n' data sets $\{Z_k, k = 1$ to $n', n' < n\}$. Each data set of mean value Z_k is defined on a support S_k which is the union of the supports v_{i_k} of the component data Z_{i_k}, i.e., $S_k = \bigcup v_{i_k}$.

Thus, on Fig. V.14, the estimation neighbourhood of the square panel V is taken as the union of nine squares of the same size (the square V itself plus the eight squares which constitute the first aureole around V). The available data are classed into nine sets S_k according to their locations within one or the other of the nine squares. Let $\{Z_k, k = 1$ to $9\}$ be the mean grades of these nine sets. The estimator considered is, thus, a linear combination of the nine mean grades Z_k, i.e.,

$$Z^* = \sum_{k=1}^{9} \lambda_k Z_k.$$

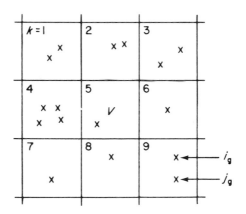

FIG. V.14. Grouping the information into nine sets.

The datum Z_9, for example, which corresponds to the square $k = 9$, consists of the union of the two elementary data Z_{i_9} and Z_{j_9}. If these two elementary data have a quasi-point support and are located at points x_{i_9} and x_{j_9}, the following relations hold:

$$Z_9 = (Z_{i_9} + Z_{j_9})/2,$$

$$\bar{\gamma}(S_9, S_k) = [\bar{\gamma}(x_{i_9}, S_k) + \bar{\gamma}(x_{j_9}, S_k)]/2, \quad \forall S_k.$$

This grouping of the available data into nine sets allows the initial dimension (17) of the kriging system to be reduced to 10. The corollary of this reduction in the dimension of the system is a loss in precision: the kriging variance resulting from the nine mean grades Z_k will be greater than that resulting from the 16 elementary grades. If the data are distributed more or less uniformly, as shown on Fig. V.14, then this loss in precision will, in general, be negligible. To evaluate this loss in precision, the two kriging systems can be solved for an average data configuration; it can then be determined whether the gain in computer time justifies the loss in precision.

However, grouping the data into sets does not always involve a loss in precision, particularly when the grouping is based upon a structural symmetry.

Structural symmetry   All data that would receive *strictly* equal weights from kriging can be grouped together. This grouping reduces the dimension of the kriging system but does not alter the kriged value or the kriging variance.

Such data must be "structurally" symmetric with respect to the support V to be estimated. This symmetry will be identical to geometrical symmetry only if the structural function $\gamma(h)$ is isotropic. For example, consider the estimation of the square panel V on Fig. V.15 by the geometrically symmetric configuration of the five quasi-point grades $Z(x_1), \ldots, Z(x_5)$. If the structural function $\gamma(h_\alpha, h_\beta)$ is isotropic, then the four exterior data will have exactly the same weights and they can, thus, be grouped into one single set called the "first aureole" with a mean grade $Z_A = [Z(x_2) + Z(x_3) + Z(x_4) + Z(x_5)]/4$. The two-weight linear estimator is then $Z_1^* = \lambda_1 Z(x_1) + \lambda_2 Z_A$.

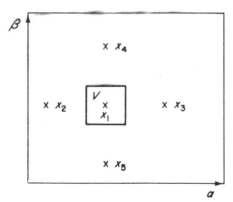

FIG. V.15. Structural and geometrical symmetries.

However, if the structural function $\gamma(h_\alpha, h_\beta)$ is anisotropic with, for example, α being a preferential direction of continuity, the grouping would then be along the two perpendicular α and β with the two mean grades $Z_\alpha = [Z(x_2) + Z(x_3)]/2$, and $Z_\beta = [Z(x_4) + Z(x_5)]/2$. The three-weight linear estimator would then be $Z_2^* = \lambda_1 Z(x_1) + \lambda_2 Z_\alpha + \lambda_3 Z_\beta$. Other examples of such symmetrical grouping were given in the case studies, cf. section V.A.5.

2. *Reducing the number of kriging systems*

In section V.A.1, remark 5, it has already been noted that if the data configuration is repeated from one panel to another, the first member matrix [K] is the same for each panel and it is necessary to take its inverse only once. Moreover, if the panels to be estimated are equal and have the same geometry, the second member matrix [M2] is also the same for each panel and, thus, there is a unique kriging system which needs to be solved only once. The system of solution weights, $[\lambda] = [K]^{-1} \cdot [M2]$ is unique and applies to all the panels to be estimated, cf. the case study of the kriging of Donoso copper mine in section V.B.5, Fig. V.27.

Systematic sampling on a regular grid is, in this sense, particularly favourable for precise and rapid kriging. In addition, such regular sampling plans facilitate the structural analysis which can be carried out precisely and without any risk of bias due to preferential positioning of the data.

In practice, it is, thus, very convenient to have a data configuration that is repeated identically from one panel to another. This may be achieved by:

(i) insisting on a systematic and regular sampling pattern when the sampling campaign is being prepared;
(ii) defining the panels to be estimated on a constant support, at least within the central homogeneous zones of the deposit;
(iii) considering a unique "mean" data configuration by using the techniques of random kriging or kriging of gaps in the data grid, cf. the following paragraphs.

Definition of the support to be estimated   A constant estimation support is very frequent in mining applications as it very often corresponds to constant mining or selection units.

The geometry of the estimation support V should be simple (rectangular in two dimensions or parallelepipedic in three dimensions) so as to make the calculation of the terms $\bar{\gamma}(v_\alpha, V)$ as easy as possible. The division of the deposit into panels to be estimated should be in geometric harmony with the regular data configuration. Thus, for the two-dimensional case shown

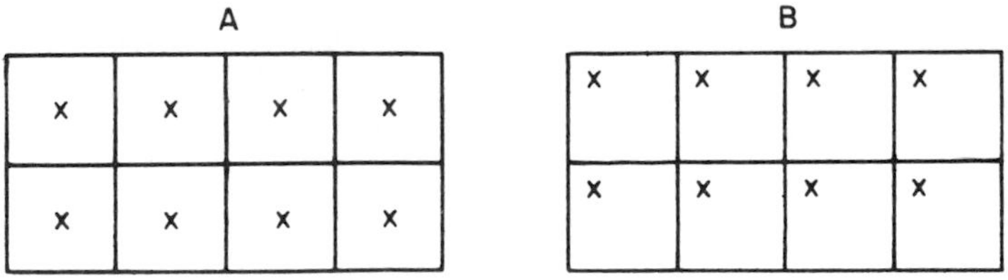

FIG. V.16. Relative locations of panel and information.

on Fig. V.16, the centred configuration A is preferable to the displaced configuration B.

For the size of the support to be estimated, the practical rule is to define a support equal to the grid size or some simple submultiple of it, cf. Fig. V.17.

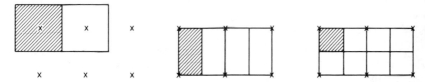

FIG. V.17. Relative size panel – grid of information.

It may sometimes happen that units have to be defined on variable supports on the borders of a deposit or a lease. The kriging of these units would then be considered separately, especially as specific problems (non-stationarity, preferential location of data sets) often arise which would require particular forms of kriging.

Kriging of gaps in the data grid It often happens that certain points on a data grid have, for various reasons, not been sampled (difficulty of access, accidental interruption of drilling, not enough time, etc.). Before considering these gaps as missing data, it should be verified (cf. section III.C.1, "Review of the data") that, in fact, they are not zero values nor preferentially located in some way. Thus, consider a regular grid with non-preferentially located gaps. There are as many different kriging systems as there are different locations of the gaps, cf. Fig. V.18. The presence of gaps in a regular data configuration thus increases the number of kriging systems to be solved.

In order to return to a unique kriging system corresponding to a complete data configuration, each gap can be replaced by its kriged estimate. It has been shown, cf. A Journel (1977, French version), that the

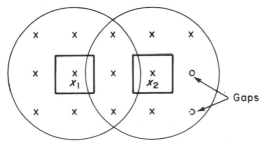

FIG. V.18. Regular data configuration with gaps.

panel kriged estimator so obtained is identical to the direct kriged estimator, using the configuration with the gaps. Of course, the kriging variance given by the completed configuration will underestimate that obtained using the configuration with the gaps. *In practice*, prior kriging of data gaps will be of interest only if there is a small number of such gaps. Then, the kriging of these gaps can often be replaced by simply taking the arithmetic mean of the neighbouring existing data, cf. the Prony case study in section V.B.4.

Random kriging   Random kriging results from a principle of random data positioning within determined domains.

   For example, let $Z(x_i)$ be one of the data considered in the estimation of a panel V. The location of $Z(x_i)$ relative to the panel may vary from one panel V to another V', cf. Fig. V.19, but the datum $Z(x_i)$ remains inside a domain β_i of fixed location relative to the panel to be estimated. It is assumed that the location x_i in β_i is uniformly random. Similarly, β_j is the domain of the uniform random location of the datum $Z(x_j)$.

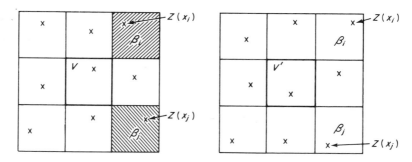

FIG. V.19. Random location of the data.

   Particular values of $\bar{\gamma}(x_i, x_j)$ and $\bar{\gamma}(x_i, V)$ correspond to each particular location of x_i in β_i and x_j in β_j of the data $Z(x_i)$ and $Z(x_j)$, i.e., a particular kriging system to be solved. The principle of random location of x_i in β_i and x_j in β_j amounts to replacing the values $\bar{\gamma}(x_i, x_j)$ *for* $i \neq j$ and $\bar{\gamma}(x_i, V)$ by their mean values when x_i and x_j describe their respective domains β_i and β_j, i.e., by $\bar{\gamma}(\beta_i, \beta_j)$ and $\bar{\gamma}(\beta_i, V)$. This results in considerably reducing the number of kriging systems to be solved. For each panel V, i.e., for each particular configuration (x_i, x_j), random kriging provides an estimator which differs from the optimal estimator provided by strict kriging using the true $\bar{\gamma}(x_i, x_j)$. What is gained in computing time is lost in precision for each particular local estimator. This loss in precision can be evaluated by comparing the estimation variance obtained by random kriging with strict kriging using a particular and simple data configuration. *In practice,* random kriging is used during the first stages of the estimation of a deposit when no great precision is required from local estimations. During this phase, local estimations are used more as an intermediate step before proceeding to a global estimation by combining all the local kriged estimates; the simplified procedure of random kriging is then entirely justified. A further justification occurs when there is a nugget effect in the structural model $\gamma(h)$: the nugget effect removes the screen effect, i.e., it tends to give the same influence to all data regardless of their distances from the panel V. In such a case, the precise location of the datum $Z(x_i)$ in β_i is not important. When there is a pure nugget effect, all data with the same support will have the same weight; this is the limiting case of random kriging in which there is only one domain β, with random locations, and β is the entire deposit or the neighbourhood of quasi-stationarity around the panel V to be estimated.

The diagonal terms in the random kriging system   Random kriging amounts to replacing the rectangular terms $\bar{\gamma}(x_i, x_j)$, $i \neq j$, by $\bar{\gamma}(\beta_i, \beta_j)$ and the terms $\bar{\gamma}(x_i, V)$ of the second member matrix by $\bar{\gamma}(\beta_i, V)$. However, this approximation does not affect the diagonal terms $\bar{\gamma}(x_i, x_i)$ which only depend on the support v_i of each datum $Z(x_i)$.

If all the data $Z(x_i)$, $Z(x_j)$, ..., have the same support v, then

$$\bar{\gamma}(x_i, x_i) = \bar{\gamma}(x_j, x_j) = \bar{\gamma}(v, v)$$

and, if the support v can be considered as a quasi-point, then $\bar{\gamma}(v, v) = 0$.

If the domain β_i contains n_i data $Z(x_{i_k})$ with respective supports v_{i_k} which are not necessarily points, and if each of these supports are located at random in β_i, then the support S_i of the data within β_i is the union of the n_i supports v_{i_k} and the diagonal term becomes

$$\bar{\gamma}(S_i, S_i) = \frac{1}{n_i^2} \sum_{k=1}^{n_i} \sum_{k'=1}^{n_i} \bar{\gamma}(v_{i_k}, v_{i_{k'}})$$

$$= \frac{1}{n_i^2} \left[\sum_k \bar{\gamma}(v_{i_k}, v_{i_k}) + \sum_k \sum_{k' \neq k} \bar{\gamma}(v_{i_k}, v_{i_{k'}}) \right]$$

with

$$\bar{\gamma}(v_{i_k}, v_{i_{k'}}) = \bar{\gamma}(\beta_i, \beta_i), \qquad \forall k \neq k',$$

and, finally,

$$\bar{\gamma}(S_i, S_i) = \frac{1}{n_i^2} \sum_{k=1}^{n_i} \bar{\gamma}(v_{i_k}, v_{i_k}) + \frac{n_i - 1}{n_i} \bar{\gamma}(\beta_i, \beta_i). \tag{V.31}$$

The diagonal terms thus depend on the number and supports of the data $Z(x_{i_k})$ in each domain β_i, and the random kriging system is not identical for each panel V to be estimated.

Some random kriging configurations   The domain β_i of random locations may be a volume, a surface, a line, or a set of discrete points, depending on the type of sampling. Thus, for example, we have the following.

(i) In three dimensions, β_i may be a polygonal envelope around the panel V to be estimated, cf. Fig. V.20(a). The corresponding data $Z(x_{i_k})$ may be the grades of the intersections of drill cores with the polygonal envelope β_i.

(ii) In two dimensions, β_i may be the surface of the basic regular data grid, cf. Fig. V.19, or the level at the top of a block, this level being sampled by drifts, cf. Fig. V.20(b).

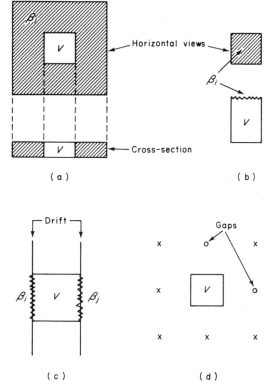

FIG. V.20. Some random kriging configurations.

(iii) In one dimension, β_i may be one of the sides of the panel V to be estimated; this side may correspond to a drift sampled by channel samples, cf. Fig. V.20(c).

(iv) In two dimensions, β_i may be the discrete set of the eight nodes of the first aureole of a regular grid. Gaps may occur at random in any of these eight locations, cf. Fig. V.20(d).

3. *Rapid calculation of the mean values $\bar{\gamma}$ or $\bar{C}$*

Before being able to solve a kriging system, the system itself must first be written down, i.e., the terms $\bar{\gamma}(v_\alpha, v_\beta)$ and $\bar{\gamma}(v_\alpha, V)$ which make up the matrices [K] and [M2] must be calculated. Algorithms for the rapid calculation of these mean values $\bar{\gamma}$ were presented in section II.E, and it is recalled that there were two possibilities: either (i) to calculate the required mean value $\bar{\gamma}(v_\alpha, v_\beta)$ by approximating each of the two supports v_α and v_β by a set of discrete points, *or* (ii) to split the calculation of the multiple integral $\bar{\gamma}(v_\alpha, v_\beta)$ into successive steps, some of which being calculated in advance and presented in the form of auxiliary functions or graphs.

Some authors, cf. J. Serra (1967*a*), Royle (1975), have constructed kriging graphs which give the weights and kriging variances for various particularly simple panel data configurations (two-dimensional and rectangular) and for various isotropic structural models. Experience has shown that such graphs are only useful as demonstration exercises; in mining applications there is no reason for any particular case to conform to the necessarily very limited frame of these graphs. It is certainly more precise, and often faster, to construct a kriging plan adapted to the particular problem rather than trying to compel the reality into the prerequisites of a preconceived graph.

The smaller the number of different mean values $\bar{\gamma}(v_\alpha, v_\beta)$ to be calculated, the faster the calculation of the $\bar{\gamma}$ terms will be. Again, the advantage of a systematic data file on a regular grid is obvious. It may very often be possible to calculate in advance all the values $\bar{\gamma}(v_\alpha, v_\beta)$ that will be required by kriging; these values could then be stored in computer memory and read out at the appropriate stage of solving each kriging system.

Note that the random kriging procedure allows a considerable reduction in the number of different relative geometries (v_α, v_β) to be considered. In fact, the random kriging system retains only the value $\bar{\gamma}(\beta_i, \beta_j)$, whatever the particular locations x_{i_k} and $x_{j_{k'}}$ of the available data are in the two distinct domains β_i and β_j.

4. *Preparation of a file for a kriging plan*

A considerable amount of the cost of kriging is spent in defining the estimation neighbourhood and searching for data within these neighbourhoods. This cost can very often be cut down by establishing a data file adapted to the particular kriging plan. The set of data would be classified and organized into a file which would permit rapid access to the data required for solving each kriging system with a minimum of intermediate calculations.

Consider the example on Fig. V.14. In the two-dimensional space, the plan for kriging the square panel V consists of grouping all the data on the nine grid squares $\{V_k, k = 1 \text{ to } 9\}$, and considering the linear combination of the nine mean grades on the nine domains V_k, i.e.,

$$Z^* = \sum_{k=1}^{9} \lambda_k Z_k.$$

In order to build a data file adapted to this kriging plan, the space is divided into a certain number of grid squares, each square being indexed by a pair of integer coordinates (iu_k, jv_k). For each domain V_k the following information is recorded in the data file.

(i) The number n_k of data (with support supposed constant and quasi-point) within V_k.

(ii) The coordinates (u_{i_k}, v_{i_k}) of the location x_{i_k} of each of these data $(i_k = 1 \text{ to } n_k)$. This information allows the calculation of the mean values $\bar{\gamma}$ of the first member kriging matrix $[K]$:

$$\bar{\gamma}(S_k, S_{k'}) = \frac{1}{n_k n_{n'}} \sum_{i_k=1}^{n_k} \sum_{j_{k'}=1}^{n_{k'}} \gamma(x_{i_k} - x_{j_{k'}}),$$

S_k being the support of the data within the domain V_k.

(iii) The mean grade of the n_k data, i.e.,

$$Z_k = \frac{1}{n_k} \sum_{i_k=1}^{n_k} Z_{i_k}.$$

This information is used to construct the kriging estimator Z^*.

The following section, V.B.3, details a typical program, CLAS, for the classification of data from any given set of drill-holes (in all directions) within a regular network of parallelepipedic blocks. This program is by no means universal, it simply prepares a data file adapted to the kriging plan used by program KRI3D. This second program for three-dimensional kriging is also given as an example.

5. *Choosing an algorithm for solving kriging systems*

The most important element of kriging costs is the time involved in solving the system of kriging equations. For a three-dimensional deposit sampled by drill-holes in all directions on an irregular grid, it is common to have to estimate a number of blocks of the order of 10 000, each of which requires a particular kriging system which may consist of five to 12 equations. Thus, it is essential to make use of a solution algorithm that is both adapted to the kriging plan and is particularly competitive in machine time.

The kriging (stationary or otherwise) system of linear equations has two important properties that should be taken into account.

   (i) The first member matrix of the system is symmetric.

  (ii) The n first equations of the system written in terms of the covariance† are such that the pivot is always on the diagonal of the first member matrix. Thus, for the first n equations (n being the number of data sets considered in the system) of the non-stationary kriging system (V.12),

$$\sum_{\beta} \lambda_\beta \bar{C}(v_\alpha, v_\beta) - \sum_{l} \mu_l b^l_{v_\alpha} = \bar{C}(v_\alpha, V), \qquad \forall \alpha = 1 \text{ to } n,$$

it is certain that

$$\bar{C}(v_\alpha, v_\alpha) \geq |\bar{C}(v_\alpha, v_\beta)|, \qquad \forall \alpha, \beta.$$

The symmetry property of the first member matrix $[K]$ suggests the use of the square roots method (the matrix $[K]$ is expressed in the form of a product of two triangular matrices, one the transpose of the other: $[K] = [T]^t \cdot [T]$), or the Cholesky method ($[K] = [R] \cdot [D] \cdot [R]^t$, where $[D]$ is a diagonal matrix). Both these methods can be found in B. Demidovitch and I. Maron (1973, pp. 287–294). These two methods were tried and rejected at the Centre de Géostatistique of Fontainebleau, because of the machine

---

† If the structural analysis has been carried out in terms of the semi-variogram $\gamma(h)$, the kriging system can be written in terms of the covariance $C(h) = C(0) - \gamma(h) = \gamma(\infty) - \gamma(h)$, provided the covariance exists. If the covariance does not exist (e.g., a semi-variogram model without a sill, i.e., $\gamma(\infty) = \infty$), a pseudocovariance $C'(h) = A - \gamma(h)$ can be considered, where A is a constant with a value greater than the greatest value of $\bar{\gamma}(v_\alpha, v_\beta)$ that can be encountered in the various kriging systems. Because of the non-bias condition $\sum_\beta \lambda_\beta = 1$, A is eliminated from the kriging equations, which are then written as

$$\sum_{\beta} \lambda_\beta \bar{C}'(v_\alpha, v_\beta) - \sum \mu_l b^l_{v_\alpha} = \bar{C}'(v_\alpha, V), \qquad \forall \alpha = 1 \text{ to } n,$$

and

$$\bar{C}'(v_\alpha, v_\alpha) \geq |\bar{C}'(v_\alpha, v_\beta)|, \qquad \forall \alpha, \beta.$$

time involved. Two other methods are used at Fontainebleau: the standard pivotal method and the method of successive approximations when the number of unknowns is greater than 20 (which is not often the case in mining applications).

Program RELMS (*due to P. Delfiner*)   The speeding up of the pivotal method is due to the fact that there is no search for the maximum element on each line of the first member matrix [K], the pivots being taken in order along the main diagonal of the matrix [K] *written in terms of the covariance.* For a stationary kriging system (with only one non-bias equation) the pivots chosen in this manner are effectively the maximum elements. For an unbiased kriging system of order k (with $n + k$ equations of which k are non-bias equations) all pivots after the nth are not necessarily the maximum elements. However, it can be shown that these pivots are never null because of the existence and uniqueness of the solution of the kriging system, cf. relations (V.15). The listing of RELMS is given in section V.B.3.

Program REITER (*due to A. Maréchal*)   The method of successive approximations, cf., in particular, B. Demidovitch and I. Maron (1973, pp. 294, 316, 447) consists of putting the matrix equation $[K] \cdot [\lambda] = [M2]$ in the form $[\lambda] = [B] + [A] \cdot [\lambda]$. Let $[\lambda^{(0)}]$ be the initial approximation; for example, the second member column matrix could be taken as the first approximation, $[\lambda^{(0)}] = [M2]$. Successive approximations are then derived as follows:

$$[\lambda^{(1)}] = [B] + [A] \cdot [\lambda^{(0)}],$$
$$[\lambda^{(2)}] = [B] + [A] \cdot [\lambda^{(1)}],$$
$$\vdots \qquad \vdots \qquad \vdots \qquad \vdots$$
$$[\lambda^{(k+1)}] = [B] + [A] \cdot [\lambda^{(k)}].$$

If this sequence of approximations has a limit, $[\lambda] = \lim_{k \to \infty} [\lambda^{(k)}]$, then this limit is the solution of the system $[\lambda] = [B] + [A] \cdot [\lambda]$, and thus also of the kriging system $[K] \cdot [\lambda] = [M2]$.

This iterative procedure will converge to a unique solution if any of the canonical norms $\|\alpha\|$ of the matrix [A] is less than unity. Considering the usual three canonical norms, this sufficient condition for convergence is written as

$$\|\alpha\| = \max_i \sum_{j=1}^{n} |a_{ij}| < 1$$

or

$$\|\alpha\| = \max_j \sum_{i=1}^{n} |a_{ij}| < 1$$

or

$$\|\alpha\|^2 = \sum_{i=1}^{n} \sum_{j=1}^{n} |a_{ij}|^2 < 1.$$

An equivalent sufficient condition is that the greatest eigenvalue ν of matrix [A] is such that $|\nu| < 1$.

The Lusternik method for improving the convergence consists in correcting the last solution $[\lambda^{(k)}]$ using this maximal eigenvalue ν:

$$[\lambda] \simeq [\lambda^{(k)}] + \frac{\nu}{1-\nu} [[\lambda^{(k)}] - [\lambda^{(k-1)}]]. \qquad (V.32)$$

The maximal eigenvalue ν can be estimated after each iteration $[\lambda^{(k)}]$ by the relation

$$\nu \simeq \frac{1}{n} \sum_{i=1}^{n} \frac{\lambda_i^{(k)} - \lambda_i^{(k-1)}}{\lambda_i^{(k-1)} - \lambda_i^{(k-2)}} \qquad (V.33)$$

or

$$\nu = \frac{\sum\limits_i [\lambda_i^{(k)} - \lambda_i^{(k-1)}]^2}{\sum\limits_i [\lambda_i^{(k-1)} - \lambda_i^{(k-2)}] \cdot [\lambda_i^{(k)} - \lambda_i^{(k-1)}]}$$

$$= \frac{\sum\limits_i [\Delta\lambda_i^{(k)}]^2}{\sum\limits_i [\Delta\lambda_i^{(k-1)}] \cdot [\Delta\lambda_i^{(k)}]},$$

if the matrix [A] is symmetric.

The $\{\lambda_i^{(k)}, i = 1 \text{ to } n\}$ are the n values of the column matrix $[\lambda^{(k)}]$. The imprecision in the estimation variance due to the approximation error $[\lambda] - [\lambda^{(k)}]$ can be evaluated using formula (V.32). Let σ_K^2 be the kriging variance corresponding to the exact solution $[\lambda]$ and $\sigma_{E_k}^2$ that corresponding to the approximation $[\lambda^{(k)}]$. The imprecision in the estimation variance, $\Delta\sigma_{E_k}^2 = \sigma_{E_k}^2 - \sigma_K^2$, is given by

$$\Delta\sigma_{E_k}^2 \simeq \frac{\nu^2}{1-\nu} \sum_i k_{ii} [\Delta\lambda_i^{(k)}]^2; \qquad (V.34)$$

the $\{k_{ii}, i = 1 \text{ to } n\}$ are the diagonal elements of the first member kriging matrix [K].

Using this last relation, the iterations can be stopped as soon as the imprecision $\Delta\sigma_{E_k}^2$ is less than a given limit.

The listing of program REITER is given in section V.B.3.

For the sake of comparison, the solution times for an n-dimensional stationary kriging system, written in terms of covariance, are given in Table V.4 for each of the following programs.

(i) GELG (standard IBM subroutine) pivot method with search for maximum elements.
(ii) RECHOL, Cholesky method.
(iii) RELMS, pivot method without search for maximum elements.
(iv) REITER, method of successive approximations with Lusternik's correction. The precision on the variance calculation is $\Delta\sigma_{E_k}^2 < 10^{-6}$ for a kriging variance close to 1. The number of iterations is 2 for $n \le 10$ and 10 for $n = 40$.

TABLE V.4. *Solution times for solving a system of linear equations*

Dimension, n	GELG	RECHOL	RELMS	REITER
5	0·02	0·02	0·01	0·01
10	0·09	0·07	0·03	0·03
15	0·24	0·18	0·08	0·06
20	0·50	0·36	0·17	0·10
30	1·57	1·20	0·48	0·20
40	3·66	2·72	1·05	0·36

The times given in this table are only relative (in IRIS-80 seconds), the same FORTRAN programs translated by another computer would result in different absolute times.

Note that, for the most common dimensions ($n \le 15$), RELMS is three times faster than the standard GELG, and, for large dimensions, ($n \ge 40$) REITER is more than 10 times faster than GELG. Moreover, the performance of RELMS and REITER can be improved if they are rewritten taking account of the particularities of the machine language of the computer used.

V.B.3 Computer subroutines

The following subroutines of data classification (CLAS), of three-dimensional kriging (KRI3D), of linear system solving (RELMS and REITER) should be considered as initial elements of a kriging package† that must be constructed according to each practitioner's requirements. It is clear that a unique kriging program, even a three-dimensional one, cannot suit (within reasonable cost limits) all the possible data configurations and kriging plans that could be considered in mining practice.

Program CLAS

This program is designed for the classification of drill-hole data within a regular network of parallelipedic blocks. The drill-holes are of any length and direction in the three-dimensional space. For each block of the network and, according to the index, when INC = 1, the number of distinct samples having their centre of gravity within the block is calculated. The characteristics of each of these samples (coordinates, length of the sample, number of the drill-hole from which it originates, mean grade, etc.) are then put into memory. Thus, for the block in Fig. V.21 which is intersected by two drill-holes, two distinct samples are considered on drill-hole no. 1, and four distinct samples on drill-hole no. 2.

When INC = 0, within each block the samples belonging to the same drill-hole are grouped, thus constituting pieces of core originating from

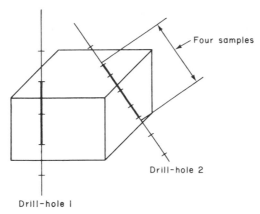

FIG. V.21. Sorting of the data in a network of blocks.

†The Centre de Géostatistique of Fontainebleau gives out (under certain conditions) its own Mining Geostatistics package, cf. Ch. Huijbregts (1975*b*).

distinct drill-holes (e.g., two pieces of core for the block of Fig. V.21). The characteristics of these pieces of core are put into memory.

If required, the program CLAS provides a listing of this classification once completed.

Locating the blocks   All coordinates are referred to orthogonal axes parallel to the three main directions of the parallelepipedic network of blocks, cf. Fig. V.22. The network is constituted of $NX * NY * NZ = NB$ blocks of identical sizes. These blocks are numbered in standard matrix form (integer coordinates IXC, IYC, IZC). The origin of the network, of coordinates XYZ(3), is located at the centre of block no. NY(IYC = NY and IXC = IZC = 1).

Each sample is defined on its drill-hole by the coordinates (XF, YF, ZF) – end of the sample.

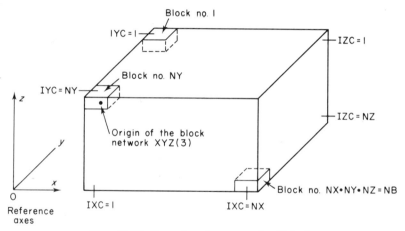

FIG. V.22. Locating the network of blocks.

Organization of the data file   Only the informed blocks (which contain at least one sample or piece of core) have a memory reservation on the file. The size of this memory reservation is given by the parameter NIMB – maximum number of data per block; each datum being characterized by nine values (the 2×3 coordinates of head and end of sample, the length of sample, the number of the drill-holes and the mean grade); there are $9 \times NIMB$ words reserved per block.

The location within the file of this memory reservation is read from the integer table NUM (NX, NY, NZ). The number of data contained in each informed block is given by the integer table NPB (NX, NY, NZ), with $1 < NPB \leqslant NIMB$. If $NPB \geqslant NIMB$, only the NIMB first data are stocked in the file.

The CLAS version given here corresponds to a direct word coding on file (CII French computer). This type of coding can easily be adapted to the particular coding required.

```
      SUBROUTINE CLAS(NS,XYZ,NXYZ,DB,INC,IL,IF1,IF2
     1,NIMB,NUM,NPB,IS)
C
C ...............................................
C                     SORTING OUT CF DRILL-HOLE DATA
C WITHIN A REGULAR SYSTEM OF BLOCKS (DEFINEC BY
C ORIGIN ANC GRID):SORTING CAN EE SAMPLE BY SAMPLE
C OR BY DRILL-HOLE INTERSECTION..RESULTS ARE WRITTEN
C ON A DIRECT ACCESS FILE..
C
C ....PARAMETERS
C  NS   NUMBER CF CRILL-HOLES TC EE SCRTEC CUT
C  XYZ(3)  COORDINATES OF CRIGIN OF BLOCK SYSTEM
C          BLOCKS ARE RANKED IN INCREASING X
C                             IN CECREASING Y
C                             FRCM TOP TO WALL IN Z
C  NXYZ(3)  GRIC OF BLOCKS ALCNG OX,OY,OZ
C           ORIGIN CF BLOCKS IS CENTRAL TC BLOCK:
C           LINE NY,COLUMN 1,LEVEL 1
C           LINE 1 CORRESPONDS TC MAXIMUM Y
C  DB(3)  DIMENSION OF BLOCK ALCNG AXIS CX,CY,CZ
C  IL   UNIT NUMBER OF CRILL-HOLE DATA FILE
C  IF1(20)  FORMAT OF DRILL-HCLE COLLAR INFCRMATICN
C           :DRILL-HOLE NUMEER,NUMBER CF SAMPLES,
C           COCRDINATES OF PCINT WHERE SAMPLING
C           STARTS
C  IF2(20)  FORMAT CF SAMPLES:ENC COORCINATE,VALUE
C  NIMB  MAXIMUM NUMBER OF INFCRMATICNS CONSIDERED
C          IN A BLOCK:DEFINES RECCRD LENGTH
C  NUM(NX*NY*NZ)  ACCRESS ON FILE IE CF BLCCK.IF
C               NUM=0 BLOCK CONTAINS NO INFCRMATICN
C  NPB(NX*NY*NZ)  INCICATES NUMEER OF DATA OF BLOCKS
C
C ....CPTIONS
C  INC.EQ.0  ALL DISTINCT CRILL-HCLE INTERSECTIONS
C            ARE COMPUTED ANC WRITTEN
C  INC.EQ.1  SCRTING AND FILING SAMPLE BY SAMPLE
C  IS.NE.1   SORTING RESULTS ARE PRINTEC BY SUBRCUTINE
C
C ....COMMONS
C  IOUT      LINE PRINTER UNIT NLMBER
C  TEST      INFERICR BCUNCARY CF EXISTING CATA
C  IF T.LE.TEST MISSING OR ELIMINATED CATA
C  /WRIT/IE,IE1,IE2  UNIT NUMBERS OF CA FILES:
C                    SORTED CATA,ARRAYS NUM AND NPB
C ...............................................
C
```

```
      DIMENSION NUM(1),NPB(1),IF1(20),IF2(20),
     1XYZ(3),NXYZ(3),DB(3)
      COMMCN ICUT,TEST
      COMMON/WRIT/IE,IE1,IE2
C
C                    INITIALIZE
C
      NX=NXYZ(1)
      NY=NXYZ(2)
      NZ=NXYZ(3)
      NB=NX*NY*NZ
      XOB=XYZ(1)
      YOB=XYZ(2)
      ZOB=XYZ(3)
      DO 10 I=1,NB
      NPB(I)=0
   10 NUM(I)=0
      ICL=0
      NECT=C
      NCUT=C
      NTUT=C
      IRE=NIMB*9
      WRITE(IOUT,2000)XYZ,NXYZ,CE
      WRITE(IOUT,2CC1)NS,NIMB
C
C                    SORTING DRILL-HCLE/DRILL-HCLE
C
      CO 4 ISON=1,NS
      READ(IL,IF1)NUS,NECH,XC,YC,ZD
      NOB1=C
      NECT=NECT+NECH
      JND=1
      DO 31 IECH=1,NECH
      READ(IL,IF2)XF,YF,ZF,T
      IF(T.LE.TEST)GO TO 3
C
      XC=(XF-XD)/2.+XD
      YC=(YF-YD)/2.+YD
      ZC=(ZF-ZC)/2.+ZD
      HC=SORT((XF-XC)*(XF-XD)+(YF-YD)*(YF-YD)+
     1(ZF-ZC)*(ZF-ZC))
C
      IXC=INT((XC-XOB)/DB(1)+1.5)
      IYC=INT((YC-YCB)/DB(2)+1.5)
      IZC=INT((ZOB-ZC)/CB(3)+1.5)
      IYC=NY-IYC+1
C
      IF(IXC*(NX+1-IXC).LE.0)GC TO 3
      IF(IYC*(NY+1-IYC).LE.0)GC TO 3
      IF(IZC*(NZ+1-IZC).LE.0)GC TO 3
C
      NOBL=IYC+NY*(IXC-1)+NY*NX*(IZC-1)
```

```
            IAD=NUM(NOBL)
            IF(IAC.GT.0)GC TO 11
            ICL=ICL+1
            NUM(NCBL)=ICL
        11 IF(INC.EC.0)GC TC 2
C
C          SORTING ANC WRITING SAMPLE/SAMPLE
C
            NPB(NCBL)=NPB(NOBL)+1
            IF(NPB(NCBL).GT.NIMB)GC TC 13
        12 NCUT=NCUT+1
C
C          WRITE ON DIRECT ACCESS FILE IE AT ACCRESS
C          IDA IN WORDS
C
            IDA=IRE*(NUM(NOBL)-1)+9*(NPB(NCBL)-1)
            WRITE DISC IE,IDA,NUS,XD,YC,ZD,XF,YF,ZF,HC,T
C
            GC TC 3
        13 WRITE(IOUT,2002)NCBL,IYC,IXC,IZC,NFB(NCBL)
            GO TO 3
C
C          COMPUTATION CF DRILL-HCLE INTERSECTICN
C
         2 IF(NOB1.EQ.0)GC TO 23
        20 IF(NOBL.NE.NOB1)GO TC 22
            IF(NPB(NCBL).GT.NIMB)GO TC 3
            NCT=NCT+1
            TV=TV+HC*T
            HL=HL+HC
            XTF=XF
            YTF=YF
            ZTF=ZF
            IF(IECH.LT.NECH)GO TC 3
C
C          END OF DRILL-HCLE
C
        21 JND=1
            IF(HL.LE.0.)GC TO 4
            TV=TV/HL
            NCUT=NCUT+NCT
C
            IDA=IRE*(NUM(NOBL)-1)+9*(NPB(NOBL)-1)
            WRITE DISC IE,IDA,NUS,XTC,YTD,ZTD,XTF,YTF,
           1ZTF,HL,TV
C
            GO TC 4
C
C          WRITE DRILL-HCLE INTERSECTION
C
        22 IF(NPB(NOB1).GT.NIMB)GC TC 23
            XTF=XC
```

```
         YTF=YC
         ZTF=ZC
         TV=TV/HL
         NCUT=NCUT+NCT
C
C        WRITE ON DIRECT ACCESS FILE IE AT ADCRESS
C        IDA IN WCRDS
C
         IDA=IRE*(NUM(NOB1)-1)+9*(NFB(NOB1)-1)
         WRITE DISC IE,IDA,NUS,XTC,YTD,ZTD,XTF,YTF,ZTF,HL,TV
C
      23 NOB1=NOBL
         NPB(NCBL)=NPB(NOBL)+1
         IF(NPB(NOBL).GT.NIMB)GO TC 13
C
C        INITIALIZE NEW INTERSECTICN
C
         NCT=1
         NTUT=NTUT+1
         JND=1
         IF(IECH.EQ.NECH)GO TO 12
         JND=0
         XTD=XC
         YTD=YD
         ZTD=ZC
         XTF=XF
         YTF=YF
         ZTF=ZF
         TV=HC*T
         HL=HC
       3 CONTINUE
         XD=XF
         YD=YF
         ZD=ZF
      31 CONTINUE
         IF(INC.EC.0.AND.JND.EQ.0)GC TO 21
       4 CONTINUE
         DO 41 IJL=1,NB
      41 NPB(IJL)=MINO(NPB(IJL),NIMB)
         IAD=0
         WRITE DISC IE1,IAD,(NUM(I),I=1,NB)
         WRITE DISC IE2,IAD,(NPB(I),I=1,NB)
         IF(INC.EQ.0)GO TO 42
         WRITE(IOUT,2003)ICL,NECT,NCUT
         GO TC 43
      42 WRITE(IOUT,2004)ICL,NECT,NCUT,NTUT
      43 IF(IS.EQ.1)GO TO 5
C
C                  PRINT RESULTS IF IS.NE.1
C
         WRITE(ICUT,2010)
         DO 47 L=1,NZ
```

```
      ZZ=ZOE-CB(3)*(L-1)
      WRITE(IOUT,2005)L,ZZ,TEST
      DO 47 I=1,NY
      DO 46 J=1,NX
      IJL=I+NY*(J-1)+NX*NY*(L-1)
      IF(NPB(IJL).EC.0)GO TO 46
      YY=YOB+DB(2)*(NY-I)
      WRITE(IOUT,2006)I,YY
      IAD=NUM(IJL)
      XX=XOB+DB(1)*(J-1)
      WRITE(IOUT,2007)J,XX,NPB(IJL),IAD
      NO=NPE(IJL)
      IDA=IRE*(IAD-1)
      DO 45 K=1,NO
      READ CISC IE,IDA,NUS,XO,YC,ZD,XF,YF,ZF,HL,T
      WRITE(IOUT,2008)NUS,XD,YC,ZD,XF,YF,ZF,HL,T
      IDA=IDA+9
   45 CONTINUE
   46 CONTINUE
   47 CONTINUE
C
 2000 FORMAT(1H1,'SCRTING CN FILE OF CRILL-HOLE ',
     1'INFCRMATIONS '/1H ,5X,'BLCCK GRID : CRIGIN ',
     2'XO= ',F10.3,' YO= ',F10.3,' ZO= ',F10.3/1H ,
     317X,' NUMBER /OX=',I4,' /CY=',I4,' /CZ=',I4
     4/1H ,17X,'  SIZE  DBX=',F1C.3,' DBY=',F10.3,
     5' DBZ=',F10.3)
 2001 FORMAT(1H ,'NUMBER OF DRILL-HOLES = ',I4,10X,
     1'MAXIMUM NUMBER OF INFORMATIONS TAKEN IN A ',
     2'BLOCK =',I2)
 2002 FORMAT(1H ,'**BLOCK ',I5,' IYC=',I3,' IZC=',
     1I3,' IZC=',I3,4X,'NUMBER CF DATA =',I3,'ABOVE ',
     2'ALLOWED MAXIMUM')
 2003 FORMAT(1HO,' NUMBER OF BLOCKS CONTAINING DATA'
     1,' =',I5/1H ,'TOTAL NUMBER OF SAMPLES =',I6,
     2' NUMBER OF SAMPLES TAKEN =',I6)
 2004 FORMAT(1HO,' NUMBER OF BLCCKS CONTAINING DATA'
     1,' =',I5/1H ,'TOTAL NUMBER OF SAMPLES =',I6,
     2' NUMBER OF SAMPLES TAKEN =',I6,' IN THE ',
     3I6,' INTERSECTIONS WRITTEN')
 2005 FORMAT(///1HO,'  **LEVEL ',I5,'*** ELEVATION ',
     1'Z= ',F10.3,'  ** LIST OF VALUES PER BLOCK**'
     2/1H ,E11.5,'=MISSING VALUES')
 2006 FORMAT(1HO,'*LINE ',I5,' Y=',F10.3)
 2007 FORMAT(1H ,'COLUMN ',I5,'  X=',F1C.3,' NUMBER'
     1,' OF DATA =',I2,5X,' FILE ADDRESS= ',I6)
 2008 FORMAT(1H ,'DRILL-HOLE ',I8,' START AND END '
     1'COORDINATES ',6(2X,F8.3),5X,'LENGTH =',F10.3
     2/1H ,'VALUE =',F10.3)
 2010 FORMAT(///1HO)
C
    5 RETURN
      END
```

Program KRI 3D

This program is designed for the kriging estimation of parallelepipedic units from three-dimensional drill-hole data, previously sorted by the program CLAS.

The unit v_k to be estimated is a submultiple of the block B of CLAS network. There are NSB units to be estimated in each block B. The location of these units v_k are deduced from the location of the block B by a system of integer coordinates similar to that used in CLAS (the unit no. 1 within B is located at the superior left corner of the upper level of B, cf. Fig. V.22).

The same estimation neighbourhood is considered for all the units v_k comprising block B. The form of this neighbourhood is parallelepipedic, centred on block B, and consisting of an integer number of blocks B_i. The data within the estimation neighbourhood are grouped into NL sets. Each of these sets consists of the data which are within a certain number NP ($L = 1$ to NL) of blocks B_i comprising the neighbourhood. A particular kriging weight will correspond to each of these NL sets of data. All the units v_k of the same block B are estimated from the same data configuration; their kriging systems differ only by their second member matrices.

The mean value $\bar{C}(L_1, L_2)$ calculations, between two sets L_1 and L_2, are carried out taking into account the precise locations of each datum (sample or piece of core) constituting L_1 and L_2. For the calculation of the mean values $\bar{C}(L_i, v_k)$, the unit v_k is approximated by a regular set of discrete points.

The algorithm used for solving the system of linear equations is RELMS, cf. following section.

The total volume to be estimated is contained in the network of blocks B defined by the program CLAS, and is constituted by an integer number of blocks B_i (those indicated by the index IEXP = 1). All the units v_k, comprising a block B to be estimated, are estimated.

Subroutines used KRIGMS + RELMS, for the writing and solving of the kriging system. When kriging systems with large dimensions (more than 20 equations) are to be considered, RELMS (pivot method) should be replaced by REITER (successive approximations method).

CBARS, for the calculation of the various mean values $\bar{C}$. The structure of this subroutine is similar to that of FUNCTION GBAR given in Section II.E.1.

VOIS, for the data search within the estimation neighbourhood of each block B.

FUNCTION COVA, for the calculation of the covariance function $C(h)$ defined as a nested sum of basic models (exponential or spherical). Except

for the definition of the nugget effect, the structure of this subroutine is
similar to that of FUNCTION GAM given in section III.B.5.

```
C
C ....................................................
C                          KRI-3D KRIGING PROGRAM
C
C KRIGING OF A THREE-DIMENSIONAL SYSTEM OF BLOCKS
C DEFINED BY ORIGIN AND GRID FROM DRILL-HOLES WITH
C ANY ORIENTATION..DATA HAVE BEEN INITIALLY SORTED
C IN BLOCK SYSTEM BY SUBROUTINE CLAS..EACH BLOCK
C MIGHT BE SUBDIVIDED IN A SYSTEM OF SUB-BLOCKS WITH
C SAME KRIGING NEIGHBOURHOOD..
C
C ....PARAMETERS
C                          :SORTING OF DATA (IN BLOCK SYSTEM)
C   XOB,YOB,ZOB ORIGIN OF BLOCK SYSTEM : CENTRE OF
C                          BLOCK LOCATED DOWN LEFT ON TOP
C   DBX,DBY,DBZ  SIZE OF BLOCKS ALONG AXIS CX,OY,OZ
C   NX,NY,NZ  BLOCK SYSTEM IS ARRANGED AS IN <CLAS>
C   NIMB  SEE SUBROUTINE CLAS
C   TEST,IS  SEE COMMON AND PRINT OPTION
C   NSB,NSY,NSZ SUB-BLOCK SYSTEM WITHIN EACH BLOCK
C                          :KRIGING PLAN
C   NL    NUMBER OF DATA GROUPINGS (KRIGING WEIGHTS)
C   IJV(3)   NUMBER OF LINES,COLUMNS,LEVELS OF
C                 PARALLELEPIPEDIC KRIGING NEIGHBOURHOOD
C   NP(NL)  NUMBER OF BLOCKS CORRESPONDING TO EACH
C                 DATA GROUPING
C   NOB(MAX.NP)  LOCATION OF THESE BLOCKS WITHIN
C                 NEIGHBOURHOOD.BLOCKS ARE NUMBERED
C                 ACCORDING TO LINE ,COLUMN,LEVEL
C                 POSITION IN NEIGHBOURHOOD
C                          :STRUCTURAL PARAMETERS
C   NST,CO  NUMBER OF STRUCTURES,NUGGET EFFECT
C   C(NST)  SILLS
C   AA(NST)  RANGES:<0. EXPONENTIAL,>0. SPHERICAL
C   CAX(NST,3),CAY(NST,3),CAZ(NST,3) SEE <CCVA>
C   NBX,NBY,NBZ  SUBDIVISION OF SUB-BLOCKS FOR
C                 COVARIANCE CALCULATION
C                          :DIRECT-ACCESS FILES UNIT NUMBERS
C   IE:   SORTED DATA(PROVIDED BY <CLAS>)
C   IEX:   ARRAY IEXP    LOCATION OF KRIGED BLOCKS
C         IEXP(NX*NY*NZ) =1 KRIGED BLOCK =0 NO KRIGING
C         IN SAME ORDER AS BLOCK SYSTEM
C   IE1:  ARRAY NUM PROVIDED BY <CLAS>
C   IE2:  ARRAY NPB PROVIDED BY <CLAS>
C   IE2:  ARRAY NPB PROVIDED BY CLAS
C   IO:   OUTPUT FILE  OF RESULTS.ONLY BLOCKS SUCH
C         THAT IEXP=1 ARE WRITTEN IN SEQUENCE.WITHIN
C         EACH BLOCK SUB-BLOCKS ARE NUMBERED ACCORDING
C         TO LINE,COLUMN,LEVEL
C
```

```
C ...MEMORY REQUIREMENTS
C    CB,CB1,CB2    NSB=NSX*NSY*NSZ
C    XOB,YOB,ZOB   NSB*NBX*NBY*NBZ
C    IP            3*NBT=3*(NP(1)+NP(2)+...+NP(NL))
C    NIN           NL+1
C    XA,YA.ZA      NBT*NIMB
C    TA            NL
C    TK,SK         NX*NY*NSB
C    NUM,NPB       NX*NY*IJV(3)
C    R,RR,X        (NL+1)*NSB
C    A             (NL+1)*(NL+2)/2
C
C ....COMMONS
C    IOUT       LINE PRINTER UNIT NUMBER
C    TEST       INFERIOR BOUNDARY OF EXISTING DATA
C    IN CASE OF BLOCK WITH NO INFORMATION
C    VALUES ARE THEN =TEST.
C           /VOISI/       PARAMETERS FOR VOIS
C           /STRUC/       PARAMETERS FOR COVA
C
C ....OPTIONS
C    IS.EQ.1  STATISTICS OF KRIGED BLOCKS ARE PRINTED
C    IS.NE.1  ALSO RESULTS BLOCK/BLOCK ARE PRINTED
C
C ....SUBROUTINES CALLED
C    KRIGMS+RELMS   KRIGING SYSTEM+SOLUTION
C    VOIS           NEIGHBOURHOOD OF ESTIMATED BLOCK
C    CBARS          AVERAGE COVARIANCE
C    COVA           COVARIANCE FUNCTION
C
C ....COMMENTS:
C    BLOCKS ARE NUMBERED AS IN SUBROUTINE CLAS:BLOCK
C NUMBER I+NY*(J-1)+NY*NX*(L-1) IS ON LINE I,
C COLUMN J,LEVEL L..LINE 1 CORRESPONDS ALWAYS TO
C MAXIMUM Y COORDINATE,COLUMN 1 TO ORIGIN,LEVEL 1
C TO TOP LEVEL..   KRIGING PLAN:NEIGHBOURHOOD IS
C DEFINED AS A PARALLELEPIPEDIC ENVELOPPE MADE UP
C OF BLOCKS..INFORMATIONS BELONGING TO BLOCKS OF
C THE ESTIMATION NEIGHBOURHOOD ARE GROUPED IN NL
C SUB-SETS EACH OF WHICH WILL HAVE A PARTICULAR
C KRIGING WEIGHT..CHOICE OF THESE SUB-SETS SHOULD
C BE MADE ACCORDING TO LOCATION,SYMMETRY,ETC...TO
C DEFINE A PARTICULAR SUB-SET L THE NUMBERS OF THE
C BLOCKS DELINEATING IT ARE GIVEN AS A LIST OF NP(L)
C VALUES (ARRAY NOB)..THESE BLOCKS ARE NUMBERED
C ACCORDING TO THEIR RELATIVE LINE (1 TO IJV(1))
C POSITION I1 (LINE 1 CORRESPONDS ALWAYS TO Y
C MAXIMUM),COLUMN (1 TO IJV(2)) AND LEVEL (1 TO
C IJV(3)) POSITIONS I2 AND I3 BY I1+IJV(1)*(I2-1)+
C IJV(2)*IJV(1)*(I3-1)..SUCH NUMBERS ARE CONVERTED
C INTO RELATIVE POSITIONS USED BY SUBROUTINE VOIS
C (SEE ARRAY IP)..
```

```
C
C   ..............................................
C
      DIMENSION NP(10),NOB(75)
      DIMENSION C(5),AA(5),CAX(5,3),CAY(5,3),CAZ(5,3)
      DIMENSION CB(8),CB1(8),CB2(8)
      DIMENSION XDB(64),YDB(64),ZDB(64)
      DIMENSION IP(225),NIN(11),TA(10)
      DIMENSION XA(150),YA(150),ZA(150)
      DIMENSION TK(800),SK(800),IEXP(100)
      DIMENSION NUM(300),NPB(300)
C
      DIMENSION R(88),RR(88),X(88),A(330)
C
      DIMENSION IJV(3),NXYZ(3),XYZ(3),DB(3),JL(3)
C
      COMMON ICUT,TEST
      COMMON/VOISI/NIMB,XB,YE,ZE,IJVL
      COMMON/STRUC/NST,CO,C,AA,CAX,CAY,CAZ
C
C   READ INPUT PARAMETERS AND PRINT TITLE
C
      INP=1
      IOUT=3
      READ(INP,1000)XOB,YOB,ZOB,CBX,CBY,CBZ,NX,NY,
     1NZ,TEST,NIMB,IS
      READ(INP,1001)NSX,NSY,NSZ
      CSX=CBX/NSX
      CSY=CBY/NSY
      CSZ=CBZ/NSZ
      NSB=NSX*NSY*NSZ
      WRITE(IOUT,2000)DBX,DBY,CBZ,XOB,YOE,ZOB,NY,NX,
     1NZ,DSX,CSY,CSZ,NSB
C
C   DEFINE NEIGHBOURHOOD AND DATA GROUPING
C
      READ(INP,1002)NL,(IJV(I),I=1,3)
      READ(INP,1003)(NP(L),L=1,NL)
      WRITE(IOUT,2001)NL
      N1=0
      NLI=IJV(1)
      NLC=IJV(1)*IJV(2)
      IJVL=IJV(3)/2+1
C
C   READ BLOCK NUMBERS DEFINING GROUPING L
C
      DO 100 L=1,NL
      N=NP(L)
      READ(INP,1003)(NOB(IJ),IJ=1,N)
      WRITE(IOUT,2002)L,(NOB(IJ),IJ=1,N)
      DO 100 I=1,N
      N1=N1+1
```

```
      N2=3*(N1-1)
      LB=(NCB(I)-1)/NLC+1
      JB=(NCB(I)-NLC*(LB-1)-1)/NLI+1
      IB=NOB(I)-NLI*(JB-1)-NLC*(LB-1)
      IP(N2+1)=IB-IJV(1)/2-1
      IP(N2+2)=JB-IJV(2)/2-1
      IP(N2+3)=LB-IJV(3)/2-1
  100 CONTINUE
C
C     READ AND PRINT STRUCTURAL PARAMETERS
C
      READ(INP,1004)NST,CO
      WRITE(IOUT,2003)NST,CO
      DO 103 N=1,NST
      READ(INP,1005)C(N),AA(N),(CAX(N,I),CAY(N,I),
     1CAZ(N,I),I=1,3)
  103 WRITE(IOUT,2004)C(N),AA(N),(CAX(N,I),I=1,3),
     1(CAY(N,I),I=1,3),(CAZ(N,I),I=1,3)
C
C     INPUT OUTPUT DIRECT ACCESS FILES
C
      READ(INP,1006)IE,IE1,IE2,IE3,IO
      NXYZ(1)=NX
      NXYZ(2)=NY
      NXYZ(3)=NZ
      XYZ(1)=XCB
      XYZ(2)=YCB
      XYZ(3)=ZCB
      DB(1)=DBX
      DB(2)=DBY
      DB(3)=DBZ
C
C     SUBDIVIDE SUB-BLOCKS IN POINTS
C
      READ(INP,1001)NBX,NBY,NBZ
      NDB=NBX*NBY*NBZ
      I1=0
      DO 14 LS=1,NSZ
      ZSO=DBZ/2.-DSZ/2.-DSZ*(LS-1)
      DO 13 IS=1,NSY
      YSO=DBY/2.-DSY/2.-DSY*(IS-1)
      DO 12 JS=1,NSX
      XSO=DBX/2.+DSX/2.+DSX*(JS-1)
      J1=-1
      J2=-1
      XO=XSO-DSX/2.+DSX/(4.*NBX)
      YO=YSO-DSY/2.+DSY/(4.*NBY)
      ZO=ZSO+DSZ/2.-DSZ/(2.*NBZ)
      L=0
      DO 11 L1=1,NBZ
      J1=-J1
      XO=XC+J1*DSX/(2.*NBX)
      DO 11 L2=1,2
      L=L+1
```

```
      IF(L.GT.NBZ)GO TO 12
      J2=-J2
      YO=YO+J2*DSY/(2.*NBY)
      DO 10 J=1,NBX
      DO 10 I=1,NBY
      I1=I1+1
      XDB(I1)=XO+(J-1)*DSX/NBX
      YDB(I1)=YO+(I-1)*DSY/NBY
   10 ZDB(I1)=ZO
      ZO=ZO-DSZ/NBZ
   11 CONTINUE
   12 CONTINUE
   13 CONTINUE
   14 CONTINUE
C
C
C     COMPUTE VARIANCE OF A POINT WITHIN SUB-BLOCK
C
      NIN(1)=1
      NIN(2)=NDB+1
      CALL CBARS(XDB,YDB,ZDB,1,0,NIN,XDB,YDB,ZDB,
     1NDB,NSB,CB)
C
C     KRIGING BLOCK PER BLOCK
C
      NBL=0
      UK=0.
      VK=0.
      NK=0
      IAD=0
      NO=NX*NY
      IF(IS.NE.1)WRITE(IOUT,2006)
      DO 1 LB=1,NZ
      ZB=ZOB-DBZ*(LB-1)
      IF(IS.NE.1)WRITE(IOUT,2007)LB,ZB
      NBK=0
      LO=MAXO(1,LB-IJV(3)/2)
      L1=MINO(NZ,LB+IJV(3)/2)
C              READ IEXP,NUM,NPB
C              ON RESPECTIVE UNITS IEX,IE1,IE2
C              ADDRESSES START AT O
      IAD1=NX*NY*(LB-1)
      READ DISC IE1,IAD1,(IEXP(IJ),IJ=1,NC)
      IAD2=NX*NY*(LO-1)
      N1=NX*NY*L1-IAD2+1
      READ DISC IE2,IAD2,(NUM(IJ),IJ=1,N1)
      READ DISC IE3,IAD2,(NPB(IJ),IJ=1,N1)
C              COMPUTE LINE BY LINE
      DO 2 IB=1,NY
      DO 3 JB=1,NX
      IJB=IB+NY*(JB-1)
      IF(IEXP(IJB).EQ.O)GO TO 3
      NBK=NBK+1
      NBL=NBL+1
      XB=XOB+DBX*(JB-1)
```

```
        YB=YOB+DBY*(NY-IB)
        JL(1)=IB
        JL(2)=JB
        JL(3)=LB
C
C       SORT CUT INFORMATICN WITHIN NEIGHBOURHOOD
C
        CALL VOIS(NL,NP,IP,JL,NXYZ,NPB,NUM,IE,XA,YA,
       1ZA,NIN,TA,NA)
        IF(NA.EQ.0)GO TO 37
        IF(NA.GT.1)GO TO 31
C               CNLY CNE INFORMATICN FCUNC
        CALL CBARS(XA,YA,ZA,1,1,NIN,XDB,YCB,ZCB,NDB,
       1NSB,CB1)
        CALL CBARS(XA,YA,ZA,1,C,NIN,XDB,YCB,ZCB,NDB,
       1NSB,CB2)
        DO 30 K=1,NSB
        J=K+NSB*(NBK-1)
        TK(J)=TA(1)
        SK(J)=CB(1)-2*CB2(K)+CB1(1)
  30 CONTINUE
        GO TO 34
C                         SOLVE KRIGING SYSTEM
  31 CALL KRIGMS(NA,XA,YA,ZA,NIN,XDB,YCB,ZCB,NDB,
       1NSB,R,RR,A,X,IER,CB1)
        IF(IER.NE.0)GO TO 36
C                         SCLUTION
        DO 33 K=1,NSB
        J=K+NSB*(NBK-1)
        TTK=0.
        IK=(NA+1)*K
        SSK=CB(1)-X(IK)
        DO 32 I=1,NA
        IK=I+(NA+1)*(K-1)
        TTK=TTK+X(IK)*TA(I)
  32 SSK=SSK-X(IK)*RR(IK)
        TK(J)=TTK
  33 SK(J)=SSK
  34 DO 35 K=1,NSB
        J=K+NSB*(NBK-1)
        UK=UK+TK(J)
        VK=VK+TK(J)*TK(J)
  35 NK=NK+1
        IF(IS.EQ.1)GO TO 3
C                     PRINT OUT IF IS.NE.1
        WRITE(IOUT,2009)IB,JB,XB,YB,NA
        DO 40 K=1,NSB
        J=K+NSB*(NBK-1)
  40 WRITE(IOUT,2010)K,TK(J),SK(J)
        GO TO 3
C                     ERRCRED SYSTEM
  36 WRITE(IOUT,2008)IB,JB
        GO TC 38
C                     NO INFORMATICN
```

```
   37 WRITE(IOUT,2012)IB,JB
   38 DO 39 K=1,NSB
      J=K+NSB*(NBK-1)
      TK(J)=TEST
   39 SK(J)=TEST
    3 CONTINUE
    2 CONTINUE
      N2=NBK*NSB
C
C      WRITE RESULTS ON FILE
C
      WRITE DISC IO,IAD,(TK(N),SK(N),N=1,N2)
      IAD=IAD+2*N2
    1 CONTINUE
C
C      PRINT STATISTICS OF KRIGED BLOCKS
C
      WRITE(IOUT,2013)DSX,DSY,DSZ,NBL
      IF(NK.EQ.0)GO TO 4
      VK=(VK-UK*UK/NK)/NK
      UK=UK/NK
    4 WRITE(IOUT,2014)UK,VK,NK
C
 1000 FORMAT(6F8.3,3I3,F8.3,I1)
 1001 FORMAT(3I2)
 1002 FORMAT(4I2)
 1003 FORMAT(40I2)
 1004 FORMAT(I2,F10.4)
 1005 FORMAT(F10.4,F8.3,9F4.2)
 1006 FORMAT(5I2)
C
 2000 FORMAT(1H1,4(/),48X,'****KRIGING OF BLOCKS****'
     1//'*DRILL-HOLE DATA SORTING GRID (IN BLOCK S',
     2'YSTEM)*'//'    SIZE  : DX= ',F8.3,' DY= ',F8.3
     3,' DZ= ',F8.3/1H ,' ORIGIN OF BLOCKS :  XO= ',
     4F11.3,' YO= ',F11.3,' ZO= ',F11.3/1H ,
     5' GRID OF ',I3,' LINES, ',I3,' COLUMNS AND ',
     6I3,' LEVELS '/' *KRIGED BLOCKS* SIZE  : DSX= ',
     7F8.3,' DSY= ',F8.3,' DSZ= ',F8.3,'   NUMBER ',
     8'OF SUB-BLOCKS= ',I3)
 2001 FORMAT(1H ,'KRIGING PLAN: ',I2,' WEIGHTS'/1H ,
     2'WEIGHT:CORRESPONDING BLOCKS WITHIN NEIGHBOUR',
     3'HOOD (NUMBERS)')
 2002 FORMAT(1H ,2X,I2,2X,':',5X,40(1X,I2))
 2003 FORMAT(1H0,' STRUCTURAL PARAMETERS OF THE ',
     1'ESTIMATED VARIABLE'/1H ,'NUMBER OF STRUCTURES'
     2,' = ',I1,' NUGGET EFFECT = ',F12.4/'   SILL ',
     3'     RANGE  |       CAX       |      CAY      ',
     4'|      CAZ')
 2004 FORMAT(1H ,F10.4,1X,F8.3,1X,9('|',F4.2))
 2006 FORMAT(1H0,4(/),48X,'**KRIGING RESULTS BLOCK ',
     1'PER BLOCK**'//)
 2007 FORMAT(1H0,/,1H ,'    **LEVEL ',I3,'  Z=',
     1F8.3//)
```

```
 2008 FORMAT(1HO,'*BLOCK IB= ',I3,' JB= ',I3,
     1'ERROR IN SYSTEM*')
 2009 FORMAT(1HO,'***BLOCK IB= ',I3,' JB= ',I3,
     1' XB= ',F11.3,' YB= ',F11.3,' NUMBER CF '
     2,'WEIGHTS NA= ',I3,'***')
 2010 FORMAT(1H ,'SUB-BLOCK ',I2,' VALUE = ',F1C.3,
     1' VARIANCE OF ESTIMATICN = ',F12.5)
 2012 FORMAT(1H ,'*BLOCK IB= ',I3,' JB= ',I3,
     1'  NO INFORMATION*')
 2013 FORMAT(1H1,4(/),25X,'****KRIGING OF BLOCKS ',
     1' DX= ',F8.3,'  DY= ',F8.3,'  DZ= ',F8.3,
     2' ****'/1HO,2X,I5,' KRIGEC BLOCKS  STATISTICS'
     3/1H ,'              AVERAGE      CISPERSION ',
     3' NUMBER OF BLOCKS')
 2014 FORMAT(1H ,9X,F10.3,2X,F1C.3,8X,I4)
C
      STOP
      END



      SUBROUTINE KRIGMS(NA,XA,YA,ZA,NIN,XCE,YCB,ZCB
     1,NDB,NSB,R,RR,A,X,IER,CB)
C
C ................................................
C        COMPUTE KRIGING MATRIX ANC SCLVE SYSTEM
C
C ....PARAMETERS
C  NA NUMBER OF EFFECTIVE DATA GROUPINGS
C  XA,YA,ZA COORDINATES OF DATA WITHIN NEIGHBOURHOOD
C  NIN  LOCATION OF CATA GROUPING IN XA,YA,ZA
C  NSB   NUMBER OF SUB-BLOCKS
C  R((NA+1)*NSB) RIGHT-HAND SITE MATRIX
C  A((NA+1)*(NA+2)/2) COEFFICIENT MATRIX
C  X((NA+1)*NSB) SOLUTION
C  KTILT      ERROR INDICATOR IF KTILT.NE.0
C
C ....SUBROUTINES CALLED
C  CBARS     COVARIANCE COMPUTATICN
C  RELMS     SYSTEM SOLUTION
C
C ................................................
C
      DIMENSION XA(1),YA(1),ZA(1),XDB(1),YCB(1),
     1ZCB(1),NIN(1)
      DIMENSION CB(1)
      DIMENSION R(1),RR(1),A(1),X(1)
      NEQ=NA+1
      NN=(NEQ+1)*NEC/2
C
C     COMPUTE COEFFICIENT MATRIX
C
```

```
      IN=0
      DO 1 J=1,NA
      DO 2 I=1,J
      IN=IN+1
      CALL CBARS(XA,YA,ZA,I,J,NIN,XDB,YCB,ZDB,NDB,
     1NSB,CB)
      A(IN)=CB(1)
    2 CONTINUE
      CALL CBARS(XA,YA,ZA,J,0,NIN,XDB,YCB,ZCB,NDB,
     1NSB,CB)
      DO 10 L=1,NSB
      K=J+NEQ*(L-1)
      R(K)=CB(L)
   10 RR(K)=R(K)
    1 CONTINUE
      DO 3 I=1,NA
      IN=IN+1
    3 A(IN)=1.
      IN=IN+1
      A(IN)=0.
      DO 4 L=1,NSB
      K=NEQ+NEC*(L-1)
      R(K)=1.
    4 RR(K)=1.
      NSM=NSB
      M=NEQ
C
C     CALL SYSTEM SOLUTION ROUTINE
C
      CALL RELMS(X,A,R,M,NSM,IER)
      RETURN
      END


      SUBROUTINE RELMS(X,A,B,M,N,KTILT)
C
C ........................................................
C                  OBTAIN SOLUTICN OF SYSTEM
C OF SIMULTANEOUS LINEAR EQUATICNS WITH SYMMETRIC
C COEFFICIENT MATRIX UPPER TRIANGULAR PART BEING
C ASSUMED TO BE STORED COLUMNWISE..
C
C ....PARAMETERS
C  X(M*N)  FINAL SOLUTION
C  A(M*(M+1)/2)  UPPER TRIANGULAR PART OF
C                COEFFICIENT MATRIX DESTROYED BY
C                TRIANGULATION
C  B(M*N)  RIGHT-HAND SIDE MATRIX DESTROYED
C  M   NUMBER OF EQUATIONS
C  N   NUMBER OF RIGHT-HAND SIDE VECTORS
C  KTILT   RESULTING ERROR PARAMETER
C          KTILT=0 NO ERROR
```

```
C                KTILT=-1 NO RESULT M<1
C                KTILT=K ZERO PIVOT WAS FOUND AT
C                   ELIMINATION STEP K
C
C ....COMMENTS:
C   SOLUTION IS DONE BY MEANS OF GAUSS-ELIMINATION
C  WITH PIVOTING IN MAIN DIAGONAL..HENCE NO SEARCH
C  OF MAXIMUM ELEMENT ..EQUATIONS ARE KEPT IN ORDER
C  AND RIGHT-HAND SIDE MATRIX IS FIRST KEPT..SYSTEM
C  IS THEN SOLVED BY SIMPLY WORKING BACKWARDS..
C  KTILT=K INDICATES HIGHEST RANK MINOR DET(G(K))
C  IS ZERO..IN GENERAL AFTER TRIANGULATION THE MAIN
C  DIAGONAL ELEMENT OF LINE I HAS THE VALUE
C  A(I*(I+1)/2)=DET(G(I))/DET(G(I-1))..
C  RELMS SHOULD BE USED ONLY FOR SYSTEMS WITH NON-
C  ZERO HIGHEST RANK MINORS..
C ...............................................
C
      DIMENSION X(1),A(1),B(1)
      IF(M.LE.0) GO TO 14
      IF(M.GT.1) GO TO 1
      IF(A(1).EQ.0.) GO TO 20
      DO 19 I=1,N
   19 X(I)=B(I)/A(1)
      KTILT=0
      RETURN
   20 KTILT=1
      RETURN
C
C      INITIALIZE
C
    1 TOL=0.0
      KTILT=0
      NM=N*M
      M1=M-1
      KK=0
C
C      START TRIANGULATION
C
      DO 7 K=1,M1
      KK=KK+K
      AK=A(KK)
      IF(AK-TOL)3,2,3
    2 KTILT=K
      RETURN
    3 PIV=1./AK
      II=KK
      LP=0
      KM1=K-1
      DO 6 I=K,M1
      LL=II
      II=II+I
```

```
       R=A(II)*PIV
       LP=LP+1
       IJ=II-KM1
       DO 4 J=I,M1
       IJ=IJ+J
       LL=LL+J
     4 A(IJ)=A(IJ)-R*A(LL)
       DO 5 LLB=K,NM,M
       IN=LLB+LP
     5 B(IN)=B(IN)-R*B(LLB)
     6 CONTINUE
     7 CONTINUE
       R=A(IJ)
       IF(R-TOL)9,8,9
     8 KTILT=M
       RETURN
C
C
C      END TRIANGULATION
C
C      START BACK SOLUTION
C
     9 PIV=1./R
       DO 10 LLB=M,NM,M
    10 X(LLB)=B(LLB)*PIV
       I=M
       KK=IJ
       DO 13 II=1,M1
       KK=KK-I
       PIV=1./A(KK)
       I=I-1
       DO 12 LLB=I,NM,M
       IN=LLB
       R=B(IN)
       IJ=KK
       DO 11 J=I,M1
       IJ=IJ+J
       IN=IN+1
    11 R=R-A(IJ)*X(IN)
    12 X(LLB)=R*PIV
    13 CONTINUE
C
C      END SOLUTION
C
C      RETURN
C
C      ERROR RETURN
C
    14 KTILT=-1
       RETURN
       END
```

```
      SUBROUTINE CBARS(XA,YA,ZA,I,J,NIN,XDB,YDB,ZDB
     1,NDB,NSB,CB)
C
C ................................................
C              COMPUTE AVERAGE COVARIANCE BETWEEN
C  SETS OF DATA I AND J OR BLOCK IF J=0
C
C ....PARAMETERS
C  CB(NSB)    AVERAGE COVARIANCE
C              OTHER PARAMETERS SEE KRIGMS
C
C ....SUBROUTINES CALLED
C  COVA  COVARIANCE FUNCTION
C ................................................
C
      DIMENSION XA(1),YA(1),ZA(1),NIN(1),CB(1)
      DIMENSION XDB(1),YDB(1),ZDB(1)
      NC=0
      DO 100 I=1,NSB
  100 CB(I)=0.
      IF(I.EC.0)RETURN
      N1=NIN(I)
      N2=NIN(I+1)-1
      IF(J.EQ.0)GO TO 3
C
C     AVERAGE COVARIANCE BETWEEN SETS I AND J
C
      M1=NIN(J)
      M2=NIN(J+1)-1
      DO 1 I2=N1,N2
      DO 2 J2=M1,M2
      CB(1)=CB(1)+COVA(XA(I2),YA(I2),ZA(I2),XA(J2),
     1YA(J2),ZA(J2))
      NC=NC+1
    2 CONTINUE
    1 CONTINUE
      CB(1)=CB(1)/MAXO(1,NC)
      RETURN
C
C     AVERAGE COVARIANCE BETWEEN SET I AND SUB-BLOCKS
C
    3 DO 4 K2=1,NSB
      NC=0
      DO 5 I2=N1,N2
      DO 5 J1=1,NDB
      J2=J1+NDB*(K2-1)
      CB(K2)=CB(K2)+COVA(XA(I2),YA(I2),ZA(I2),
     1XDB(J2),YDB(J2),ZDB(J2))
      NC=NC+1
    5 CONTINUE
      CB(K2)=CB(K2)/MAXO(1,NC)
    4 CONTINUE
      END
```

```
      SUBROUTINE VCIS(NL,NP,IP,JL,NXYZ,NPB,NUM,IE,XA
     1,YA,ZA,NIN,TA,NA)
C
C ...............................................
C                  SORTS OUT INFORMATIONS
C BELONGING TO NEIGHBOURHOOD OF ESTIMATED BLOCK
C AND GROUPS THESE BY SETS WITH SAME KRIGING WEIGHT
C ACCORDING TO KRIGING PLAN..DATA ARE READ ON FILE
C PROVIDED BY <CLAS>
C
C ....PARAMETERS
C  IP(3*NBT) LOCATION OF BLOCKS WITHIN NEIGHBOURHOOD
C  XA(NIMA),YA(NIMA),ZA(NIMA)  COORDINATES OF DATA
C                              NIMA=NBT*NIMB
C  NIN(3*NBT)  LOCATION OF EACH GROUPING IN ARRAYS
C              XA,YA,ZA
C  TA(NL)  AVERAGE VALUE OF EACH GROUPING
C ....COMMONS
C   /VOISI/    SEE MAIN PROGRAM
C ...............................................
C
      DIMENSION NP(1),JL(1),IP(1)
      DIMENSION NXYZ(1),NPB(1),NUM(1)
      DIMENSION XA(1),YA(1),ZA(1),NIN(1),TA(1)
C
      COMMON/VOISI/NIMB,XB,YB,ZB,IJVL
C
      NX=NXYZ(1)
      NY=NXYZ(2)
      NZ=NXYZ(3)
      NA=0
      IA=0
      J=0
      NIN(1)=1
      N=0
C
C     NEW GROUPING CONSIDERED
C
      DO 1 IL=1,NL
      NB=NP(IL)
      IA=NA+1
      HA=0.
      TA(IA)=0.
C                  NEW BLOCK OF THIS GROUPING
      DO 2 M=1,NB
      N=N+1
      N1=3*(N-1)
      IO=JL(1)+IP(N1+1)
      JO=JL(2)+IP(N1+2)
      LO=JL(3)+IP(N1+3)
      LB=IP(N1+3)+IJVL
      IF(IO*(NY+1-IO).LE.0)GO TO 2
      IF(JO*(NX+1-JO).LE.0)GO TO 2
```

```
      IF(LO*(NZ+1-LO).LE.0)GC TC 2
      IJL=IC+NY*(JO-1)+NX*NY*(LB-1)
      IAD=NPB(IJL)
      IF(IAC.EQ.0)GO TO 2
      IDA=9*(NUM(IJL)-1)*NIMB
      CO 3 I=1,IAD
C
C     READ CATA ON FILE IE AT ACCRESS ICA IN WCRDS
C
      READ CISC IE,ICA,NUS,XO,YC,ZO,XF,YF,ZF,HL,T
C
      IF(T.LE.TEST)GC TC 3
      J=J+1
      XA(J)=XO+(XF-XO)/2.-XB
      YA(J)=YO+(YF-YC)/2.-YB
      ZA(J)=ZO+(ZF-ZC)/2.-ZB
      HA=HA+HL
      TA(IA)=TA(IA)+T*HL
      IDA=ICA+9
    3 CONTINUE
    2 CONTINUE
      IF(HA.EQ.0)GO TO 1
C                     GROUPING WITH CATA
      NIN(IA+1)=J+1
      TA(IA)=TA(IA)/HA
      NA=NA+1
    1 CONTINUE
      RETURN
      END




      FUNCTICN CCVA(X1,Y1,Z1,X2,Y2,Z2)
C
C ...............................................
C                  COVARIANCE FLNCTICN
C DEFINEC AS THE SUM OF EXPONENTIAL CR SPHERICAL
C STRUCTURES WHICH ARE ISOTRCPIC AFTER LINEAR
C TRANSFORMATICN OF THE INITIAL COORDINATES CF
C THE VECTOR (X1-X2,Y1-Y2,Z1-Z2)..
C
C ....COMMCNS   /STRUC/
C NST  NUMBER OF ELEMENTARY STRLCTURES
C CO   NUGGET EFFECT
C C(NST)   SILLS
C A(NST)   RANGES <C. EXPONENTIAL
C                 >C. SPHERICAL
C CAX(NST,3),CAY(NST,3),CAZ(NST,3)  LINEAR
C          TRANSFORMATICN CF INITIAL CCCRCINATES
C ...............................................
C
```

```
      DIMENSICN A(1),C(1),CAX(1),CAY(1),CAZ(1)
      COMMON/STRUC/NST,CO,C,A,CAX,CAY,CAZ
      COVA=0.
      HX=X2-X1
      HY=Y2-Y1
      HZ=Z2-Z1
      H=SQRT(HX*HX+HY*HY+HZ*HZ)
      IF(H.GT.1.E-03)GO TO 2
      COVA=CO
C
  2 DO 1 IS=1,NST
      IJS=IS
      IJS1=IJS+NST
      IJS2=IJS1+NST
      DX=HX*CAX(IJS)+HY*CAX(IJS1)+HZ*CAX(IJS2)
      DY=HX*CAY(IJS)+HY*CAY(IJS1)+HZ*CAY(IJS2)
      DZ=HX*CAZ(IJS)+HY*CAZ(IJS1)+HZ*CAZ(IJS2)
      H=SQRT(DX*DX+CY*CY+DZ*DZ)
      IF(A(IS))10,12,11
 10 COVA=COVA+C(IS)*EXP(H/A(IS))
      GO TO 1
 11 IF(H.GE.A(IS))GO TO 12
      COVA=COVA+C(IS)*(1.-1.5*H/A(IS)+
    10.5*H*H*H/(A(IS)*A(IS)*A(IS)))
      GO TC 1
 12 CONTINUE
  1 CONTINUE
      RETURN
      END



      SUBROUTINE REITER(X,A,R,M,EPS,IER)
C
C ...........................................
C       OBTAIN APPROXIMATE SOLUTICN
C OF SIMULTANEOUS LINEAR EQUATICNS AX=R..UFPER
C TRIANGULAR PART CF COEFFICIENT MATRIX IS ASSUMED
C TC BE STORED COLUMNWISE..
C
C ....PARAMETERS
C  X(M)   VECTOR OF FINAL SCLUTICNS
C  A(M*(M+1)/2)  UPPER TRIANGULAR PART OF
C                COEFFICIENT MATRIX (UNMCCIFIED)
C  R(M)   RIGHT-HANC SIDE VECTOR
C  EPS    TOLERANCE CN SOLUTION (GIVEN AS FRECISION
C         OF (KRIGING) VARIANCE)
C  IER    DIVERGENCE INDICATCR
C         IER=0  GOCD SOLUTICN
C         IER=-1 DIVERGING SYSTEM
C
C ....COMMENTS:
C   THE METHOD IS BY SUCCESSIVE APPRCXIMATICNS
```

```
C   IN WHICH THE FINAL SOLUTION IS IMPROVED (REF.
C   CEMIDOVITCH/MARON ELEMENTS DE CALCUL NUMERIQUE
C    ED. DE MOSCOU 1973)..THIS METHOD HAS BEEN
C   ACAPTED TO THE SOLUTION OF A KRIGING SYSTEM..THE
C   SCLUTION BEING CNLY APPROXIMATE THE ITERATIVE
C   PROCESS IS STOPPED WHEN THE KRIGING VARIANCE
C   IS OBTAINED WITH A GOOD PRECISION (PARAMETER
C   EPS  E.G.:A GOOD LEVEL OF PRECISION IS CBTAINED
C   WITH EPS=1.E-04 OR 1.E-06 FCR A KRIGING VARIANCE
C   ABOUT 1.)..MAXIMUM NUMBER OF ITERATICNS IS 20.
C
C   ...CAPACITY:
C   50 EQUATIONS  (SEE ARRAYS T AND DT)
C   ........................................
C
      DIMENSION T(50),CT(50)
C
      DIMENSION A(1),R(1),X(1)
C
C     INITIAL SOLUTION
C
      DO 1 I=1,M
      T(I)=C.
   1  X(I)=R(I)
C
C     NEW ITERATION
C
      IER=0
      NITER=0
      IT=4
   2  CO 5 K=1,IT
      L=0
C
C                 PRECEDING SCLLTION
C
      DO 3 I=1,M
      L=L+I
      V=X(I)/A(L)
      CT(I)=V-T(I)
      T(I)=V
   3  X(I)=R(I)
C
C                 CCMPUTE NEW SCLUTICN
C
      L=0
      DO 4 J=2,M
      LEND=J-1
      L=L+1
      DO 4 I=1,LEND
      L=L+1
      X(I)=X(I)-A(L)*T(J)
   4  X(J)=X(J)-A(L)*T(I)
   5  CONTINUE
```

```
C
C      TEST OF ACCURACY AND CONVERGENCE
C
       L=0
       NITER=NITER+IT
       CNMU=0.
       DDMU=0.
       CSIG=0.
       DO 6 I=1,M
       L=L+I
       V=X(I)/A(L)-T(I)
       CNMU=CNMU+V*V
       CDMU=CDMU+V*DT(I)
     6 DSIG=CSIG+V*V*A(L)
       ANU=DCMU/DNMU
C                    DIVERGING SYSTEM
       IF(ABS(ANU).LT.1.01)GO TO 9
       DS=DSIG/(AMU*(ANU-1)*EPS)
C                    GIVEN ACCURACY ACHIEVED
       IF(DS.LT.1.)GO TO 7
       IF(NITER.GE.20)GO TO 9
       IT=2
       GO TO 2
C
C      FINAL SOLUTION
C
     7 ANU=1./(ANU-1.)
       L=0
       DO 8 I=1,M
       L=L+I
       V=X(I)/A(L)
     8 X(I)=V+(V-T(I))*ANU
       RETURN
     9 IER=-1
       RETURN
       END
```

V.B.4. Typical kriging plans

Numerous complete geostatistical studies of actual deposits, from structural analysis to kriging, can be found in the bibliography. In this section, some of these studies have been selected because they are typical examples of kriging plans and their presentation will be limited to this kriging aspect.

Three types of deposit can roughly be distinguished: stratiform deposits proven by vertical drill-holes; massive deposits (e.g. porphyry-copper) proven by drilling on an irregular grid and in all directions; and vein-type deposits proven by drilling or by drifts with channel samples.

This rough classification, based on geostatistical practice, does not correspond to any precise metallogenic limits. From a geostatistical point of view, a thick vein (subhorizontal or not) can be treated as a stratiform deposit in the plane of a vein. A porphyritic mass may also be regarded as a stratiform deposit on each horizontal bench. Conversely, for a deep sedimentary deposit sampled by drives and percussion-drill fans it may be advisable to use kriging plans analogous to those used for three-dimensional porphyritic masses, etc.

1. *Stratiform deposits*

In geostatistical applications, the techniques of stratiform deposit are relevant when the structure of the mineralization and the layout of the available data, in the three-dimensional space, favour one dimension with respect to the other two. The most characteristic example is a subhorizontal sedimentary deposit sampled by vertical drilling: the vertical direction is the best known, and it is often the direction of preferential variability (due to surface leaching by water for example).

The study of these deposits can very often be reduced to a two-dimensional problem, by considering, for instance, the horizontal regionalizations of variables accumulated over the total vertical thickness.

Prony nickeliferous laterites (New Caledonia), after Ch. Huijbregts and A. Journel (1972), A. Journel (1973b).

The deposit was sampled by vertical drills on a regular grid. The drill cores were cut into constant 1-m lengths and analysed for Ni grade and MgO impurity. From the experimental profile of the Ni grades of each drill-hole S_i, a new, more regular and representative grade file of a small panel centred on the drill S_i was defined, using an unbiased kriging of order 3 (with a quadratic vertical drift), cf. Fig. V.23 and Fig. II.1 in section II.A.1.

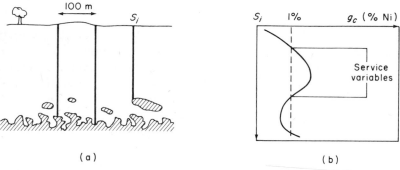

FIG. V.23. Prony deposit (vertical cross-section).

The horizontal dimensions of this small unit panel are $10 \, \text{m} \times 10 \, \text{m}$. Various nickel cut-off grades were applied to the small panel profile, cf. Fig. V.23(b): $g_c = 0$, 0.8, 1, 1.2% Ni. A thickness $t(x)$ and two accumulations in Ni and in MgO impurity correspond to each cut-off grade g_c. An accumulation $a(x) = t(x).z(x)$ is defined as the product of the thickness $t(x)$ at point x and the corresponding mean grade $z(x)$. These variables, called service variables, are regionalized in the horizontal two-dimensional space. These horizontal regionalizations are sampled on a systematic regular grid of $100 \, \text{m} \times 100 \, \text{m}$ and by two 20-m in-fill drilling crosses to detect small-scale variability structures. The fitted variogram models are isotropic and consist of the nested sum of a nugget effect and two spherical models.

The data on the regular $100 \, \text{m} \times 100 \, \text{m}$ grid were used to estimate by kriging the service variables at each node of a $10 \, \text{m} \times 10 \, \text{m}$ grid, cf. Fig. V.24. This kriging is not intended to provide estimates of such small $10 \, \text{m} \times 10 \, \text{m}$ unit panels, since the relative estimation standard deviation would vary between 25% and 95% depending on the variable and the location of the unit. The object of this kriging is to condition a simulation of the deposit, cf. section VIII.B.1, or to provide estimates of larger panels of various shapes by combining the kriged estimates of the $10 \, \text{m} \times 10 \, \text{m}$ unit panels.

The kriging plan   The zone that was studied covered 176 squares of the $100 \, \text{m} \times 100 \, \text{m}$ grid, i.e., 17 600 small $10 \, \text{m} \times 10 \, \text{m}$ units. It is, of course, out of the question to solve 17 600 kriging systems for each of the 12 studied service variables (three variables for each of the four cut-off grades). First, note that provided the regular data grid is complete, the data configuration is the same for all 100-m grid squares. This limits the number of kriging systems to be solved to 100 for each service variable; there are 100 unit panels within each 100-m grid square, to which correspond 100 kriging systems with a unique first member matrix [K] and 100 second member matrices [M2] (ν), $\nu = 1$ to 100. Furthermore, because of the two symmetries with respect to the medians and diagonals of the 100-m grid square, the 100 second member matrices [M2](ν) can be reduced to 21, cf. Fig. V.24. Thus, for each service variable, the problem of solving 17 600 kriging systems is reduced to that of solving a *single* system with 21 different second member matrices. In such a case, kriging proves no more costly than any of the more traditional estimation methods.

There are several non-preferential gaps in the 100-m data grid due to nine accidental breaks in the drilling. These gaps were filled in, before kriging was undertaken, by simply taking the arithmetic mean of the eight data values surrounding each of them on the 100-m grid, cf. kriging of gaps in section V.B.2. All 100 unit panels within the 100-m grid square were estimated considering the 16 data $\{Z_\alpha, \alpha = 1$ to $16\}$ of the first and second aureola, cf. Fig. V.24. The kriging estimator is thus a linear combination of these 16 data,

$$Z^*(V) = \sum_{\alpha=1}^{16} \lambda_\alpha(\nu) . Z_\alpha$$

for each of the 100 unit panels ($\nu = 1$ to 100). The dimension of the stationary kriging system is, thus, 17 (16 weights λ_α plus the Lagrange parameter μ).

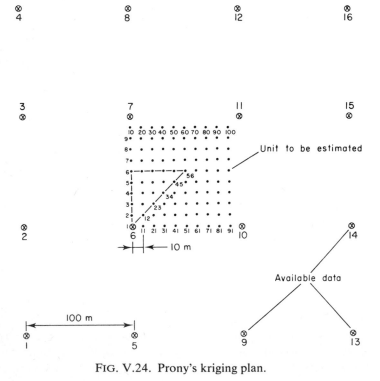

FIG. V.24. Prony's kriging plan.

Togo sedimentary phosphates (West Africa), after A. Journel and H. Sans (1974)

The deposit was sampled by vertical pits on a regular $50\ \text{m} \times 50\ \text{m}$ grid and analysed in $0{\cdot}5$-m sections for percentage recovery $p(x)$ of a granulometric fraction and for the tricalcic phosphate grade $z(x)$ of this fraction. The two studied regionalized variables are the percentage $p(x)$ and the accumulation $a(x) = p(x) . z(x)$. These variables are defined on the constant support, 0.5 m × horizontal section of the pit, and are regionalized in the three-dimensional space. In addition to the 50-m regular square grid, data were also available on two diagonal 10-m in-fill grids to detect micro-structures.

The three-dimensional variograms that were fitted consist of three nested structures:

$$\gamma(h_u, h_v, h_w) = C_0 + \gamma_1(\sqrt{(h_u^2 + h_v^2)}, h_w) + \gamma_2(\sqrt{h_u^2 + h_v^2})$$

where C_0 is the nugget effect.

γ_1 is a structure characterized by a marked geometric anisotropy: the range in the vertical direction h_w of sedimentation is 1.5 m, while the isotropic range is 25 m in the horizontal directions (h_u, h_v).

γ_2 is an isotropic horizontal macro-structure with a 300-m range corresponding to the horizontal silting of beds. The problem consisted of kriging a 25 m × 25 m × 0.5 m panel slice P, cf. Fig. V.25. The object of this kriging was to produce a vertical mean grade curve for each mining unit of 25 m × 25 m horizontal dimensions. Cut-off grades would then be applied to this curve so as to define optimal roof–wall profiles, taking into account various constraints such as ore quality control and accessibility to mining equipment.

The kriging plan was essentially determined by the structure of the three-dimensional model γ adopted.

Only the structure γ_1 differentiates the vertical distance h_w and, therefore, the 0.5-m sections in any given pit. The horizontal range of this structure γ_1 is 25 m, which is not greater than half the distance between the pits. Thus, when kriging the slice P which is wedged against the pit S_5, it serves no purpose to differentiate between any of the vertical pit section other than those of S_5. Furthermore, since the vertical range of γ_1 is 1.5 m, it is only necessary to distinguish those data in S_5 which are less than 1.5 m from P; in the example shown on Fig. V.25, only the first seven sections of pit S_5 should be distinguished.

As the horizontal structure γ_2 is isotropic, the data from S_2 and S_6, S_1 and S_9, S_4 and S_8 can be grouped together because of their symmetrical locations in relation to P. It was shown that, in regard to the structure γ_2, the first aureole (pits S_1, S_2, S_3, S_4, S_6, S_7, S_8, S_9) forms an almost total screen again the influence of more distant data.

Eight different weights were distinguished in the kriging plan, as shown on Fig. V.26: the five weights β_1, β_2, β_3, β_4, β_5 for the eight pits in the first aureole (grouped into five sets, S_3, $S_2 + S_6$, $S_1 + S_9$, $S_4 + S_8$, S_7); and three weights, α_1, α_2, α_3, for the vertical data of pit S_5 which runs down on one side of the panel slice P.

Of course, if any of the data corresponding to the weights β_i or α_j are missing, the dimension of the kriging system will be reduced accordingly.

The eight weights resulting from kriging the accumulation variable for the particular configuration considered are also given on Fig. V.26. Note that the value $\alpha_1 = 0.11$ is significantly different from the value 1 which

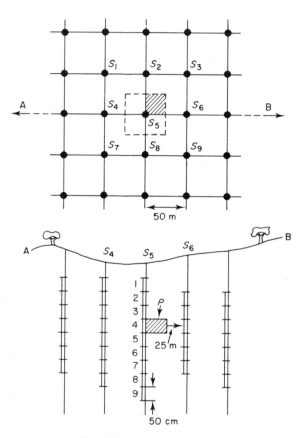

FIG. V.25. The kriging data configuration in Togo.

would have been obtained if the value of each vertical section had been assigned to the corresponding slice P (polygons of influence method). Note also the significant influence of the eight pits of the first aureole, $\sum \beta = 0.67$; kriging of slices P results in little vertical differentiation. Although kriging causes a smoothing effect (cf. Chapter VI, relation (VI.1)), it is normal for the mean characteristics of slices P of $25\,\text{m} \times 25\,\text{m} \times 0.5\,\text{m} =$ $312.5\,\text{m}^3$ to be much less dispersed than the characteristics of data defined on pit sections of $0.5 \times \varnothing\text{-pit} \simeq 0.4\,\text{m}^3$, and, indeed, production data did verify this remark.

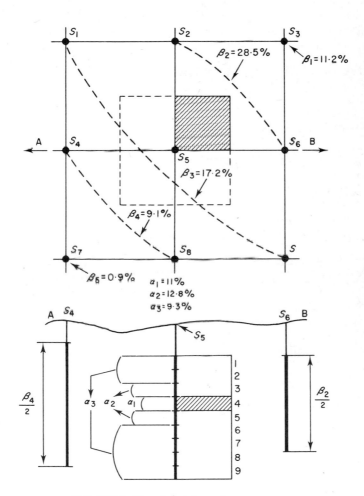

FIG. V.26. Togo's kriging plan.

2. *Massive deposits*

In geostatistical applications, a massive deposit is one in which the geometry of the mineralized mass has no preferential direction of elongation or flattening. In general, the mineralization is diffused and the scale of the heterogeneities is small when compared to the scale of mining. Estimation usually consists of evaluating a mean grade or a mean recovery rather than determining the exact geometry of each ore lode. The most characteristic examples of this type of deposit are porphyry-type brecciated masses (copper or molybdenite). Such masses may be proven on any sort of grids with drills in all directions.

Los Bronces copper deposit (Chile), after J. Damay and M. Durocher (1970)

The first campaign consisted of vertical drilling and horizontal drilling from drives. All drill cores were cut into constant 2-m lengths and analysed for copper.

The structural analysis, given in section IV.A, Case Study 2, resulted in an isotropic structural model consisting of an exponential model with a nugget effect:

$$\gamma(h_u, h_v, h_w) = C_0 + C_1[1 - e^{-r/a_1}], \qquad \forall r \in [2, 60 \text{ m}]$$

with $r^2 = |h|^2 = h_u^2 + h_v^2 + h_w^2$.

A second campaign was then carried out with vertical drilling on a regular 25 m × 25 m grid determined by the previous practical range $3a_1 \approx 25$ m. The object of this second campaign was to estimate the upper part of the deposit (Donoso zone) which was to be mined by open-pit. The mining units were cubic 12 m × 12 m × 12 m blocks, the horizontal dimension being approximately equal to one quarter of the 25 m × 25 m square grid, cf. Fig. V.27. *A single data configuration* was used for the kriging of these blocks. This configuration consisted of the following.

  (i) The estimated global mean grade of the Donoso zone.
  (ii) the 4 × 24 grades of core lengths of 2-m support, located on the four drill-holes S_1, S_2, S_3, S_4 in the estimation neighbourhood of each block. Taking symmetry and structural isotropy into account, the dimension of the single kriging system can be reduced to 38 and consists of:
  one weight for the global mean grade;
  12 weights for each of the three cores S_1, S_2, S_3 (because of symmetry, core S_4 receives the same weights as S_2);
  one Lagrange parameter required for the non-bias condition,

$$\sum_{\alpha=1}^{37} \lambda_\alpha = 1.$$

Figure V.27 shows the weights of each core as a function of the distance to the median horizontal plane of the block to be estimated. Note the following:

(i) the rapid decrease of the curve S_1;
(ii) the screen effect of cores S_2, S_4 on S_3, which means that enlarging the estimation neighbourhood would serve no purpose.
(iii) the relative importance of the weight assigned to the global mean grade, $\lambda_m = 0 \cdot 143$, which is due to the presence of the relative nugget effect $C_0/C_1 = 0 \cdot 28$.

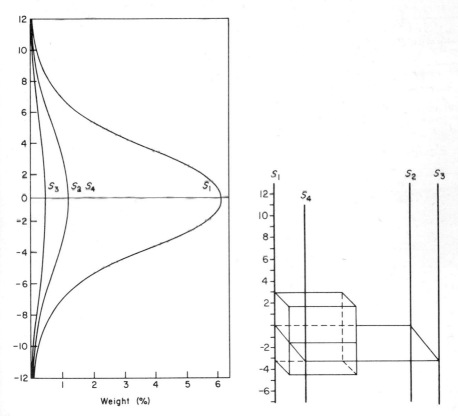

FIG. V.27. Donoso's kriging plan. Global mean grade weight $= 0 \cdot 143$.

Chuquicamata copper deposit (Chile), after I. Ugarte (1972)

This deposit is mined by open-pit and estimated by two types of sampling.

(i) Drilling in all directions. Drill cores are cut and analysed on 1·5-m lengths. These samples are used essentially for long term estimation.
(ii) Vertical blast-holes analysed over the thickness of 13-m benches. These samples are used for short-term estimation of mining units.

The regularized variograms calculated from the 1·5-m core lengths were compared to the horizontal variograms graded over the constant 13-m bench thickness calculated from the blast-hole data. The three-dimensional regionalization of the copper grades, with the 1·5-m support assimilated to a point, is characterized by a model of anisotropic, nested variograms, specifying two horizontal directions and the vertical direction. The presence of a proportional effect means that neighbourhood of quasistationarity must be considered.

Short-term estimation of a mining unit (20 m × 20 m × 13 m block) by blast-hole data on previously mined neighbouring blocks is illustrated in Fig. V.28. Note that the blast-hole data available do not completely surround the block V to be estimated. On the Western front, with mining advancing from East to West, all the blast-hole data is located on the Eastern side of block V, and the reverse is true for blocks on the Eastern front.

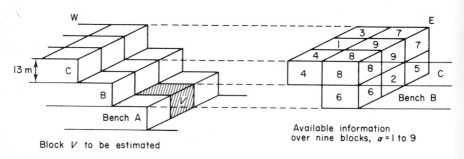

FIG. V.28. Short-term kriging plan at Chuquicamata.

For kriging purposes, the blast-hole data are grouped together inside each of the nine neighbouring blocks already mined, cf. Fig. V.28. Since the number and exact arrangement of the blast-holes within each of these blocks are variable, it is assumed when calculating the mean values $\bar{\gamma}$ of the kriging system that each block contains, on average, six blast-holes on a regular 6 m × 10 m grid, the grid being positioned in a uniform random manner within the block, cf. Fig. V.29(a).

Thus the terms $\bar{\gamma}(\alpha, \beta)$ between two blocks α and β for which data is available, each with six blast-holes, are approximated by the mean value

$$\bar{\gamma}(\alpha, \beta) = \frac{1}{36} \sum_{i=1}^{6} \sum_{j=1}^{6} \gamma(x_i - x_j).$$

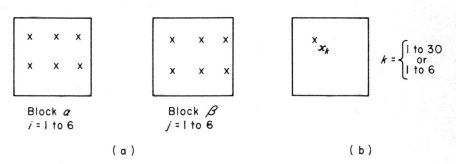

FIG. V.29. Discrete approximation of the blocks (a) blocks for which data is known; (b) block V to be estimated.

If the object is to estimate the mean grade of the entire block V, then the terms $\bar{\gamma}(\alpha, V)$ are calculated using a discrete approximation of V consisting of 30 blast-holes on a regular centred grid within V, cf. Fig. V.29(b). On the other hand, if the object is to estimate the arithmetic mean of the six blast-holes within V, then the terms $\bar{\gamma}(\alpha, V)$ are calculated using a discrete approximation of V consisting of six blast-holes on a regular $6\,\text{m} \times 10\,\text{m}$ grid positioned at random within V:

$$\bar{\gamma}(\alpha, V) = \frac{1}{6} \sum_{i=1}^{6} \bar{\gamma}(x_i, V) \quad \text{with} \quad \bar{\gamma}(x_i, V) = \begin{cases} \dfrac{1}{30} \sum_{k=1}^{30} \cdot \gamma(x_i - x_k), \\ \\ \dfrac{1}{6} \sum_{k=1}^{6} \gamma(x_i - x_k). \end{cases}$$

Using the blast-hole data file from the benches already mined, the experimental arithmetic mean Z_V^* of the blast-holes within V can be calculated and compared with the kriging estimator Z_K^* corresponding to the discrete approximation ($k = 1$ to 6). Note that the same experimental value Z_V^* can be compared to two kriging estimators Z_{KW}^* and Z_{KE}^* depending on whether the block V is considered as coming from the Western front (the data α are then located East of the block) or from the Eastern front (the data α are located West of the block). The comparisons

(Z_{KW}^*, Z_V^*) and (Z_{KE}^*, Z_V^*) are given by the histograms of errors $(Z_V^* - Z_{KW}^*)$, and $(Z_V^* - Z_{KE}^*)$ on Fig. V.30.

If the results corresponding to the East and West data configurations are grouped together, the following remarkable verifications can be made:

(i) the non-bias of kriging, the mean error being 0·00% Cu;

(ii) the representativity of the kriging variance, $\sigma_K^2 = 0\cdot117$, which is very close to the experimental variance of 0·118.

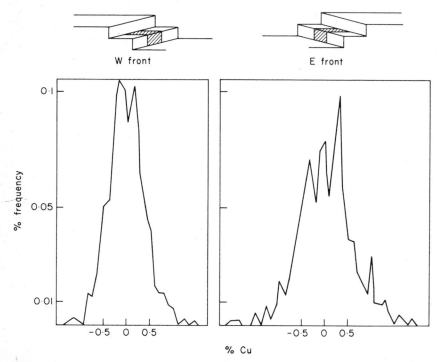

FIG. V.30. Histograms of errors for the stationary case in which the kriging system is of dimension 10.

For the Western front, mean error $= -0\cdot11\%$ Cu; experimental estimation variance $= 0\cdot125$ (% Cu)^2; number of values $= 410$. For the Eastern front, mean error $= +0\cdot11\%$ Cu; experimental estimation variance $= 0\cdot111$ (% Cu)^2; number of values $= 400$.

Grouping the two fronts, mean error $= 0\cdot00\%$ Cu; number of values $= 810$; experimental estimation variance $= 0\cdot118$ (% Cu)^2; kriging variance $\sigma_K^2 = 0\cdot117$ (% Cu)^2; mean grade of the considered zone $= 2\%$ Cu.

On the other hand, if a distinction is made between the East and West configurations, there is a significant bias of modulus 0·11% Cu, the sign of which depends on whether the block is considered as coming from the Western front or from the Eastern front. This bias can be explained by a local East–West drift that cannot be corrected because of the anisotropic data configuration. More precisely, as shown on Fig. V.31, the Chuqui-camata deposit is bounded by a large fault oriented N–S, and the variability of the grades is related to a network of small faults with the same main direction. On average, over the whole deposit, this E–W variability appears on the mean stationary variogram as a marked geometric anisotropy in the horizontal plane. Locally, at the scale at which the comparison (Z_K^*, Z_V^*) is made, this E–W variability may appear as a local drift of the copper grades (non-stationarity). In fact, as all the data are located on only one side of block V, any separation of these data by a mineralized fault will result in a locally biased estimate; if all the data are located on the other side of the block, this bias will be reversed.

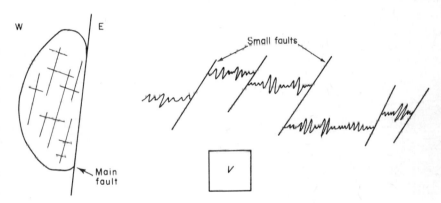

FIG. V.31. Local East–West drift.

The solution consists in taking this local E–W drift into account in the short-term kriging system, which means introducing more non-bias conditions than the single condition,

$$\sum_{\alpha=1}^{9} \lambda_\alpha = 1,$$

present for the stationary case. The variogram model γ remains unchanged, and a linear drift is considered which, in the three-dimensional space, amounts to introducing three additional non-bias conditions:

$$\sum_{\alpha=1}^{9} \lambda_\alpha u_\alpha = u_V, \qquad \sum_\alpha \lambda_\alpha v_\alpha = v_V, \qquad \sum_\alpha \lambda_\alpha w_\alpha = w_V,$$

where $(u_\alpha, v_\alpha, w_\alpha)$ and (u_V, v_V, w_V) are the three-dimensional coordinates of the centre of gravity of the block α for which data is known and the block V to be estimated, cf. relation (V.10) in section V.A.2.

Using the same data file coming from the same already-mined benches considered above, the comparison of the kriging estimates Z^*_{KW}, Z^*_{KE} and the experimental values Z^*_V yields (cf. Fig. V.32):

(i) an almost null bias, whether the East or West data configuration is used;

(ii) an excellent agreement between the kriging variance $\sigma^2_K = 0\cdot261$ and the experimental estimation variance. Note that the value $0\cdot261$ is greater than the value $0\cdot117$ previously obtained for the stationary

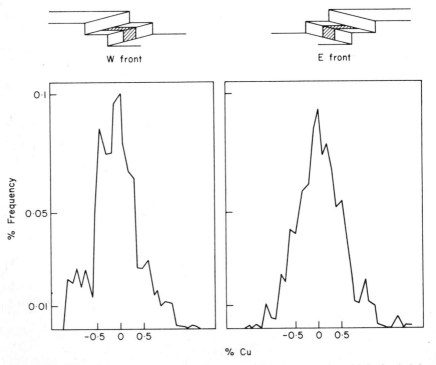

FIG. V.32. Histograms of errors for the non-stationary case in which the kriging system is of dimension 13.

For the Western front, mean error $= -0\cdot02\%$ Cu, experimental estimation variance $= 0\cdot247$ (% Cu)², number of values $= 397$. For the Eastern front, mean error $= +0\cdot01\%$ Cu; experimental estimation variance $= 0\cdot273$ (% Cu)²; number of values $= 397$.

Grouping the two fronts, mean error $= 0\cdot00\%$ Cu; number of values $= 794$; experimental estimation variance $= 0\cdot261$ (% Cu)², kriging variance, $\sigma^2_K = 0\cdot261$ (% Cu)²; mean grade of the considered zone $= 2\%$ Cu.

case: what is gained in correcting the bias is lost in the kriging variance.

An important aspect of this short-term estimation is the simple manner in which the local drift was taken into account. It was enough to introduce several additional non-bias conditions in the stationary kriging system without modifying the structural model. It is advisable to use such a procedure when the data is anisotropically positioned and if a local drift is suspected, e.g., at the edges of a deposit.

Long-term estimation of blocks B (60 m $\times$ 60 m $\times$ 13 m) from the core samples only, is illustrated in Fig. V.33. The estimation neighbourhood consists of a set of $3 \times 9 = 27$ blocks of the same size, centred on the block B to be estimated, i.e., nine blocks on each of the three horizontal benches. Each intersection of the drill cores with each of the three blocks of the vertical column containing B is considered separately; on Fig. V.33, these intersections correspond to the five weights 1, 2, 3, 5, 6. Each intersection

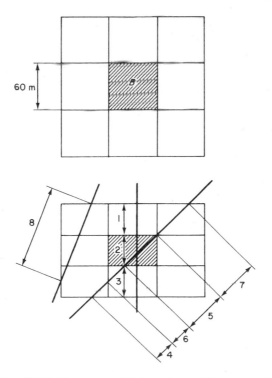

FIG. V.33. Long-term kriging plan at Chuquicamata.

of the drill core with the aureole of $27 - 3 = 24$ remaining blocks is also considered separately; on Fig. V.33 these interesections correspond to the three weights 4, 7, 8. A length l_α corresponds to each of these inter-sections, identified by its centre of gravity x_α and a mean grade Z_α. The stationary kriging system can be written in the standard form:

$$\begin{cases} \sum_{\beta=1}^{n} \lambda_\beta \bar{\gamma}(l_\alpha, l_\beta) + \mu = \bar{\gamma}(l_\alpha, B), & \forall \alpha = 1 \text{ to } n, \\ \sum_{\beta} \lambda_\beta = 1, \end{cases}$$

where there are n intersections l_α ($n = 8$ in the example of Fig. V.33).

For the calculation of the terms $\bar{\gamma}(l_\alpha, l_\beta)$, $\alpha \neq \beta$, the lengths l_α are assimilated to their centres of gravity: $\bar{\gamma}(l_\alpha, l_\beta) \simeq \gamma(x_\alpha - x_\beta)$, $\forall \alpha \neq \beta$. Similarly, $\bar{\gamma}(l_\alpha, B) \simeq \bar{\gamma}(x_\alpha, B)$ where the block B is represented by a set of discrete points. On the other hand, the length l_α must be taken into account when calculating the diagonal terms

$$\bar{\gamma}(l_\alpha, l_\alpha) = F(l_\alpha),$$

where F is the standard auxiliary function.

A certain number of blocks already mined, and thus containing blast-hole data, were estimated by this long-term kriging. The resulting kriging estimates Z_K^* were then compared to the corresponding arithmetic mean Z_B^* of the blast-hole data within B. The resulting histogram of errors $(Z_B^* - Z_K^*)$ has a mean of 0.004% Cu (non-bias) and a variance 0.093, which falls within the interval of kriging variances $[0.045$ to 0.209 $(\% \text{ Cu})^2]$.

Michiquillay copper deposit (Peru), after D. Guibal, (1976)

This massive porphyry–copper deposit was estimated by 135 vertical drills on a pseudo-regular $100 \text{ m} \times 100 \text{ m}$ grid. The variable studied was the grade of total copper defined on the support of a core sample of 15 m length, which is equal to the height of a bench. The zone studied here is the upper enriched zone (secondary sulphur mineralization).

The structural analysis resulted in the following structural model:

$$\gamma(h_u, h_v, h_w) = C_0 + C_1 \gamma_1(r) + C_2 \gamma_2(\sqrt{(h_u^2 + h_v^2)}),$$

where γ_1 is a three-dimensional isotropic model with a range $a_1 = 105$ m and γ_2 represents an horizontal macro-structure with a range $a_2 = 300$ m.

As the data grid was not regular, but only of constant density (on average one drill per $100 \text{ m} \times 100 \text{ m}$ square and assumed to be located at random within this square), a random kriging plan was adopted for this estimation of 100 m × 100 m × 15 m blocks. The estimation neighbourhood consisted of 27 blocks: the block B to be estimated and the 26 blocks of the first aureole surrounding it, cf. Fig. V.34. These 27 blocks were divided into six domains $\{D_\alpha, \alpha = 1$ to $6\}$ within which it was assumed that the data were randomly located within each 100 m × 100 m square.

D_1 is the domain consisting of the block B to be estimated; D_2 and D_3 are the blocks located above and below B, respectively; D_4 is made up of the eight blocks of the first aureole surrounding B on the same level; D_5 and D_6 consist of the eight blocks located, respectively, above and below the eight blocks of D_4.

The data inside each domain D_α consist of n_α drill intersections, all vertical, 15 m long and of mean grade Z_α. In the general case, the six domains D_α will all contain some data $(n_\alpha > 0, \forall \alpha)$, and the kriging estimator will be the linear combination of the six mean grades Z_α:

$$Z_K^* = \sum_{\alpha=1}^{6} \lambda_\alpha Z_\alpha.$$

The stationary kriging system is then written as

$$\begin{cases} \sum_{\beta=1}^{6} \lambda_\beta \bar{\gamma}(D_\alpha, D_\beta) + \mu = \bar{\gamma}(D_\alpha, B), & \forall \alpha = 1 \text{ to } 6, \\ \\ \sum_{\beta} \lambda_\beta = 1. \end{cases}$$

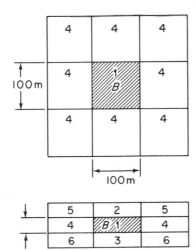

FIG. V.34. Six domains with random positioning of the data.

As the three intersections S_1, S_2, S_3 come from the same vertical drill intersecting the block B, cf. Fig. V.35, then $\bar{\gamma}(D_1, D_2) = \bar{\gamma}(S_1, S_2)$, and similarly for the other two rectangular terms $\bar{\gamma}(D_1, D_3)$ and $\bar{\gamma}(D_2, D_3)$. This relation holds, strictly, only when there is a single vertical drill intersecting each block B. For this reason, only one intersection per block was retained in the classifying program of the data $(n_1, n_2, n_3 \leqslant 1)$; in any case, only rarely did more than one drill intersect a block.

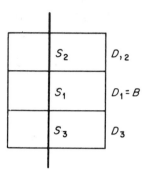

FIG. V.35. Data within domains D_1, D_2, D_3.

For the calculation of the other rectangular terms $\bar{\gamma}(D_\alpha, D_\beta)$, $\alpha \neq \beta$, the hypothesis of random locations was used. Thus, for example $\bar{\gamma}(D_1, D_4)$ is the arithmetic mean of the eight mean values $\bar{\gamma}(B, B_k)$, $k = 1$ to 8, between the block $B = D_1$ and the eight blocks B_k of the first aureole that make up D_4.

The diagonal terms are

$$\bar{\gamma}(D_1, D_1) = \bar{\gamma}(S_1, S_1) = F(15).$$

Similarly,

$$\bar{\gamma}(D_2, D_2) = \bar{\gamma}(D_3, D_3) = F(15).$$

The domain D_4 consists of eight blocks $\{B_k, k = 1$ to 8$\}$ and n_4 drill intersections $(0 < n_4 \leqslant 8)$ and, using a formula similar to (V.31),

$$\bar{\gamma}(D_4, D_4) = \frac{1}{n_4}F(15) + \sum_{k=1}^{n_4}\sum_{\substack{k'=1 \\ k' \neq k}}^{n_4} \bar{\gamma}(B_k, B_{k'}).$$

The terms $\bar{\gamma}(B_k, B_{k'})$ correspond to the blocks in D_4 that are intersected by drills. The same applies for $\bar{\gamma}(D_5, D_5)$ and $\bar{\gamma}(D_6, D_6)$. The hypothesis of random locations is again used for the calculation of the terms $\bar{\gamma}(D_\alpha, B)$ of the second member matrix. Thus, for example, $\bar{\gamma}(D_1, B) = \bar{\gamma}(B, B)$ and

$$\bar{\gamma}(D_4, B) = \frac{1}{8} \sum_{k=1}^{8} \bar{\gamma}(B_k, B).$$

In the case of a complete configuration ($n_\alpha > 0$, $\forall \alpha = 1$ to 6, and $n_4 = n_5 = n_6 = 8$), the following results are obtained:

$$\lambda_1 = 0 \cdot 147, \qquad \lambda_2 = \lambda_3 = 0 \cdot 128,$$

$$\lambda_4 = 0 \cdot 167, \qquad \lambda_5 = \lambda_6 = 0 \cdot 215$$

and $\sigma_K^2 = 0 \cdot 0212$ (% Cu)^2, for a mean estimated grade of $0 \cdot 7$ (% Cu).

Note that while the three intersections S_1, S_2, S_3 of Fig. V.35 represent only 11% (3/27) of the volume of the total data used in kriging, they receive 40% of the total weight ($\lambda_1 + \lambda_2 + \lambda_3 = 0 \cdot 403$).

El Salvador copper deposit (Chile), after M. Didyk and A. Maréchal quoted by J. Walton (1972)

The deposit (porphyry-copper mass) is mined by block caving and, for the structural analysis, two types of data were available.

(i) Drilling in all directions. The drill cores were cut and analysed in constant 3-m lengths.
(ii) Continuous channel samples along drifts, which were also analysed in constant 3-m lengths.

The structural analysis resulted in a single structural model for both sets of data. The model adopted is isotropic with a nugget effect and a hole effect.

$$\gamma(h_u, h_v, h_w) = C_0 + C\left[p(1 - e^{-ar}) + (1 - p)\left(1 - \frac{\sin br}{br}\right) \right], \qquad \forall r = |h| > \lambda$$

Two types of mineralization can be distinguished from the geology of the deposit. The first one is a massive breccia which corresponds to the exponential model $(1 - e^{-ar})$, cf. section IV.A, Case Study 2, the breccia structure of the Los Bronces deposit. The second mineralization is due to intensive enrichment within fractures, having a pseudo-periodic character corresponding to the hole effect $(1 - (\sin br)/br)$.

In the kriging of a cubic $100 \, \text{m} \times 100 \, \text{m} \times 100 \, \text{m}$ mining block B, the following data sets were used, cf. Fig. V.36:

(i) the 10 clusters of drill intersections within the 10 horizontal slices of the block, each having a thickness of 10 m;
(ii) the 10 clusters within the 10 slices of the aureole of a $300 \, \text{m} \times 300 \, \text{m} \times 100 \, \text{m}$ volume, surrounding the block B;

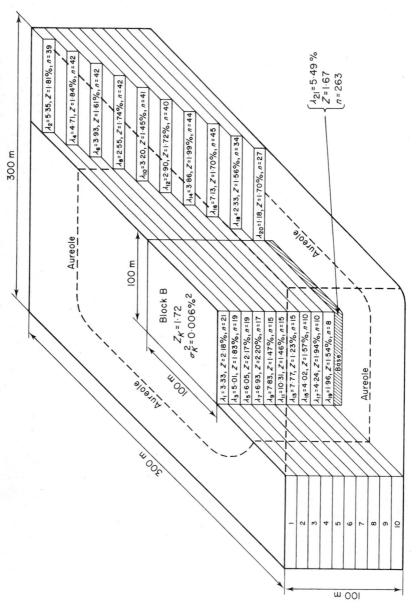

FIG. V.36. El Salvador's kriging plan. n = number of core samples within each slice, Z = mean grade of each slice, λ = weight in %.

(iii) the set of channel samples taken along the drifts which run along the base of block *B*.

This provides a total of 21 weights for the 21 data sets $\{G_\alpha, \alpha = 1 \text{ to } 21\}$, and a stationary kriging system of dimension 22. As the results of this kriging were to be used for production planning, it was essential that they be as accurate as possible. For this reason, the number and exact location of each intersection S_i constituting the cluster G_α were taken into account when calculating the terms $\bar{\gamma}(G_\alpha, G_\beta)$ rather than assuming random locations of the clusters G_α in their respective domains:

$$\bar{\gamma}(G_\alpha, G_\beta) = \frac{1}{n_\alpha n_\beta} \sum_{i=1}^{n_\alpha} \sum_{j=1}^{n_\beta} \bar{\gamma}(S_i, S_j).$$

Similarly,

$$\bar{\gamma}(G_\alpha, B) = \frac{1}{n_\alpha} \sum_{i=1}^{n_\alpha} \bar{\gamma}(S_i, B),$$

where the cluster G_α consists of n_α intersections S_i of length l_i. This precision in calculating the various terms $\bar{\gamma}$ of the kriging matrix will result in an increase in computing time over the method of random positioning; all these terms must be recalculated for each block B to be estimated. However, for block-caving mining, the number of such blocks B is limited.

The kriging results for a particular block are given on Fig. V.36: $Z_K^* = 1 \cdot 72 \pm (2\sigma_K = 0 \cdot 15)\%$ Cu. Note the small weight, $\lambda_{21} = 0 \cdot 055 = 5 \cdot 5\%$, given to the data on the level at the base of the block, even though this level is extremely well sampled by drifts and channel samples. The number of channel samples was shown to be excessive by studying their influence on the kriging variance σ_K^2; nothing is gained by increasing the number of channel samples and drifts if more drilling is not carried out within the mass of block B. This is a common characteristic of massive deposits proven by drilling and channel samples along drifts, cf. the similar study of the El Teniente data configuration in A. Journel and R. Segovia (1974).

3. *Vein-type deposits*

As far as geostatistical estimation is concerned, the difference between vein deposits and stratiform or massive deposits is the significant influence of the geometric error on the error of estimation of the mean ore characteristics. Thus, in a deposit which consists of a single vein affected by tectonics, the estimation will consist more in locating the vein and determining its walls than in estimating the mean quality of intra-vein ore. These geometric problems often cannot be formalized probabilistically, which explains why geostatistical applications to vein deposits are not as developed as those for stratiform or massive deposits.

The first problem is to determine when, from a geostatistical point of view, a deposit should be considered as vein-type. A brecciated deposit with multiple interlaced veins can be regarded as a massive deposit. The estimation of the mean width of a single vein estimated by drifts can sometimes be reduced, in the plane of the vein, to a two-dimensional problem of the stratiform type, cf. the following case study of Eagle vein. If the vein is thick enough, intra-vein estimations can be made: the geometric problem vanishes and kriging plans used for massive-type deposits can be applied.

Eagle copper vein (Canada), after A. J. Sinclair and J. Deraisme (1974)

This single quasi-vertical vein forms an undulating surface of small amplitude. The curvilinear distances h were measured along the vein, which amounts to flattening it out, cf. Fig. V.37. We can thus speak of the plane of the vein, this plane being vertical.

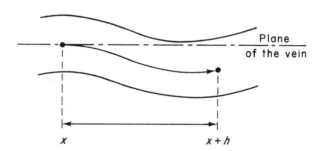

FIG. V.37. Plane of the vein.

Data for this study consist of vein width and copper assay values taken at approximately 10-ft intervals along three main horizontal development drifts located in the plane of the vein. The two regionalized variables used in the study were the vein width and the corresponding copper accumulation.

Nested structures, $\gamma(h) = \gamma_1(h) + \gamma_2(h)$, were fitted to the variograms of these two regionalizations. The two structures are spherical with respective ranges for the width of $a_1 = 20$ ft and $a_2 = 100$ ft. These two ranges can be interpreted as pseudoperiods of the successive bulges in the vein, as shown by the thickness profile obtained on one of the development levels cf. Fig. V.38.

In the plane of the vein, the three levels are about 600 ft apart, which is a much greater distance than the largest range $a_2 = 100$ ft. Thus, local estimations of ore resources can be made only within 100 m of workings for which samples are available, i.e., the zone of influence of the levels. This

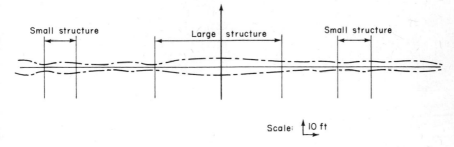

Scale: 10 ft

FIG. V.38. Thickness profile on part of level 6930.

zone of influence is a narrow 200-ft band centred on each level, cf. Fig. V.39. Beyond these three bands of influence, no differentiated local estimations are possible.

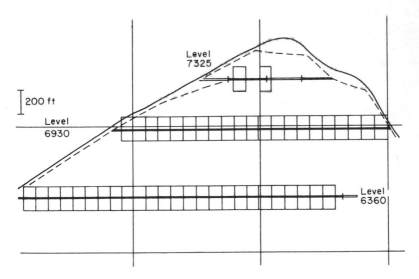

FIG. V.39. Local estimation of Eagle.

The three vertical bands of influence are split into panels of vertical dimension 100 ft × 200 ft. For the estimation by kriging of the mean horizontal thickness and the corresponding mean copper accumulation of each panel, the information on the central level is grouped into three data sets, S_1, S_2, S_3, cf. Fig. V.40. S_1 is the data set within the panel to be estimated; S_2 and S_3 are the next data sets taken along the central level.

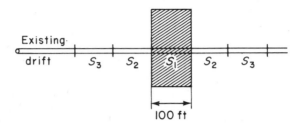

FIG. V.40. Data configuration for the estimation of a 100 ft × 200 ft panel.

As the mining company planned to carry out additional sampling for the local estimation of all the mining panels, several kriging plans were established to predict the improvement to be gained by different additional sampling plans, cf. Fig. V.41. It was shown that, with an additional level between each of the existing ones, the relative standard deviation σ_E/m, of the global estimation of the ore-tonnage, would be reduced from 4·8% to 2·3%, the relative standard deviation σ_K/m, of the local kriging of the ore-tonnage of 100 ft × 200 ft unit panels, would be reduced from 17% to 15%. This latter decrease is not especially significant; however, all the units in the deposit could be differentiated by kriging, while, with the three existing levels, only 40% of the panels were differentiated (had different

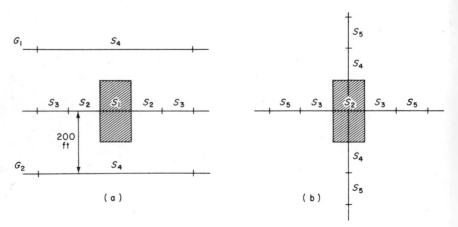

FIG. V.41. Kriging plans with additional data. (a) Two additional drifts; (b) one additional vertical raise.

kriging estimates). Similar results were obtained for the estimation of the quantity of metal and the mean copper grade.

Uchucchacua argentiferous vein (Peru) after E. Tulcanaza (1972)

The data for this study consist of samples taken along drifts and raises. The samples were taken at 1-m intervals perpendicularly to the direction of the workings on an approximately constant support. The mineralization essentially consists of argentiferous pyrite with a little galena, and the problem was to determine the intra-vein variability of the silver grade. The regionalization of this grade is characterized by a marked zonal anisotropy: the horizontal variability as determined on the drifts is greater than the vertical variability as determined on the raises. This is explained by the vertical direction of the main faults. The chosen structural model is anisotropic and consists of the nested sum of two horizontal structures (with 4-m and 70-m ranges) and a vertical structure (with a vertical 20-m range).

The problem was to estimate the mean silver grades of blocks of dimensions 50 m × 50 m or 40 m × 20 m in the plane of the vein (the constant thickness of the vein is fairly well known). The kriging configuration simply consists of two data sets, cf. Fig. V.42: (i) for the 50 m × 50 m block sampled by drifts and raises, the two mean grades Z_d and Z_r of the drifts and the raises were used; (ii) for the 40 m × 20 m block sampled by a single central drift, the two mean grades Z_{d_1} and Z_{d_2} of the samples in the section of the drift inside the block and the two symmetric sections outside the block were used.

For the calculation of the $\bar{\gamma}$ terms of the kriging matrix, the supports of the channel samples were assimilated to line segments; thus, for the third configuration, cf. Fig. V.42, the support of the datum Z_{d_1} is the line d_1 of 40 m length. Thus, $\bar{\gamma}(d_1, d_1) = F(40)$ and $\bar{\gamma}(d_1, B) = \chi(20; 40)$, where F and χ are the standard auxiliary functions.

This assimilation of the support of the discrete set of channel samples to a continuous line amounts to assuming that the mean grade of the line

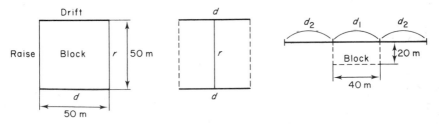

FIG. V.42. Kriging configuration at Uchacchacua.

segment is perfectly estimated by the mean grade of the channel samples within it. It is then necessary to add a correction term to the preceding kriging variance. This correction term corresponds to the estimation of the line segments by their component channel samples; thus, for the third kriging configuration, $Z_K^* = \lambda Z_{d_1} + (1-\lambda)Z_{d_2}$, the three segments of 120 m total length are estimated by 120 channel samples spaced at 1-m intervals, and the variance correction is $\sigma_e^2/120$, where σ_e^2 is the extension variance of a central point to its segment of influence of 1 m length, cf. section V.C.2, the principle of independence of elementary errors. For this third configuration, $\sigma_K^2 = 11 \cdot 8$ and $\sigma_e^2/120 = 0 \cdot 3$, which gives an estimation variance $E\{[Z_B - Z_K^*]^2\}$ of the mean grade Z_B of block B equal to $11 \cdot 8 + 0 \cdot 3 = 12 \cdot 1$.

V.C. VARIANCE OF GLOBAL ESTIMATION

V.C.1. Definition of the global estimation variance

A global estimation differs from a local estimation in two important aspects.

(i) The distances involved in global estimation are usually greater than the limits of quasi-stationarity defined by the structural analysis, and sometimes include several heterogeneous mineralizations.

(ii) The global estimation does not have the repetitive character of the local estimation. As the deposit is unique, there is only one global estimation error and it is, thus, difficult to probabilize it.

The first aspect means that a global estimator should be built by combining various local or partial estimators, as shown in section V.A.3. In general, kriging techniques can be used to provide local kriged estimates within each zone or neighbourhood of quasi-stationarity; the global estimator is then defined by the weighted sum of these local estimators, according to formula (V.18).

The problem of the definition and calculation of a norm of the quality of a global estimation remains. In local estimation, the estimation configuration for a particular block V_i can be considered as repeating itself identically for a large number of blocks V within a zone of homogeneous mineralization. In such a case, the probabilistic interpretation of the local error $(z_{V_i} - z_{V_i}^*)$ as a particular realization of the random error $(Z_{V_i} - Z_{V_i}^*)$ has a physical meaning, and the local estimation variance $E\{[Z_{V_i} - Z_{V_i}^*]^2\}$ of a block V was then defined. In global estimation, the deposit D is unique as well as its mean grade z_D. The probabilization† of the global error $(z_D - z_D^*)$ is obtained, no longer by considering various particular deposits D_i, but by considering various locations of the data configuration as a whole over the unique and fixed deposit D: a particular global estimate z_{iD}^* corresponds to each particular position i of the origin of this data configuration, cf. Fig. V.54, the example of two surface estimates $S^*(x_0)$ and $S^*(x_0')$ linked to two particular locations x_0 and x_0' of the data grid origin. Thus, the global estimation error $(z_D - z_{iD}^*)$ has been probabilized, and the global estimation variance is defined as $E\{[Z_D - Z_D^*]^2\}$.

However, the distribution of the error $(z_D - z_{iD}^*)$ will always remain inaccessible since, in practical applications, only one particular location i_0 of the data configuration over D is available. Hence, all confidence intervals (e.g., the $\pm 2\sigma$ Gaussian standard) attached to the global estimation variance will have but a mere conventional meaning.

Choice of the structural model $\gamma(h)$ or $C(h)$, used in the calculation of the global estimation variance

Consider the "local variogram" on D, i.e., the variogram $2\gamma_{(D)}(h)$ that would be available if the domain or deposit D were perfectly known, cf. section III.B.7, formula (III.40):

$$2\gamma_{(D)}(h) = \frac{1}{D'} \int_{D'} [z(x+h) - z(x)]^2 dx.$$

In practice, this local variogram is estimated by the experimental variogram, cf. formula (III.39),

$$2\gamma_D^*(h) = \frac{1}{N'} \sum_{i=1}^{N'} [z(x_i+h) - z(x_i)]^2,$$

where N' is the number of pairs of data (distant of vector h) available over the *whole* domain D.

As this experimental variogram $2\gamma_D^*(h)$, and, thus, its fitted model $2\gamma_D(h)$ are defined over the *whole* deposit D, they may differ from the quasi-stationary models $\gamma_i(h)$ used in the local estimation of each particular zone or block V_i of the deposit D. Often, in mining practice, these

---

† *Note* The global estimation error probabilization as well as the conceptual basis of the global estimation variance calculation is related to an approach different from the approach of regionalized variables considered in this book. This approach is called "transitive theory", cf. G. Matheron (1965, p. 17) and (1971, p. 9), and results in calculation algorithms and solutions remarkably similar to those provided by the regionalized variables theory. In the present text, this transitive theory approach is briefly outlined in section V.C.4, which is devoted to the surface or volume estimation problem.

quasi-stationary local models $\gamma_i(h)$ are linked by a proportional effect, cf. section III.B.6: the local model $\gamma_{m^*}(h)$ corresponding to the estimator $m^* = z_D^*$ of the mean grade of D should then be taken as the global model to be used in the calculation of the global estimation variance.

If the deposit D comprises various zones of truly heterogeneous mineralizations, it is advisable to proceed to a global estimation in each homogeneous zone (homogeneous at the mining scale, not at the scale of the geologist's pick), and then to combine the various estimation variances assumed to be independent.

In general, two types of error are present in global estimation:

   (i) a *geometric* error, due to the uncertainty of the exact limits of the mineralization or of the zone considered as the deposit;
  (ii) a *quality* error, due to the estimation of grades or other characteristics of the ore within a geometrically and volumetrically defined domain.

In the following two sections, V.C.2 and V.C.3, the study will be limited to that of the quality error, assuming that the geometry and volume of the domain or deposit to be estimated are known. In section V.C.4, the geometric error and its influence on the global estimation of ore and metal quantities and mean grades will be quantified.

V.C.2. Global estimation on a domain of known geometry

Consider a deposit D of known geometry and volume, e.g., D consists of N unit blocks V_i,

$$D = \sum_{i=1}^{N} V_i.$$

The global estimator of the mean grade Z_D is obtained by a weighted sum of the N local estimators $Z_{V_i}^*$, i.e., $Z_D^* = (1/D)\sum_i V_i Z_{V_i}^*$. The error of global estimation is then the weighted sum of the N local errors, i.e.,

$$(Z_D - Z_D^*) = \frac{1}{D}\sum_i V_i[Z_{V_i} - Z_{V_i}^*]$$

and the global estimation variance is given by formula (V.19):

$$\sigma_{ED}^2 = E\{[Z_D - Z_D^*]^2\}$$

$$= \frac{1}{D^2}\left[\sum_i V_i^2 \sigma_{KV_i}^2 + \sum_i \sum_{j \neq i} V_i V_j E\{[Z_{V_i} - Z_{V_i}^*][Z_{V_j} - Z_{V_j}^*]\}\right].$$

If Z_{V_i} was estimated by kriging, the local estimation variance, $\sigma_{KV_i}^2 = E\{[Z_{V_i} - Z_{V_i}^*]^2\}$, could then be obtained directly from the kriging system.

The difficulty lies in the calculation of the $N(N-1)$ error covariances $E\{[Z_{V_i} - Z^*_{V_i}][Z_{V_j} - Z^*_{V_j}]\}$. Such a calculation quickly becomes tedious when N is great. The practical solution consists in expressing the global estimation error $[Z_D - Z^*_D]$ as the sum of n (with, in general, $n \leqslant N$) elementary independent errors:

$$Z_D - Z^*_D = \sum_{k=1}^{n} \lambda_k (Z_k - Z^*_k). \qquad (V.35)$$

The global estimation variance then takes the simplified form

$$\sigma^2_{ED} = E\{[Z_D - Z^*_D]^2\} = \sum_{k=1}^{n} \lambda^2_k E\{[Z_k - Z^*_k]^2\}. \qquad (V.36)$$

The problem thus consists in distinguishing n successive stages of estimation, the errors of which are approximately independent. These n successive stages will depend on the available data configurations; in mining applications, there are essentially two cases, depending on whether the elementary errors come from the same estimation configuration (cf. subsection 1, "Direct combining of elementary errors", immediately following) or from different configurations (cf., hereafter, subsection 2, "Combining the terms of a global estimation error").

1. *Direct combining of elementary errors*

First, consider the two-dimensional example of the regionalization on the horizontal plane of the vertical thickness of a sedimentary bed. The deposit is sampled by vertical drilling, and three cases will be considered, according to whether the drills are on regular, random stratified (i.e., constant density) or irregular grids.

Regular grid　Consider the regular square grid of side l shown on Fig. V.43. The surface S of the deposit is divided into N elementary grid surfaces s_i, each of which has a positive central drill-hole (i.e., with a non-zero mineralized thickness $Z(x_i)$). The mean thickness Z_{s_i} of each of these elementary units is estimated by kriging from the thicknesses measured on the central drill-hole and, for example, the eight holes of the first aureole surrounding s_i. If the deposit is assimilated to the surface S of known geometry, then the estimator of the mean thickness Z_S of the deposit is the average of the N local kriging estimators, i.e., $Z^*_S = (1/N) \sum_i Z^*_{s_i}$.

Any two units s_i and s_j which are separated by a distance less than $3l$ will have common drill-holes in their estimation neighbourhood and, as a consequence, the two estimation errors $[Z_{s_i} - Z^*_{s_i}]$ and $[Z_{s_j} - Z^*_{s_i}]$ cannot be considered as being independent. On the other hand, if the global estimation error $[Z_S - Z_S^*]$ is considered as the sum of the $n = N$ elementary errors of extending the thicknesses $Z(x_i)$ of the central drill-holes to their respective surfaces of influence s_i, then the simplified formula (V.35) can be used:

$$Z_S - Z_S^* \simeq Z_S - \frac{1}{N} \sum_i Z(x_i) = \frac{1}{N} \sum_i [Z_{s_i} - Z(x_i)].$$

As a first approximation, the N elementary errors $[Z_{s_i} - Z(x_i)]$ with a variance σ_{Es}^2, can be considered as independent† because none of them have data in common. The global estimation variance can then be expressed simply as

$$E\{[Z_S - Z_S^*]^2\} \simeq \frac{1}{N^2} \sum_{i=1}^{N} E\{[Z_{s_i} - Z(x_i)]^2\} = \frac{1}{N} \sigma_{Es}^2 \qquad \text{(V.37)}$$

with $\sigma_{Es}^2 = 2H(l/2; l/2) - F(l; l)$, according to the formula for $\sigma_{E_3}^2$ corresponding to Fig. II.51 in Section II.E.3.

Note that this sum of elementary independent errors, $Z_{s_i} - Z(x_i)$, is related to the approximation

$$Z_S^* = \frac{1}{N} \sum_i Z_{s_i}^* \simeq \frac{1}{N} \sum_i Z(x_i).$$

This approximation improves as the number N of elementary units s_i increases. For a local kriging neighbourhood limited to the first aureole, $N > 50$ gives an excellent approximation.

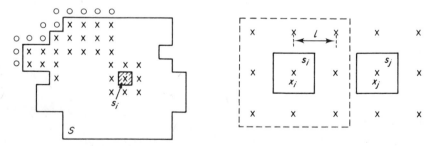

FIG. V.43. Direct combining of elementary errors from a regular grid.

---

† The N elementary errors $[Z_{s_i} - Z(x_i)]$ are independent but only as a first approximation, although they do not involve any common data. This is because the spatial regionalization entails a certain degree of correlation between the values Z_{s_i}, Z_{s_j} and their estimators $Z(x_i)$, $Z(x_j)$. However, formula (V.37) provides a good approximation for the estimator variance, as will be shown later, in the addendum, on Figure V.47.

Random stratified grid   A random stratified grid, cf. Fig. V.44, is one in which the sample grid, while not regular, has a constant density with an average of one sample in each elementary grid surface s; the location of this sample within s is assumed to be random and uniform (i.e., the sample is likely to be on any point of s).

Consider the preceding sedimentary deposit with the N positive drill-holes on a random stratified grid. As before, the global estimator Z_S^* is the weighted sum of a certain number of local estimators obtained, for example, by local kriging, and the global estimation error $[Z_S - Z_S^*]$ is considered as the sum of the N independent elementary errors of extending the thickness of each positive drill-hole, $Z(x_i)$, to its zone of influence s_i (in this case, s_i is the grid unit square):

$$Z_S - Z_S^* \simeq \frac{1}{N} \sum_i [Z_{s_i} - Z(x_i)] \quad \text{and} \quad E\{[Z_S - Z_S^*]^2\} \simeq \frac{1}{N^2} \sum_i \sigma_{Es_i}^2.$$

In this case, each of the elementary errors $[Z_{s_i} - Z(x_i)]$ has its own estimation variance $\sigma_{Es_i}^2$ which depends on the particular position x_i of the drill-hole in the grid square s_i. Over a large number N of such grid squares, x will take all possible locations inside the grid square s and the elementary estimation variances $\sigma_{Es_i}^2$ will have the following mean value:

$$\frac{1}{N} \sum_{i=1}^{N} \sigma_{Es_i}^2 = \frac{1}{s} \int_s E\{[Z_s - Z(x)]^2\} \, dx = D^2(0/s).$$

Indeed, from formula (II.33), the mean value over s of the variance of extension $E\{[Z_s - Z(x)]^2\}$ of a point datum $Z(x)$ to the mean value Z_s is, by definition, the dispersion variance $D^2(0/s)$ of this point datum in s. The global estimation variance is then written as

$$E\{[Z_S - Z_S^*]^2\} \simeq \frac{1}{N^2} \sum_i \sigma_{Es_i}^2 = \frac{1}{N} D^2(0/s). \qquad (V.38)$$

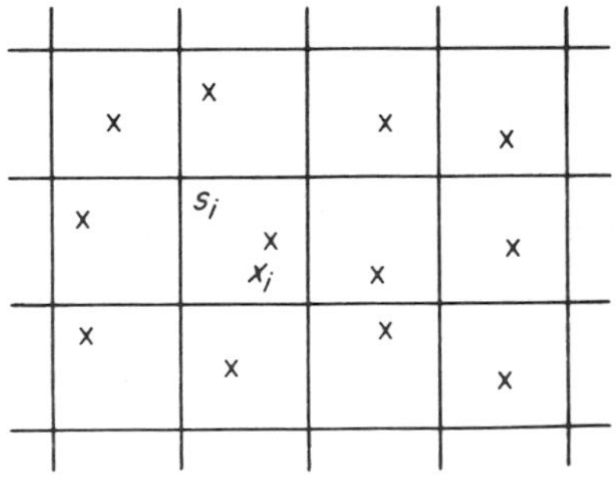

FIG. V.44. Random stratified grid.

The dispersion variance $D^2(0/s)$ is expressed very simply by means of the auxiliary function F:

$$D^2(0/s) = \bar{\gamma}(s, s) = F(l; l),$$

if s is a square of side l.

The principle of uniform random location of the data within the grid unit can be applied in a similar manner to data configurations in one, two or three dimensions. In three dimensions for example, for a parallelepipedic grid unit v containing, on average, one datum of support $v' \ll v$ and located in a uniform random manner within v, the global estimation variance of a large volume D consisting of N grid units v can be obtained by direct combination:

$$E\{[Z_D - Z_D^*]^2\} \simeq \frac{1}{N} D^2(v'/v) = \frac{1}{N}[\bar{\gamma}(v, v) - \bar{\gamma}(v', v')]. \qquad (V.39)$$

Irregular grid   Consider the same sedimentary deposit, this time with the N positive data on an irregular grid, cf. Fig. V.45 (the N surfaces of influence s_i are unequal). Note that under no circumstances can a preferential grid, with all its associated risks of bias, be assimilated to an irregular grid, cf. section III.C.1.

The global estimator Z_S^* is the weighted sum of a certain number of local estimators obtained, for instance, by local kriging; note that these local estimators may not be defined on the surface of influence s_i of each positive hole. By expressing the global estimation error $[Z_S - Z_S^*]$ as the sum of the N independent elementary errors $[Z_{s_i} - Z(x_i)]$, of extending the thicknesses $Z(x_i)$ to their respective surface of influence s_i, the following formula is obtained:

$$E\{[Z_S - Z_S^*]^2\} \simeq \frac{1}{S^2} \sum_{i=1}^{N} s_i^2 E\{[Z_{s_i} - Z(x_i)]^2\} = \frac{1}{S^2} \sum_i s_i^2 \sigma_{E s_i}^2, \qquad (V.40)$$

where $S = (1/N)\sum_i s_i$, and $\sigma_{E s_i}^2$ is the variance of extension of a central point datum to its surface of influence. This variance is expressed by means of the two auxiliary functions H and F, cf. section II.E.3, if the surface s_i is a rectangle. If s_i is not rectangular, it can, for instance, be assimilated to a square of the same surface area $s_i = l_i^2$, and, thus,

$$\sigma_{E s_i}^2 = H(l_i/2; l_i/2) - F(l_i; l_i).$$

Note that the previous expression (V.40) of the global estimation variance is related to the approximation

$$Z_S^* \simeq \frac{1}{S} \sum_i s_i Z_{s_i}^* = \frac{1}{S} \sum_i s_i Z(x_i),$$

i.e., to the assimilation of the global estimator Z_S^*, weighted sum of local kriging estimators, to the polygons of influence estimator. When N is large, there is very little difference between these two global estimators. Nevertheless, at the local level, the kriging estimators should always be preferred to the polygons of influence estimators (because the local kriging variances are, by definition, minimum).

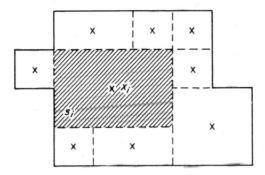

FIG. V.45. Direct combining of elementary errors from an irregular grid.

Other configurations for directly combining elementary errors

In one dimension, let L be a line sampled by N points regularly spaced at intervals l, cf. Fig. V.46(a). This line may represent, for example, a drift of length L with a negligible transverse section which has been sampled by vertical channel samples spaced at regular intervals l. The line L thus consists of N segments l each having a central, centred point datum. The global estimation error $[Z_L - Z_L^*]$ of the mean characteristic on L is considered as the sum of the N elementary errors of extension of the central data $Z(x)$ to their segments of influence l. As a first approximation, these errors can be considered independent, resulting in the following expression similar to (V.37):

$$E\{[Z_L - Z_L^*]^2\} \simeq \frac{1}{N}\sigma_{E_l}^2 \qquad (V.41)$$

with

$$\sigma_{E_l}^2 = E\{[Z_l - Z(x)]^2\} = 2X(l/2) - F(l).$$

The same line $L = Nl$ may be sampled by $(N+1)$ points in a closed arrangement, each segment l_i having two extreme data $Z(x_i)$ and $Z(x_{i+1})$, as shown on Fig. V.46(b). Two contiguous segments l_i and l_j have one datum in common and, as a consequence, the two corresponding local estimation errors can no longer be considered as independent. It can be shown that, for $N \geq 6$, a closed arrangement of $(N+1)$ points can be assimilated, with an excellent approximation, to a centred arrangement of N points, and formula (V.41) can then be applied. This practical rule can also be applied for closed arrangements in two or three dimensions (with $N \geq 36$).

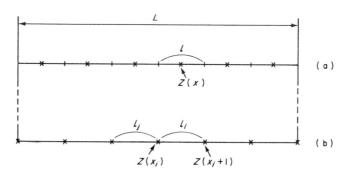

FIG. V.46. Direct combining of elementary errors on a line. (a) Centred arrangement; (b) closed arrangement.

If the N points are in a centred arrangement, but on an irregular grid ($l_i \neq l_j$), each segment l_i being informed by a central point datum $Z(x_i)$, then the weighted average of the elementary errors of extension of the $Z(x_i)$ to their respective segments of influence l_i results in a formula similar to (V.40):

$$E\{[Z_L - Z_L^*]^2\} \simeq \frac{1}{L^2} \sum_i l_i^2 \sigma_{El_i}^2 \qquad (V.42)$$

where $L = \sum_i l_i$ and

$$\sigma_{El_i}^2 = E\{[Z_{l_i} - Z(x_i)]^2\} = 2X(l_i/2) - F(l_i).$$

Addendum

On this simple example of centred arrangement in one dimension, it is advisable to test the robustness of the approximation underlying formula (V.41), i.e., the influence of the hypothesis of independence of the N elementary errors of extending the central data to their segments l of influence. For this, a spherical scheme with range a and unit sill and without nugget effect, was considered, cf. formula (III.15). The estimation variance of a line $L = Nl$ surveyed by N point data regularly spaced on a central arrangement is then calculated from the exact formula (II.27) and compared with the approximation given by formula (V.41). These compared results are shown on Fig. V.47 (by courtesy of Michel David) for various values of N and the quotient l/a. The approximation is excellent for all small values of the spacing ($l/a < 0.5$, which is a common case in practice); for greater values of the spacing ($l/a > 0 \cdot 5$, the number N must be great ($N > 30$) to provide a good enough approximation.

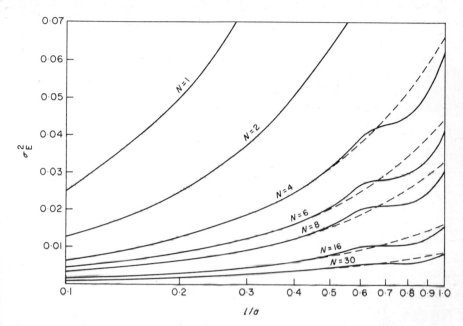

FIG. V.47. Estimation variance of a line $L = Nl$ by N data on a centred arrangement. Spherical model, sill 1, range a. ———, Exact values; ----, approximate values.

In two dimensions, consider the estimation of the mean grade over a surface S divided into N rectangles $s_i = L_i \times l_i$, each containing a median drift assimilated to a line of length L_i, as on Fig. V.48. The variance $\sigma^2_{Es_i}$ of

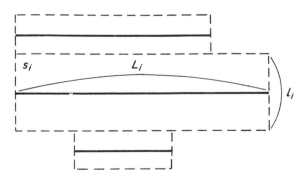

FIG. V.48. Direct combination of elementary errors from a configuration of median drifts.

the elementary error of extension of the mean grade of a drift L_i to its rectangle of influence $s_i = L_i \times l_i$ is calculated using the auxiliary functions χ and F, or read directly from charts no. 8 and 18, cf. section II.E.3. The global estimation variance over the surface

$$S = \sum_{i=1}^{N} s_i$$

can then be obtained by directly combining the elementary estimation variances $\sigma_{Es_i}^2$. The result is a formula similar to (V.40):

$$E\{[Z_S - Z_S^*]^2\} \simeq \frac{1}{S^2} \sum_i s_i^2 \sigma_{Es_i}^2, \qquad (V.43)$$

where

$$\sigma_{Es_i}^2 = 2\chi(l_i/2; L_i) - F(l_i; L_i) - F(L_i).$$

In three dimensions, consider the estimation of the mean grade over a volume V divided into N rectangular parallelepipeds $v_i = s_i \times h_i$, each of which being informed by its median section of surface s_i, cf. Fig. V.49. The variance $\sigma_{Ev_i}^2$ of the elementary error of extension of the mean grade of the median level s_i to its parallelepiped of influence $v_i = s_i \times h_i$ is calculated using the auxiliary functions χ and F, or read directly from charts no. 9 and 19, cf. section II.E.3. The global estimation variance over the volume

$$V = \sum_{i=1}^{N} v_i$$

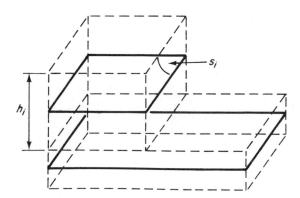

FIG. V.49. Direct combining of elementary errors from a configuration of median sections.

can then be obtained by directly combining the elementary estimation variances $\sigma^2_{Ev_i}$:

$$E\{[Z_V - Z^*_V]^2\} \simeq \frac{1}{V^2} \sum_i v_i^2 \sigma^2_{Ev_i}, \qquad (V.44)$$

where, if the median section s_i is assimilated to a square of the same area, $l_i^2 = s_i$,

$$\sigma^2_{Ev_i} = 2\chi(h_i/2; l_i^2) - F(h_i; l_i^2) - F(l_i; l_i).$$

Note In the last two examples, corresponding to Fig. V.48 and V.49, the elementary extensions of the mean grade of a median drift or a median level to its rectangle or parallelepiped of influence were considered. But both the median drift and the median level have been estimated with a certain estimation error. Thus, a term corresponding to this estimation error must be added to the estimation variances given by formulae (V.43) and (V.44); this results in combining the various terms of the global estimation error.

2. Combining the terms of a global estimation error

The data available for the estimation of a three-dimensional phenomenon is very often concentrated along alignments (e.g., pieces of core along drill-holes, sampling along drifts) and planes (e.g., planes of vertical sections along profiles of vertical drill-holes or horizontal levels of the drifts). Thus, the data available for the estimation of the mass shown on Fig. V.50 are vertical channel samples taken along horizontal drifts on each of the three horizontal levels.

The procedure for estimating the mean grade of this mass can be considered as consisting of three steps corresponding to three terms of the global estimation variance.

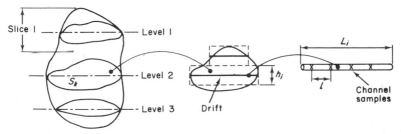

FIG. V.50. Combining the terms of the estimation of a mass.

(i) The estimation of the drifts by the channel samples. The estimation variance corresponding to this step is called the "line term", T_l.

(ii) The estimation of the levels by the drifts which are supposed to be perfectly known. The estimation variance corresponding to this step is called the "section term", T_S.

(iii) The estimation of the slices of the mass by their median levels, which are supposed to be perfectly known. The estimation variance corresponding to this last step is called the "slice term", T_V.

As a first approximation, the three estimation errors corresponding to these three steps can be considered as independent, and the global estimation variance of the mass then consists of the sum of the three preceding terms:

$$\sigma_E^2 = T_l + T_S + T_V. \tag{V.45}$$

The following example of the calculation of these three terms is based on Fig. V.50. Another example is given in the study of Ity Mount Flotouo gold deposit in section V.C.5.

Line term If each drift is assimilated to a line L_i along which channel samples are taken at regular intervals, then the estimation variance of this line can be expressed using the direct combining formula (V.41):

$$\sigma_{EL_i}^2 \simeq \frac{1}{N_i} \sigma_{El}^2,$$

where $L_i = N_i \times l$ and $\sigma_{El}^2 = 2X(l/2) - F(l)$.

The estimation of all the drifts (of total length $L = \sum_i L_i = Nl$) by the set of $N = \sum_i N_i$ channel samples, regularly spaced at intervals l along these drifts, is thus characterized by the line term

$$T_l = \sigma_{EL}^2 = \frac{1}{N} \sigma_{El}^2.$$

Section term The three levels are assimilated to the sum of n rectangles $s_i = h_i \times L_i$, each estimated by a median drift L_i supposed to be perfectly known. The estimation variance of these three sections (of a total surface $S = \sum_{i=1}^n s_i$), known as the section term, is then obtained from the direct combining formula (V.43):

$$T_S = \sigma_{ES}^2 = \frac{1}{S^2} \sum_i s_i^2 \sigma_{Es_i}^2,$$

where $\sigma_{Es_i}^2$ is the variance of extension of the median L_i to its rectangle of influence $S_i = h_i \times L_i$.

Slice term   The final term is the error of estimation of the three slices of the mass by their median levels supposed to be perfectly known. Each slice may, for example, be assimilated to a rectangular parallelepiped estimated by its median level S_k, and the slice term is then obtained from the direct combining formula (V.44):

$$T_V = \sigma_{EV}^2 = \frac{1}{V^2} \sum_k v_k^2 \sigma_{Ev_k}^2,$$

where

$$V = \sum_{k=1}^{3} v_k,$$

and $\sigma_{Ev_k}^2$ is the extension variance of the median rectangle of surface S_k to its parallelepiped of influence v_k.

Remarks   The terms combining to give formula (V.45), $\sigma_E^2 = T_l + T_S + T_V$, express the intuitive notion that the estimation would not be improved by increasing the density of channel samples in the drifts (T_i decreases) without also increasing the number of drifts on each level (T_S decreases) and the number of levels within the mass (T_V decreases). The line term is often negligible with respect to the section (T_S) and slice (T_V) terms. If the costs of channel sampling and drift development are known, the two-term formula, $T_l + T_S$, can be used to make optimum use of the sampling budget on any particular level.

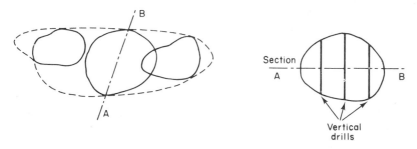

FIG. V.51. Estimation of a mass from vertical drills along sections.

The term-combining formula used in the preceding example can be applied to configurations other than channel samples in drifts on levels. For instance, it could be applied to the global estimation of a deposit from vertical drilling along vertical sections, cf. Fig. V.51.

V.C.3. Estimation of a quotient or a product

The quantity to be estimated is often in the form of a quotient or a product of two quantities estimated separately. Consider, for instance, a sub-horizontal sedimentary deposit of surface S and variable vertical thickness $t(x)$ as shown on Fig. V.52. For each vertical drill at point x, two additive variables are defined: mineralized thickness $t(x)$ and accumulation $a(x) = t(x) . z(x)$, where $z(x)$ is the grade measured along $t(x)$. This grade $z(x)$ is not an additive variable because it is defined on a variable support $t(x)$, and so the regionalization of $t(x)$ cannot be studied directly, cf. section III.C.2. The mean grade Z_S of the deposit is estimated by the quotient of the two global estimators of mean accumulation and mean thickness:

$$Z_S = \frac{1/S \int_S a(x)\,dx}{1/S \int_S t(x)\,dx} \quad \text{estimated by} \quad Z_S^* = \frac{A_S^*}{T_S^*}.$$

Similarly, if the mineralized surface S is estimated by S^*, the estimator of the global volume of ore is in the form of the product of the two global estimators S^* and T_S^*:

$$V = S \times T_S = \int_S t(x)\,dx \quad \text{estimated by} \quad V^* = S^* \times T_S^*.$$

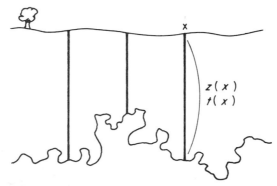

FIG. V.52. Definition of the accumulation $a(x) = t(x) . z(x)$.

The global estimation variances of mean accumulation and thickness can be calculated by means of the regionalization of the factor variables $a(x)$ and $t(x)$. The problem is to deduce the global estimation variance of the mean grade estimated by the quotient $Z_S^* = A_S^*/T_S^*$. More generally, the estimation variances of the product $P = A . B$ and of the quotient $Q = A/B$, estimated by $P^* = A^* . B^*$ and $Q^* = A^*/B^*$, respectively, are to be evaluated knowing the estimation variances, $\sigma_A^2 = E\{[A - A^*]^2\}$ and $\sigma_B^2 = E\{[B - B^*]^2\}$, of the two factors A and B.

Warning   The product and quotient of two random functions underline difficult problems of existence and convergence. Thus, the quotient of two normally-distributed variables is a variable with a Cauchy distribution which has no moments (no expected value, no variance); the expectation of a limited expansion is not always the limit of the sum of the expectations of the terms, etc. The following demonstrations are more symbolic than rigorous, which should be sufficient for the practitioner of mining estimation. All the following results can be established rigorously for the product and quotient of two lognormally distributed variables.

Estimation variance of the product $P = AB$

The true global values P, A, B are considered to be known, and their estimators P^*, A^*, B^* are interpreted as random variables:

$$A^* = A + \varepsilon_A, \qquad B^* = B + \varepsilon_B, \qquad P^* = P + \varepsilon_P,$$

where ε_A, ε_B, ε_P are the random estimation errors.

The estimations of the factors A and B are assumed to be unbiased, i.e., $E\{\varepsilon_A\} = E\{\varepsilon_B\} = 0$, and their estimation variances, $\sigma_A^2 = E\{\varepsilon_A^2\}$ and $\sigma_B^2 = E\{\varepsilon_B^2\}$, are known.

Thus,

$$P^* = A^*B^* = AB + A\varepsilon_B + B\varepsilon_A + \varepsilon_A\varepsilon_B$$

and

$$E\{P^*\} = P + A \cdot E\{\varepsilon_A\} + B \cdot E\{\varepsilon_A\} + E\{\varepsilon_A\varepsilon_B\} = P + E\{\varepsilon_A\varepsilon_B\}.$$

Therefore there is a bias, $E\{P^* - P\} = E\{\varepsilon_A\varepsilon_B\}$, in the estimation of the product P by $P^* = A^*B^*$. This bias is null if the estimation of the two factors, A and B, can be considered as independent, i.e., if $E\{\varepsilon_A\varepsilon_B\} = E\{\varepsilon_A\} \cdot E\{\varepsilon_B\} = 0$. In practice, the term $\varepsilon_A\varepsilon_B$ corresponds to a second-order error, and the bias term $E\{\varepsilon_A\varepsilon_B\}$ is negligible with respect to the value AB.

By neglecting the third- and fourth-order errors, the quadratic estimation error of the product P is written as

$$(P^* - P)^2 = A^2\varepsilon_B^2 + B^2\varepsilon_A^2 + 2AB\varepsilon_A\varepsilon_B + \ldots$$

and the relative estimation variance of P by P^* is

$$\frac{\sigma_P^2}{P^2} = \frac{E\{[P - P^*]^2\}}{P^2} = \frac{\sigma_A^2}{A^2} + \frac{\sigma_B^2}{B^2} + \frac{E\{[A - A^*][B - B^*]\}}{AB} + \ldots.$$

Putting $\rho_{AB} = (E\{[A - A^*][B - B^*]\}/\sigma_A \cdot \sigma_B)$ for the correlation coefficient of the two estimation errors of the two factors, the formula can be written as

$$\frac{\sigma_P^2}{P^2} = \frac{\sigma_A^2}{A^2} + \frac{\sigma_B^2}{B^2} + 2\rho_{AB} \cdot \frac{\sigma_A}{A} \cdot \frac{\sigma_B}{B} + \ldots \qquad (V.46)$$

In practice, the true values A and B in formula (V.46) are replaced by their unbiased estimators A^* and B^*. The calculation of ρ_{AB} will be detailed later.

Estimation variance of the quotient $Q = A/B$

Using the preceding notations, the estimation of the quotient can be expressed as

$$Q^* = \frac{A^*}{B^*} = \frac{A + \varepsilon_A}{B + \varepsilon_B} = \frac{A + \varepsilon_A}{B} \cdot \frac{1}{1 + \varepsilon_B/B} \cdot$$

If the relative error ε_B/B is small with respect to 1 – and this is generally the case for global estimation – the following limited expansion is obtained:

$$Q^* = \left(Q + \frac{\varepsilon_A}{B}\right) \cdot \left[1 - \frac{\varepsilon_B}{B} + \frac{\varepsilon_B^2}{B^2} - \frac{\varepsilon_B^3}{B^3} + \ldots\right] \qquad (V.47)$$

and

$$E(Q^*) = Q\left[1 + \frac{\sigma_B^2}{B^2} - \frac{E\{\varepsilon_A \varepsilon_B\}}{AB} + \ldots\right].$$

There is, thus, a relative bias, $(\sigma_B^2/B^2) - (E\{\varepsilon_A \varepsilon_B\}/AB) + \ldots$, which persists even when the two estimations are independent, i.e., even when $E\{\varepsilon_A \varepsilon_B\} = E\{\varepsilon_A\} \cdot E\{\varepsilon_B\} = 0$. In mining applications, this relative bias is negligible when the relative estimation variance σ_B^2/B^2 is less than 0.01, which is often the case in global estimation. Using (V.47), the relative estimation variance of the quotient Q can be expanded up to the second order as

$$\frac{\sigma_Q^2}{Q^2} = \frac{E\{[Q - Q^*]^2\}}{Q^2} = \frac{\sigma_A^2}{A^2} + \frac{\sigma_B^2}{B^2} - 2\frac{E\{[A - A^*][B - B^*]\}}{AB} + \ldots,$$

i.e.,

$$\frac{\sigma_Q^2}{Q^2} = \frac{\sigma_A^2}{A^2} + \frac{\sigma_B^2}{B^2} - 2\rho_{AB}\frac{\sigma_A}{A} \cdot \frac{\sigma_B}{B} + \ldots \qquad (V.48)$$

Calculation of the coefficient $\rho_{AB} = E\{[A - A^*][B - B^*]\}/(\sigma_A \cdot \sigma_B)$

Note first that if the two estimations of the two factors A and B are independent, the covariance $E\{[A - A^*][B - B^*]\}$ of the two errors is null, and the approximate formulae (V.46) and (V.48) reduce to the standard formula of addition of relative variances:

$$\frac{\sigma_Q^2}{Q^2} \quad \text{or} \quad \frac{\sigma_P^2}{P^2} = \frac{\sigma_A^2}{A^2} + \frac{\sigma_B^2}{B^2} + \ldots \tag{V.49}$$

If the two variables $a(x)$ and $b(x)$ are intrinsically coregionalized, i.e., if the two structural functions $\gamma_a(h)$ and $\gamma_b(h)$ are proportional, cf. section III.B.3, formula (III.30), *and* if the two configurations for the estimation of their mean values A and B are identical (similar in practice), then it can be shown that the coefficient ρ_{AB} is equal to the standard correlation coefficient ρ_{ab} between the two variables $a(x)$ and $b(x)$.

If the two variables $a(x)$ and $b(x)$ are not intrinsically coregionalized, then the covariance of the two errors $E\{[A-A^*][B-B^*]\}$ must be calculated, i.e., the coefficient ρ_{AB} is calculated from the cross-structural function

$$\gamma_{ab}(h) = \tfrac{1}{2}E\{[a(x+h)-a(x)][b(x+h)-b(x)]\}.$$

When the two configurations for the estimation of the two factors A and B are identical (or very similar), a "reduced difference" variable can be defined, $d(x) = a(x)/A - b(x)/B$. Its structural function can then be determined:

$$2\gamma_d(h) = E\{[d(x+h)-d(x)]^2\} = E\left\{\left[\frac{a(x+h)}{A} - \frac{a(x)}{A}\right]\left[\frac{b(x+h)}{B} - \frac{b(x)}{B}\right]\right\}$$

$$= \frac{2\gamma_a(h)}{A^2} + \frac{2\gamma_b(h)}{B^2} - \frac{4\gamma_{ab}(h)}{AB}.$$

If the two configurations for the estimations of the two factors A and B are supposed to be identical, then the two estimation variances σ_A^2 and σ_B^2 are obtained by applying the same linear operator $\mathscr{E}$ to the two structural models γ_a and γ_b, respectively, cf. formula (III.2):

$$\sigma_A^2 = E\{[A-A^*]^2\} = \mathscr{E}(\gamma_a) \quad \text{and} \quad \sigma_B^2 = E\{[B-B^*]^2\} = \mathscr{E}(\gamma_b);$$

similarly,

$$E\{[A-A^*][B-B^*]\} = \mathscr{E}(\gamma_{ab}).$$

The relative estimation variance of the quotient A/B can then be written as

$$\frac{\sigma_Q^2}{Q^2} = \frac{1}{A^2}\mathscr{E}(\gamma_a) + \frac{1}{B^2}\mathscr{E}(\gamma_b) - \frac{2}{AB}\mathscr{E}(\gamma_{ab}) + \ldots \tag{V.50}$$

$$= \mathscr{E}(\gamma_d) + \ldots.$$

The relative variance σ_Q^2/Q^2 is obtained directly by applying the operator $\mathscr{E}$ to the structure γ_d of the "reduced difference" variable.

In the case of the estimation of the product AB, the "reduced sum" variable, $s(x) = a(x)/A + b(x)/B$ would be defined, its semi-variogram $\gamma_s(h)$ would be calculated, and the relative estimation variance of the product could then, similarly, be written as

$$\frac{\sigma_P^2}{P^2} = \frac{1}{A^2}\mathscr{E}(\gamma_a) + \frac{1}{B^2}\mathscr{E}(\gamma_b) + \frac{2}{AB}\mathscr{E}(\gamma_{ab}) + \ldots \tag{V.51}$$

$$= \mathscr{E}(\gamma_s) + \ldots .$$

V.C.4. Global estimation on a domain of unknown geometry

Up to this point, the geometric error has not been considered, i.e., the error due to the ignorance of the exact limits of the mineralization. The volume of the area to be estimated was assimilated to the sum of N positive blocks or zones of influence v_i, i.e., the estimator of the total mineralized volume V was taken as the sum

$$V^* = \sum_{i=1}^{N} v_i$$

of the volumes of these blocks. In doing so, an error $(V - V^*)$ was incurred and must be evaluated, for example, through a variance of the geometric error. This geometric error $(V - V^*)$ not only affects the volume and the ore tonnage, but also the estimations of metal tonnage and mean grade of the deposit D. Thus, it is equally important to combine the influence of the geometric error with that of the different quality errors, e.g., the quality error due to the regionalization of the grade variables in the total volume V.

1. *Variance of the geometric error*

The determination of the limits of the mineralization is a constant problem in global estimation of a deposit. It sometimes happens that some of these limits are known *a priori*, e.g., the limit of a lease, a fixed depth for drilling and estimation, clearly defined vein walls, etc. Other limits require estimation by one of the following methods:

(i) all-or-nothing-type drilling, e.g., the limits of an underground reservoir or a mineralized lens;

(ii) economic cut-off grade, e.g., in deposits such as porphyry-copper, all zones with grades greater than 0·4% Cu will be considered as mineralized zones;

(iii) with the aid of geological interpretations made on planes or cross-sections.

When this estimation of the geometry of the deposit is directly related to geological hypotheses (metallogenesis, tectonics, etc.), the geometric error is of a deterministic type not favourable to its quantification (this does not mean that the error will be large or that such geological methods of estimation should be disregarded). In such a case, it is advisable to examine these geological hypotheses and possibly consider different ones. There will be a different global estimator of the ore tonnage for each interpretation of the geology which returns a certain random character to the estimation. Thus, even if the variance of the geometric error cannot be evaluated, at least its limits can be.

However, the geometric error may intrinsically have a purely or partially random character which can be modelled.

(i) From the experimental knowledge of similar deposits (e.g., random location in space of mineralized lenses of random volume and geometry). The morphology of the deposit can be simulated several times. The available data configuration can be applied to these various simulated deposits, and an estimation variance can, thus, be determined. It should be stressed that such a variance depends heavily on the model adopted for the morphology of the deposit.

(ii) There are, however, some cases, frequently encountered in mining applications, which allow an approach entirely free from any *a priori* hypothesis or model. This is the "transitive theory" approach, cf. G. Matheron (1971, p. 9), the main aspects of which are outlined in the following subsection.

The transitive approach   Let $k(x)$ be the indicator variable of the mineralized volume V to be estimated. This indicator is an all-or-nothing variable defined as

$$k(x) = \begin{cases} 1 & \text{if the point } x \text{ is in } V, \quad x \in V, \\ 0 & \text{if not,} \quad x \notin V. \end{cases} \tag{V.52}$$

The mineralized volume V is then the triple integral, in the three-dimensional space, of this indicator variable:

$$V = \int k(x)\, dx = \int_{-\infty}^{+\infty} dx_u \int_{-\infty}^{+\infty} dx_v \int_{-\infty}^{+\infty} k(x_u, x_v, x_w)\, dx_w. \tag{V.53}$$

In the transitive approach, the structural function "*geometric covariogram*" of the volume V is defined as

$$K(h) = \int k(x)k(x+h)\,dx, \qquad (V.54)$$

where h is the standard notation for a vector in the three-dimensional space for the estimation of a volume V, or in the two-dimensional space for the estimation of a surface S.

The real unknown volume (or surface) to be estimated is simply the value of the geometric covariogram at the origin: $V = K(0)$.

In the two-dimensional case, when the surface S is estimated by a regular grid (a_1, a_2), the origin of which is located at the point x_0 of coordinates (x_{0u}, x_{0v}), the estimator S^* of the surface area can be written, cf. Fig. V.53, as

$$S^*(x_0) = S^*(x_{0u}, x_{0v}) = a_1 a_2 \sum_{p_u=-\infty}^{+\infty} \sum_{p_v=-\infty}^{+\infty} k(x_{0u} + p_u \cdot a_1, x_{0v} + p_v \cdot a_2),$$

p_u and p_v being two integers defining the various nodes of the regular grid. If $n(x_0)$ is the number of positive drills, the estimator $S^*(x_0)$ is then written as $S^*(x_0) = a_1 a_2 \cdot n(x_0)$. Note that this estimator depends on the choice of the origin within the elementary grid surface $a_1 a_2$, cf. Fig. V.53.

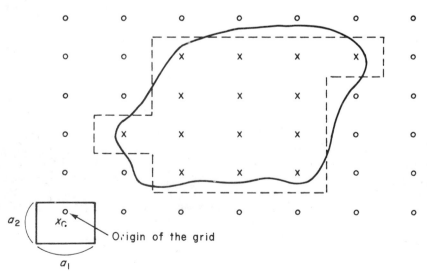

FIG. V.53. Estimation of a surface from a regular grid. ×, Positive hole; ○, negative hole; ——, real unknown limit; – – – –, polygonal estimated limit.

For each new position x'_0 of the grid, there is a corresponding estimator $S^*(x'_0)$ and, thus, a new realization $[S - S^*(x'_0)]$ of the estimation error of the surface S, cf. Fig. V.54. One possible way of defining the quality of the estimator S^* is to take the mean value of the quadratic error $[S - S^*(x_0)]^2$ when the origin x_0 of the grid takes all possible locations inside the elementary grid surface a_1a_2. This mean quadratic error is expressed as

$$E^2(a_1, a_2) = \frac{1}{a_1a_2} \int_{a_1} dx_{0u} \int_{a_2} [S - S^*(x_{0u}, x_{0v})]^2 dx_{0v}.$$

Note the similarity of this expression for the mean quadratic error to that of the estimation variance $E\{[S - S^*]^2\}$. Using this similarity, even though no probabilistic interpretation is involved in this "transitive" definition, the notation σ_E^2 for the estimation variance can be used, i.e., in one-dimensional notation:

$$\sigma_E^2(a) = \frac{1}{a} \int_a [S - S^*(x_0)] \, dx_0. \tag{V.55}$$

This mean quadratic error, called the "*variance of the geometric error*", is a function of the available data grid; here, $a = (a_1, a_2)$.

If $S^*(x_0)$ is replaced in the integral of (V.55) by its one-dimensional expression

$$S^*(x_0) = a \sum_{p=-\infty}^{+\infty} k(x_0 + p_a),$$

then, after some calculations, the following formula is obtained:

$$\sigma_E^2(a) = a \sum_{p=-\infty}^{+\infty} K(pa) - \int_{-\infty}^{+\infty} K(h) \, dh. \tag{V.56}$$

The variance of the geometric error linked to the grid spacing a can be expressed uniquely with the geometric covariogram $K(h)$ of the surface S (or volume V) to be estimated. However, this covariogram is just as unknown as the geometry and surface area of S (or V). It is at this point that an important property of geometric covariograms is introduced, which allows both the formula (V.56) and the transitive approach to geometric errors to be of practical use.

Whatever the geometry and the extent of the domain to be estimated (S or V) are, their geometric covariogram $K(h)$ exhibits a linear behaviour at the origin:

$$K(h) = K(0) - \bar{\omega}_\alpha|h| + \ldots, \quad \text{when } |h| \to 0.$$

The direction subscript α indicates that this linear behaviour may not be isotropic.

Using this property, the first term in the limited expansion of the expression $\sigma_E^2(a)$ can be evaluated when a tends towards zero. This first term is sufficient to give the order of magnitude of the geometric error variance.

For the two-dimensional problem of the estimation of a surface from a regular, rectangular grid (a_1, a_2), it can be shown (cf. G. Matheron (1965, p. 108)), that the relative estimation variance can be expanded up to the first order as

$$\frac{\sigma_S^2}{S^2} = \frac{1}{n^2}\left[\tfrac{1}{6}N_2 + 0\cdot06\frac{(N_1)^2}{N_2}\right] + \ldots, \qquad (V.57)$$

where $n \geqslant 10$ is the number of positive drills (the estimator is then $S^* = n \cdot a_1 a_2$), and $N_2 \leqslant N_1$; $2N_1$ and $2N_2$ being the number of elements parallel to the two sides of the grid (a_1, a_2) which make up the perimeter of the reunion S^* of the n positive zones of influence. The perimeter of any possible gaps in the mineralization must be taken into account in the calculation of the values $2N_1$ and $2N_2$.

Taking the example of Fig. V.53, with $n = 14$ positive drills, the values for the perimeter of the estimated surface S^* are:

$2N_1 =$ ten elements of length a_1 parallel to the side a_1 of the grid;

$2N_2 =$ eight elements of length a_2 parallel to the side a_2 of the grid.

Thus, the order of magnitude of the relative estimation variance of the surface S is

$$\frac{\sigma_S^2}{S^2} = \frac{1}{14^2}\left[\frac{4}{6} + 0\cdot06\frac{25}{4}\right] = 0\cdot005,$$

i.e., a relative standard deviation of 7%, $\sigma_S/S \approx 0\cdot07$, and a standard confidence interval equal to $S \in [14 \pm 2]a_1 a_2$. It will be recalled from section V.C.1 that this interval should be considered more as a comparative norm of the quality of the estimation (S, S^*) than as the true 95% confidence interval. Figure V.54 gives two further examples of the use of formula (V.57); the same surface S is estimated from the same grid (a_1, a_2), but with two different locations, x_0 and x_0', of the origin of the grid. Note that there are gaps in the mineralized surface.

Remarks The relative estimation variance of the surface, given by formula (V.57), is proportional to the number of elements N_1 and N_2 which make up the perimeter. This means that, for a constant surface area

S, the more jagged (i.e., the longer) the perimeter, the more difficult is the surface estimation.

When a portion of the perimeter is perfectly known (e.g., the limit of a lease), this portion must not be considered in the calculation of the values $2N_1$ and $2N_2$, and this will reduce the relative variance σ_S^2/S^2 by approximately the ratio of known perimeter to total perimeter.

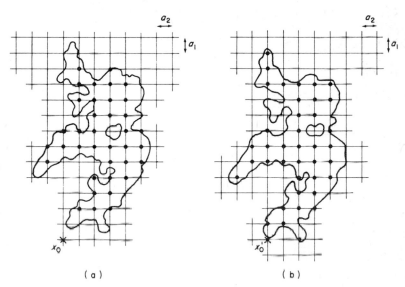

FIG. V.54. Estimation of the same surface from two distinct origins of the data grid.

(a) $n = 37$, $2N_1 = 26$, $2N_2 = 22$, $s^* = 37a_1a_2$, $\sigma/s = 4\cdot5\%$, $s = (37 \pm 3\cdot3)a_1a_2$.

(b) $n = 39$, $2N_1 = 32$, $2N_2 = 28$, $s^* = 39a_1a_2$, $\sigma/s = 4\cdot7\%$, $s = (39 \pm 3\cdot7)a_1a_2$.

In practice, S is estimated either by the sum of positive zones of influence, or by drawing a plausible contour, taking into account any knowledge of the morphology of the domain under study.

Three-dimensional case   For the estimation of volumes by a parallelepipedic grid of sides (a_1, a_2, a_3) and when the distances (a_1, a_2, a_3) tend towards zero, there is a first-order expansion formula similar to (V.57), cf. G. Matheron (1965, p. 105). In mining applications, such a parallelepipedic grid is rarely available and volumes are usually estimated by combining the surface estimations of different sections, or sometimes by the sum of prisms of influence of positive drill-holes.

First, let $V = \sum_i l_i S_i$ be the volume estimated by $V^* = \sum_i l_i S_i^*$, where l_i is the thickness of the section S_i (vertical or horizontal), cf. Figs V.51 and 55.

The surface S_i is estimated by S_i^* with an estimation variance of $\sigma_{S_i}^2 = E\{[S_i - S_i^*]^2\}$.

The elementary estimations of the surfaces S_i are usually independent and the estimation variance of the volume V can then be expressed as $\sigma_V^2 = E\{(V - V^*)^2\} = \sum_i l_i^2 \sigma_{S_i}^2$. The relative estimation variance is

$$\frac{\sigma_V^2}{V^2} = \frac{1}{V^2} \sum_i v_i^2 \frac{\sigma_{S_i}^2}{S_i^2}, \qquad (V.58)$$

where $v_i = l_i . S_i$.

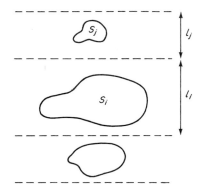

FIG. V.55. Estimations of a volume by combining sections.

Second, let V be a volume containing n positive vertical drill-holes of mineralized length l_i. These n drill-holes are on a regular horizontal grid of sides (a_1, a_2), cf. Fig. V.56. The horizontal projection S of the mineralized volume V is estimated by S^*, e.g., the union $S^* = n . a_1 a_2$ of the n positive zones of influence. The mean vertical mineralized thickness L is estimated by

$$L = \sum_{i=1}^{n} l_i.$$

The estimator of the volume V is then $V^* = L^* \times S^*$.

The first term of the limited expansion at the origin (when a_1 and a_2 tend toward zero) of the relative estimation variance of V by V^* is given by the formula

$$\frac{\sigma_V^2}{V^2} = \frac{S^*}{L^{*2}} \cdot \frac{1}{n^2} \left[0.06\lambda + \frac{\pi}{90} \frac{1}{\lambda^2} \right] + \ldots, \qquad (V.59)$$

where $\lambda = a_2/a_1 \leqslant 1$. For such a three-dimensional estimation of a volume, this formula only provides a rough order of magnitude of the estimation variance. In mining applications, if there are sufficient data to characterize the regionalization of the vertical thickness $l(x)$ at each point x, it is better to calculate the estimation variance of the mean thickness $E\{[L-L^*]^2\}$, then to combine this variance with the variance of estimation of the surface $E\{[S-S^*]^2\}$, using formula (V.63).

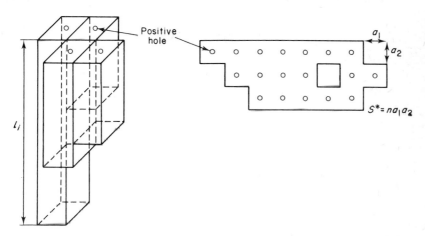

FIG. V.56. Estimation of a volume by combining vertical prisms of influence.

2. *Influence of the geometric error on global estimation*

Up to this point, a distinction has been made between the geometric error due to the determination of the limits of the mineralization and the quality error due to the estimation of a mean characteristic (e.g., grade) within a field of known limits. In practice, these two errors simultaneously affect the global estimations.

Consider, for instance, a subhorizontal sedimentary deposit sampled by vertical drill-holes, cf. Fig. V.57. A contour of the estimated surface S^* is defined, using the positive drill-holes. By means of the transitive approach,

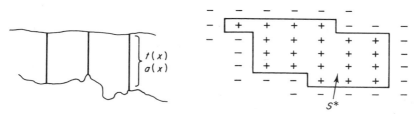

FIG. V.57. Estimation of a subhorizontal deposit from a regular grid of vertical drill-holes.

the order of magnitude of the relative estimation variance of the surface can be evaluated from formula (V.57), i.e., σ_S^2/S^2, with $\sigma_S^2 = E\{[S - S^*]^2\}$.

On the *estimated* field S^* of known limits, the following unbiased estimations can be made:

(i) the mean mineralized thickness

$$T_{S*} = \frac{1}{S^*} \int_{S*} t(x)\,dx,$$

estimated by T_{S*}^*;

(ii) the mean metal accumulation

$$A_{S*} = \frac{1}{S^*} \int_{S*} a(x)\,dx,$$

estimated by A_{S*}^*;

(iii) the mean grade $Z_{S*} = A_{S*}/T_{S*}$, estimated by $Z_{S*}^* = A_{S*}^*/T_{S*}^*$. (The relative bias due to the quotient of the factors A_{S*}^* and T_{S*}^* is neglected, cf. section V.C.3, formula (V.47)).

By means of the study of the two regionalizations, the thickness $t(x)$ and the accumulation $a(x) = t(x) \cdot z(x)$, defined as the product of the thickness $t(x)$ by the corresponding grade $z(x)$, the corresponding estimation variances can be calculated, i.e.,

for the estimator T_{S*}^*, $\sigma_{ET}^2 = E\{[T_{S*} - T_{S*}^*]^2\}$;

for the estimator A_{S*}^*, $\sigma_{EA}^2 = E\{[A_{S*} - A_{S*}^*]^2\}$.

These variances correspond to the estimation of the mean characteristics (mean thickness and accumulation) over the estimated field S^*. In fact, the real field is $S = S^* + [S - S^*]$, and it is the mean accumulation (or thickness) on this true field S that is required, i.e.,

$$A_S = \frac{1}{S} \int_S a(x)\,dx.$$

The difference between A_S and A_{S*}, i.e., $d_S A = A_S - A_{S*}$, corresponds to the impact of the surface error.

Thus, estimating the mean accumulation A_S on the true field S by A_{S*}^*, the error involved is $[A_S - A_{S*}^*] = d_S A + [A_{S*} - A_{S*}^*]$, the variance of which is

$$E\{[A_S - A_{S*}^*]^2\} = E\{(d_S A)^2\} + \sigma_{EA}^2 + 2E\{d_S A \cdot [A_{S*} - A_{S*}^*]\}.$$

The hypothesis of internal independence   Often (but not always†) in mining applications, the surface or volume estimation is made independently of the estimation of the mean quality; the estimation error $[A_{S^*} - A_{S^*}^*]$ on the field S^* of known limits can then be assumed to be independent of the impact $d_S A$ of the geometric error. The covariance of these two errors is null and, therefore,

$$E\{[A_S - A_{S^*}^*]^2\} = E\{(d_S A)^2\} + \sigma_{EA}^2.$$

The global estimation variance of the mean accumulation A_S is then reduced to the sum of two terms:

(i) a term σ_{EA}^2 for the estimation on the known field S^*, provided by the standard study of the regionalization of the accumulation $a(x)$;
(ii) a "border" term $E\{(d_S A)^2\}$ due to the geometric error.

It can be shown that, under the hypothesis of internal independence,

$$E\{(d_S A)^2\} = \frac{\sigma_S^2}{S^2} \cdot D_a^2(0/S), \qquad (V.60)$$

where σ_S^2/S^2 is the relative estimation variance of the surface, and $D_a^2(0/S)$ is the dispersion variance in S of the accumulation $a(x)$ of quasi-point support. In practice, provided that the data $a(x_i)$ are distributed in an approximately uniform manner over the field S^*, this dispersion variance $D_a^2(0/S)$ can be estimated by the experimental dispersion variance of the n data $\{a(x_i), i = 1$ to $n\}$ available in S^*, cf. section II.C.1.

Finally, the global estimation variance of the mean accumulation A_S can be expressed as

$$\sigma_A^2 = E\{[A_S - A_{S^*}^*]^2\} = \sigma_{EA}^2 + \frac{\sigma_S^2}{S^2} \cdot D_a^2(0/S). \qquad (V.61)$$

Similarly, the global estimation variance of the mean thickness T_S over the true surface S by the estimator $T_{S^*}^*$, defined on the estimated surface S^*, can be expressed as

$$\sigma_T^2 = E\{[T_S - T_{S^*}^*]^2\} = \sigma_{ET}^2 + \frac{\sigma_S^2}{S^2} \cdot D_t^2(0/S).$$

---

† When the limits of the mineralization are economic limits linked to the quality of the ore (e.g., cut-off grade), the geometric error and the quality error cannot be considered as being independent, particularly at the borders of the deposit. But if the extent of these border zones is negligible with respect to the extent of the deposit (of surface S), an hypothesis of internal independence can be made as a first approximation.

The estimation variance of the mean grade $Z_S = A_S/T_S$ by the quotient $Z_{S*}^* = A_{S*}^*/T_{S*}^*$ of the previous estimators defined on the known surface S^* is obtained from formula (V.48):

$$\frac{\sigma_Z^2}{Z^2} = \frac{\sigma_A^2}{A^2} + \frac{\sigma_T^2}{T^2} - 2\rho_{AT}\frac{\sigma_A}{A}\cdot\frac{\sigma_T}{T} + \ldots$$

The quantity of metal in the deposit is the product $Q = A_S . S$ of the mean accumulation A_S and the true mineralized surface S. The global estimation variance of Q by $Q^* = A_{S*}^* . S^*$ is obtained from formula (V.49):

$$\frac{\sigma_Q^2}{Q^2} = \frac{\sigma_A^2}{A^2} + \frac{\sigma_S^2}{S^2} + \ldots$$

Under the hypothesis of internal independence, the correlation ρ_{AS} between the estimation errors of the mean accumulation and the surface is null. If σ_A^2 is replaced by its expression (V.61), the following formula is obtained:

$$\frac{\sigma_Q^2}{Q^2} = \frac{\sigma_{EA}^2}{A_S^2} + \frac{\sigma_S^2}{S^2}[1 + D_a^2(0/S)] + \ldots \tag{V.62}$$

Similarly, the relative variance of global estimation of the ore tonnage $Vd = T_S \times S \times d$ by $V^*d = T_{S*}^* . S^* . d$ is

$$\frac{\sigma_{Vd}^2}{(Vd)^2} = \frac{\sigma_V^2}{V^2} = \frac{\sigma_{ET}^2}{T_S^2} + \frac{\sigma_S^2}{S^2}[1 + D_i^2(0/S)] + \ldots, \tag{V.63}$$

where the density d of the ore is supposed to be constant and known. If this density, $d(x)$, is regionalized and sampled along the drill-holes, it should then be studied by means of a new accumulation variable $b(x) = a(x) . d(x)$.

Note   All the preceding formulae correspond to a geometric error of surface $(S - S^*)$. If a volume is involved, the border term $E\{(d_VA)^2\} = E\{[A_V - A_{V*}]^2\}$ is written similarly to (V.60):

$$E\{(d_VA)^2\} = \frac{\sigma_V^2}{V^2}\cdot D_a^2(0/V), \tag{V.64}$$

where σ_V^2/V^2 is the relative estimation variance of the volume, and $D_a^2(0/V)$ is the dispersion variance of the point accumulation $a(x)$ in V.

V.C.5. Case study

Global estimation of Ity Mt Flotouo deposit, after Ph. Formery (1963)

This case study is well known to geostatisticians and has been used in courses at the Centre de Géostatistique of Fontainebleau since 1968. It concerns the Ity Mt Flotouo deposit in the Ivory Coast, which consists of a gold-bearing crust of laterite and clay. The thickness of this crust varies from 3 m to 5 m with a slight overburden. The deposit was sampled by vertical pits on a regular 20 m × 30 m grid, cf. Fig. V.58. Channel samples were taken down each pit along the entire mineralized thickness.

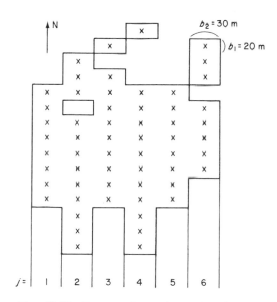

FIG. V.58. Data configuration on Ity deposit.

Estimation of the mineralized surface   The locations of the positive pits (non-null mineralized thickness) are shown on Fig. V.58. The estimated mineralized surface consists of the sum of the grids of influence of the $n = 58$ positive pits. Note the gap in the mineralization at the NW corner and the rather jagged outline of the polygonal contour of the Northern end. The influence zones of the positive pits towards this Northern boundary are debatable, but on such a small deposit, no better estimation of the Northern boundary can be made from geology. The transitive approach by means of formula (V.57) is used to characterize the surface error:

(i) the estimated mineralized surface is $S^* = nb_1b_2 = 58 \times 20 \times 30 = 34\,800\ m^2$;

(ii) the polygonal contour of S^*, including the NW gap, consists of $2N_1 = 44$ and $2N_2 = 20$ grid elements (b_1, b_2);

(iii) the order of magnitude of the relative estimation variance of the surface is then

$$\frac{\sigma_S^2}{S^2} \simeq \frac{1}{58^2}\left[\frac{10}{6} + 0\cdot06\frac{\overline{22}^2}{10}\right] = 0\cdot0014,$$

i.e., a relative standard deviation of $3\cdot7\%$, $\sigma_S/S \simeq 0\cdot037$.

Estimation of the mean accumulation   The regionalized variable (in the horizontal two-dimensional space) is the accumulation $a(x) = t(x)\,.\,z(x)$, defined at each point x by the product of the mineralized thickness $t(x)$ and the mean grade $z(x)$ measured over this thickness. If A^* is the mean accumulation measured over the 58 positive pits,

$$A^* = \frac{1}{n}\sum_{i=1}^{n} a(x_i),$$

then the relative accumulation $a(x)/A^*$ can be defined. By defining this relative regionalized variable, the value of A^* can be left unknown (for proprietary reasons). By means of this relative variable $a(x)/A^*$, the various relative variances can be calculated directly; indeed,

$$D^2\{a(x)/A^*\} = D^2\{a(x)\}/A^{*2}.$$

The N–S and E–W semi-variograms of the variable $a(x)/A^*$, calculated from the 58 positive pits, showed structural isotropy, and the two experimental curves were combined to give a single mean semi-variogram characteristic of the isotropic horizontal regionalization of the accumulation. A spherical model with a range $a = 66$ m and a nugget was fitted:

$$\gamma(h) = C_0 + C_1\gamma_1(r), \qquad \forall r = |h| > 0,$$

where $C_0 = 0\cdot30$ and $C_1 = 0\cdot75$, giving a sill value of $C_0 + C_1 = 1\cdot05$, which is very close to the experimental dispersion variance of the 58 positive accumulation values, $s^2 = 1\cdot0$.

The estimation variance of the mean accumulation A_{S^*} over the estimated known field S^*, $\sigma_{EA}^2 = E\{[A_{S^*} - A^*]^2\}$, can be calculated either by direct combining of elementary errors, assumed to be independent, or by combining terms of the error $[A_{S^*} - A^*]$, cf. section V.C.2.

Direct combining of elementary errors. As the estimator A^* is simply the arithmetic mean of the $n = 58$ positive accumulations, the global estimation errors $[A_{S^*} - A^*]$ is made up of the 58 elementary extension errors of a central accumulation datum to its grid of influence (b_1, b_2). Thus, from formula (V.37), the global estimation variance of the mean accumulation on the known field S^* is

$$\sigma_{EA}^2 = E\{[A_{S^*} - A^*]^2\} \simeq \frac{1}{n} \sigma_{Eb_1b_2}^2$$

with

$$\sigma_{Eb_1b_2}^2 = 2H(b_1/2; b_2/2) - F(b_1; b_2).$$

For a spherical model of range a and sill 1, the auxiliary functions H and F are given on Charts no. 3 and 4 in section II.E.4:

$$\sigma_{Eb_1b_2}^2 = C_0 + C \times 0 \cdot 11 = 0 \cdot 38 \quad \text{and} \quad \sigma_{EA}^2 \simeq 0 \cdot 38/58 = 0 \cdot 0066.$$

Combining terms. In this case, the global estimation variance is calculated as the sum of a line term and section term. The line term (corresponding to the N–S lines which have the greatest data density, $b_1 < b_2$) measures the extension error of the 58 central accumulations to their segments ($b_1 = 20$ m) of influence. This line term is expressed by formula (V.41) as

$$T_l \simeq \frac{1}{n} \sigma_{Eb_1}^2 = \frac{1}{n} [2\chi(b_1/2) - F(b_1)].$$

The value of the variance of the elementary extension $\sigma_{Eb_1}^2$ is read from Chart no. 7 and gives $\sigma_{Eb_1}^2 = C_0 + C \times 0 \cdot 077 = 0.36$. (Note the preponderance of the nugget effect $C_0 = 0 \cdot 30$.) The line term is then $T_l \simeq 0 \cdot 36/58 = 0 \cdot 0062$.

The section term measures the extension of the mean accumulations of the lines L_j, $j = 1$ to 6, to their sections of influence (rectangle $L_j \times b_2$). The elementary extension variance of a median line L_j to its rectangle of influence $L_j \times b_2$ is calculated using Chart no. 8, and, for the first line ($L_1 = 160$ m), this gives

$$\sigma_{EL_1b_2}^2 = C \times 0 \cdot 01 = 0 \cdot 0075.$$

In this case the nugget effect has no influence, cf. formula (III.9); this is due to the fact that the nugget constant C_{0L_1}, regularized over the length L_1, can be considered as zero because L_1 is much greater than the diameter of the sampling pit, which is the support on which the constant $C_0 = 0 \cdot 3$ was defined.

Table V.5 gives the variances $\sigma^2_{EL_j b_2}$ as well as the variances weighted by the square L_j^2, of their length of influence, for each of the six lines L_j. The section term is expressed by formula (V.43):

$$T_S \simeq \frac{1}{S^{*2}} \sum_{j=1}^{6} (L_j \cdot b_2)^2 \cdot \sigma^2_{EL_j b_2},$$

where $S^* = b_2 \cdot \sum_j L_j$, i.e.,

$$T_S \simeq \frac{\sum_j L_j^2 \sigma^2_{EL_j b_2}}{\left(\sum_i L_j\right)^2} = \frac{1534}{(1160)^2} = 0 \cdot 0011.$$

TABLE V.5. *Calculation of the section term*

j	L_j	$\sigma^2_{EL_j} \times b_2$	$L_j^2 \sigma^2_{EL_j} \times b_2$
1	160	0·0075	192
2	240	0·0060	345
3	200	0·0067	268
4	240	0·0060	345
5	160	0·0075	192
6	160	0·0075	192
	$\sum = 1160$ m		$\sum = 1534$

The global estimation variance of the mean accumulation on the *known* field S^* is the sum of the line and section terms:

$$\sigma^2_{EA} \simeq T + T_S = 0 \cdot 0062 + 0 \cdot 0011 = 0 \cdot 0073.$$

Thus, the combining-terms method provides the same order of magnitude for the global estimation variance as the direct combining method (0·0066). To be on the safe side, the higher value $\sigma^2_{EA} = 0 \cdot 0073$ is retained.

Calculation of the border term The preceding variance, $\sigma^2_{EA} = E\{[A_{S^*} - A^*]^2\}$, corresponds to an estimation on a known field S^* and takes no account of the surface error $(S - S^*)$. The impact of this surface error on the estimation of the mean accumulation A_S over the true surface S is characterized by the border term, cf. formula (V.60),

$$E\{(d_S A)^2\} = \frac{\sigma^2_S}{S^2} \times D_a^2(0/S),$$

where $\sigma^2_S/S^2 = 0 \cdot 0014$ is the relative estimation variance of the surface, and $D_a^2(0/S)$ is the dispersion variance of the point accumulation in S, and is estimated by the *a priori* experimental variance of the 58 positive accumulation data, i.e., $D_a^2(0/S) = 1\cdot0$. The border term is then $E\{d_SA)^2\} = 0\cdot0014$.

Taking the surface error into account, the relative† estimation variance of the mean accumulation is finally given by formula (V.61):

$$\frac{\sigma_A^2}{A^{*2}} = \sigma_{EA}^2 + E\{(d_SA)^2\} = 0\cdot0014 + 0\cdot0073 = 0\cdot0087,$$

i.e., a relative standard deviation of $9\cdot3\%$, $\sigma_A/A^* = 0\cdot093$.

Estimation variance of the quantity of metal   The quantity of metal available in the deposit of surface S is equal to the product $Q = A_S . S$. This quantity of metal is estimated by the product $Q^* = A^* . S^*$. Under the hypothesis of internal independence, the estimation variance of Q by Q^* is given by formula (V.49):

$$\frac{\sigma_Q^2}{Q^2} = \frac{\sigma_A^2}{A^2} + \frac{\sigma_S^2}{S^2} + \ldots = 0\cdot0087 + 0\cdot0014 \approx 0\cdot01,$$

i.e., a relative standard deviation of 10%, $\sigma_Q/Q = 0\cdot1$.

Note that the accumulation error σ_A^2/A^2 is quite significant with respect to the surface error σ_S^2/S^2.

---

† Recall that, since the regionalized variable is the relative accumulation, the calculated variances are also relative.

VI

Selection and Estimation of Recoverable Reserves

SUMMARY

This chapter is intended as a study of the influence of the regionalization of the selection variable on the recovery of the resources of a deposit.

Section VI.A is concerned with the respective influences of the support of the selection unit (mining selection is made on blocks and not on core samples) and of the level of available information (in general, selection is made on estimates while real grades are recovered and treated in the mill).

Section VI.B is a study of hypotheses of permanence of distribution laws, by means of which one histogram – e.g., the histogram of block grades used for mining selection – can be deduced from another with a different variance – e.g., the experimental histogram of core-sample grades.

VI.A. *IN SITU* RESOURCES AND RECOVERABLE RESERVES

Because of the variability of mining and treatment costs added to the spatial variability of ore quality, the entire resources of a deposit are rarely mineable. Out of these *"in situ resources"*, the mining company must define a certain tonnage of mineable ore, or *"recoverable reserves"*, within a given economic and technical context.

Whilst these *in situ* resources are determined by the geological environment of the deposit, the recoverable reserves will depend, not only on the *in situ* resources with their particular characteristics (spatial variabilities of grades, mineralogy, tectonics, etc.) but also, and essentially, on the economic and technical context of the mining project, i.e., as follows.

(i) On the criterion of selection, e.g., maximization of some benefit function or supply of a market.

444

(ii) On the cut-off parameters adopted to achieve this selection, e.g., a cut-off grade or a limiting mineable thickness.

(iii) On the technological constraints of the mining project, e.g., in an open-pit operation, a block V can only be mined if all the blocks above it, in the cone with its vertex on V, have been previously mined.

(iv) On the support (size and geometry) of the selection unit (block of several hundred or thousand tonnes). In general, this support will be considerably different from the support of the exploration unit (core samples for example).

(v) On the information available at the time of selection. In general, selection is made on estimates (kriged grades, for example) and, once the selection is made, real values are recovered (the mill receives and treats real grades, not estimates).

The influence on selection of these last two factors, support and level of information, is expressed by the geostatistical variances of dispersion and estimation, and this represents one of the main contributions of geostatistics to the techniques of mining estimations. The influence of economic and technical factors on the recovery of the resources of a deposit, while essential, will not be treated in this work. These factors are extremely variable from one deposit to another, and they involve a wide range of fields which have little or nothing to do with regionalized variables, such as sociology, politics, fiscal studies, mining techniques, haulage, etc. Since the object of this book is the study of regionalized variables, we shall limit ourselves to the study of the effect of regionalization of the selection variables on recovery. It is the ultimate job of the mining economist and mining engineer to conclude the feasibility study of the deposit by using information such as grade/tonnage curves supplied by the geostatistician.

VI.A.1. Influence of the selection support

Consider a deposit G sampled, for example, by drills, the cores of which are cut into constant lengths c and analysed. Let v be the mining unit on which future selection will be made.

The *in situ* resources are often characterized by histograms of dispersion, and the histogram of the grades $z_c(x)$ of the core lengths of support c is shown on Fig. VI.1; the abscissa values are for the grades and the ordinates are the frequency of occurrence of these grades. It is supposed that this histogram is representative of the entire deposit or zone G, which, in practice, amounts to saying that the sampling of G by the set of drill cores is unbiased – no particular zone of G is preferentially sampled.

Representativity of an experimental histogram

It is very important to ensure that the experimental histogram of the grades of the available core samples is truly representative of the histogram (called "local over G" from the terminology of section III.B.7) that would result from the set of core samples obtained by dividing the entire deposit G into drill cores. Thus, the experimentally available core samples should be uniformly distributed in the volume G; this will be approximately so if the drilling grid is regular over G. But if the drilling grid is irregular, with, for example, significant gaps or clusters of drills, it is advisable to assign an adequate weight to each grade value when constructing the histogram. This will prevent a non-representative mode appearing in the rich-grade area of the histogram if the drills tend to be concentrated in rich zones. The weighting of the grade values will, of course, influence the mean and variance of the histogram. In particular, the mean of the weighted histogram will no longer be the arithmetic mean of the data set available over G, but it will be closer to the true mean grade over G (correction of the bias due to the preferential concentration of drills). It should be ensured that the new variance agrees with the theoretical variance of dispersion $D^2(c/G)$ of the grades of support c in G; if the mineralization is globally homogeneous over G, then $D^2(c/G)$ can be calculated from the structural function (covariance or variogram), cf. section II.C.2. For each of the data values z_c, the weight can be taken as either

(i) its volume of influence in G, in which case the mean of the histogram is considered as the polygonal influence estimator (unbiased but not optimal) of the mean grade over G; or

(ii) the weight attributed to it in the global kriging of G, in which case the mean of the histogram is considered as the kriged mean grade of G.

Selection of units of support v

Consider the histogram of Fig. VI.1, which is supposed to be representative of the dispersion in G of the grades z_c of core samples of support c. This histogram has

(i) an experimental mean m^*, which for the moment is supposed to be equal to the true mean m of the deposit; this amounts to neglecting the global estimation error of the deposit;

(ii) an experimental dispersion variance $s^2(c/G)$;
(iii) a certain shape, asymmetric for example, which could be fitted to a
   lognormal distribution.

Consider the grade z_0 on the abscissa: the hatched area represents the
proportion of core samples with grades $z_c \geqslant z_0$.

Suppose now that the true grades $z_v(x)$ of all the mining units of support
v are known; the histogram of dispersion of the grades $z_v(x)$ can then be
constructed as on Fig. VI.1. This histogram has

(i) a mean equal to the mean of the core sample grades;
(ii) an experimental dispersion variance $s^2(v/G)$. (Note that $s^2(v/G) <
   s^2(c/G)$ since, in general, the support v is much larger than the
   support c.);
(iii) a certain shape, symmetric for example.

Consider the cut-off grade z_0; the dotted area represents the proportion
of panels v with true grades $z_v \geqslant z_0$.

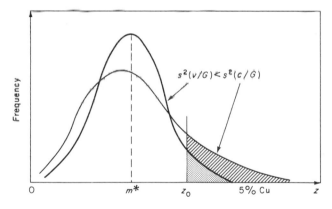

FIG. VI.1. Influence of support on recovery.

Just as the semi-variograms γ_c and γ_v regularized on core-sample
lengths c and panels v, respectively, are different, so are the two histograms
and the two hatched and dotted areas of these histograms. There may be a
non-negligible proportion of core samples c with grade $z_c \geqslant z_0 = 5\%$ Cu,
but there are no panels v with a mean grade $z_v \geqslant 5\%$ Cu. For high grades,
$z_0 > m$, the hatched area of the experimental histogram of core grades may
seriously overestimate the possible mined recovery in both tonnage and
mean grade and, correspondingly, underestimate the tonnage of metal left
in the waste, i.e., underestimate the area corresponding to the panels with
true grades $z_v < z_0$.

In mining, the actual selection is made on panels v and not on core samples c, and *for the estimation of reserves, it is essential to take the support v of the selection unit into account.* Otherwise there is a risk of bias with serious economic consequences.

This remark is not so trivial as it may sound. For example, consider a sedimentary deposit sampled by vertical drills analysed over the total mineralized thickness. If the mean mineralized thickness is $\bar{t}$, the support c of the data is a core of section ϕ and mean length $\bar{t}$ as shown on Fig. VI.2.

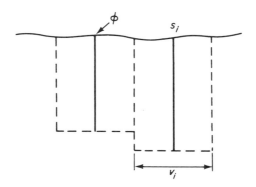

FIG. VI.2. Estimation using polygons of influence.

If the polygonal influence method is used, then the mean grade of the central drill core S_i is attributed to each polygon v_i. Now, if the cut-off grade z_0 is applied to the estimates obtained by this method, the estimated recovered tonnage will correspond exactly to the hatched area of Fig. VI.1, with all the associated risks of bias. The results of the case study given in section VI.A.4 are particularly good illustrations of this point.

Another typical example concerns the method of estimation by inverse squared distances and, more generally, all estimation methods which do not take account of the particular geometry of the panel to be estimated. For a given data configuration, such a method would give the same estimated value regardless of the estimation support v. Accordingly, if selection is made on these estimated values, the result will not depend on the size v of the selection unit. Now, it is obvious that mining a deposit with a small hammer (very selective mining with a very small support c) will yield results very different from block-caving (the mining unit v being very large), see P. Switzer and H. Parker (1975).

Within any homogeneously mineralized zone G (at the scale of mining), geostatistics can be used to calculate the variance of dispersion $D^2(v/G)$ of the grades Z_v of the mining units of support v (cf. section II.C.2), either formally by the general formula (II.36), $D^2(v/G) = \bar{\gamma}(G, G) - \bar{\gamma}(v, v)$, or experimentally by means of Krige's additivity relation (II.37), $D^2(v/G) = D^2(c/G) - D^2(c/v)$. The two dispersion variances of the core grades in G and v can be obtained experimentally.

The mean over G of the grades of the units v can be estimated, for example, by the mean m^* of the grades of available core samples. But the first two moments, mean m and dispersion variance $D^2(v/G)$, are not sufficient to determine the distribution of the grades Z_v. It is also necessary to know the type of function.

(i) In some cases, this distribution may be known *a priori*; for example, as a result of mining, or from blast-holes on benches, or from zones that have already been mined, cf. section II.C.3, the case study of the block grade distribution at Chuquicamata.

(ii) In other cases, an hypothesis of permanence of law can be postulated, and the shape of the histogram of experimental data Z_c can be adopted for the distribution of Z_v. Thus, if a lognormal distribution can be fitted to the histogram of the Z_c, the Z_v are supposed to be lognormally distributed with the variance $D^2(v/G)$. The problems involved in determining one distribution (Z_v) from another (Z_c) defined on a different support by means of permanence-of-law hypotheses will be treated later, in section VI.B.

VI.A.2. Influence of the level of information

In the preceding section, VI.A.1, we were concerned with the estimation of the histogram of the true mean grades Z_v, defined on the support v on which selection will actually be made. But is it really useful to know this histogram of *true* grades?

In fact, except in rare cases, these true grades z_v are unknown at the time of selection, and selection is made on the estimates z_v^*. This means that the actual recovery will consist not of the units with true grades $z_v \geq z_0$, but of the units with estimated grades $z_v^* \geq z_0$ corresponding to the dotted area in Fig. VI.3.

This dotted area, corresponding to the units actually recovered (i.e., those for which $z_v^* \geq z_0$), differs from the hatched area corresponding to perfect selection made on the true grades $(z_v \geq z_0)$. This difference becomes more important as the deviation between the estimators z_v^* and the true values z_v becomes greater, i.e., as the estimation variance $E\{[Z_v - Z_v^*]^2\}$ becomes larger. Now kriging minimizes this estimation variance.

Influence of technological conditions

In the preceding selection, it was supposed that it was possible to recover all units v with estimated grade z_v^* greater than the cut-off grade z_0, regardless of the position of v in the field G. In practice, this is not always possible; the actual recovery of a unit also depends on its accessibility and the proposed mining method. In an open-pit operation, for example, a block can be mined only if all the blocks inside a cone with its vertex on this block have already been mined; in underground mining, a new shaft will be sunk only if sufficient global reserves are accessible from it. In fact, technological considerations define a first rough selection, from which a second and more precise selection (by cut-off grade, for instance) can be made.

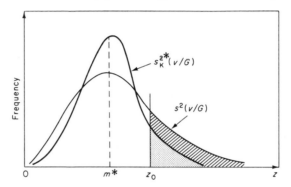

FIG. VI.3. Distributions of true and estimated grades of units v.

Thus, on each level, the pit design defines the accessible surfaces; one then proceeds to a second stage of selection which consists in distinguishing waste from ore on the basis of the mining unit.

   In this chapter, we limit ourselves to a one-stage selection only, made on the basis of a mean characteristic (grade, for example) of the unit v, independent of its location in the deposit or in the zone G. The difficult problems of estimation of the recoverable reserves after a two-stage selection are still in the domain of geostatistical research. One approach to these problems may consist of simulating the deposit and applying a simulation of the mining process to this numerical model: the influence of technological conditions on the recovery of reserves can then be observed, cf. J. Deraisme (1977) and section VII.B.3.

Smoothing effect

If the estimator Z_v^* used in the actual selection is a kriging estimator, then the variances $D^2(v/G)$ and $D_K^{2*}(v/G)$ of the true and estimated values can be related by one or the other of the following two relations, called "smoothing relations".

If the dimensions of the field G are large and if the available information is sufficient to provide a global estimation variance σ_{KG}^2 which is negligible with respect to the mean local estimation variance $\overline{\sigma_{Kv}^2}$, of a unit v, then

$$D^2(v/G) \simeq D_K^{2*}(v/G) + \overline{\sigma_{Kv}^2} - 2\bar{\mu} \qquad (VI.1)$$

(the notations $\overline{\sigma_{Kv}^2}$ and $\bar{\mu}$ are defined in the following paragraphs).

If the kriging procedures that provide the estimators Z_{Kv}^* can be assimilated to a complete kriging procedure, i.e., to a unique kriging procedure using the same data configuration comprising all the data available in G, then

$$D^2(v/G) = D_K^{2*}(v/G) + \overline{\sigma_{Kv}^2} - \sigma_{KG}^2 \simeq D_K^{2*}(v/G) + \overline{\sigma_{Kv}^2}, \qquad (VI.2)$$

where $D^2(v/G)$ is the dispersion variance of the true grades $Z_v(x)$ in the deposit or zone G, and $D_K^{2*}(v/G)$ is the dispersion variance of the corresponding kriged values $Z_{Kv}^*(x)$. From section II.C.1, it is known that these two variances, $D^2(v/G)$ and $D_K^{2*}(v/G)$, are estimators of the two experimental dispersion variances $s^2(v/G)$ and $s_K^{2*}(v/G)$ of the two histograms of the true and kriged grades of Fig. VI.3.

$$\overline{\sigma_{Kv}^2} = \frac{1}{N} \sum_{i=1}^{N} \sigma_{Kv_i}^2 = \frac{1}{N} \sum_i E\{[Z_{v_i} - Z_{Kv_i}^*]^2\}$$

is the mean value of the kriging variance of a unit v. There are N such units v in the zone G. If the grid of available data over G is not regular, then the various local kriging variances $\sigma_{Kv_i}^2$ will be different. $\bar{\mu} = (1/N) \sum_i \mu_i$ is the mean value of the Lagrange parameter used in the kriging system of each unit v, cf. system (V.1).

$\sigma_{KG}^2 = E\{[Z_G - Z_G^*]^2\}$ is the kriging variance of the mean grade Z_G over the zone G. In the case of a complete kriging system, it can be shown (cf. section V.A.3, "Theorem of combining kriged estimates") that the global kriged estimator Z_{KG}^* is the average of the N local kriged estimators $Z_{Kv_i}^*$, i.e.,

$$Z_{KG}^* = \frac{1}{N} \sum_i Z_{Kv_i}^*.$$

The proofs of these two smoothing relations ((VI.1) and (VI.2)) are given in the addendum at the end of this section, VI.A.2.

The smoothing relation (VI.2) implies a smoothing effect: $D_K^{2*}(v/G) \leqslant D^2(v/G)$. Indeed, the global kriging variance is always less than the mean local kriging variance: $\sigma_{KG}^2 \leqslant \overline{\sigma_{Kv}^2}$. On the other hand, in the local kriging system ((VI.1 or (V.2)), there is no restriction that $\sigma_{Kv_i}^2 \ge 2\mu_i$, so that it is not necessary that $\overline{\sigma_{Kv}^2} \ge 2\bar{\mu}$. However, the smoothing effect $D_K^{2*}(v/G) \le D^2(v/G)$ has always been observed in all data configurations encountered in mining applications and for the standard variogram or covariance models.

The kriged dispersion is smoothed with respect to the true dispersion and the amount of smoothing increases as the mean $\overline{\sigma_{Kv}^2}$ of the local kriging variance increases, i.e., as the estimators $Z_{Kv}^*(x)$ of $Z_v(x)$ become worse. This smoothing effect means that the histogram of the kriged values $Z_{Kv}^*(x)$ will display more values around the mean (close to m^*) but less extreme values than the histogram (unknown in practice) of the true grades $Z_v(x)$, cf. Fig. VI.3. This may be awkward when the cut-off grade is high, $z_0 > m^*$.

Note that the smoothing effect expressed in relations (VI.1) and (VI.2) is characteristic of the kriging estimator. The linear estimator obtained from the polygonal influence method, for instance, has a tendency to over-evaluate the real dispersion $D^2(v/G)$, cf. Fig. VI.9 in section VI.A.4.

Practical estimation of the recovery

At the time of selection, the true values z_0 are generally unknown, and the actual selection is made on the estimators z_v^* (preferably on the kriging estimator z_{Kv}^*). It is, thus, *imperative* to evaluate the recoverable tonnage from the histogram of the estimators z_v^*, i.e., by the dotted area ($z_v^* \ge z_0$) on Fig. VI.3.

1. If, at the time of the evaluation of the reserves, the ultimate estimators z_v^*, upon which selection will be based, are known, then the histogram of the z_v^* is experimentally available and the recoverable tonnage ($z_v^* \ge z_0$) can be estimated without any problem. In the next section, VI.A.3, it will be shown that the kriging of the selected units (those for which $z_{Kv}^* \ge z_0$) provides a fair estimator of the mean grade of this recoverable tonnage.

From the histogram of the kriging estimators z_{Kv}^*, it is possible to obtain an estimator of the histogram of the true values z_v by the following means.

(i) Use of a permanence-of-law hypothesis to extend the type of law observed on the distribution of the z_{Kv}^* to the distribution of the z_v.

(ii) Use of a correction of variance: the variance of the distribution of the z_v is $D^2(v/G)$. This variance can be deduced from $D_K^{2*}(v/G)$ by one of the smoothing relations (VI.1) or (VI.2), or calculated from the general formula (II.36).

It is then possible to evaluate the increase in the recovery of the resources of a deposit by improving local estimation: a decrease in σ^2_{Kv} will result in the histogram of the z^*_{Kv} (used by the actual selection) becoming closer to the histogram of the true values z_v.

2. Very often, at the time of the study of the evaluation of reserves, the ultimate data (e.g., blast-holes in an open pit) from which the estimators z^{**}_v used in selection will be constructed, are not yet available. This means that the only histogram available at this stage is that of the estimators z^*_{Kv}, constructed from the data available at the time of the study, rather than the histogram of the z^{**}_v on which the actual selection will be made. This first set of data is often very inferior to the ultimate set of data available for selection, e.g., the study of evaluation must be performed from drill-hole core data, while actual selection is based on a dense grid of blast-hole data, plus important visual information from the mining front. As a consequence, the two estimation variances $\sigma^{2*}_K = E\{[Z_v - Z^*_{Kv}]^2\}$ and $E\{[Z_v - Z^{**}_v]^2\}$, as well as the two histograms, may differ considerably.

Note that from the fact that $\sigma^{2*}_K > E\{[Z_v - Z^{**}_v]^2\}$, the dispersion variance $D^{2**}(v/G)$ of the ultimate estimators Z^{**}_v will generally be greater than the dispersion variance $D^{2*}_K(v/G)$ of the kriging estimator Z^*_{Kv} corresponding to the primary information.

The tonnage recoverable from future selection ($z^{**}_v \geq z_0$) corresponds to the dotted area on Fig. VI.4.

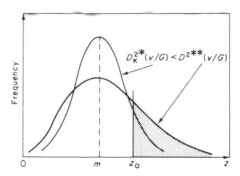

FIG. VI.4. Distributions of estimated values corresponding to two levels of information.

In practice, at the time of the study, both the histogram of the core grades z_c and that of the kriged grades z^*_{Kv} are available, these two histograms having respective variances $D^2(c/G)$ and $D^{2*}_K(v/G)$. The problem is to obtain an estimator of the histogram of the ultimate estimators z^{**}_v (to be used for selection) from these two experimentally available histograms. To do this, it is necessary to act as follows.

(i) To evaluate the level of ultimate information which will be used for future selection and to calculate the corresponding mean kriging variance $\overline{\sigma_{Kv}^{2**}}$. Remember that a kriging variance can be calculated in advance, once the corresponding data configuration is known, cf. section V.A.1, remark 4. However, the prior calculation of $\overline{\sigma_{Kv}^{2**}}$ is often approximate, since it must take into account future information which may be difficult to quantify, e.g., the experience of the shovel operator or information about the stope face.

(ii) Using one of the smoothing relations (VI.1) or (VI.2), to deduce the dispersion variance $D^{2**}(v/G)$ of the ultimate estimators Z_v^{**}. Thus, for example, if the following approximate smoothing relation is used:

$$D^2(v/G) \simeq D^{2**}(v/G) + \overline{\sigma_{Kv}^{2**}},$$

the variance $D^{2**}(v/G)$ is written as

$$D^{2**}(v/G) \simeq D_K^{2*}(v/G) + [\overline{\sigma_{Kv}^{2*}} - \overline{\sigma_{Kv}^{2**}}].$$

(iii) To deduce an estimator of the histogram of the Z_v^{**} from one of the two experimentally available histograms by a variance correction and assuming permanence of law.

If the core grade histogram is considered, the variance correction is a smoothing, $D^{2**}(v/G) < D^2(c/G)$. It is then advisable to verify that the same permanence-of-law hypothesis will also reproduce the histogram of the Z_v^* by correcting the variance $D^2(c/G) \rightarrow D_K^{2*}(v/G)$, with $D_K^{2*}(v/G) < D^{2**}(v/G) < D^2(c/G)$.

If the histogram of the kriged grades Z_{Kv}^* is considered as the starting point, correcting the variance then amounts to increasing the variability: $D^{2**}(v/G) \geq D_K^{2*}(v/G)$, since, in general, $\overline{\sigma_{Kv}^{2*}} \geq \overline{\sigma_{Kv}^{2**}}$.

The case study of section VI.B.4 gives an example of the development of such permanence-of-law procedures.

Note that, sometimes the sought after histogram of the ultimate selection estimator Z_v^{**} can be obtained directly from data coming from zones or benches already mined.

Addendum. Demonstration of the smoothing relations

(This demonstration does not provide any new insight into the problems of selection and can be omitted on a first reading.)

Let V be a stationary field made up of the union of N units, all with the same support v, centred on the points $\{x_i, i = 1$ to $N\}$. The global mean grade over V can be written as a sum of the N true grades of the units v:

$$Z_V(x) = \frac{1}{N} \sum_{i=1}^{N} Z_v(x_i).$$

The dispersion variance of the true grades of the units v in V can be written according to formulae (II.31) and (II.35):

$$D^2(v/V) = E\left\{ \frac{1}{N} \sum_i [Z_v(x_i) - Z_V(x)]^2 \right\} = \bar{C}(v, v) - \bar{C}(V, V)$$

Consider now the N unbiased estimators $Z_v^*(x_i)$ of the grades of the units v. The global mean grade estimated over V is then

$$Z_V^*(x) = \frac{1}{N} \sum_i Z_v^*(x_i).$$

In a similar way, we can define the dispersion variance of the estimated grades of the units v in V:

$$D^2(v/V) = E\left\{ \frac{1}{N} \sum_i [Z_v^*(x_i) - Z_V^*(x)]^2 \right\}$$

$$= \frac{1}{N} \sum_i E\{Z_v^{*2}(x_i)\} - E\{Z_V^{*2}(x)\}$$

The case of kriged estimators If the estimators used above are kriged estimators, then $Z_v^*(x_i) = Z_{Kv}^*(x_i)$ is a linear combination of the n_i data $\{Z_\alpha(x_i), \alpha = 1 \text{ to } n_i\}$ in the neighbourhood of estimation, i.e., $Z_{Kv}^*(x_i) = \sum_\alpha \lambda_\alpha Z_\alpha(x_i)$, and the kriging system (V.1) is written as

$$\begin{cases} \sum_\beta \lambda_\beta \bar{C}(\alpha, \beta) = \bar{C}(v_i, \alpha) + \mu_i, & \forall \alpha = 1 \text{ to } n_i, \\[2mm] \sum_\beta \lambda_\beta = 1 \end{cases}$$

and

$$\sigma_{Kv_i}^2 = E\{[Z_v(x_i) - Z_{Kv}^*(x_i)]^2\}$$

$$= \bar{C}(v_i, v_i) + \mu_i - \sum_\alpha \lambda_\alpha \bar{C}(v_i, \alpha).$$

Considering that $C(h)$ is a centred covariance, and that the stationary expectation is $m = E\{Z_{Kv}^*(x)\} = E\{Z_{KV}^*(x)\}$, the following relation can be expressed:

$$E\{[Z_{Kv}^*(x_i) - m]^2\} = \sum_\alpha \sum_\beta \lambda_\alpha \lambda_\beta \bar{C}(\alpha, \beta) = \sum_\alpha \lambda_\alpha \bar{C}(v_i, \alpha) + \mu_i$$

$$= \bar{C}(v_i, v_i) + 2\mu_i - \sigma_{Kv_i}^2 \qquad\qquad (VI.3)$$

Now, by putting

$$\bar{\mu} = \frac{1}{N} \sum_i \mu_i \quad \text{and} \quad \overline{\sigma_{Kv}^2} = \frac{1}{N} \sum_i \sigma_{Kv_i}^2,$$

and noting that $\bar{C}(v_i, v_i) = \bar{C}(v, v)$, $\forall i$ (all units v_i have the same support v), the following relation is obtained for the dispersion variance of the kriged grades:

$$D_K^{2*}(v/V) = \bar{C}(v, v) + 2\bar{\mu} - \overline{\sigma_{Kv}^2} - E\{[Z_{KV}^*(x) - m]^2\}.$$

By substituting $D^2(v/V) + \bar{C}(V, V)$ for $\bar{C}(v, v)$, this relation becomes

$$D^2(v/V) = D_K^{2*}(v/V) + \overline{\sigma_{Kv}^2} - 2\bar{\mu} - [\bar{C}(V, V) - E\{[Z_{KV}^*(x) - m]^2\}]. \quad \text{(VI.4)}$$

The term between the brackets is of the order of magnitude of the global estimation variance over the field V, and can generally be disregarded when compared to mean local estimation variance $\overline{\sigma_{Kv}^2}$. The first approximate smoothing relation (VI.1) is then obtained:

$$D^2(v/V) \simeq D_K^{2*}(v/V) + \overline{\sigma_{Kv}^2} - 2\bar{\mu}.$$

The case of complete kriging The term "complete kriging" is used when the N krigings $Z_{Kv}^*(x_i)$ have been carried out using the same data configuration comprising all[†] the available data in V. From section V.A.3, "Theorem of combining kriged estimates", it can then be shown that the sum $Z_V^* = (1/N) \sum_i Z_{Kv}^*(x_i)$ corresponds exactly to the direct kriging of $Z_V(x)$ from the previous data configuration. For $E\{[Z_{KV}^*(x) - m]^2\}$ we then obtain an expression similar to (VI.3):

$$E\{[Z_{KV}^*(x) - m]^2\} = \bar{C}(V, V) + 2\mu_V - \sigma_{KV}^2,$$

where

$$\mu_V = \frac{1}{N} \sum_i \mu_i = \bar{\mu},$$

cf. relation (V.17).

Substituting $(\sigma_{KV}^2 - 2\mu_V)$ for the term between brackets of formula (VI.4), relation (VI.4) gives the second smoothing relation (VI.2):

$$D^2(v/V) = D_K^{2*}(v/V) + \overline{\sigma_{Kv}^2} - \sigma_{KV}^2$$

The mean local kriging variance $\overline{\sigma_{Kv}^2}$ is always much greater than the global kriging variance σ_{KV}^2, and, thus, the previous relation reduces to

$$D^2(v/V) \simeq D_K^{2*}(v/V) + \overline{\sigma_{Kv}^2},$$

which characterizes the smoothing effect of kriging:

$$D_K^{2*}(v/V) \leqslant D^2(v/V).$$

[†] In practice, the information used for the kriging of a unit v is limited to a neighbourhood surrounding v. However, and provided there is no strong nugget effect (since the nugget effect removes the screen, cf. section V.A.1, remark 7), this nearby information screens the influence of more distant data and, thus, the kriging of v can be assumed to be complete and taking all the available information over V into account.

VI.A.3. Bivariate distribution of the estimated and true values

In the previous section, it was shown that the actual selection is carried out on the estimators z_v^* and not on the unknown true values z_v. The recoverable tonnage $(z_v^* \geq z_0)$ must thus be estimated from the distribution of the estimator Z_v^*. But once this selection is made, the mill receives the true values, not the estimates; more precisely, it receives the true values z_v *conditioned* by the fact that their estimators z_v^* are greater than the cut-off z_0. The set of recovered values is denoted by $\{z_v/z_v^* \geq z_0\}$.

By interpreting the true and estimated grades z_v and z_v^*, as realizations of the random functions $Z_v(x)$ and $Z_v^*(x)$, the bivariate distribution of the two variables Z_v and Z_v^* can be conveniently represented by the scatter diagram (or correlation cloud) on Fig. VI.5. The true grades appear along the ordinate axis and the estimated grades along the abscissa axis. The distribution of the true grades, with variance $D^2(v/G)$, can thus be seen as the projection of this diagram on to the ordinate axis; similarly, the distribution of the estimated grades, with variance $D^{2*}(v/G)$, can be seen as the projection of the diagram onto the abscissa axis.

A regression curve $f(z)$, giving the mean of the true grades of the units for which the estimated grades are equal to a fixed value $Z_v^* = z$, is also shown on the same figure. In probabilistic terms, $f(z)$ is the "conditional expectation" and is denoted by $f(z) = E\{Z_v/Z_v^* = z\}$. Note that the recoverable tonnage for a cut-off z_0, $\{Z_v/Z_v^* \geq z_0\}$, corresponds to the dotted area on Fig. VI.5.

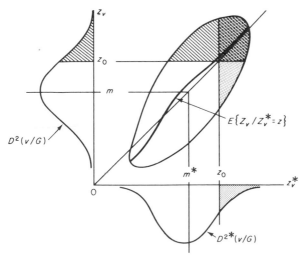

FIG. VI.5. Bivariate distribution of the true and estimated grades, Z_v and Z_v^*.

Conditional non-bias

The true grades $\{Z_v/Z_v^* = z_0\}$ of the selected units with estimated grades $Z_v^* = z_0$ are distributed along the vertical chord with abscissa z_0, and their mean conditional expectation is $f(z_0) = E\{Z_v/Z_v^* = z_0\}$. In practice, this true mean grade is unknown and is estimated by the cut-off value z_0. It is very important that this estimation z_0 be as close as possible to the true value $f(z_0)$. The ideal case would be $f(z_0) = z_0$, $\forall z_0$, i.e., an estimator Z_v^* such that its conditional expectation curve $E\{Z_v/Z_v^* = z\}$ is identical to the first bisector. The conditional non-bias relation can then be defined as

$$E\{Z_v/Z_v^* = z\} = z, \qquad \forall z, \qquad\qquad (VI.5)$$

which can also be written as

$$E\{[Z_v - Z_v^*]/Z_v^* = z\} = 0, \qquad \forall z.$$

Note that this conditional non-bias condition is more severe than the global non-bias condition $E\{Z_v - Z_v^*\} = 0$. Note also that the conditional non-bias (VI.5) entails the unbiasedness of the estimation of the true mean grade of the recoverable tonnage by the mean of the estimated grades of the selected unit, i.e.,

$$E\{(Z_v - Z_v^*)/Z_v^* \geqslant z_0\} = 0, \qquad \forall z_0$$

If selection is made from the estimators Z_v^* verifying the conditional non-bias relation (VI.5), the problem of the estimation of the recoverable reserves is thus resolved, The recoverable tonnage, i.e., the number of units with estimated grades $z_v^* \geqslant z_0$, is evaluated from the histogram of the estimated values z_v^*; the true mean grade of this recoverable tonnage is estimated without bias by the mean of the estimated grades of the selected units.

In practice, there is no reason why the estimator used, Z_v^*, should verify the conditional non-bias relation (VI.5); the conditional expectation $E\{Z_v/Z_v^* = z\} = f(z)$ will be different from the first bisector: $f(z) \neq z$, cf. Fig. VI.5.

An estimator Z_v^* should then be found, which would minimize the mean quadratic deviation $E\{[f(z) - z]^2\} = E\{[f(z) - Z_v^*]^2\}$. This criterion characterizes the accuracy of the effective selection; this selection is said to be accurate if the mean recovered grade $f(z)$ is close to the value z of the selection parameter.

Accuracy, however, is not sufficient: the selection made on the estimator $(Z_v^* \geqslant z_0)$ must also be as close as possible to the ideal selection carried out on the true values $(Z_v \geqslant z_0)$. The tonnage recovered by this ideal selection corresponds to the hatched area $\{Z_v/Z_v \geqslant z_0\}$ on Fig. VI.5. This hatched area differs in two ways from the dotted area $\{Z_v/Z_v^* \geq z_0\}$ corresponding to the tonnage which can be recovered in practice from the information Z_v^*.

(i) A certain number of units estimated as being too poor when they are, in fact, above the cut-off z_0; these units are wrongly rejected and correspond to the hatched area only.

(ii) A certain number of units estimated above the cut-off grade when they are, in fact, too poor; these units are wrongly accepted and correspond to the dotted area only.

The problem is to minimize these two areas corresponding to wrongly estimated units by reducing the length of the vertical chords of the scatter diagram (z_v, z_v^*), i.e., to minimize the dispersion of the selected true grades around their mean $f(z) = E\{Z_v/Z_v^* = z\}$, which amounts to minimizing *the conditional variance* $E\{[Z_v - f(z)]^2\}$.

The desired estimator Z_v^* is, thus, the one which minimizes both the accuracy criterion $E\{[f(z) - Z_v^*]^2\}$ and the conditional variance $E\{[Z_v - f(z)]^2\}$.

Consider the variance of estimation of Z_v by Z_v^*:

$$E\{[Z_v - Z_v^*]^2\} = E\{[Z_v - f(z)]^2\} + E\{[f(z) - Z_v^*]^2\}$$
$$+ 2E\{[Z_v - f(z)][f(z) - Z_v^*]\}.$$

From the definition of the conditional expectation $f(z) = E\{Z_v/Z_v^* = z\}$, the cross-term of the preceding expansion is zero and, thus,

$$E\{[Z_v - Z_v^*]^2\} = E\{[Z_v - f(z)]^2\} + E\{[f(z) - Z_v^*]^2\}. \qquad \text{(VI.6)}$$

Now, the kriging estimator Z_{Kv}^* is precisely the linear estimator which minimizes the estimation variance, i.e., the sum of the accuracy criterion and the conditional expectation. By minimizing the sum of these two criteria, kriging represents a compromise between satisfying one or the other, and, because of this, it is preferred to any other linear estimator for selection problems.

VI.A.4. Case study on a simulated deposit

A simulated deposit, as opposed to a real one, is perfectly known; in particular, all the distributions necessary for the solution of selection problems can be readily obtained (distribution of sampling grades, bivariate distribution of true and estimated grades of selection units, etc.). On a simulated deposit, it is thus possible to determine exactly the effect of any estimation procedure on the recoverable reserves.

The deposit

This simulation can be considered either as a zone of sedimentary deposit of variable vertical thickness (such as the lateritic or silicated nickel deposits of New Caledonia, cf. section IV.E, Case Study 13), or as a horizontal mining bench of constant vertical thickness in a massive deposit mined by open pit (such as a porphyry – copper deposit).

The simulated zone is a rectangle of dimensions $50u \times 10u$, made up of 500 square blocks v_i (of dimensions $u \times u$), cf. Fig. VI.6. A total of 60 500 samples were simulated on a regular $u/11 \times u/11$ grid over this rectangle, giving a total of $11 \times 11 = 121$ samples per block. These 60 500 samples are considered as the real deposit G to be studied.

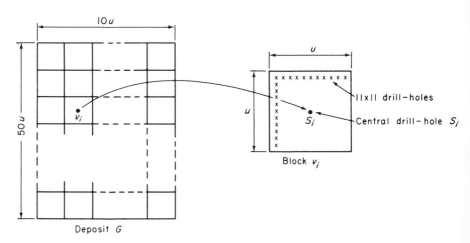

FIG. VI.6. The simulated data basis.

If G is a subhorizontal sedimentary deposit, then these 60 500 simulated values represent the accumulation $a(x) = t(x) \cdot z(x)$ at each point x, i.e., the product of the thickness $t(x)$ by the corresponding mean grade $z(x)$. If G is a bench in an open pit, then the simulated values represent the grade $z(x)$ over the constant vertical height of the bench. There are, thus, 500 true mean grades $z_v(x_i)$ of the blocks v_i, available over G (the true mean grade of each block being taken as the arithmetic mean of the 121 grades of the samples that fall within it). The samples S_i at the centre of each block are assumed to be the result of a sampling campaign providing 500 grades $z(x_i)$ on a regular $u \times u$ grid, cf. Fig. VI.6.

The grades $z(x)$ were simulated according to the techniques of simulation presented in Chapter VII, i.e., in such a way that they reproduce the following.

(i) A given distribution – with mean 9·57% and variance $\sigma_Z^2 = 22 \cdot 4\%^2$ – which is, in fact, the experimental histogram of the nickel accumulations (thickness × grade) of the Mea deposit (New Caledonia). Obviously, if $z(x)$ is interpreted as a grade, the value $m = 9 \cdot 57\%$ is closer to a lead grade, while it is closer to a copper grade when divided by 10.

(ii) A given model of spatial variability; in this case, a spherical model with a nugget effect and isotropic in the horizontal plane:

$$\tfrac{1}{2}E\{[Z(x+h)-Z(x)]^2\} = \gamma(h) = C_0 + C_1\gamma_1(r), \qquad \forall r = |h| > u/11,$$

where $C_0 = 6\%^2$, $C_1\gamma_1(r)$ is a spherical model with sill $C_1 = 18\%^2$ and range $a_1 = 10u$.

Kriging of selection units

A model $\gamma(h)$ was fitted to the experimental semi-variogram $\gamma^*(h)$ of the 500 samples from the sampling campaign of the $u \times u$ grid. These samples were then used to estimate each block v_i by kriging, using the estimation configuration on Fig. VI-7, i.e.: one weight λ_1 for the grade of the central sample; one weight λ_2 for the mean grade of the four median samples of the first aureole; one weight λ_3 for the mean grade of the four corner samples of the first aureole.

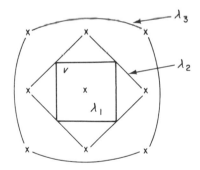

FIG. VI.7. Kriging configuration of the selection units.

This kriging configuration is complete only for the 384 blocks v_i inside the zone G' of dimensions $8u \times 48u$, defined as the erosion of G by a unit distance u. The following study will be confined to these 384 blocks for which the kriging system is identical.

The results $\lambda_1, \lambda_2, \lambda_3, \sigma_K^2$ of the above kriging system are given in Table VI.1. Since the true grades of these 384 blocks are available, the kriging variance $\sigma_K^2 = 1 \cdot 32\%^2$ can be compared to the experimental estimation variance $\sigma_E^{2*} = 1\cdot35\%^2$. The kriging system thus provides an excellent prediction of the error variance.

If each of these 384 blocks had been estimated by the polygons of influence method (Poly), by assigning the grade of the central sample to each unit v, i.e., $\lambda_1 = 1$ and $\lambda_2 = \lambda_3 = 0$, the experimental estimation variance would have been $\sigma_E^{2*} = 6\cdot92\%^2$. Using geostatistics, this variance would have been predicted, according to formula (II.28), as $\sigma_E^2 = 7\cdot31\%^2$.

TABLE VI.1. *Compared results of the kriging and polygons of influence methods of estimation*

	Krig	Poly
λ_1	0·204	1
λ_2	0·458	0
λ_3	0·338	0
σ_E^2, theoretical	1·323	7·307
σ_E^2, experimental	1·348	6·919

Estimation of recoverable reserves

Suppose, now, that at the time of mining, the only information available on which to carry out selection is the $u \times u$ grid of samples. Recovery consists of all units v_i which have an estimated grade $z_{v_i}^*$ greater than the cut-off grade z_0.

The tonnage of recovered ore $T(z_0)$, i.e., the number of units $N(z_0)$ recovered, is determined from the histograms VI.9(a) and (b) of the 384 estimated values, $Z_{Kv_i}^*$ by kriging or Z_i^* by polygons of influence. Note that histogram VI.9(b) of the estimators Z_i^* is that of the 384 grades of the central samples; thus, using the polygons of influence estimator Z_i^* amounts to ignoring the influence of the selection support v, cf. section VI.A.1. The histogram of the 384 true grades of the blocks v_i is given on Fig. VI.9(c).

Once the selection is made, $Z_{Kv_i}^* \geqslant z_0$ or $Z_i^* \geqslant z_0$, the conditional non-bias condition must be verified, cf. section VI.A.3, formula (VI.5), i.e., the true mean grade $E\{Z_v/Z_v^* \geqslant z_0\}$ and the estimated mean grade $E\{Z_v^*/Z_v^* \geqslant z_0\}$ of the recovered tonnage must agree. For each of the two estimators considered, Fig. VI.8(a) and (b) give the two curves $E\{Z_v/Z_v^* \geqslant z_0\}$ and $E\{Z_v^*/Z_v^* \geqslant z_0\}$ as functions of the cut-off value z_0. Note that kriging ensures conditional non-bias, whereas selection carried out on polygonal influence estimators is subject to a serious bias (systematic overestimation of the actually recovered grades); this bias increases with the selectivity (i.e., as z_0 increases).

The global quantity of metal actually recovered, $Q(z_0) = T(z_0) . E\{Z_v/Z_v^* \geqslant z_0\}$, is proportional to $N(z_0) . E\{Z_v/Z_v^* \geqslant z_0\}$. The difference between the two quantities of metal recovered, corresponding to the two types of selection estimator considered, is a direct measure of the economic benefit to be obtained from a geostatistical approach to selection problems, i.e., an approach which takes into account the two fundamental influences of the selection support and the level of information.

The results of a hypothetical selection based on the true grades Z_{v_i} of the 384 selection units (i.e., perfect knowledge of the deposit) can also be given. This ideal selection will obviously lead to the best results.

For the three selection hypotheses (based, respectively, on true grade, kriged grade and grade estimated by polygons of influence) Table VI.2 gives, for the three cut-off grades $z_0 = 4$, 9 and 14%:

1. the number $N(z_0)$ of recovered blocks;
2. the true mean grade of the recovered tonnage, $E\{Z_v/Z_v^* \geqslant z_0\}$;
3. the estimated mean grade of the recovered tonnage, $E\{Z_v^*/Z_v^* \geqslant z_0\}$;
4. the conditional bias, $E\{Z_v - Z_v^*/Z_v^* \geqslant z_0\}$;
5. the experimental estimation variance of the grades of the recovered blocks, $E\{[Z_v - Z_v^*]^2/Z_v^* \geqslant z_0\}$;
6. the global quantity of metal actually recovered, $N(z_0) . E\{Z_v/Z_v^* \geqslant z_0\}$;
7. the estimation of this quantity of metal, $N(z_0) . E\{Z_v^*/Z_v^* \geqslant z_0\}$.

TABLE VI.2. *Comparative results of selection based on various estimators*

	z_0	(1)	(2)	(3)	(4)	(5)	(6)	(7)
True grade selection, Z_v	4	370	10·08		0	0	3730	
	9	203	12·97		0	0	2633	
	14	67	16·19		0	0	1085	
Kriging, Z_v^*	4	365	10·13	10·05	0·08	1·36	3698	3668
	9	205	12·82	12·66	0·16	1·36	2628	2593
	14	65	16·04	15·91	0·13	1·56	1043	1034
Poly, Z^*	4	337	10·51	10·68	−0·17	6·94	3542	3599
	9	199	12·52	13·52	−1·00	7·08	2491	2690
	14	67	14·93	16·95	−2·02	7·81	1000	1136

Note that the selection made on kriged estimates results in a better recovery from the deposit, in terms of the global quantity of metal actually recovered. Also note that the standard polygons of influence estimator always overestimates the effectively recoverable quantity of metal, which is particularly dangerous during a feasibility study. Selection based on the standard estimators of inverse distances (ID) or inverse squared distances (ID2) would have yielded better results than selection based on polygons of influence estimators (Poly), but these results are always strictly inferior to those obtained from selection based on a kriged estimator.

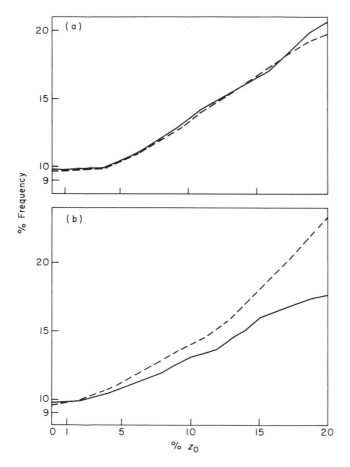

FIG. VI.8. Conditional bias of the estimators. (a) Kriging. ———, $E\{Z_v/Z^*_{Kv} \geqslant z_0\}$; ––––, $E\{Z^*_{Kv}/Z^*_{Kv} \geqslant z_0\}$. (b) Polygons of influence. ———, $E\{Z_v/Z^*_v \geqslant z_0\}$; ––––, $E\{Z^*_v/Z^*_v \geqslant z_0\}$.

If, instead of a regular grid, the 384 blocks were estimated from data on a pseudo-regular grid with clusters of data and gaps (which is often the case in mining applications) the results obtained from kriging would be still even better than those obtained from ID, ID2, and Poly.

The smoothing relation due to kriging

The smoothing relation (VI.2) due to kriging can be verified from the experimental dispersion variances of Fig. VI.9(a) and (c):

$$D^2(v/G') \simeq D_K^{2*}(v/G') + \sigma_{Kv}^2$$

(the global estimation variance $\sigma_{KG'}^2$ is disregarded). Indeed, the experimental values show

$$16 \cdot 14 \simeq 14 \cdot 71 + 1 \cdot 35 = 16 \cdot 06.$$

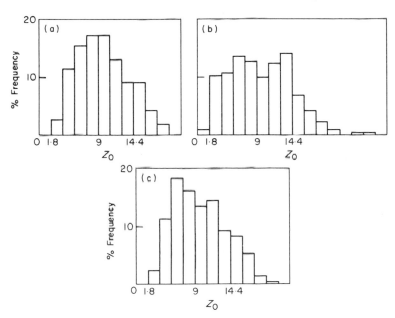

FIG. VI.9. Histograms of true grades and of their estimators. (a) Kriging: $m^* = 9 \cdot 72$, $\sigma^{2*} = 14 \cdot 71$. (b) Polygons of influence: $m^* = 9 \cdot 70$, $\sigma^{2*} = 23 \cdot 13$. (c) True grades: $m^* = 9 \cdot 82$, $\sigma^2 = 16 \cdot 14$.

The bivariate distribution of true and estimated grades

It was shown on Fig. VI.5 of section VI.A.3 that the conditional non-bias $E\{Z_v/Z_v^* = z\} = z$, $\forall z$, is verified when the conditional expectation curve $f(z) = E\{Z_v/Z_v^* = z\}$ is identical to the first bisector. The best linear estimator of this regression curve $f(z)$ is the regression line D_1 of the true values Z_v given the estimated values Z_v^*. It is thus advisable to verify the proximity of D_1 to the first bisector on the two scatter diagrams (z_v, z_v^*) corresponding to the two types of estimator used.

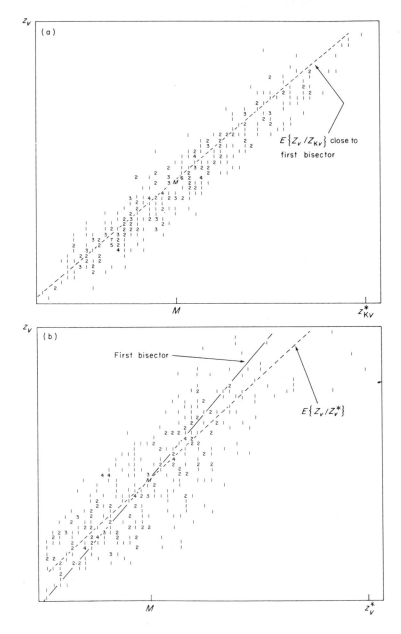

FIG. VI.10. Scatter diagrams, true grade/estimator. (a) Kriging; (b) polygons of influence.

For the kriging estimator, the equation of D_1 is $Z_v = 0\cdot074 + 1\cdot003\, Z^*_{Kv}$, with a residual variance approximately equal to the kriging variance $1\cdot348$. Thus, there is no significant difference between this line D_1 and the first bisector, cf. Fig. VI.10(a).

For the polygons of influence estimator, the equation of D_1 is $Z_v = 3\cdot040 + 0\cdot699 Z^*_v$, with a much greater residual variance ($4\cdot829$). This regression line D_1 is significantly different from the first bisector, as can be seen on Fig. VI.10(b).

On the other hand, if we consider the second regression line D_2 of the estimated values Z^*_v, given the true values Z_v (i.e., the best linear estimator of the second regression curve $g(z) = E\{Z^*_v / Z_v = z\}$), then kriging gives the equation of D_2 as $Z^*_{Kv} = 0\cdot744 + 0\cdot911 Z_v$, with a residual variance of $1\cdot229$.

The polygons of influence estimator gives the equation of D_2 as $Z^*_v = -0\cdot144 + 1\cdot002 Z_v$, with a residual variance approximately equal to the estimation variance $6\cdot919$, cf. Table VI.1. In this case, line D_2 corresponding to the Poly estimator is closer to the first bisector. In fact, the mean of the grades Z^* of the samples within a block v of true grade $Z_v = z$ is equal to z. However, since the actual selection is made on the estimator Z^*_v and not on the unknown true grade Z_v, regression line D_1 is the line of interest and, thus, the kriging estimator Z^*_{Kv} is retained and conditional non-bias is ensured (D_1 is close to the first bisector).

It was D. G. Krige (1951) who first distinguished the practical influence on mining selection of these two regression lines, and for this reason the preceding scatter diagrams (Z_v, Z^*_v) are sometimes referred to as "Krige's ellipse" the term ellipse is used because the experimental scatter diagram takes an elliptical shape when the bivariate distribution is Gaussian.

VI.B. PERMANENCE OF DISTRIBUTION

It was seen in the preceding section that the estimation of the recoverable tonnage requires the inference of the histogram of the estimators Z^{**}_v which are ultimately used in the future selection. For example, this inference is made from the histogram of the grades Z_c of the available core samples, or from the histogram of the estimators Z^*_v available at the time of the study.

More generally, we have the experimental histogram of a stationary variable Z, from which it is required to deduce the distribution of a variable Y that has the *same* expectation, but a different, *known* variance: $E\{Y\} = E\{Z\} = m$ and $\text{Var}\{Y\} \neq \text{Var}\{Z\}$. This problem as posed has no solution, since the type of the distribution of Y is undetermined. However, the problem can be solved if we assume that the distributions of Y and Z have the same "shape". The following section will show how this permanence of shape, or of law, can be interpreted in practice.

VI.B.1. Simple hypotheses of permanence

1. *Permanence of normality*

Let Z be a Gaussian stationary variable (in practice, a variable the histogram of which can be approximated by a normal distribution), with expectation m and variance σ_Z^2. Let U be a standard normal variable with zero expectation and unit variance. These two variables are related by the standard formula $Z = m + \sigma_Z . U$.

The hypothesis[†] of permanence of normality means that Y is considered to follow a normal distribution with expectation $m = E\{Y\} = E\{Z\}$ and a variance $\sigma_Y^2 \neq \sigma_Z^2$. Y can be expressed in terms of the standard normal law as

$$Y = m + \sigma_Y . U. \tag{VI.7}$$

The distribution of Y can then be readily determined from tables of standard normal distribution.

2. *Permanence of lognormality*

Let Z be a lognormal variable with expectation m and variance σ_Z^2. The Gaussian transform function φ_Z which transforms a standard normal variable U into this lognormal variable Z is

$$Z = \varphi_Z(U) = e^{m' + \sigma_Z' U} \tag{VI.8}$$

or, conversely,

$$U = \varphi_Z^{-1}(Z) = (\log Z - m')/\sigma_Z',$$

where m' and $\sigma_Z'^2$ are the expectation and variance of the normal variable $\log Z$. These logarithmic parameters are deduced from the known arithmetic parameters m and σ_Z^2 by the standard relations

$$\left\{ \begin{matrix} m = E\{Z\} = e^{m' + \sigma_Z'^2/2}, \\ \sigma_Z^2 = \text{Var}\{Z\} = m^2(e^{\sigma_Z'^2} - 1). \end{matrix} \right\} \tag{VI.9}$$

[†] If Y can be expressed as a linear combination $Y = \sum_\alpha \lambda_\alpha Z_\alpha$ of normal variables, and if the distribution of the RF $Z(x)$ is multivariate Gaussian, then its distribution is normal, and the permanence of normality is a property of Y and not an hypothesis. However, normal-type symmetrical distributions are not very frequent in mining applications; the histograms of grades in particular are usually asymmetric of the lognormal type.

The preceding transform function φ_Z can then be written as

$$Z = \varphi_Z(U) = m \cdot e^{-\sigma_Z'^2/2} \cdot e^{\sigma_Z' \cdot U}. \tag{VI.10}$$

The hypothesis[†] of permanence of lognormality amounts to considering that the variable Y follows a lognormal distribution with the same expectation $m = E\{Y\} = E\{Z\}$ and a variance $\sigma_Y^2 \neq \sigma_Z^2$. The Gaussian transform function φ_Y which expresses Y as a function of the standard normal variable U is thus written as

$$Y = \varphi_Y(U) = m \cdot e^{-\sigma_Y'^2/2} \cdot e^{\sigma_Y' \cdot U}. \tag{VI.11}$$

The logarithmic parameter σ_Y' is deduced from the known arithmetic variance σ_Y^2 by the relation

$$\sigma_Y^2 = \text{Var}\{Y\} = m^2(e^{\sigma_Y'^2} - 1).$$

It is thus possible to construct the required distribution of Y from the standard normal distribution of U. The distribution function is then

$$F_Y(y) = \text{Prob}\{Y < y\} = \text{Prob}\{U < u\} = G(u), \tag{VI.12}$$

where

$$y = \varphi_Y(u) \quad \text{or} \quad u = \varphi_Y^{-1}(y),$$

and $G(u)$ is the standard normal distribution function.

3. *Indirect correction by permanence of a two-parameter distribution*

The standard two-parameter distributions, such as the normal, lognormal or gamma, have the advantage of being tabulated, and an immediate correction of the variance $\sigma_Z^2 \to \sigma_Y^2$ can be made. However, as these distributions depend on two parameters only, they are often unable to take the major characteristics of an experimental histogram into account; a bimodal histogram, for instance, requires at least a three-parameter function in order to obtain a good fit. The lognormal distribution fitted to the experimental histogram on Fig. VI.11(a) cannot take account of the two significant modes. However, the major interest lies in an estimator of

[†] Contrary to the normal case, a linear combination $Y = \sum_\alpha \lambda_\alpha Z_\alpha$ of lognormal variables Z_α does not follow a lognormal distribution. Thus, if the grades Z_c of core samples follow a lognormal distribution, then, to be theoretically rigorous, the grades Z_v of blocks cannot follow a lognormal distribution. In mining applications, however, the histogram of the block grades z_v very often has the same lognormal type of asymmetry as the histogram of core sample grades z_c. For this reason, a working hypothesis of permanence of lognormality is adopted. A case study involving permanence of lognormality is presented in section II.C.3, "Grade dispersions at Chuquicamata mine", cf. Figs. II.22 to II.24.

the distribution of the variable Y with the same expectation but different variances $\sigma_Y^2 \neq \sigma_Z^2$, and not in the quality of the fit of the histogram of Z.

If the two modes of the Z histogram are to be exactly reproduced on the estimated histogram of Y, then, under certain conditions,[†] the following procedure can be used.

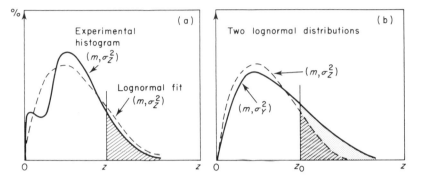

FIG. VI.11. Indirect correction of variance through a lognormal model. (a) Lognormal fit; (b) correction of variance.

(i) Using the simple two-parameter distribution which fits best the experimental histogram of Z, make the variance correction $\sigma_Z^2 \rightarrow \sigma_Y^2$, cf. the two lognormal distributions on Fig. VI.11(b).

(ii) From these two two-parameter distributions, read the recoveries corresponding to the cut-off grade z_0. The quotient of these two recoveries gives an estimate ρ of the correction coefficient, to be applied to the recoveries deduced from the experimental histogram of Z to obtain the recoveries corresponding to the histogram of Y.

Thus, if the recovery read on Fig. VI.11(b) from the lognormal distribution (m, σ_Z^2) is 20% (hatched area) and if the recovery read from the lognormal distribution (m, σ_Y^2) is 25% (dotted area), the correction coefficient is $25/20 = 1 \cdot 25$. The hatched area on Fig. VI.11(a) gives the recovery read from the experimental histogram of Z, say, for example, 18%. The estimator of the recovery of Y for this cut-off grade z_0 is, then,

$$\text{Prob}^* \{ Y \geq z_0 \} = 18 \times \rho = 22 \cdot 5\%.$$

[†] This indirect correction procedure retains all the secondary modes observed on the histogram of Z. As a consequence, and in mining applications, this procedure will be used if these secondary modes are significant and therefore should be reproduced on the histogram of Y. For this, the variance correction $\sigma_Z^2 \rightarrow \sigma_Y^2$ must not be significant; in practice, the relative differences $|\sigma_Z^2 - \sigma_Y^2|/\sigma_Z^2$ must be less than 30%. In fact, if this variance correction represents a significant smoothing $(\sigma_Y^2 \ll \sigma_Z^2)$, this strong smoothing should mask the secondary modes, which should not then appear on the Y-histogram.

This procedure has the advantage of being simple and of reproducing the major characteristics of the experimental histogram of Z on the estimated histogram of Y. However, the histogram of Y estimated in this way will not have the exact required values (m, σ_Y^2) of the mean and variance. In practice, the deviations from the required values (m, σ_Y^2) can usually be corrected by a slight modification of the few extreme values of the histogram of Y. In any case, no permanence-of-law procedure is robust enough to be used for these extreme values (y greater than $m + 3\sigma_Y$ for instance).

4. Affine correction of variance

Instead of the preceding indirect correction, it is also possible to make a direct variance correction $\sigma_Z^2 \to \sigma_Y^2$ by means of the very simple affine relation

$$Y = \frac{\sigma_Y}{\sigma_Z}(Z - m) + m. \tag{VI.13}$$

This simple variance correction on the standard variable $(Z - m)/\sigma_Z$ results in a variable Y with the required mean m and variance σ_Y^2.

Knowing the distribution of Z (or its experimental histogram), it is a simple matter to deduce the distribution of Y:

$$F_Y(y) = \text{Prob}\{Y < y\} = \text{Prob}\{Z < z\} = F_Z(z), \tag{VI.14}$$

where

$$z = \frac{\sigma_Z}{\sigma_Y}(y - m) + m.$$

This affine variance correction results in a distribution for Y, the shape of which is very similar to that of Z, i.e., all secondary modes, significant or not, of the Z-histogram are reproduced on the Y-histogram, cf. section VI.B.4, Fig. VI.16. However, if σ_Y/σ_Z is greater than 1, there is a non-zero probability of obtaining negative values of Y, which can be awkward if Y represents a grade. If they are few, these negative values can be corrected by hand (set equal to zero, for example) on the Y-histogram.

In practice, this affine variance correction produces results very close to those provided by the indirect correction by permanence of a two-parameter distribution. Either one of these two procedures can be used when the variance correction is small, in practice when $|\sigma_Y^2 - \sigma_Z^2|/\sigma_Z^2 < 30\%$, and when a great deal of confidence can be placed in the experimental Z-histogram. Note that the indirect correction, as opposed to the affine correction, guarantees that no value of Y will be negative.

VI.B.2. Generalized permanence

As the simple two-parameter distributions are often insufficient to take the various characteristics of an experimental histogram into account, a more general solution would be to use n-parameter distributions or, more precisely, order n expansions with Hermite polynomials.

Let Z be a stationary variable with any distribution, represented by an experimental histogram with mean m and variance σ_Z^2. We can then define the Gaussian transform function φ_Z which expresses the variable Z in terms of the standard normal variable U, i.e., $Z = \varphi_Z(U)$. This transform function is fitted to the experimental Z-histogram, and the variance σ_Z^2 appears as one of the parameters defining the function φ_Z.

One possible hypothesis of generalized permanence of law consists in considering that the Gaussian transform function φ_Y of the variable Y is identical to φ_Z, with the variance parameter σ_Z^2 replaced by σ_Y^2. The application of generalized permanence then consists in:

(i) determining the transform function φ_Z of a variable Z from a representative experimental histogram;

(ii) determining how this function φ_Z depends on the variance σ_Z^2 so as to make the variance correction $\sigma_Z^2 \to \sigma_Y^2$.

Fitting φ_Z by a Hermite expansion

In general, the distribution of Z is given in the form of an experimental histogram which is often too complex to be correctly represented by a simple two- or three-parameter model. A way around this problem is to fit an expansion up to an order n to this histogram; more precisely, to its Gaussian transform function φ_Z. The order n increases with the complexity of the histogram. It can be shown that the Gaussian transform function φ_Z of a variable Z which follows any distribution with a *finite* variance can be expressed in the form

$$\varphi_Z(u) = \sum_{i=0}^{+\infty} \frac{\psi_i}{i!} H_i(u),$$

where the $H_i(u)$ are the Hermite polynomials. The definition and main properties of these polynomials will be presented in the following section, VI.B.3.

In practice, the Hermite expansion is carried out up to an order n, i.e.,

$$\varphi_Z(u) \simeq \sum_{i=0}^{n} \frac{\psi_i}{i!} H_i(u). \tag{VI.15}$$

The order n is such that the distribution of the variable $\varphi_Z(U)$ correctly fits the experimental histogram of the data z. The method of fitting the $(n+1)$ coefficients ψ_i of the expansion of φ_Z from the experimental Z-histogram is given in the following section, VI.B.3. It can be shown that

$$\psi_0 = m, \qquad \sum_{i=1}^{n} \frac{\psi_i^2}{i!} = \sigma_Z^2, \tag{VI.16}$$

where m and σ_Z^2 are the mean and variance of the experimental Z-histogram.

Applying the hypothesis of permanence to this Hermite expansion consists of considering that the Hermite expansion of the transform function φ_Y of the variable Y can be written as

$$\varphi_Y(u) \simeq \sum_{i=0}^{n} \frac{\psi_i c_i}{i!} H_i(u) \tag{VI.17}$$

where

$$\psi_0 \cdot c_0 = m, \qquad \text{i.e., } c_0 = 1, \qquad \sum_{i=1}^{n} \frac{\psi_i^2 c_i^2}{i!} = \sigma_Y^2.$$

Thus, φ_Z and φ_Y have the same Hermite expansion, except for the variance correction coefficients c_i. The variance correction is written $c_i = a^i$ (a to the power i) which entails

$$\sigma_Y^2 = \sum_{i=1}^{n} \frac{\psi_i a^{2i}}{i!} = f(a). \tag{VI.18}$$

The justification for adopting the form $c_i = a^i$ for the variance correction is given later, in remark 1.

Since the n parameters $\{\psi_i, i = 1 \text{ to } n\}$ are known, the function $f(a)$ can be plotted as on Fig. VI.12. The known value σ_Y^2 then determines the parameter a of the variance correction. Formula (VI.17) gives the expression of the Hermite expansion of φ_Y, i.e.,

$$\varphi_Y(u) \simeq \sum_{i=0}^{n} \frac{\psi_i a^i}{i!} H_i(u). \tag{VI.19}$$

Remark 1   If Z is a normally distributed variable with expectation m and variance σ_Z^2, the Hermite expansion of its Gaussian transform function is reduced to two terms:

$$z = \varphi_Z(u) = \psi_0 + \psi_1 H_1(u) = m + \sigma_Z \cdot u.$$

Knowing that $H_1(u) = -u$, relations (VI.16) can be verified:

$$\psi_0 = m \quad \text{and} \quad \psi_1 = -\sigma_Z, \quad \text{i.e., } \psi_1^2 = \sigma_Z^2.$$

The variance correction factor is given by formula (VI.18):

$$\sigma_Y^2 = \psi_1^2 a^2, \quad \text{i.e., } a = \sigma_Y / \sigma_Z,$$

and formula (VI.7) for the permanence of normality is obtained:

$$y = \varphi_Y(u) = \psi_0 + \psi_1 a \cdot H_1(u) = m + \sigma_Y \cdot u.$$

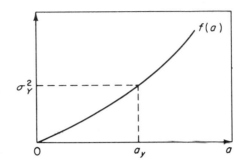

FIG. VI.12. Determination of the variance correction factor a_Y.

If Z is a lognormal variable with expectation m and variance σ_Z^2, its Gaussian transform function can be written, according to (VI.10), as

$$z = \varphi_Z(u) = m \cdot e^{-\sigma_Z'^2/2} \cdot e^{\sigma_Z' \cdot u},$$

where $\sigma_Z'^2$ is the variance of the normal variable log Z. It can be shown that the Hermite expansion of this function φ_Z can be written as

$$\varphi_Z(u) = \sum_{i=0}^{\infty} \frac{\psi_i}{i!} H_i(u) = m \sum_{i=0}^{\infty} \frac{(-\sigma_Z')^i}{i!} H_i(u);$$

thus,

$$\psi_i = m(-\sigma_Z')^i, \quad \text{i.e., } \psi_0 = m = E\{Z\}$$

and

$$\sigma_Z^2 = \text{Var}\{Z\} = \sum_{i=1}^{\infty} \frac{\psi_i^2}{i!} = m^2 \sum_{i=1}^{\infty} \frac{(\sigma_Z'^2)^i}{i!} = m^2(e^{\sigma_Z'^2} - 1).$$

The variance correction factor is given by formula (VI.18):

$$\sigma_Y^2 = \text{Var}\{Y\} = \sum_{i=0}^{\infty} \frac{\psi_i a^{2i}}{i!} = m^2(e^{\sigma_Z'^2 a^2} - 1) = m^2(e^{\sigma_Y'^2} - 1),$$

i.e., $a = \sigma_Y'/\sigma_Z'$, the ratio of the standard deviations of the variables log Y and log Z.

The Gaussian transform function φ_Y of the variable Y is, thus,

$$y = \varphi_Y(u) = \sum_{i=0}^{\infty} \frac{\psi_i a^i}{i!} H_i(u) = m \sum_{i=0}^{\infty} \frac{(-\sigma_Y')^i}{i!} H_i(u)$$

which is the Hermite expansion of the Gaussian transform function φ_Y of a lognormal variable Y with expectation $E\{Y\} = m$ and variance

$$\sigma_Y^2 = m^2(e^{\sigma_Y'^2} - 1).$$

Thus, formula (VI.18) adopted for the variance correction assures strict permanence of normality and, above all, of lognormality, which is a frequently observed experimental fact. By extension, this variance correction formula is used for all variables Z, whatever their distribution.

Remark 2   The variance correction by a Hermite model must always be made in the direction of smoothing, i.e., $\sigma_Z^2 \to \sigma_Y^2$, where $\sigma_Y^2 < \sigma_Z^2$. This decrease of the variance corresponds to a convolution, and, in general, it is much easier to convolute a curve than to deconvolute it. The variance correction procedure, with $c_i = a^i$, is particularly unstable for an increase in the variance ($\sigma_Y^2 > \sigma_Z^2$) corresponding to a correction factor $a > 1$, and this instability increases with the order n of the Hermite expansion. This is illustrated by the example on Fig. VI.20.

The decrease in the variance, $\sigma_Y^2 < \sigma_Z^2$, corresponds to $a < 1$, which entails a smoothing of the various secondary modes observed on the Z-histogram; the Y-histogram tends, at the limit (when $\sigma_Y^2 \ll \sigma_Z^2$), to the normal distribution (m, σ_Y^2).

As a consequence, if the Y-histogram must faithfully reproduce the shape of the Z-histogram with all its secondary modes, an affine variance correction procedure should be used, cf. section VI.B.1, formula (VI.13).

The problem consists of determining whether or not the Y-histogram (with the lesser variance $\sigma_Y^2 < \sigma_Z^2$) should smooth the secondary modes observed on the experimental Z-histogram. In mining applications, Z usually represents the grade of a core sample, for instance, and Y the grade of a panel on a considerably larger support, which would involve a smoothing of the secondary modes, and the Hermite procedure would thus be preferable. However, this is not an absolute rule, cf. the case study in section VI.B.4. The choice between methods will be made on the basis of physical information about the studied phenomenon.

VI.B.3. Fitting Hermite models

(This section can be omitted at first reading. For mathematical theory, see A. Angot (1965, p. 480); for geostatistical applications see G. Matheron (1975c, d) and A. Maréchal (1975a).)

A Hermite model of the Gaussian transform function φ_Z will be fitted to the experimental histogram of the variable Z and, by permanence, this model will be extended to the variable Y. The experimental Z-histogram must thus represent the distribution of Z, i.e., the unbiasedness of this histogram should be verified beforehand, cf. section VI.A.1, the remark on the representativeness of an experimental histogram, and section III.C.1, "Critical review of data".

Graphical transform

The transform function φ_Z can be determined graphically from the two cumulative distribution curves, $G(u) = \text{Prob}\,\{U < u\}$ corresponding to the standard normal variable U, and $F_Z^*(z) = \text{Prob}\,\{Z < z\}$ corresponding to Z, cf. Fig. VI.13. A class (u_k, u_{k+1}) of the theoretical Gaussian curve $G(u)$ corresponds graphically to each class (z_k, z_{k+1}) of the experimental cumulative Z-histogram. An interpolation (linear, for example) can be used within each class interval (z_k, z_{k+1}). This way, a one-to-one correspondence $z \rightleftarrows u$ can be defined graphically, i.e., not only the transform $z = \varphi_Z(u)$, but also its inverse $u = \varphi_Z^{-1}(z)$.

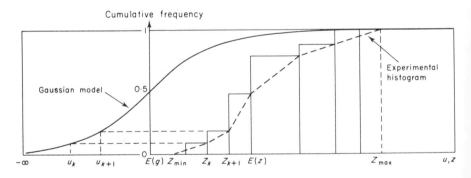

FIG. VI.13. Graphical transform.

Hermite expansion of φ_Z

Let $g(x)$ be the density of the standard normal distribution; the cumulative distribution is then written as

$$G(u) = \text{Prob}\,\{U < u\} = \int_{-\infty}^{u} g(x)\,dx = \frac{1}{\sqrt{2\pi}} \int_{-\infty}^{u} e^{-x^2/2}\,dx. \quad (VI.20)$$

The Hermite polynomials $H_i(x)$ are defined as the derivatives of the Gaussian density:

$$H_i(x) = e^{x^2/2} \frac{d^n}{dx^n} e^{-x^2/2}. \qquad (VI.21)$$

These Hermite polynomials are orthogonal for the Gaussian density $g(x)$, i.e.,

$$\int_{-\infty}^{+\infty} H_i(x) \cdot H_j(x) g(x) \, dx = i! \delta_{ij},$$

where

$$\delta_{ij} = \begin{cases} 1 & \text{if } i = j, \\ 0 & \text{if } i \neq j. \end{cases}$$

An orthogonal basis can then be defined by considering the standardized Hermite polynomials $\eta_i(x) = H_i(x)/\sqrt{i!}$, which are such that

$$\int_{-\infty}^{+\infty} \eta_i(x) \cdot \eta_j(x) g(x) \, dx = \delta_{ij}.$$

The Hermite polynomials can be expressed by the recurrence relation:

$$H_{i+1}(x) = -x H_i(x) - l H_{i-1}(x),$$

which entails

$$H_0(x) = 1; \quad H_1(x) = -x; \quad H_2(x) = x^2 - 1; \quad H_3(x) = -x^3 + 3x; \quad \text{etc.}$$

Consider the condition

$$E\{Z^2\} = E\{\varphi_Z^2(U)\} = \int_{-\infty}^{+\infty} \varphi_Z^2(u) \cdot g(u) \, du < \infty.$$

This condition is satisfied for any variable Z with finite variance, which is always the case in practice since this variance is deduced from the experimental histogram of Z. The following property then results: any function $\varphi(x)$, such that

$$\int_{-\infty}^{+\infty} \varphi^2(x) g(x) \, dx < \infty$$

can be expressed as a linear combination of the vectors $\eta_i(x)$ of the Hermitian basis, i.e.,

$$\varphi(x) = \sum_{i=0}^{\infty} k_i \eta_i(x) = \sum_{i=0}^{\infty} \frac{\psi_i}{i!} H_i(x). \qquad (VI.22)$$

The coefficients ψ_i of this expansion are given by

$$\psi_i = \int_{-\infty}^{+\infty} \varphi(x) H_i(x) g(x) \, dx, \qquad \forall i. \tag{VI.23}$$

From this, it follows that the expansion (VI.22) is the best approximation, up to any order n, of $\varphi(x)$ by a polynomial of degree n, in the weighted (by $g(x)$) least squares sense. Indeed, with

$$I = \int_{-\infty}^{+\infty} \left[\varphi(x) - \sum_{i=0}^{n} \frac{\psi_i}{i!} H_i(x) \right]^2 g(x) \, dx,$$

the equations $\partial I / \partial \psi_i = 0$, $\forall i$, providing the minimum of I, give the relations (VI.23) which define the coefficients ψ_i.

Now, let φ_Z be the Gaussian transform function of Z; the moments of Z are written as

$$E\{Z\} = E\{\varphi_Z(U)\} = \int_{-\infty}^{+\infty} \varphi_Z(u) g(u) \, du = \psi_0,$$

$$E\{Z^2\} = E\{\varphi_Z^2(U)\} = \sum_{i=0}^{\infty} \sum_{j=0}^{\infty} \frac{\psi_i \psi_j}{i! j!} \int_{-\infty}^{+\infty} H_i(u) \cdot H_j(u) \cdot g(u) \, du$$

$$= \sum_{i=0}^{\infty} \frac{\psi_i^2}{i!}.$$

There are, thus, two relations relating the coefficients of the Hermite expansion of φ_Z to the mean and variance of Z:

$$\left. \begin{array}{l} E\{Z\} = m = \psi_0, \\[2mm] \mathrm{Var}\{Z\} = \sigma_Z^2 = \displaystyle\sum_{i=0}^{\infty} \frac{\psi_i^2}{i!} - \psi_0^2 = \sum_{i=1}^{\infty} \frac{\psi_i^2}{i!}. \end{array} \right\} \tag{VI.24}$$

In practice, the Hermite expansion of φ_Z is stopped when $\sum_{i=1}^{n} \psi_i^2 / i!$ is close enough to the required variance σ_Z^2. The contributions $\psi_i^2 / i!$ of the terms of higher degree $(i > n)$ tend rapidly toward zero.

φ_Z being expanded up to the order n, it must be verified graphically that the distribution of the variable

$$Z' = \sum_{i=0}^{n} \frac{\psi_i}{i!} H_i(u)$$

is a good enough approximation of the experimental cumulative histogram of Z, cf. Fig. VI.14.

Practical determination of the coefficient ψ_i

The coefficient ψ_i of the Hermite expansion of φ_Z are given by the integral (VI.23). In practice, function $\varphi_Z(u)$ is determined by graphical transform using linear interpolation, cf. Fig. VI.13. Thus, for an experimental cumulative Z-histogram discretized into K classes, $[z_{min}, z_2], [z_2, z_3], \ldots, [z_k, z_{k+1}], \ldots, [z_K, z_{max}]$, the transform function φ_Z is defined as follows:

$$\text{for } u \in [-\infty, u_2], \qquad z = \varphi_Z^{(1)}(u) = z_{min} = z_1,$$

$$\text{for } u \in [u_k, u_{k+1}], \qquad z = \varphi_Z^{(k)}(u) = a_k + b_k . u,$$

where a_k and b_k are the coefficients of linear interpolation for the kth class $[z_k, z_{k+1}]$ of the cumulative Z-histogram;

$$\text{for } u \in [u_K, +\infty], \qquad z = \psi_Z^{(K)}(u) = z_{max}.$$

z_{max} and z_{min} are the maximum and minimum values of the variable Z. In general, for a grade variable, $z_{min} = 0$ and z_{max} is fixed by the mineralogy, e.g., $z_{max} = 70\%$ Fe for pure hematite iron ore.

Thus, the preceding integral (VI.23) giving the coefficients ψ_i, is approximated by the sum of the K contributions $\{\psi_i^{(k)}, k = 1 \text{ to } K\}$:

$$\psi_i = \sum_{k=1}^{K} \psi_i^{(k)}, \qquad \forall i, \qquad (VI.25)$$

where

$$\psi_i^{(k)} = \int_{u_k}^{u_{k+1}} \varphi_Z^{(k)}(u) . H_i(u) . g(u) \, du$$

and, in particular, for $k = 1$

$$\psi_i^{(1)} = \int_{-\infty}^{u_2} \varphi_Z^{(1)} . H_i(u) . g(u) \, du$$

$$= z_{min} \int_{-\infty}^{u_2} H_i(u) . g(u) \, du,$$

$$\psi_0^{(1)} = z_{min} \int_{-\infty}^{u_2} g(u) \, du = z_{min} . G(u_2)$$

and

$$\psi_i^{(1)} = z_{min} . H_{i-1}(u_2) . g(u_2).$$

For $k = K$,

$$\psi_i^{(K)} = \int_{u_K}^{+\infty} \varphi_Z^{(K)}(u) \cdot H_i(u) \cdot g(u)\,du$$

$$= z_{\max} \int_{u_k}^{+\infty} H_i(u) \cdot g(u)\,du,$$

$$\psi_0^{(K)} = z_{\max} \cdot [1 - G(u_K)]$$

and

$$\psi_i^{(K)} = -z_{\max} \cdot H_{i-1}(u) \cdot g(u_K).$$

For any k, a linear interpolation in the class $[z_k, z_{k+1}]$ gives

$$\psi_i^{(k)} = \int_{u_k}^{u_{k+1}} (a_k + b_k \cdot u) \cdot H_i(u) \cdot g(u)\,du.$$

By putting

$$\Delta_i = H_i(u_{k+1}) \cdot g(u_{k+1}) - H_i(u_k) \cdot g(u_k), \quad \forall i \geq 0.$$

$$\Delta_0 = g(u_{k+1}) - g(u_k),$$

$$\Delta_{-1} = G(u_{k+1}) - G(u_k),$$

$$\Delta_{-2} = 0,$$

the expression used in practice becomes

$$\psi_i^{(k)} = a_k \Delta_{i-1} - b_k [\Delta_i + i\Delta_{i-2}]. \tag{VI.26}$$

Inverse transform function

Once the nth-order Hermite expansion of the transform function $Z = \varphi_Z(U)$ has been determined, it is often useful to determine the inverse transform function $U = \varphi_Z^{-1}(Z)$. For example, this inverse function is required to provide the distribution function $F_Z(z) = \mathrm{Prob}\{Z < z\}$ from that of U, i.e.,

$$F_Z(z) = G(u) \quad \text{with} \quad u = \varphi_Z^{-1}(z).$$

This inverse transform can be achieved iteratively using the standard Newton inversion method, cf. B. Demidovitch and I. Maron (1973, p. 120); the required inverse value $u = \varphi^{-1}(z)$ is the limit of the series (when l tends towards infinity):

$$u_l = u_{l-1} - \frac{[\varphi(u_{l-1}) - z]}{\varphi'(u_{l-1})}. \tag{VI.27}$$

The derivative $\varphi'(u)$ at any point u of the nth-order Hermite expansion of the function $\varphi(u)$ is given by

$$\varphi'(u) = -\sum_{i=1}^{n} \frac{\psi_i}{(i-1)!} H_{i-1}(u). \qquad (VI.28)$$

If the transform function φ is monotonic in the region of z, then the series (VI.27) rapidly converges (for $l < 50$ iterations). This will always be so in practice, except sometimes for extreme values of u in the tails of the histogram. For these tail values, it may happen that the Hermite expansion does not ensure that $\varphi(u)$ is monotonic, then a simple linear interpolation is used instead of the expansion.

VI.B.4. Case study. Selection at Prony

The nickeliferous laterite deposit at Prony in New Caledonia has been described in section V.B.4, Fig. V.23. This deposit was the subject of a thorough geostatistical evaluation and simulation study, cf. A. Journel (1973b). The present case study is limited to the section of the work related to the permanence of distribution by means of which recovery can be forecast.

The variable studied was total mineralized thickness measured along each vertical drill-hole and regionalized in the two-dimensional space, cf. Fig. V.23. Let $z(x)$ be this thickness. At the time of the study, 405 experimental values $z(x_\alpha)$ were available on a regular square 200 m grid. Figure VI.14 shows the histogram of these quasi-point support experimental values (the horizontal section of the drill core is assimilated to a point). This histogram has a mean $m = 12\cdot09$ meters and a variance $D^2(0/G) = 48\cdot98$ m^2. There is a large mode at the origin (zero mineralized thickness) corresponding to rises of the karstic bedrock. There is also a main mode around the 12-m mean and several secondary modes; the modes around 20 m and 26 m correspond to deep mineralized pockets in the karstic bedrock.

Fitting an Hermite expansion

Hermite expansions of order $n = 15$ and $n = 30$ were fitted to the Gaussian transform function φ_Z of the experimental histogram on Fig. VI.14. Note that on that figure, both expansions take the mode at the origin and the main mode (around 12 m) into account, but take no account of the various other secondary modes. For this purpose, it would be necessary to use Hermite expansions up to order $n = 60$ or 100, which would involve prohibitive calculations. The mean $m = 12{\cdot}09$ and variance $D^2(0/G) = 48{\cdot}98$ are, by construction, exactly reproduced.

The fact that the secondary modes are not represented is not especially important, since the interest lies in recoveries defined on mining panels (horizontal support of the order of $100\,\text{m} \times 100\,\text{m}$), and not recoveries defined on drill cores (quasi-point support). It can be reasonably assumed that the histogram of the mean mineralized thicknesses of panels as large as $100\,\text{m} \times 100\,\text{m}$ has no secondary modes. It is physically known that while there are a few panels with a very small average thickness (rise of the bedrock) there are certainly no deep karstic pockets with an horizontal section as large as $100\,\text{m} \times 100\,\text{m}$.

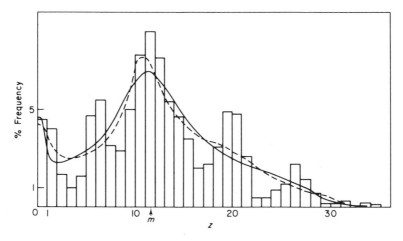

FIG. VI.14. Thickness histogram (core section support). Experimental values for $m = 12{\cdot}09$, $D^2(0/G) = 48{\cdot}98$ are shown by the histogram. ———, Hermite fit, $n = 15$; – – – –, Hermite fit, $n = 30$.

Kriging of mining panels

The experimental data $z(x_\alpha)$ on the 200-m sampling grid were used to estimate (by kriging) the mean mineralized thickness $z_v(x)$ of 2214 panels v with horizontal dimensions $100\,\text{m} \times 100\,\text{m}$. The mean of these 2214 kriged values $z^*_{Kv}(x)$ is $m = 11{\cdot}83\,\text{m}$, the dispersion variance is $D^{2*}_K(v/G) = 24{\cdot}1\,\text{m}^2$, and the histogram of these kriged values is shown on Fig. VI.15.

Note that the secondary modes for the large thicknesses are no longer present, but the secondary mode around $8\,\text{m}$ remains. The following models were fitted to this experimental histogram, cf. Fig. VI.15.

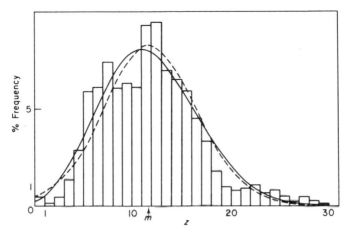

FIG. VI.15. Kriged thicknesses histogram (100 m × 100 m support, 200-m sampling grid). Experimental values for $m = 11\cdot83$ and $D_K^{2*} = 24\cdot1$ are shown by the histogram. ———, Hermitian permanence, $n = 15$; – – – –, Gaussian fit.

(i) A normal distribution with the same mean and variance. There is a slight probability that this normal distribution will result in negative thicknesses.

(ii) An Hermite expansion, deduced by variance correction $D^2(0/G) \rightarrow D_K^{2*}(v/G)$ from the $n = 15$ order Hermite expansion fitted to the histogram of thicknesses measured along the drill cores (cf. Fig. VI.14). The correction factor a used in this variance correction is given by formula (VI.18) and $a = 0\cdot7064$. Note that this histogram, deduced by Hermitian permanence, tends at the limit towards the normal distribution with the same mean $11\cdot83$ and variance $24\cdot1$. Indeed, the decrease in the variance is particularly significant here, the difference between $D^2(0/G) = 48\cdot98$ and $D_K^{2*}(v/G) = 24\cdot1$ being more than 50%. Note that if an Hermite model were fitted directly to the experimental histogram on Fig. VI.15, the secondary mode around 8 m would show up more clearly, cf. Fig. VI.20.

Selection study

The 100 m × 100 m panel v is considered as the selection unit on which future production will be carried out. The selection variable is the mean mineralized thickness.

If the only information available at the time of study is the data from the 200-m sampling grid, the cut-off will be made on the preceding kriged values $z^*_{Kv}(x)$, i.e., on the experimental histogram on Fig. VI.15, the results of which are shown in column 1 of Table VI.3. From this table, for example, a cut-off thickness $z_0 = 5$ m provides a recovery of 94·5% of the panels, i.e., 94·5% of the kriged values $z^*_{Kv}(x)$ are greater than the cut-off $z_0 = 5$ m.

In fact, much more data are available at the time actual selection is made; this future information is assimilated to a regular, square 25-m grid, i.e., 16 samples per panel. The actual future selection is thus assumed to be made on the ultimate estimators $z^{**}_{Kv}(x)$ deduced by kriging from the data on the 25-m grid. Therefore, the problem is to deduce an estimator of the histogram of these 2214 ultimate selection values z^{**}_{Kv} from the histogram available at the time of study, i.e., the histogram of the kriged values z^*_{Kv}. To do this, the procedure outlined in section VI.A.2 is followed.

(i) and (ii). The kriging variance σ^{2**}_{Kv} corresponding to the ultimate 25-m grid is calculated, and then the dispersion variance of the ultimate kriged values z^{**}_{Kv} is calculated by means of the smoothing relation (VI.1). This results in a value $D^{2**}_K(v/G) = 32·8$ m^2.

(iii) Using an hypothesis of permanence of law, an estimator of the future histogram of the z^{**}_{Kv} is deduced, either from the sample thicknessses histogram by decreasing the variance $D^2(0/G) \to D^{2**}_K(v/G)$, or from the histogram of the now available kriged values z^*_{Kv} by increasing the variance $D^{2*}_K(v/G) \to D^{2**}_K(v/G)$. Applying the cut-off thickness z_0 to this estimated histogram of the z^{**}_{Kv} results in the required prediction of actually recoverable reserves.

Two procedures based on two different permanence-of-law hypotheses were considered, namely:

   (i) the affine variance correction, cf. section VI.B.1, point no. 4, the correction by increasing the variance $D^{2*}_K \to D^{2**}_K$ is made directly on the z^*_{Kv}-histogram, according to formula (VI.14);

(ii) the correction by decreasing the variance $D^2(0/G) \to D^{2**}_K$ is made on the Hermite expansion fitted to the sample thicknesses histogram.

1. *Affine variance correction*   Let Z^*_{Kv} be the variable kriged from the 200-m grid, with mean $m = 11·83$ and variance $D^{2*}_K = 24·1$. The future kriged variable Z^{**}_{Kv} with the same mean and a variance $D^{2**}_K = 32·8$ is then considered as having a distribution identical to that of the variable

$$Y = \frac{D^{**}_K}{D^*_K}(Z^*_{Kv} - m) + m,$$

cf. relation (VI.13), i.e.,

$$Y = 1·17(Z^*_{Kv} - 11·83) + 11·83.$$

The distribution function (cumulative histogram) of Z_{Kv}^{**} is then given by the relation (VI.14):

$$\text{Prob}\{Z_{Kv}^{**} < y\} = \text{Prob}\{Z_{Kv}^{*} < z\}$$

with

$$z = 0.86(y - 11.83) + 11.83.$$

The histogram of the future values z_{Kv}^{**} thus estimated is shown on Fig. VI.16. The corresponding recoveries for different cut-off thicknesses z_0 are given in columns 4, 5 and 6 of Table VI.3.

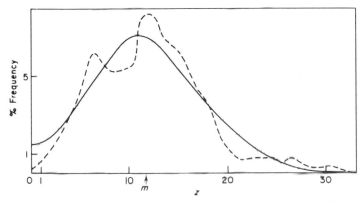

FIG, VI.16. Estimated histograms of the future kriged thicknesses (100 m × 100 m support, 25-m sampling grid). ———, Hermitian permanence; – – – –, affine correction. $m = 11.83$, $D_K^{2**} = 32.8$.

TABLE VI.3. *Estimations of recoveries*

	Z_{Kv}^{*}			Z_{Kv}^{**}				
z_0 (1)	(2)	(3)	(4)	(5)	(6)	(7)	(8)	(9)
0 100	100	11·83	99·3	99·97	11·91	100	100	11·83
5 94·5	98·7	12·4	88·6	96·7	12·9	88·3	97·1	13·0
7 82·5	92·3	13·2	77·3	91·0	13·9	79·2	92·5	13·8
11 56·6	72·7	15·2	55·6	74·3	15·8	53·4	72·6	16·1
15 22·5	35·4	18·6	25·9	42·0	19·2	27·8	44·7	19·0
20 6·0	12·1	23·8	7·3	15·4	24·9	8·9	17·2	22·9

Z_{Kv}^{*}, Selection made using the data on the 200-m grid.
Z_{Kv}^{**}, Future selection to be made using the data on the 25-m grid.

For $z_0 = 5$ m, the estimators are:

> 88·6%   for percentage of recovered panels;
> 96·7%   for percentage of recovered volume;
> 12·9 m   for mean thickness of the recovered panels.

2. *Variance correction on the Hermite models* After fitting an $n = 15$ order Hermite expansion to the experimental histogram of the sample thicknesses (cf. Fig. VI.14), variance correction was used to deduce the 15th-order Hermite expansion of the variable Z_{Kv}^{**}. In this case, the variance is decreased: $D^2(0/G) = 48·98 \rightarrow D_K^{2**} = 32·8$. The correction factor a is found from formula (VI.18) to be $a = 0·823$. The resulting estimated histogram of the Z_{Kv}^{**} is shown on Fig. VI.16. The corresponding recoveries are given in columns 7, 8 and 9 of Table VI.3.

For a cut-off thickness $z_0 = 5$ m the estimator of the percentage of recovered panels is 88·3%, which is very close to the value 88·6% resulting from the affine variance correction.

If $f(z)\,dz = \text{Prob}\{z - dz < Z_{Kv}^{**} < z + dz\}$ is the frequency density of the future values Z_{Kv}^{**}, and if z is the mean value of Z_{Kv}^{**} in the corresponding class $[z \pm dz]$, then

(i) the percentage of recovered panels, or recovered surface, cf. columns 1, 4 and 7, is given by

$$\% \ S(z_0) = \text{Prob}\{Z_{Kv}^{**} > z_0\} = \int_{-\infty}^{z_0} f(z)\,dz,$$

expressed in per cent;

(ii) the percentage of the recovered volume, cf. columns 2, 5 and 8, is given by

$$\% \ V(z_0) = \left[\int_{-\infty}^{z_0} zf(z)\,dz \right] \Big/ \int_{-\infty}^{+\infty} zf(z)\,dz,$$

expressed in per cent, and the total *in situ* volume (for $z_0 = 0$) is equal to

$$2214 \times 100 \times 100 \times 11·83 = 261·9 \times 10^6 \ \text{m}^3;$$

(iii) the mean thickness of the recovered panels, cf. columns 3, 6 and 9, is given by

$$p(z_0) = \left[\int_{-\infty}^{z_0} zf(z)\,dz \right] \Big/ \int_{-\infty}^{z_0} zf(z)\,dz,$$

expressed in metres.

Remark 1 The practical values of the cut-off thickness are between $z_0 = 3$ m and $z_0 = 7$ m; thus, the results of Table VI.3 appear to be robust with respect to the adopted permanence hypothesis. In the absence of any test data which could be used to choose between the affine variance correction and the correction on the Hermite model, the more pessimistic results should be retained; thus, for $z_0 = 5$ m, 96·7% of recovered volume with a mean thickness of 12·9 m.

This robustness should not be observed on the histograms (relative frequency curve) on Fig. VI.16, but on the corresponding cumulative histograms on Fig. VI.17; this robustness is even more evident on the curves on Fig. VI.18 and VI.19, which give the cumulative percentage of volume recovery and the mean thickness of the recovered volume, respectively. By eliminating the cut-off thickness z_0 from Figs VI.18 and VI.19, a curve giving the recovered volume as a function of the corresponding mean thickness can be obtained.

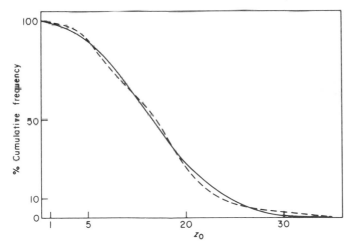

FIG. VI.17. Cumulative percentage of panel recovery. ———, Hermitian permanence; – – – –, affine correction.

The effect of the adopted hypothesis of permanence on the two histograms on Fig. VI.16 is twofold:

(i) the affine correction reproduces the secondary modes of the experimental histogram of the kriged values $z_{K_v}^*$ (compare Fig. VI.15 and VI.16);

(ii) the secondary modes of the experimental histogram of the sample values (cf. Fig. VI.14) are smoothed by decreasing the variance of the Hermite model.

However, this smoothing $D^2(0/G) = 48 \cdot 98 \to D_K^{2**} = 32 \cdot 8$ is not enough to eliminate the mode at the origin (zero thickness corresponding to rises of the bedrock). For the greater decrease of the variance, $D^2(0/G) \to D_K^{2*} = 24 \cdot 1$, the smoothing is sufficient to almost entirely mask this mode at the origin, cf. Fig. VI.15.

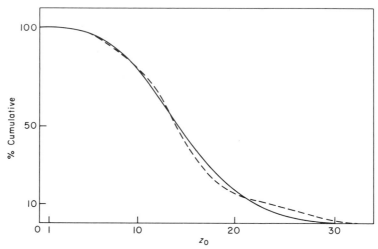

FIG. VI.18. Cumulative percentage of volume recovery. ———, Hermitian permanence; – – – –, affine correction.

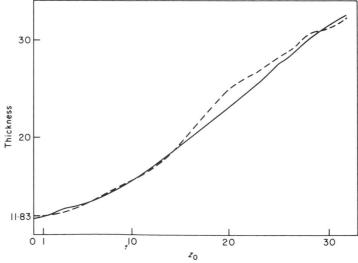

FIG. VI.19. Mean thickness of the recovered volume. ———, Hermitian permanence; – – – –, affine correction.

The affine correction does not guarantee a zero probability of obtaining negative thicknesses Z_{Kv}^{**}; thus, from Table VI.3, 0·7% of the Z_{Kv}^{**} values are estimated to be negative. However, in practice, these 0·7% aberrant values do not affect the estimation of the recoveries in the practical range $z_0 \in [3 \text{ to } 7]$ m.

Fitting an Hermite model directly to the histogram of the Z_{Kv}^{*}   It might be thought possible to fit an Hermite model directly to the available histogram of the values Z_{Kv}^{*} kriged from the data on the 200-m sampling grid and then, by increasing the variance $D_K^{2*} \to D_K^{2**}$, to deduce therefrom the Hermite model corresponding to the histogram of the Z_{Kv}^{**}, kriged from the data on the 25-m grid. It has already been stated that formula (VI.18) for the variance correction of an Hermite model is particularly unstable for an

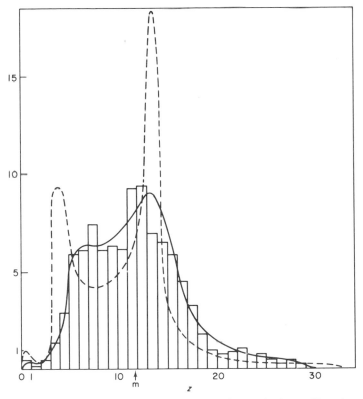

FIG. VI.20. Unstable variance correction (by increasing) on Hermite model. Experimental values for $m = 11·83$ and $D_K^{2*} = 24·1$ are shown by the histogram. ———, Hermitian fit, $n = 15$, $D_K^{2*} = 24·1$; ————, Hermitian permanence, $D_K^{2**} = 32·8$.

increase in the variance. In fact, if an extremely good direct fit up to order $n = 15$ of the Hermite model to the experimental histogram of the Z^*_{Kv} is obtained, cf. Fig. VI.20, the model obtained by increasing the variance will considerably exaggerate the two modes, and cannot be used as an estimator of the histogram of the future estimated thicknesses Z^{**}_{Kv}, cf. Fig. VI.20. The correction factor a corresponding to this variance increase is greater than 1 in this case: $a = 1 \cdot 13$. Thus, when using Hermite models, the permanence-of-law hypothesis must always be applied in the sense of decreasing the variance, which in the present case means to decrease from $D^2(0/G) = 48 \cdot 98$ to $D^{2**}_K = 32 \cdot 8$ or to $D^{2*}_K = 24 \cdot 1$.

VII

Simulation of Deposits

SUMMARY

A regionalized variable $z(x)$ is interpreted as one realization of a certain random function $Z(x)$. This RF – more precisely this class of RF's – is characterized by a distribution function and a covariance or variogram model. The idea of simulations consists in drawing other realizations $z_s(x)$ from this class of RF's, and to retain only those realizations $z_{sc}(x)$ which meet the experimental values at the data locations x_α, i.e., $z_{sc}(x_\alpha) = z(x_\alpha)$, $\forall x_\alpha \in$ data set. The realizations $z_{sc}(x)$ are called "conditional simulations" of the regionalized phenomenon $z(x)$.

The theory of conditional simulations is outlined in section VII.A together with the original "turning bands" method, which provides three-dimensional simulated realizations from one-dimensional realizations simulated on lines rotating in space.

The practical application of conditional simulations is given in section VII.B, through two-case studies. A basic FORTRAN program is given, which should allow each user to write his own simulation package. Then several typical uses of simulated deposits in the field of mine project and planning are presented.

THE OBJECTIVES OF SIMULATION

Local and global estimations of recoverable reserves are often insufficient at the planning stage of a new mine or a new section of an operating mine. For the mining engineer, as well as the metallurgist and chemist, it is often essential to be able to predict the variations of the characteristics of the recoverable reserves at various stages in the operation, e.g., after mining, hauling, stockpiling, etc.

For instance, the choice of mining and haulage methods depends in part on the spatial dispersions of the various ore characteristics. Conversely, the potential recovery rate of the *in situ* resources will depend, in part, on the mining technology. The choice of mining equipment or of a method of removal of the mined products in a subhorizontal sedimentary deposit may

491

depend on the daily variations in overburden and mineralized thickness. The blending process and the flexibility of the plant will depend on the dispersion of the grades received at all scales (daily, monthly, yearly).

If the *in situ* reality were perfectly known, the required dispersions, and, thus, the most suitable working methods, could be determined by applying various simulated processes to this reality. Unfortunately, the perfect knowledge of this *in situ* reality is not available at the planning stages of an operation. The information available at this stage is usually very fragmentary, and limited to the grades of a few samples. The estimations deduced from this information – even through kriging – are far too imprecise for the exact calculations of dispersions that are required.

As it is impossible to estimate the *in situ* reality correctly in sufficient detail, one simple idea is to simulate it on the basis of a model. In a way, a real deposit and simulations of this deposit are different variants of the same mineralized phenomenon, as characterized by a given model.

Consider, for example, the true grade variable $z_0(x)$ at each point x of a deposit. The geostatistical approach consists in interpreting the spatial distribution of the grade $z_0(x)$ as a particular realization of a random function $Z(x)$. This RF is characterized by its first two moments and its distribution function, which are estimated from the experimental data. This model is, thus, suitable for the practical problem of determining various dispersions of the grades in the deposit, since the dispersion variances of $Z(x)$ can be expressed as a function of the second-order moment only – covariance or variogram – cf. formula (II.36). A simulation thus consists of drawing another realization $z_s(x)$ of this RF $Z(x)$. The two realizations, real and simulated, differ from each other at given locations but come from the same RF $Z(x)$, the first two moments and the univariate distribution function of which are fixed.

As far as the dispersion of the simulated variable is concerned, there is no difference between the simulated deposit $\{z_s(x)\}$ and the real deposit $\{z_0(x)\}$. The simulated deposit has the advantage of being known at all points x and not only at the experimental data points x_α. This simulated deposit is also called a "numerical model" of the real deposit.

Conditioning

There are an infinite number of possible realizations $\{z_s(x), s = 1 \text{ to } \infty\}$ of a RF $Z(x)$. From among this infinity of realizations, the simulations $\{z_{sc}(x)\}$ that are chosen are those that meet the experimental data values at the actual data locations x_α, i.e., those simulations for which

$$z_{sc}(x_\alpha) = z_0(x_\alpha), \qquad \forall x_\alpha.$$

This is known as conditioning the simulation to the experimental data: the simulated deposits and the real one have the same clusters of rich and poor data at the same locations. This conditioning confers a certain robustness to the simulation $\{z_{sc}(x)\}$ with respect to the characteristics of the real data $\{z_0(x)\}$ which are not explicitly modelled by the RF $Z(x)$, cf. A. Journel (1975d). If, for example, a sufficient number of data show a local drift, then the conditional simulations, even though based on a stationary model, will reflect the local drift in the same zone, cf. the left-hand section of Fig. VII.1. These conditional simulations can be further improved by adding all sorts of qualitative information available from the real deposit, such as geometry of the main faults, intercalated waste seams, etc.

Simulation or estimation?

Simulated deposits have the same values at the experimental data locations, and have the same dispersion characteristics (at least up to order 2) as the real deposit. In what way then do such conditional simulations differ from an estimation? Their difference lies in their objectives.

(i) The object of an estimation is to provide, at each point x, an estimator $z^*(x)$ which is as close as possible to the true unknown grade $z_0(x)$. The criteria for measuring the quality of an estimation are unbiasedness and minimal mean quadratic error, or estimation variance $E\{[Z(x)-Z^*(x)]^2\}$. There is no reason, however, for these estimators to reproduce the spatial variability of the true grades $\{z_0(x)\}$. In the case of kriging, for instance, the minimization of the estimation variance involves a smoothing of the true dispersions, cf. the smoothing relations (VI.1) and (VI.2). Similarly, the polygonal influence method of estimation would consider the grade as constant all over the polygon of influence of a sample, and, thus, underestimate the local variability of true grades. The estimated deposit $\{z^*(x)\}$, is, thus, a biased base on which to study the dispersions of the true grades $\{z_0(x)\}$.

(ii) On the other hand, the simulation $(z_s(x)\}$, or better the conditional simulation $\{z_{sc}(x)\}$, has the same first two experimentally found moments (mean and covariance or variogram, as well as the histogram) as the real grades $\{z_0(x)\}$, i.e., it identifies the main dispersion characteristics of these true grades. On the contrary, at each point x, the simulated value $z_s(x)$ or $z_{sc}(x)$ is not the best possible estimator of $z_0(x)$. In particular, it will be shown that the variance of estimation of $z_0(x)$ by the conditionally simulated value $z_{sc}(x)$ is exactly twice the kriging variance, cf. relation (VII.6).

In general, the objectives of simulation and estimation are not compatible. It can be seen on Fig. VII.1 that, even though the estimation curve

$z^*(x)$ is, on average, closer to the real curve $z_0(x)$, the simulation curve $z^*_{sc}(x)$ is a better reproduction of the fluctuations of the real curve. The estimation curve is preferable to locate and estimate reserves, while the simulation curve is preferred for studying the dispersion of the characteristics of these reserves, remembering that in practice the real curve is known only at the experimental data points x_α.

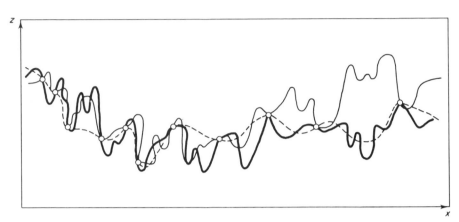

FIG. VII.1. Real, simulated and estimated profiles. ———, Reality; ———, conditional simulation; – – – –, Kriging; $\bigcirc$, conditioning data.

The right-hand side of the profiles on Fig. VII.1 illustrates the fact that a simulation cannot be used to replace sampling, which is always necessary for a good local estimation of the deposit. In any case, the better the deposit is known, the better the structure of variability can be modelled and the denser the grid of conditional data will be, all of which will result in a simulation closer to, and more representative of, reality.

VII.A. THE THEORY OF CONDITIONAL SIMULATIONS

VII.A.1. The principle of conditioning

Consider the regionalization of a variable $z_0(x)$, which may be, for example, the grade at point x. This regionalized variable is interpreted as a realization of a *stationary* RF $Z_0(x)$, with expectation m and centred covariance $C(h)$ or variogram $2\gamma(h)$. The problem is to build a realization of a RF $Z_{sc}(x)$ isomorphic to $Z_0(x)$, i.e., a RF which has the same expectation and the same second-order moment, $C(h)$ or $\gamma(h)$. Moreover, the realization $z_{sc}(x)$ must be conditional to the experimental data, i.e., at the experimental data points the simulated values and the experimental data values must be the same:

$$z_{sc}(x_\alpha) = z_0(x_\alpha), \qquad \forall x_\alpha \in \text{data set } I.$$

Consider the true value $z_0(x)$ and its kriged value $z_{OK}^*(x)$ deduced from the available data $\{z_0(x_\alpha), x_\alpha \in I\}$. These two values differ by an unknown error:

$$z_0(x) = z_{OK}^*(x) + [z_0(x) - z_{OK}^*(x)] \qquad \text{(VII.1)}$$

or, in terms of random functions,

$$Z(x) = Z_{OK}^*(x) + [Z_0(x) - Z_{OK}^*(x)]. \qquad \text{(VII.2)}$$

Now, a characteristic property of kriging is that the kriging error $[Z_0(x) - Z_{OK}^*(x)]$ is orthogonal to the kriged values, i.e.,

$$E\{Z_{OK}^*(y) \cdot [Z_0(x) - Z_{OK}^*(x)]\} = 0, \qquad \forall x, y.$$

(The proof of this result is given in the addendum at the end of this section).

To obtain the desired conditional simulation, it is thus enough to replace the unknown kriging error $[Z_0(x) - Z_{OK}^*(x)]$ in formula (VII.2) by an isomorphic and independent kriging error $[Z_s(x) - Z_{sK}^*(x)]$. More precisely, we consider a RF $Z_s(x)$ which is *isomorphic* to, but independent of $Z_0(x)$. Given a realization $z_s(x)$ of this RF $Z_s(x)$, the kriging procedure, when applied to the same data configuration $\{z_s(x_\alpha), x_\alpha \in I\}$ will result in a kriging error $[z_s(x) - z_{sK}^*(x)]$ isomorphic to the true error $[z_0(x) - z_{OK}^*(x)]$ and independent of $z_{OK}^*(x)$. The required conditional simulation is then written as

$$\boxed{z_{sc}^*(x) = z_{OK}^*(x) + [z_s(x) - z_{sK}^*(x)]} \qquad \text{(VII.3)}$$

or, in terms of random functions,

$$Z_{sc}(x) = Z_{OK}^*(x) + [Z_s(x) - Z_{sK}^*(x)].$$

The requirements of a conditional simulation are satisfied since we have the following.

(i) The RF $z_{sc}(x)$ has the same expectation as $Z_0(x)$. This follows from the unbiasedness of the kriging estimator:

$$E\{Z_{OK}^*(x)\} = E\{Z_0(x)\} \quad \text{and} \quad E\{Z_{sK}^*(x)\} = E\{Z_s(x)\}$$

which entails

$$E\{Z_{sc}(x)\} = E\{Z_0(x)\} = m, \qquad \forall x.$$

(ii) The RF $Z_{sc}(x)$ is isomorphic to $Z_0(x)$. This follows from the fact that the simulated kriging error $[Z_s(x) - Z_{sK}^*(x)]$ is isomorphic to the true error $[Z_0(x) - Z_{0K}^*(x)]$ and independent of $Z_{0K}^*(x)$. In the addendum, it will be shown that this isomorphism applies rigorously to the increments of the RF's $Z_{sc}(x)$ and $Z_0(x)$ only. In other words, the variograms of these two RF's are identical, but not necessarily their covariances. This does not pose any problem since, in geo-statistics, the variogram structural function is always preferred to the covariance.

(iii) The simulated realization $z_{sc}(x)$ is conditional to the experimental data. This follows from the fact that, at the experimental data points, the kriged values are equal to the real values, i.e.,

$$z_{0K}^*(x_\alpha) = z_0(x_\alpha) \quad \text{and} \quad [z_s(x_\alpha) - z_{sK}^*(x_\alpha)] = 0, \qquad \forall x_\alpha \in I,$$

which entails $z_{sc}(x_\alpha) = z_0(x_\alpha)$.

Remark 1   The kriging weights $\{\lambda_\alpha, \forall \alpha \in I\}$ are the same for $Z_{0K}^*(x)$ and $Z_{sK}^*(x)$, since the two RF's $Z_0(x)$ and $Z_s(x)$ are isomorphic and the two kriging configurations are identical.

Remark 2   The conditioning principle provides both values, kriged $z_{0K}^*(x)$ and conditionally simulated $z_{sc}^*(x)$, at each point x, cf. formula (VII.3). By taking into account the independence of the two kriging errors $[Z_0(x) - Z_{0K}^*(x)]$ and $[Z_s(x) - Z_{sK}^*(x)]$, the variance of estimation of the real value $z_0(x)$ by the conditional simulation $z_{sc}(x)$ can be written as

$$E\{[Z_0(x) - Z_{sc}(x)]^2\} = E\{[Z_0(x) - Z_{0K}^*(x)]^2\} + E\{[Z_s(x) - Z_{sK}^*(x)]^2\}$$
$$= 2E\{[Z_0(x) - Z_{0K}^*(x)]^2\} = 2\sigma_K^2.$$

In terms of estimation variance, kriging is twice as good an estimator as conditional simulation: estimation is not the object of a simulation.

Addendum. Proof to the property of orthogonality of kriging errors to kriged values

(This addendum can be omitted at a first reading.)

Consider the expansions of the type (VII.2) of the RF on any two points x and y:

$$Z_0(x) = Z_{0K}^*(x) + [Z_0(x) - Z_{0K}^*(x)] = Z_{0K}^*(x) + R(x),$$
$$Z_0(y) = Z_{0K}^*(y) + [Z_0(y) - Z_{0K}^*(y)] = Z_{0K}^*(y) + R(y),$$

with, for the kriged value,

$$Z_{0K}^*(y) = \sum_\alpha \lambda_\alpha(y) \cdot Z_0(x_\alpha).$$

In the case of a kriging system without the unbiasedness conditions – the stationary expectation of the RF $Z_0(x)$ being known (for example, zero) – the kriging system (VI.1) reduces to, cf. section VIII.A.1, system (VIII.3),

$$\sum_\beta \lambda_\beta(x)\sigma_{\alpha\beta} = \sigma_{\alpha x}, \qquad \forall x_\alpha \in I,$$

$\forall$ the location x of the value $Z_0(x)$ to be estimated, with

$$\sigma_{\alpha\beta} = E\{Z_0(x_\alpha)Z_0(x_\beta)\}.$$

The covariance between the kriging error and the kriged values then becomes

$$E\{Z^*_{0K}(y)R(x)\} = \sum_\alpha \lambda_\alpha(y)\left[\sigma_{\alpha x} - \sum_\beta \lambda_\beta(x)\sigma_{\alpha\beta}\right] = 0, \qquad \forall x, y,$$

and the kriging error $R(x)$ is thus orthogonal to the kriged values $Z^*_{0K}(y)$.

In the usual case of a kriging system with a non-bias condition, the kriging system (VI.1) is written as

$$\sum_\beta \lambda_\beta(x)\sigma_{\alpha\beta} - \mu = \sigma_{\alpha x}, \qquad \forall x_\alpha \in i, \forall x,$$

$$\sum_\beta \lambda_\beta(x) = 1.$$

Consider, then, the following covariance:

$$E\left\{\sum_\alpha \nu_\alpha Z_0(x_\alpha) \cdot R(x)\right\} = \sum_\alpha \nu_\alpha \sigma_{\alpha x} - \sum_\alpha \sum_\beta \nu_\alpha \lambda_\beta(x)\sigma_{\alpha\beta} = \mu \sum_\alpha \nu_\alpha, \qquad \forall x.$$

This covariance becomes zero when $\sum_\alpha \nu_\alpha = 0$, i.e., for all linear combinations $\sum_\alpha \nu_\alpha Z_0(x_\alpha)$ known as "authorized" linear combinations, cf. section II.A.3. In this case, the kriging error $R(x)$ is orthogonal only to authorized linear combinations of the data $Z_0(x_\alpha)$.

Now, the difference between any two kriged values is such an authorized linear combination; indeed,

$$Z^*_{0K}(y) - Z^*_{0K}(y') = \sum_\alpha [\lambda_\alpha(y) - \lambda_\alpha(y')]Z_0(x_\alpha)$$

with

$$\sum_\alpha [\lambda_\alpha(y) - \lambda_\alpha(y')] = 0$$

and, thus,

$$E\{R(x) \cdot [Z^*_{0K}(y) - Z^*_{0K}(y')]\} = 0, \forall x, y, y'.$$

The variogram of the RF $Z_0(x)$ can then be written as

$$2\gamma(x - y) = E\{[Z_0(x) - Z_0(y)]^2\} = E\{[Z^*_{0K}(x) - Z^*_{0K}(y)]^2\} + E\{[R(x) - R(y)]^2\}.$$

From expression (VII.3) for conditional simulation, this variogram is exactly that of the RF $Z_{sc}(x)$, since the error $[Z_s(x) - Z^*_{sK}(x)]$ is isomorphic to $R(x)$ and independent of $Z^*_{0K}(x)$. Thus, the conditional simulation $Z_{sc}(x)$ ensures the reproduction of the variogram of the RF $Z_0(x)$, but not necessarily its covariance.

VII.A.2. Non-conditional simulation in three dimensions

The first step in carrying out the conditional simulation $z_{sc}(x)$ given in formula (VII.3) is to simulate a realization $z_s(x)$ of a RF $Z_s(x)$ isomorphic to $Z_0(x)$, i.e., with the same mean and covariance $C(h)$. The simulation $z_s(x)$ is non-conditional, i.e., it does not necessarily take on the experimental values $z_0(x_\alpha)$ at the data location x_α.

It will be recalled that the notations $z_s(x)$ and $z_0(x)$ represent regionalized variables spread in the three-dimensional space, $x \in R^3$.

Many methods are available for simulating one-dimensional realizations of a stationary stochastic process with a given covariance. However, when these procedures are extended to the three-dimensional space, they are often inextricable or prohibitive in terms of computer time. The originality of the method known as the "turning bands" method, originated by G. Matheron, derives from reducing all three-dimensional (or more generally, n-dimensional) simulations to several independent one-dimensional simulations along lines which are then rotated in the three-dimensional space R^3 (or R^n). This turning band method provides multidimensional simulations for reasonable computer costs, equivalent, in fact, to the cost of one-dimensional simulations.

The turning bands method

Consider the line D_1 in the three-dimensional space R^3, as shown on Fig. VII.2. Consider the one-dimensional RF $Y(x_{D_1})$ defined on line D_1; this RF is second-order stationary, with a zero expectation, $E\{Y(x_{D_1})\} = 0$, and a one-dimensional covariance $C^{(1)}(h_{D_1})$.

Let x_{D_1} be the projection of any point x onto the line D_1, and consider the three-dimensional RF defined by $Z_1(x) = Y(x_{D_1})$, $\forall x \in R^3$. This RF $Z_1(x)$ is second-order stationary, with a zero expectation and a three-dimensional covariance equal to

$$E\{Z_1(x)Z_1(x+h)\} = E\{Y(x_{D_1}) . Y(x_{D_1}+h_{D_1})\} = C^{(1)}(h_{D_1}),$$

where h_{D_1} is the projection of the vector h onto the line D_1.

In practice, to produce a realization $z_1(x)$, the value $z(x_{D_1})$, simulated at the point x_{D_1} on the line D_1, is given to all points inside the slice (or the band) centred on the plane $\{x_{D_1} = \text{constant}\}$ perpendicular at x_{D_1} to D_1, as illustrated on Fig. VII.2. The thickness of this slice is the equidistance between the simulated values on the line D_1.

We then consider N lines $D_1, D_2, \ldots, D_N$ corresponding to the directions of the unit vectors $k_1, k_2, \ldots, k_N$ uniformly distributed over the unit sphere. On each line D_i a realization $y(x_{D_i})$ of a RF $Y(x_{D_i})$ is generated isomorphic to $Y(x_{D_1})$, the N RF's $\{Y(x_{D_i}), i = 1 \text{ to } N\}$ being independent. Using the method outlined above, a three-dimensional realization $z_i(x) = y(x_{D_i})$, $\forall x \in R^3$, is made to correspond to each one-dimensional realization $y(x_{D_i})$.

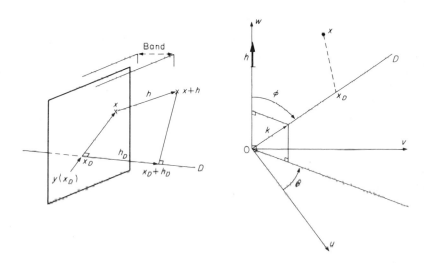

FIG VII.2. The turning bands.

A final value is then assigned to each point x by taking the sum of the N contributions from the N lines:

$$z_s(x) = \frac{1}{\sqrt{N}} \sum_{i=1}^{N} z_i(x).$$

(As the N lines can be deduced from each other by rotation, so can the slices (or bands) which they define in the three-dimensional space, whence the name "turning bands" method.)

The resulting realization $z_s(x)$ is a realization of a three-dimensional RF $Z_s(x) = Z_s(u, v, w)$ which is second-order stationary, has a zero expectation and a covariance

$$E\{Z_s(x)Z_s(x+h)\} = C(h)$$

which, when $N \to \infty$, tends toward the following isotropic covariance:

$$C(r) = \frac{1}{2\pi} \int_{1/2 \text{ unit sphere}} C^{(1)}(\langle h, k \rangle) \, dk, \qquad \text{(VII.5)}$$

where $\langle h, k \rangle$ is the projection of the vector h onto the axis k. $r = |h| = \sqrt{(h_u^2 + h_v^2 + h_w^2)}$ is the modulus of the vector h.

The preceding integral can be written in terms of spherical coordinates, cf. Fig. VII.2:

$$C(r) = \frac{1}{\pi} \int_0^{2\pi} d\theta \int_0^{\pi/2} C^{(1)}(|r \cos \varphi|) \sin \varphi \, d\varphi = \frac{1}{r} \int_0^r C^{(1)}(s) \, ds.$$

In practice, the three-dimensional covariance $C(r)$ is fixed and the one-dimensional covariance $C^{(1)}(s)$ to be simulated on each of the N lines is thus given by the derivative

$$C^{(1)}(s) = \frac{\partial}{\partial s} s \, C(s). \qquad \text{(VII.6)}$$

From Bochner's theorem, it can be shown that the function $C^{(1)}(s)$ given by formula (VII.6), is a positive definite function in one dimension and can, thus, be used as a covariance function.

It is thus always possible to determine the one-dimensional covariance $C^{(1)}(s)$ to be imposed on the RF $Y(u_i)$ simulated on the lines D_i. By turning these lines and their corresponding bands in R^3, the required three-dimensional simulation $z_s(u, v, w)$, with the specified isotropic covariance $C(r)$ is obtained.

Remark 1 Specified expectation. In general, the expectation of the three-dimensional RF $Z(x)$, a realization of which is to be simulated, is not zero: $E\{Z(x)\} = m \neq 0$. To obtain the desired expectation, it is enough to add a constant, equal to the specified mean m, to each of the previously simulated values $z_s(x)$, which constitutes a realization of the zero-expectation RF $Z_s(x)$. This additive constant does not alter the second-order moments (covariance or variogram) of the simulated RF $Z_s(x)$.

Note that this additive procedure can also be used for the simulation of a non-stationary RF interpreted as the sum of a drift and a stationary residual, $Z(x) = m(x) + R(x)$, cf. relation (V.8) in section V.A.2. The stationary residual $R(x)$, with zero expectation and known covariance, is simulated by means of the turning bands method. At each point of this simulation, a simulation $m_s(x)$ of the drift $m(x)$ is then added. The simulation $m_s(x)$ of the drift may be based, for example, on the expansion of $m(x)$ as a series of known functions:

$$m(x) = \sum_l a_l f^l(x),$$

cf. formula (V.9). One such simulation of a functional drift will be presented in section VII.B.1.

However, if the drift $m(x)$ is evident only in a restricted section of the zone, to be simulated, a sufficient representation of this local drift may be obtained by conditioning the simulation to true or fictitious data in this section, cf. the left-hand side of the simulated profile on Fig. VII.1. More precisely, a large number of conditioning data $z_0(x_\alpha)$ are placed in the section in which the drift is evident; these data may be real $(x_\alpha \in I)$ or, if the real data are insufficient, fictitious. These conditioning data will force the initially stationary simulation to follow the drift $m(x)$ within the considered section. The conditioning confers robustness to the simulation with respect to the local unmodelled characteristics of reality, and for this reason, it is always advisable to use conditional simulations $z_{sc}(x)$ rather than nonconditional ones $z_s(x)$. The robustness and hence the representativity of the conditional simulation $z_{sc}(x)$ will improve as the number of real conditioning data $\{z_0(x_\alpha), x_\alpha \in I\}$ increases: simulation is not a substitute for a good survey of the real deposit.

Remark 2   Simulation of anisotropic regionalization. The three-dimensional covariance $C(r)$ of formulae (VII.5) and (VII.6) is supposed to be isotropic, whereas, in practice, the three-dimensional RF's to be simulated often have various anisotropies. However, as was seen in section III.B.5, any anisotropic covariance $C(h_u, h_v, h_w)$ can be modelled as a nested sum of isotropic models in spaces of dimension $n \le 3$. It is, thus, sufficient to simulate independently the realizations corresponding to each of the component models and then to sum up these realizations at each point.

Take, for instance, an anisotropic covariance of the type

$$C(h_u, h_v, h_w) = K_0 C_0(r) + K_1 C_1(h_w) + K_2 C_2(\sqrt{(h_u^2 + h_v^2)}) + K_3 C_3(r),$$

where $r = |h|$; K_0, K_1, K_2, K_3 are positive constants; and C_0, C_1, C_2, C_3 are isotropic covariances defined in three, one, two and three dimensions, respectively. These covariances are, for example, spherical models with ranges a_0, a_1, a_2, a_3 and unit sills.

In this case, the RF $Z(x)$ can be interpreted as the sum of four independent RF's:

$$Z(x) = \sum_{i=0}^{3} T_i(x),$$

where $T_0(x)$ characterizes an isotropic nugget effect, i.e., a micro-structure with covariance $C_0(r)$ and a range a_0 very small with respect to the distances involved experimentally or in the simulation. All the random variables $\{T_0(x)$, for each $x\}$ are spatially uncorrelated with zero expectation and a variance $E\{T_0^2(x)\} = K_0$. The simulation of a realization of $T_0(x)$ can, thus, be carried out very simply by drawing numbers at random from a uniform distribution with a variance of K_0; there is no need in this case to use the turning bands simulation method.

$T_1(x)$ characterizes a structure of one-dimensional variability with a covariance $K_1 C_1(h_w)$. This structure may correspond to a vertical variability of grades (for example, surface weathering phenomenon) that is particular to this direction. $T_1(x)$ varies only in the vertical direction, i.e., on any given horizontal plane all the realizations $t_1(x)$ are equal, and can, thus, be simulated directly on a single vertical line. Alternatively, this simulated line may be drawn from a previous three-dimensional simulation (simulation file on tape or disc) with the isotropic covariance $C_1(r)$. All the simulated values along this line are then assigned to their respective horizontal planes, providing a three-dimensional realization with covariance $K_1 C_1(h_w)$.

$T_2(x)$ characterizes a structure of two-dimensional, isotropic variability with covariance $K_2 C_2(\sqrt{(h_u^2 + h_v^2)})$. This structure may correspond to a large-scale variability of grades specific to the horizontal directions (e.g., stratifications); the range a_2 will thus generally be much larger than a_1 and a_3. $T_2(x)$ varies only in horizontal directions, i.e., on any given vertical line, all the realizations $t_2(x)$ will be equal, and $T_2(x)$ can be simulated directly on a single horizontal plane, or the plane can be drawn from a file of three-dimensional simulation with isotropic covariance $K_2 C_2(r)$.

$T_3(x)$ characterizes a structure of three-dimensional, isotropic variability with covariance $K_3 \, C_3(r)$. This structure may correspond to a small distance (small range a_3), average, isotropic variability within each bed or mineralized lens. $T_3(x)$ varies in all directions of the space R^3 and its simulation requires the turning bands method.

A *simulation file* – or a simulation data base – can be created by storing on tape or disc different realizations of three-dimensional RF's having standard types of isotropic covariances with variable parameters, e.g., spherical or exponential covariance with unit sill and variable ranges. As an experimental anisotropic covariance can be represented as the sum of isotropic covariances, it is, thus, possible to build a simulation with the desired anisotropic covariance by selecting appropriate realizations from the simulation file.

The sill K of a transition-type covariance $KC(h)$ appears simply as a multiplicative factor. Thus, to obtain a simulation corresponding to a particular spherical or exponential model (with specified range a and sill K), it is enough to

(i) select from the simulation file the realization, spherical or exponential, that has a range closest to the specified value;

(ii) multiply all values of this realization by the constant $\sqrt{K}$.

Remark 3 The icosahedron approximation. Formula (VII–5) of the turning bands method shows that a three-dimensional isotropic covariance is only attained as a limit for an infinite number N of lines D_i. In practice, of course, this number N will have to be finite. For example, we could take a large number (N around 100) of lines distributed uniformly over the unit sphere. An equally good but much less costly approximation can be obtained by taking the $N = 15$ lines joining the mid-points of the opposite edges of a regular icosahedron, which is the regular polyhedron with the maximum number of faces (20).

The sum $z_s = (\sum_{i=1}^{15} z_i)/\sqrt{15}$, corresponds to the covariance

$$C^*(h) = \frac{1}{15} \sum_{i=1}^{15} C^{(1)}(\langle h, k_i \rangle),$$

which is the icosahedron approximation of the limit covariance $C(r)$ given in formula (VII.5). The small difference between $C^*(h)$ and $C(r)$ can be theoretically corrected using the geometric properties of the regular icosahedron. this refinement is unnecessary in mining applications.

The geometric properties of the regular icosahedron are detailed in A. Journel (1974b). For practical applications, it is sufficient to know that the 15 lines defined on the icosahedron can be divided into five groups of three, each group constituting a trirectangular trihedron. Each group of three lines can be obtained from the others by five successive rotations. More precisely, let (Ou, Ov, Ow) be any reference rectangular axes, and (Ou_1, Ov_1, Ow_1) the trirectangular trihedron with the three base vectors defined by the following direction coordinates:

$$\left. \begin{array}{l} Ou_1, \quad (k, 1, 1+k); \\ Ov_1, \quad (1, -1-k, k); \\ Ow_1, \quad (1+k, k, -1); \end{array} \right\} \qquad \text{(VII.7)}$$

where k is either one of the roots of the equation $k^2 + k + 1 = 0$, i.e., $k = 0\cdot618$ or $k = -1\cdot618$. Note that the number $(1+\sqrt{5})/2 \approx 1\cdot618$ is the "golden number" well known in geometry and architecture.

This trihedron (Ou_1, Ov_1, Ow_1) is rotated four times, with the same rotation matrix $[\mathcal{R}]$ thus defining five trirectangular trihedrons, the axes of which are the preceding 15 lines that form a regular partition of the three-dimensional space

$$
\begin{Bmatrix} Ou_1 \\ Ov_1 \\ Ow_1 \end{Bmatrix} \xrightarrow{[\mathscr{R}]} \begin{Bmatrix} Ou_2 \\ Ov_2 \\ Ow_2 \end{Bmatrix} \xrightarrow{[\mathscr{R}]} \begin{Bmatrix} Ou_3 \\ Ov_3 \\ Ow_3 \end{Bmatrix} \xrightarrow{[\mathscr{R}]} \begin{Bmatrix} Ou_4 \\ Ov_4 \\ Ow_4 \end{Bmatrix} \xrightarrow{[\mathscr{R}]} \begin{Bmatrix} Ou_5 \\ Ov_5 \\ Ow_5 \end{Bmatrix}.
$$

The matrix of rotation is

$$
[\mathscr{R}] = \begin{bmatrix} 1 & -(k+1) & k \\ k+1 & k & -1 \\ k & 1 & k+1 \end{bmatrix} \tag{VII.8}
$$

It can be verified that $[\mathscr{R}]^5$ is the unit matrix

$$
\begin{bmatrix} 1 & 0 & 0 \\ 0 & 1 & 0 \\ 0 & 0 & 1 \end{bmatrix}.
$$

VII.A.3. Non-conditional simulation in one dimension

The turning bands method uses the simulation along lines of the one-dimensional RF $Y(u)$, with covariance $C^{(1)}(s)$, to provide realizations of a three-dimensional RF with the required covariance $C(r)$.

By means of the theorem of harmonic analysis of a second-order stationary process, one-dimensional realizations $y(u)$ with any specified covariance $C^{(1)}(s)$, can be simulated, cf. A. Journel (1974b). However, this method requires the Fourier analysis of the function $C^{(1)}(s)$, which can sometimes be difficult; in addition, there are problems with the convergence of the results.

In mining applications, the usual three-dimensional models $C(r)$, e.g., spherical or exponential, are such that the corresponding one-dimensional covariances $C^{(1)}(s)$ can be expressed in the form of a convolution product of a function $f(u)$ and its transpose $\check{f}(u) = f(-u)$:

$$
C^{(1)}(s) = f * \check{f} = \int_{-\infty}^{+\infty} f(u) \cdot f(u+s)\, du. \tag{VII.9}
$$

This means that the one-dimensional simulations can be carried out as moving averages with weighting function $f(u)$, which is a much simpler method to use.

The method of moving averages

Consider a stationary random measure $T(dr)$ with a Dirac covariance measure, e.g., the differential forms of a Poisson process or of a Brownian motion. The discrete version of this process $T(dr)$ is simply a succession of independent random variables T_i with the same distribution function.

Each random measure $T(dr)$ is then regularized by the weighting function $f(u)$, and this value defines the following one-dimensional RF $Y(u)$:

$$Y(u) = \int_{-\infty}^{+\infty} f(u+r)T(dr) = T * \check{f}. \qquad (VII.10)$$

$Y(u)$ is a stationary RF with covariance $C^{(1)}(s) = f * \check{f}$. This is a standard probabilistic result, cf., in particular, G. Matheron (1971, p. 14 and exercise 10, p. 104). A symbolic and brief demonstration of this result follows:

$$C^{(1)}(s) = E\{Y(u) \cdot Y(u+s)\} = E\left\{ \iint_{-\infty}^{+\infty} f(u+r)f(u+r'+s) \cdot T(dr) \cdot T(dr') \right\}$$

with

$$E\{T(dr)T(dr')\} = \begin{cases} 0 & \text{if } r \neq r', \\ \sigma^2 \, dr & \text{if } r = r'. \end{cases}$$

Indeed, the two random measures $T(dr)$ and $T(dr')$ are orthogonal if $r \neq r'$. The preceding double integral then reduces to

$$C^{(1)}(s) = \int_{-\infty}^{+\infty} f(u+r)f(u+r+s)E\{[T(dr)]^2\} = \sigma^2 \cdot f * \check{f}$$

which, except for a multiplicative factor which is the density of variance σ^2, is simply the convolution product $f * \check{f}$ of formula (VII.9).

Discrete approximation In practice, the random measure $T(dr)$ is approximated by a series of random variables located at discrete intervals along the line, as on Fig. VII.3. These RV's T_i are independent of each other and have the same distribution with zero expectation and a specified variance $E\{T^2\} = \sigma^2$.

Realizations t_i of these RV's T_i are assigned to points at regular interval b on the line D, as on Fig. VII.3. The values of t_i could be, for example, independent realizations of the random variable with uniform distribution function as provided by the IBM subroutine RANDU. Let $t_{i-k}, \ldots, t_i, \ldots, t_{i+k}$ be the realizations at points $i - k, \ldots, i, \ldots, i + k$.

The next step is to take the moving average y_i weighted by the function $f(u)$ and defined at each point i as

$$y_i = \sum_{k=-\infty}^{+\infty} t_{i+k} \cdot f(kb). \qquad (VII.11)$$

This formula is practical only when the function $f(kb)$ decreases rapidly towards zero as $|k|$ increases, so that the discrete sum does not have to be taken over the range $k = -\infty$ to $+\infty$.

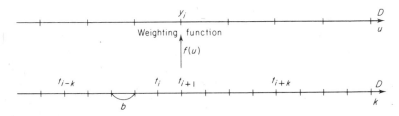

FIG. VII.3. Moving average on the line.

These values y_i can be viewed as a realization of a stationary one-dimensional RF $Y_i = Y(u)$, with zero expectation and a covariance

$$C_d^{(1)}(s) = E\{Y_i Y_{i+s}\} = E\left\{ \sum_{k=-\infty}^{+\infty} T_{i+k} f(kb) \cdot \sum_{k'=-\infty}^{+\infty} T_{i+s+k'} f(k'b) \right\}$$

After changing the variable $l = s + k'$, this covariance can be written:

$$C_d^{(1)}(s) = E\left\{ \sum_{k=-\infty}^{+\infty} \sum_{l=-\infty}^{+\infty} T_{i+k} T_{i+l} f(kb) f[(l-s)b] \right\}.$$

As the RV's T_{i+k} and T_{i+l} are independent if $k \neq l$, and have a variance $E\{T^2\} = \sigma^2$, the covariance reduces to

$$C_d^{(1)}(s) = \sigma^2 \cdot \sum_{k=-\infty}^{+\infty} f(kb) \cdot f(kb - bs). \qquad (VII.12)$$

This discrete sum $C_d^{(1)}(s)$ is simply the discrete approximation of the integral in (VII.9), which defines the specified one-dimensional covariance $C^{(1)}(s)$. This discrete approximation $C_d^{(1)}(s)$ can be calculated for each weighting function $f(u)$ and for each value b of the discrete interval. The bias $C^{(1)}(s) - C_d^{(1)}(s)$ caused by the discrete approximation can then be corrected. In practice, a multiplicative correction is enough, and this can be achieved by manipulating the parameter $\sigma^2 = E\{T^2\}$ to ensure the equality of the variances: $C^{(1)}(0) = C_d^{(1)}(0)$. An example of the correction of this bias is given in the case study of the simulation of the Prony deposit, cf. section VII.B.1.

Application to the usual covariance models All that remains is to show that the usual three-dimensional models can be expressed, by means of formula (VII.5), in terms of one-dimensional covariances of the form $C^{(1)}(s) = f * \check{f}$.

The *spherical model* is expressed by formula (III.15):

$$C(r) = \begin{cases} K\left[1 - \dfrac{3r}{2a} + \dfrac{r^3}{2a^3}\right], & \forall r \in [0, a], \\ 0, & \forall r \geqslant a \text{ (range).} \end{cases}$$

The constant $K = C(0) = \text{Var}\{Z(x)\}$ is the *a priori* variance of the three-dimensional RF $Z(x)$ and is also the sill of the corresponding semi-variogram $\gamma(h) = C(0) - C(h)$.

The corresponding one-dimensional covariance function is given by formula (VII.6):

$$C^{(1)}(s) = \begin{cases} K\left[1 - \dfrac{3s}{a} + \dfrac{2s^3}{a^3}\right], & \forall s \in [0, a], \\ 0, & \forall s \geqslant a. \end{cases} \tag{VII.13}$$

This covariance can be expressed as a convolution product $C^{(1)}(s) = f * \check{f}$, with the following weighting function:

$$f(u) = \begin{cases} \sqrt{(12K/a^3)} \cdot u, & \forall u \in [-a/2, +a/2], \\ 0, & \text{if not.} \end{cases} \tag{VII.14}$$

This result can be established by means of Fourier transforms or demonstrated by direct integration. This odd weighting function $f(u) = -f(-u)$ differs from zero only on the interval $[-a/2, +a/2]$.

In practice, an odd integer number $(2R + 1)$ of elementary values t_{i+k} is used for each moving average y_i in the discrete formula (VII.11), i.e.,

$$y_i = \sum_{-R}^{+R} t_{i+k} f(kb),$$

in which k progresses by integer values from $-R$ to $+R$. The central value t_i corresponding to $k = 0$, is assigned to the same point as the value y_i, cf. Fig. VII.3. The interval b of separation of the random variables T_i on line D is, thus, equal to $b = a/2R$. Program SIMUL, given in section VII.B.2, uses a number greater than or equal to 20. In practice, this number can be reduced $(R < 20)$, particularly when the range a is one of the ranges a_i of a nested model $C(h) = \sum_i C_i(h)$, in which case the object is to reproduce globally the nested sum $C(h)$, rather than precisely to reproduce each of its component structures $C_i(h)$, cf. the Prony case study in section VII.B.1.

The *exponential model* is expressed, according to formula (III.16), as

$$C(r) = K e^{-\lambda r}, \qquad \forall r \geqslant 0, \qquad \text{with } \lambda = 1/a.$$

The corresponding semi-variogram $\gamma(r) = C(0) - C(r)$ reaches its sill $\gamma(\infty) = C(0)$ asymptotically, cf. Fig. III.5. A practical range $a' = 3a$ is used, for which $\gamma(a) = 0 \cdot 95K$, or $C(a') = 0 \cdot 05K$.

The corresponding one-dimensional covariance is given by formula (VII.6) as

$$C^{(1)}(s) = K(1 - \lambda s) e^{-\lambda s}, \qquad \forall s \geqslant 0. \qquad (VII.15)$$

This covariance can be expressed as a convolution product, $C^{(1)}(s) = f * \check{f}$, with the following weighting function:

$$f(u) = \begin{cases} 2(\sqrt{(K\lambda)})(1 - \lambda u) e^{-\lambda u}, & \forall u \geqslant 0, \\ 0, & \forall u < 0. \end{cases} \qquad (VII.16)$$

The weighting function $f(u)$ approaches zero asymptotically as u tends to infinity. In practice, the function is limited to the interval $[0, 4a]$ and, on this interval, the random measure $T(dr)$ is approximated by $(8R+1)$ discrete random variables T_{i+k}. The separation interval b is thus equal to $b = 4a/8R = a/2R$, with $R \geqslant 20$. This high density discrete approximation is required to reproduce the asymptotic tendency of the covariance $C^{(1)}(s)$ towards its sill.

The *Gaussian covariance* is expressed by formula (III.17) as

$$C(r) = K e^{-r^2/a^2}, \qquad \forall r \geqslant 0,$$

cf. Fig. III.5. the corresponding one-dimensional covariance is

$$C^{(1)}(s) = K\left(1 - \frac{2s^2}{a^2}\right) e^{-s^2/a^2}, \qquad \forall s \geqslant 0, \qquad (VII.17)$$

which can be expressed as a convolution product, $C^{(1)}(s) = f * \check{f}$, with the following weighting function:

$$f(u) = \lambda u \, e^{-2u^2/a^2}, \qquad \forall u \in [-\infty, +\infty], \qquad (VII.18)$$

where
$$\lambda = 16 \, K/a^3 \sqrt{\pi}$$

In practice, this function is limited to the interval $[-2a, +2a]$ and, on this interval, the random measure $T(dr)$ is approximated by $(8R+1)$ discrete random variables T_{i+k}, corresponding to a separation interval $b = 4a/8R = a/2R$, with $R \geqslant 20$.

VII.A.4. Gaussian transform

Because the turning bands method of simulation consists in adding a large number (in theory infinite but in practice $N = 15$) of independent realizations defined on lines, this method will generate three-dimensional realizations $z_s(x)$ of a RF $Z_s(x)$ with a Gaussian (normal) multivariate distribution (because of the central limit theorem). Assuming the hypothesis of stationarity, upon which the method is based, the univariate distribution of the RF $Z_s(x)$ is Gaussian and independent of x. Thus, the histogram of the simulated values $\{z_s(x), x \in \text{deposit}\}$ is also Gaussian.

In practice, however, the histogram of the experimental data $\{z_0(x_\alpha), x_\alpha \in I\}$ and, more generally, the univariate distribution of the stationary RF $Z_0(x)$ is not restricted to the Gaussian type. In particular, the real values $\{z_0(x), x \in \text{deposit}\}$ are often non-negative, as in the case of grades or thicknesses, whereas the simulation $\{z_s(x), x \in \text{deposit}\}$, having a Gaussian distribution, may have a significant proportion of negative values. Thus, it is particularly important to ensure that the simulated values $z_s(x)$ have the same histogram as the experimental histogram of the available data $z_0(x_\alpha)$. Of course, it must be assumed that this experimental histogram is representative of the distribution of the variable $z_0(x)$ in the deposit, cf., in section VI.A.1, the remark on the representativeness of a histogram.

Let $Z_0(x)$ be a stationary random variable with a distribution represented by the experimental histogram of the available data $\{z_0(x_\alpha) \in I\}$. By means of a Gaussian transform function, this variable $Z_0(x)$ can be converted to a standard normal variable $U(x)$, i.e.,

$$Z_0 = \varphi_{Z_0}(U) \quad \text{and} \quad U = \varphi_{Z_0}^{-1}(Z_0).$$

Instead of simulating the grades $z_0(x)$, the transformed grades $u(x)$ are simulated. This simulation $\{u_s(x), x \in \text{deposit}\}$, obtained by the turning bands method, has a Gaussian histogram and the transform φ_{Z_0} results in a simulation $\{z_s(x) = \varphi_{Z_0}(u_s(x))\}$ with the required histogram, i.e., a histogram similar to that of the experimental data $z_0(x_\alpha)$. The covariance imposed on the simulation $u_s(x)$ must, of course, be the covariance deduced from the transformed experimental data, $u(x_\alpha) = \varphi_{Z_0}^{-1}[z_0(x_\alpha)]$, $\forall x_\alpha \in I$. Although not rigorously, the inverse transformed simulation $z_s(x) = \varphi_{Z_0}[u_s(x)]$ will then, in practice, have the required covariance $C(r)$.

Thus, the simulation $z_s(x)$ will reproduce not only the first two moments – expectation and covariance or variogram – but also the univariate distribution or, more precisely, the experimental histogram of the data $z_0(x_\alpha)$.

If, in addition, the conditioning of the simulation to the experimental data $z_0(x_\alpha)$ is desired, it is advisable first to condition the simulation $u_s(x)$ to the transformed data $u(x_\alpha) = \varphi_{Z_0}^{-1}[z_0(x_\alpha)]$ through relation (VII.3), i.e.,

$$u_{sc}(x) = u_K^*(x) + [u_s(x) - u_{sK}^*(x)],$$

and then to take the inverse transform $z_{sc}(x) = \varphi_{Z_0}[u_{sc}(x)]$. The kriged values $u_K^*(x)$ and $u_{sK}^*(x)$ are deduced, respectively, from the transformed data $u(x_\alpha)$ and from the simulated values $u_s(x_\alpha)$ located at the same points x_α. These krigings are carried out using the covariance $C_u(r)$ of the transformed RF $U(x)$.

By way of example, consider an experimental histogram of real grades $\{z_0(x_\alpha),\ \forall x_\alpha \in I\}$, to which a lognormal distribution can be fitted. The transformed variable which follows a normal distribution is: $U(x) = \log Z_0(x)$, and, thus, $Z_0(x) = \exp[U(x)]$. After determining the covariance $C_u(r)$, which characterizes the spatial regionalization of the transformed grades $\{u(x_\alpha) = \log z_0(x_\alpha),\ \forall x_\alpha \in I\}$, the turning bands method is used to provide a simulation $u_s(x)$ which has a Gaussian histogram and the imposed covariance $C_u(r)$. This simulation is then conditioned to the transformed data $u(x_\alpha)$ by formula (VII.3). Upon taking the inverse transform $z_{sc}(x) = \exp[u_{sc}(x)]$, the required conditional simulation is obtained.

In practice, however, the Gaussian tranform function $Z_0 = \varphi_{Z_0}(U)$ is not always so simple or so obvious. The most general methods are a graphical correspondence between the cumulative histogram of the data $z_0(x_\alpha)$ and the standard normal distribution of $U(x)$, cf. section VI.B.3, Fig. VI.13, or the approximation of function φ_{Z_0} by an Hermite polynomial expansion, cf. section VI.B.3, formula (VI.22) and following.

A fundamental remark

The above procedure reproduces the required univariate distribution of the RF $Z_0(x)$, but this does not mean that the 2-, 3-, ..., n-variate distributions are also reproduced, i.e., the spatial law of the RF $Z_0(x)$ is not reproduced in its entirety. As far as geostatistical applications are concerned, this is not very important since:

(i) most geostatistical techniques (kriging, estimation of recoverable reserves and study of their characteristic fluctuations) require only the univariate distribution of $Z_0(x)$ and its covariance or variogram, function;

(ii) the conditioning of the simulation $z_{sc}(x)$ to the experimental data $z_0(x_\alpha)$ induces a certain robustness with respect to the features of the reality $z_0(x)$, which are not specifically known or imposed on the simulated model. This robustness increases with the amount of conditioning data.

However, it should be stressed that the turning bands method produces simulations with a Gaussian *spatial* law and, as such, these simulations cannot correctly represent mineralized phenomena having spatial laws which do not have the very continuous character of the Gaussian spatial law. Consider, for example, a non-stationary phenomenon consisting of two states, one highly mineralized and the other waste, cf. Fig. VII.4. Each of these two states considered separately may have a Gaussian spatial law, but when considered together they definitely do not. A straightforward simulation of such a process, using the turning bands method, would result in a profile (the bold line on Fig. VII.4), which, over large zones, may have the expectation and covariance of the combined states of the real phenomenon, but nevertheless would be a poor representation of this reality.

In such a real case, the two states could be differentiated by structural analysis, and they would then be simulated independently; the transition from one state to another being simulated separately. Alternatively, one could consider a regularized phenomenon, taking averages within volumes sufficiently large to include a great number of different states; this regularization would then provide moving average profiles with a more continuous character open to simulation by the turning bands method.

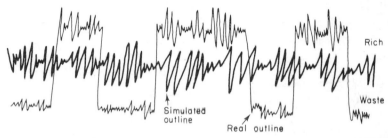

FIG. VII.4. The turning bands method is not suited for the straightforward simulation of a two-state process.

VII.A.5. Simulation of blocks

Up to this point, it has been supposed that the support of the simulated values $z_s(x)$ or $z_{sc}(x)$ is the same as that of the available data. However, it is often required to simulate the grades of blocks $Z_v(x)$, e.g., mining unit grades, while the available information $Z_0(x)$ is defined on the support of a core sample (this core sample being approximated to a quasi-point support).

1. *Averaging simulated points*

One method for simulating block values is to simulate point grades $z_{sc}(x)$ on a very dense grid and then average all simulated point grades that fall within the required block dimension. This will result in an approximate conditional simulation $z_{v,sc}(x)$ of the block grades. As the point simulation

$z_{sc}(x)$ is conditional to the data $z_0(x_\alpha)$, the block simulation will also be conditional to the same data.

The block is approximated by n interior points $(x_i, i = 1$ to $n)$, and the true grade of the block

$$Z_v(x) = \frac{1}{v} \int_{v(x)} Z(y)\,dy$$

is approximated by the arithmetic mean of the n interior point grades

$$Z_v^* = \frac{1}{n} \sum_{i=1}^{n} Z(x_i).$$

It is this arithmetic mean that is simulated. The density of the discrete representation of the block v, i.e., the number of interior points, depends on the error $(z_v - z_v^*)$ that is acceptable. The error should not mask the structural variability of the true block grades. The number n of interior points should, thus, be determined prior to each simulation of blocks, taking into account the structural function (covariance $C_v(h)$ or semivariogram $\gamma_v(h)$) which characterizes the spatial variability of support v. This structural function, $\gamma_v(h)$, for example, which is a regularization on the support v, is calculated from the point model $\gamma_0(h)$ by means of the usual regularization formulae, cf. section II.D.1, formula (II.41). In practice, it is enough to ensure that the error caused by the discrete approximation, $E\{[Z_v - Z_v^*]^2\}$, is less than the nugget constant C_{0v} corresponding to the support v. If this support v is large, C_{0V} will be close to zero, then the variance $E\{[Z_v - Z_v^*]^2\}$ should be much smaller than the *a priori* dispersion variance of the grades Z_v, i.e., $E\{[Z_v - Z_v^*]^2\} \ll D^2(v/\infty)$. As an example, for a point spherical model with no nugget effect and a range a, a sufficient discrete approximation of a cubic support v with sides $a/4$ can be obtained with eight interior points on a regular grid, i.e., the regular square grid spacing of the simulated points would be $a/8$.

This method of simulating blocks by averaging simulated point values has several advantages.

(i) It allows a simultaneous simulation of block grades and the grades of interior samples. Such a double simulation is essential for a study of selection of mining units based on available samples. It will be recalled from section VI.A.3, Fig. VI.5, that selection is always based on estimates (i.e., linear combination of sample grades), whereas the recovered grades are those of blocks.

(ii) No assumption is made about the structural variability or the univariate distribution of the regularized RF $Z_v(x)$. The entire simulation is based only on the available information, i.e., the information obtained from samples.

The disadvantage of the method is that a conditional simulation of N blocks requires a non-conditional simulation of nN points and, more significantly, the conditioning process requires kriging these nN points. If there are $N = 100\,000$ blocks and $n = 10$ points per block, then problems could arise with storage and calculation time. Cuts in computer time can be obtained, first by kriging only the N blocks, and second by simulating and kriging directly these N blocks, as shown in the following section.

2. Non-conditional simulation of points and conditioning on blocks

This procedure consists in carrying out a non-conditional point simulation, and then conditioning the simulation by kriging the blocks. More precisely, it consists of the following eight successive steps.

A. Carry out a *non*-conditional simulation of Gaussian distributed point values $u_s(x)$; $u_s(x)$ is a realization of the Gaussian RF $U(x)$ deduced from the initial point RF $Z_0(x)$ by the transform function fitted to the experimental histogram of the point data $z_0(x_\alpha)$, i.e., $z_0(x) = \varphi_{Z_0}\{u(x)\}$ and $u(x) = \varphi_{Z_0}^{-1}\{z_0(x)\}$.

B. Using the transform function φ_{Z_0}, transform this non-conditional simulation, i.e.,

$$z_s(x) = \varphi_{Z_0}\{u_s(x)\}.$$

C. Average this non-conditional simulation $z_s(x)$ within the required blocks, i.e., $z_{v,s}(x) = (1/n)\sum_i \phi_{Z_0}\{u_s(x_i)\}$, where the unit v is made up of the n interior points x_i.

(These first three steps constitute the standard procedure for the non-conditional simulation of blocks by averaging non-conditionally simulated points.)

D. From the histogram of the simulated values $z_{v,s}(x)$, determine experimentally the Gaussian transform function for a support v, i.e., $z_{v,s}(x) = \varphi_{Z_v}\{u_{v,s}(x)\}$. Once φ_{Z_v} is known, the Gaussian simulated values can be deduced from $u_{v,s}(x) = \varphi_{Z_v}^{-1}\{z_{v,s}(x)\}$. Note that, since the function $\varphi_{Z_v}^{-1}$ is, in general, non-linear, $u_{v,s}(x) \neq (1/n)\sum_i u_s(x_i)$.

E. Using the Gaussian simulated values $u_{v,s}(x)$ and $u_s(x_\alpha)$, simulate the kriging error $[u_{v,s}(x) - u_{v,sK}^*(x)]$. The kriged value $u_{v,sK}^*(x)$ is a linear combination of the simulated values $u_s(x_\alpha)$ at the data points $x_\alpha \in I$, i.e., $u_{v,sK}^*(x) = \sum_\alpha \lambda_\alpha(x) \cdot u_s(x_\alpha)$. In the kriging system, the first member matrix consists of the experimental covariances $E\{u_s(x)u_s(y)\}$ and the second member matrix consists of $E\{u_s(x)u_{v,s}(y)\}$.

F. Calculate the kriged values of the blocks which are linear combinations of the initial transformed values $u(x_\alpha) = \varphi_{Z_0}^{-1}\{z_0(x_\alpha)\}$, i.e., $u_v^*(x) = \sum_\alpha \lambda_\alpha(x) \cdot u(x_\alpha)$. Note that the kriging weights $\lambda_\alpha(x)$ are identical to those in the preceding step E.

G. Condition by these Gaussian block values, i.e.,

$$u_{v,sc}(x) = u_v^*(x) + [u_{v,s}(x) - u_{v,sK}^*(x)].$$

H. Finally, the required conditional simulation of block grades is obtained by means of the previously determined (in step D) function φ_{Z_v} for the support v, i.e.,

$$z_{v,sc}(x) = \varphi_{Z_v}\{u_{v,sc}(x)\}.$$

This procedure requires a non-conditional simulation of nN point values $u_s(x)$, n values for each of the N blocks. However, the conditioning process requires kriging only N blocks v, rather than kriging all nN points. If, however, the nonconditional simulation of the nN point values is still too costly, then the third method of direct conditioning of the N blocks should be used.

3. Direct conditional simulation of blocks

By means of the geostatistical regularization techniques, the structural model $\gamma_v(h)$ defined on the support v can be deduced from the quasi-point structural model $\gamma_0(h)$ fitted, for example, to core sample lengths. Using various permanence of distribution techniques outlined in section VI.B, the univariate distribution of the regularized RF $Z_v(x)$ can be deduced from the univariate distribution of the point RF fitted, for instance, to the experimental histogram of core sample lengths. This distribution of $Z_v(x)$ can also be deduced from production data or from control data at the entry to the milling process.

Once the structural function ($C_v(h)$ or $\gamma_v(h)$) and the distribution of Z_v are known, a *non*-conditional simulation $z_{v,s}(x)$ of the block grades can be carried out directly. However, the subsequent conditioning process poses a few problems.

The conditional data are, in general, defined on a support different from v, such as the quasi-point support of core sample lengths for instance. Now, consider the terms required in the conditioning formula (VII.3):

$$z_{v,sc}(x) = z_{vK}^*(x) + [z_{v,s}(x) - z_{v,sK}^*(x)].$$

The non-conditional block simulation $Z_{v,s}(x)$ can be immediately obtained, and each block can be kriged from the real data $z_0(x_\alpha)$ to provide the term $z_{vK}^*(x) = \sum_\alpha \lambda_\alpha z_0(x_\alpha)$. But the final term $z_{v,sK}^*(x)$ requires a point simulation $z_s(x_\alpha)$ on the actual data locations x_α and this point simulation must be coherent with that of the blocks $z_{v,s}(x)$, i.e., it cannot be carried out independently.

One solution to this problem consists in conditioning the block simulation to constructed data $z_v'(x_\alpha)$, defined on the same support v and located at the same points $x_\alpha \in I$, rather than to the real point data $z_0(x_\alpha)$. To do this, the dimensions of the support v must be very small with respect to the grid (x_α, x_β) of the conditioning data. In practice, the dimensions of v must be less than one tenth the mean distance between pairs of conditioning data (x_α, x_β). However, it is always possible to restrict the number of locations x_α to be considered in the construction of the v-support data $z_v'(x_\alpha)$. These constructed data $z_v'(x_\alpha)$ may be, for example, the kriged value of the block $v(x_\alpha)$ centred on x_α, plus an additional term to correct the smoothing effect of kriging:

$$z_v'(x_\alpha) = z_{vK}^*(x_\alpha) + \varepsilon_v, \qquad \text{(VII.19)}$$

where $z_{vK}^*(x_\alpha)$ is the kriged value of the block $v(x_\alpha)$ using the neighbouring real data $\{z_0(x_\alpha), x_\alpha \in I\}$, and ε_v is a realization of a random variable with zero expectation and a variance $E\{\varepsilon_v^2\}$ such that the variance of the constructed data $z_v'(x_\alpha)$ is equal to the dispersion variance $D^2(v/\infty)$ of the true grades of support v, cf. the kriging smoothing relation (VI.1) or (VI.2). This latter variance $D^2(v/\infty)$ is calculated from the point structural model $\gamma_0(h)$, cf. formula (II.35) or (II.36).

This method of constructing the v-support conditioning data is an approximation, since the variable ε_v is supposed to be independent of the kriged value $Z_{vK}^*(x)$, whereas, in fact, the kriging error $[Z_v(x_\alpha) - Z_{vK}^*(x_\alpha)]$ is only orthogonal to the kriged value $Z_{vK}^*(x)$. This approximation alters the spatial variability of the true values of support v, and can be justified only when the variance $E\{\varepsilon_v^2\}$ is small, i.e., when the kriging variance of a unit v centred on a data location x_α is itself small with respect to the dispersion variance $D^2(v/\infty)$.

Once these v-support conditioning data $z_v'(x_\alpha)$ are constructed, the conditioning formula (VII.3) can be used in exactly the same way as in the point case:

$$z_{v,sc}(x) = z_{vK}^*(x) + [z_{v,s}(x) - z_{v,sK}^*(x)].$$

The kriged values $z_{vK}^*(x)$ and $z_{v,sK}^*(x)$ are calculated from the conditioning v-support data $z_v'(x_\alpha)$ and $z_{v,s}'(x)$. The covariance used in this kriging is the regularized covariance $C_v(h)$.

VII.A.6. Simulation of coregionalization

In mining applications, there are often several spatially intercorrelated variables of interest, e.g., overburden and mineralized thickness, and the corresponding grades of saleable metals and of impurities.

The spatial coregionalization of the L stationary RF's $\{Z_k(x), k = 1 \text{ to } L\}$ is characterized by the covariance matrix $[C_{kk'}(h)]$, cf. section II.A.4, with

$$C_{kk'}(h) = E\{Z_k(x)Z_{k'}(x+h)\} - m_k m_k, \quad \text{and} \quad m_k = E\{Z_k(x)\}.$$

This covariance matrix must be positive definite and the linear model proposed for it in section III.B.3, formula (III.25), is

$$C_{kk'}(h) = \sum_{i=1}^{n} b_{kk'}^i K_i(h). \qquad (\text{VII.20})$$

In this model, all cross- and direct covariances $C_{kk'}(h)$ are expressed as linear combinations of n basic direct covariances $K_i(h)$. (It will be recalled here that, in the above notation, i in the superscript position represents an index and not a power.)

Consider now n stationary RF's $\{Y_i(x), i = 1 \text{ to } n\}$ with direct covariances $K_i(h)$, respectively, and all independent of each other. The L stationary RF's $\{Z_k(x), k = 1 \text{ to } L\}$, defined by the L linear combinations

$$Z_k(x) = \sum_{i=1}^{n} a_{ki} Y_i(x) \qquad (\text{VII.21})$$

have a covariance matrix $[C_{kk'}(h)]$ given by

$$\breve{C}_{kk'}(h) = \sum_{i=1}^{n} a_{ki} a_{k'i} K_i(h).$$

Thus, the simulation of the coregionalization $\{Z_k(x), k = 1 \text{ to } L\}$ with the (VII.20) covariance matrix can be reduced to the simulation of independent realizations of n RF's $Y_i(x)$ with covariances $K_i(h)$. These n realizations are then linearly combined by means of formula (VII.21) with the following parameters:

$$b_{kk'}^i = a_{ki} a_{k'i}, \quad \forall i = 1 \text{ to } n, \quad \text{and} \quad \forall k, k' = 1 \text{ to } L. \quad (\text{VII.22})$$

If the number of parameters a_{ki}, $a_{k'i}$ is insufficient to reproduce each of the coefficients $b_{kk'}^i$ of the linear model (VII.20), it is advisable to use the more complete linear combinations of formula (III.24).

Remark 1 To reproduce the univariate distribution functions of each of the component RF's $Z_k(x)$, the covariance matrix of the Gaussian RF's $U_k(x) = \varphi_K^{-1}[Z_k(x)]$ should be considered. First, the coregionalization

$\{U_k(x),\ k = 1$ to $L\}$ is simulated, and then the required coregionalization is obtained by applying the Gaussian transform:

$$Z_k(x) = \varphi_k[U_k(x)], \qquad \forall k = 1 \text{ to } L.$$

Remark 2 The conditioning of the experimental data simulation is carried out, variable by variable, by direct kriging and not by cokriging, which would be tedious and too costly:

$$z_{k,sc}(x) = z_{kK}^*(x) + [z_{k,s}(x) - z_{k,sK}^*(x)],$$

where $z_{kK}^*(x)$ and $z_{k,sK}^*(x)$ are kriged from the real data $z_k(x_\alpha)$ and the simulated data $z_{k,s}(x_\alpha)$, respectively.

VII.B. THE PRACTICE OF SIMULATIONS

VII.B.1. Case studies

1. *Conditional simulation of the Prony deposit*

The nickeliferous laterite deposit of Prony (New Caledonia) is a surface-weathered subhorizontal formation of basic rocks (peridotites). The deposit was sampled by several successive campaigns of vertical drilling on a regular horizontal grid; each core sample was analysed in 1-m lengths for nickel grade and the grade of impurity MgO. Global and local resources and recoverable reserves were evaluated by geostatistical techniques after each sampling campaign was completed.

The next step was to tackle some of the mining problems, such as deciding on the type of mining and the corresponding equipment, the need for blending, and the degree of flexibility of the mill. For this, it was necessary to predict the fluctuations at various scales of such parameters as overburden, mineralized thickness and the recovered Ni and MgO grades. Thus, it was decided to simulate the deposit and test the various methods considered for mining by applying them to the simulated deposit.

The simulation was limited to the first production zone, and consists of 17 600 simulated vertical drills on a square 10-m grid. The simulation was based on a previous geostatistical structural analysis of 176 real drill-holes on a square 100-m grid within the first production zone, cf. Fig. V.24, and was conditioned to these real data.

A conditional simulation involves three steps: structural analysis, which provides a model of the deposit; conditional simulation itself; and, finally, various verifications. Simulated deposits can be used for various purposes, depending on each particular mining project, and some of these typical uses are presented in section VII.B.3.

Drift and cut-off grades   The vertical profiles of the Ni grades measured along the drill cores all have a similar shape, cf. Fig. II.1. There is a progressive enrichment in nickel from the upper surface, followed by a rapid impoverishment, and, finally, a further enrichment at the bedrock contact. This is translated into geostatistical terms as a vertical drift of the Ni grades. The drift curve (or mean tendency) can be estimated by universal kriging from the vertical profile of the grades of the core sample lengths, and it can be shown that this drift curve is an estimator of the vertical profile of the mean grades of small panels centred on the drill holes, cf. Ch. Huijbregts and A. Journel (1972) and Figs.II.1 and VII.5.

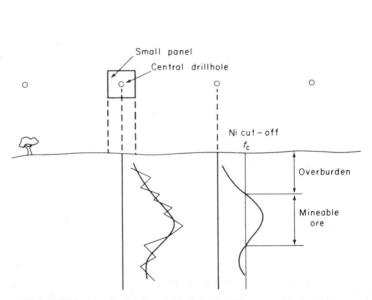

FIG. VII.5. Vertical cut and definition of the service variables.

The entire mineralized thickness intersected by each drill cannot be economically mined. Only a certain mineable thickness, defined by a cut-off grade t_c in nickel, is retained within each mining unit of 10 m × 10 m. It is supposed that future selection will be made on the basis of information *equivalent* to a 10 m × 10 m grid of samples, i.e., one central drill-hole on each mining unit. Thus, the cut-off grade t_c must be applied to the previously defined drift curve estimating the vertical profile of mean Ni grades of the mining unit, cf. Fig. VII.5. For each value t_c, a certain number of "service variables" are defined. These variables, regionalized in the two-dimensional horizontal space, are: overburden thickness P_0, mineralized thickness P_L and the corresponding accumulations A_{Ni} and A_{MgO}. The simulation of the deposit thus consists of generating a realization (coregionalization) of the corresponding service variables on a $10 \, m \times 10 \, m$ grid for each value of the cut-off grade t_c.

Horizontal structural analysis At the time of this study, the first production zone was sampled on a systematic $100 \, m \times 100 \, m$ grid, and by two cross-shaped drilling sets spaced at $20 \, m$ to detect small structures. This information was used to characterize the horizontal regionalizations of the previously defined service variables. In the actual study, four cut-off grades were considered; six service variables for each cut-off. But only the results obtained for the mineralized thickness P_L, the overburden P_0, and the Ni accumulation A_{Ni} for the two cut-off grades $t_c = 0$ (no vertical selection) and $t_c = 1 \cdot 0\%$ will be given here.

For proprietary reasons, all values given here have been multiplied by a constant factor, which does not affect the relative curves (variograms divided by the squares of the means), nor the relative dispersions.

Table VII.1 gives the means m and the relative dispersion variances D^2/m^2 of the various service variables within the zone of interest. These experimental values were deduced from the 176 drills covering the first production zone on a $100 \, m \times 100 \, m$ grid. Table VII.2 gives the correlation coefficients between these service variables for each cut-off grade.

Figure VII.6(a), (b), (c) gives the experimental relative variograms of the variables P_0 for $t_c = 1 \cdot 0$ and P_L and A_{Ni} for $t_c = 0$. Figure VII.7(a), (b) shows the scatter diagram for P_0 and P_L and the corresponding cross-variogram for $t_c = 1 \cdot 0$.

TABLE VII.1. *Experimental means and dispersion variances*

	t_c	P_0	P_L	A_{Ni}
m	0	–	23·7	26·9
D^2/m^2	0	–	0·08	0·17
m	1·0	11·7	11·7	17·6
D^2/m^2	1·0	0·29	0·47	0·51

TABLE VII.2. *Experimental correlation coefficients*

t_c	$\rho_{P_0-P_L}$	$\rho_{P_L-A_{Ni}}$
0	–	0·87
1·0	−0·45	0·98

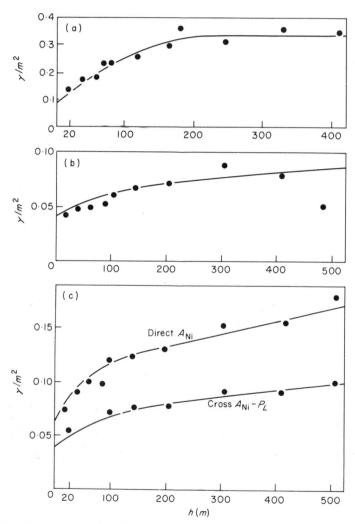

FIG. VII.6. Relative semi-variograms (experimental points and fitted models). (a) P_0 for $t_c = 1·0$; (b) P_L for $t_c = 0$; (c) A_{Ni} for $t_c = 0$.

For $t_c = 0$ there is no vertical selection and thus no overburden; the thickness P_L and accumulation A_{Ni} for $t_c = 0$ thus characterize the *in situ* resources before any selection is made.

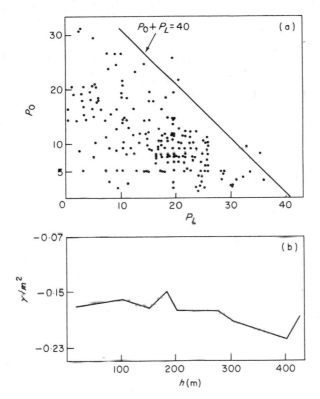

FIG. VII.7. Correlation $P_0 - P_L$ for $t_c = 1 \cdot 0$. (a) Scatter diagram; (b) relative cross-variogram.

Correlation $P_0 - P_L$. For $t_c = 1 \cdot 0$, the correlation $\rho_{P_0 - P_L} = -0 \cdot 45$ is small and negative. As the total thickness (P_L for $t_c = 0$) is not very variable, the negative correlation can be explained by the fact that, on average, the mineralized thickness decreases as the overburden increases. On the $P_0 - P_L$ scattergram, the following inequalities can be observed for $t_c = 1 \cdot 0$:

$$P_0 + P_L \leqslant 40; \quad 0 \leqslant P_0 \leqslant 32; \quad 0 \leqslant P_L \leqslant 35. \quad \text{(VII.23)}$$

As there is no evident structure on the experimental cross-semi-variogram $\gamma_{P_0 - P_L}(h)$, cf. Fig. VII.7(b), the two variables P_0 and P_L for $t_c = 1 \cdot 0$ can be simulated independently as a first approximation, but the inequalities given in (VII.23) should be verified. In practice, all conditionally simulated values z_{sc} which do not satisfy these inequalities are replaced by the corresponding kriged value z_{OK}^*, cf. the conditioning relation (VII.3). Such an approximation will slightly modify the structural variability of the simulation z_{sc}, but is quite acceptable provided that the number of altered simulated values is small.

Correlation $P_L - A_{Ni}$. Except for their nugget constants, the direct and cross-variograms on Fig. VII.6(b and c) can be deduced one from the other by a multiplicative factor. This suggests a linear coregionalization model with a single basic model, cf. formula (III.25), i.e.,

$$\left.\begin{aligned} \gamma_{P_L}/\bar{P}_L^2(r) &= C_0 + \gamma(r), \\ \gamma_A/\bar{A}^2(r) &= \alpha_a C_0 + \beta_a \gamma(r), \qquad \forall r = |h| > 0, \\ \gamma_{A-P_L}/\bar{A} \cdot \bar{P}_L(r) &= \alpha_{ap} C_0 + \beta_{ap}\gamma(r), \end{aligned}\right\} \qquad \text{(VII.24)}$$

with relative nugget constants $C_0 = 0.04$, $\alpha_a = 1.5$, $\alpha_{ap} = 1$, and sill values $\beta_a = 2.67$, $\beta_{ap} = 1.34$. The values of these parameters ensure the positivity of the coregionalization model (VII.24), since, from formula (III.26),

$$\alpha_{ap}^2 \leq \alpha_a \quad \text{and} \quad \beta_{ap}^2 \leq \beta_a.$$

The basic model $\gamma(r)$ consists of two nested spherical models: $\gamma(r) = C_1 \gamma_1(r) + C_2 \gamma_2(r)$ with

$$\text{ranges} \begin{cases} a_1 = 80\,\text{m}, \\ a_2 = 800\,\text{m}, \end{cases} \text{and sills} \begin{cases} C_1 = 0.015, \\ C_2 = 0.036. \end{cases}$$

As the experimental histogram of the variable P_L for $t_c = 0$, cf. Fig. VII.9, appears symmetric and Gaussian, P_L can be simulated directly without any prior Gaussian transformation. The same holds true for A_{Ni}; because of the very high correlation between A_{Ni} and P_L ($\rho_{P_L-A_{Ni}} = 0.81$), if the P_L-histogram is satisfied, so will the A_{Ni} histogram.

Regionalization of P_0 The experimental histogram of P_0, given on Fig. VII.8, is clearly asymmetrical. The simulation of P_0 will, therefore, require first a simulation of U, the Gaussian transform of P_0; the required simulation will then be given by $P_0 = \varphi_{P_0}(U)$. The Gaussian transform function φ_{P_0} can be fitted graphically to the experimental histogram on Fig. VII.8. It can also be assumed that P_0 follows a lognormal distribution, in which case φ_{P_0} is given directly by formula (VI.8). The model γ_V for the Gaussian variable V has been deduced directly from the structural model γ_{P_0} of the initial variable P_0. Because U has to be simulated, a simple standard model (spherical with a nugget effect) is adopted for γ_U, from which the model γ_{P_0} is deduced and compared to the experimental semi-variogram $\gamma_{P_0}^*$: the parameters can then be determined by obtaining a good approximation between γ_{P_0} and $\gamma_{P_0}^*$.

The details of these correspondance calculations $\gamma_{P_0} \to \gamma_U$ are given in Addendum A. The structural model used in the simulation of the Gaussian variable U is a spherical model with range $a_1 = 260$ m, a sill $C_1 = 0.215$ and a nugget effect $C_0 = 0.07$. The theoretical model γ_{P_0} deduced from this model is a good approximation of the experimental semi-variogram $\gamma_{P_0}^*$, cf. Fig. VII.6(a).

Results of the simulation   The simulations corresponding to each value t_c of the cut-off grade were carried out independently. However, for a given value of t_c, the various service variables were simulated on the basis of their coregionalization model.

The details of the calculation of the various parameters of the non-conditional simulation are given in Addendum B. The kriging of the 17 600 groups of service variables, required for the conditioning process, was described in section V.B.4, Fig. V.24. The remainder of this section consists of the verification of the final conditional simulations.

Tables VII.3 and VII.4 give the means, relative dispersion variances and correlation coefficients of the 17 600 groups of conditionally simulated values. These results can be compared with those of the corresponding Tables VII.1 and VII.2 for the 176 real conditioning values.

Figures VII.8 and VII.9 give the histograms and isotropic semi-variograms of the conditionally simulated values, along with – for comparison – those of the conditioning data values.

TABLE VII.3. *Simulated means and dispersion variances*

	t_c	P_0	P_L	A_{Ni}
m	0	–	23·8	26·2
D^2/m^2		–	0·08	0·175
m	1·0	11·7	11·2	16·7
D^2/m^2		0·30	0·46	0·51

TABLE VII.4. *Simulated correlation coefficients*

t_c	$\rho_{P_0 - P_L}$	$\rho_{P_L} - A_{Ni}$
0	–	0·83
1·0	−0·29	0·95

Analysis of the results   There is an excellent agreement between the global statistical characteristics, as can be seen from Tables VII.1 and VII.3. The slight differences between the simulated and experimental values is largely accounted for by the fluctuation variances of these experimental characteristics on the zone considered.

There is an equally good agreement between the strong correlations, cf. Tables VII.2 and VII.4. On the other hand, the weak correlation ($\rho_{P_0-P_L}$) = -0.45), which was not taken into account (P_0 and P_L were simulated independently) is poorly reproduced.

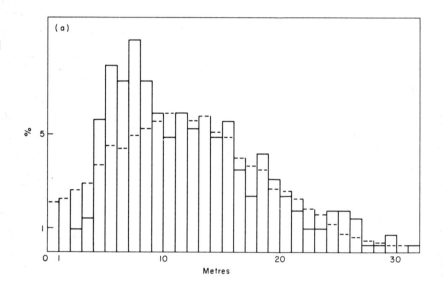

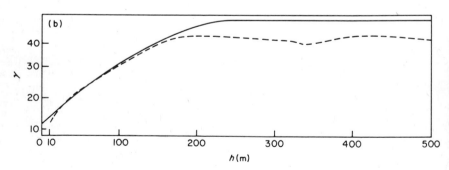

FIG. VII.8. Conditional simulation P_0, $t_c = 1.0\%$ Ni. (a) ———, Experimental values, $m = 11.7$, $D^2/m^2 = 0.29$; – – – –, conditional simulation values, $m = 11.7$, $D^2/m^2 = 0.30$. (b) ———, Fitted model, – – – –, conditional simulation.

There is a satisfactory agreement between the variograms of the simulations and the theoretical models fitted to the experimental variograms, cf. Figs VII.8(b) and VII.9(b).

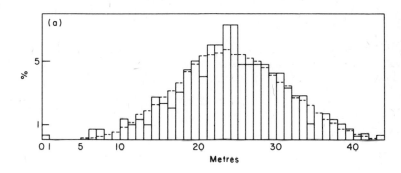

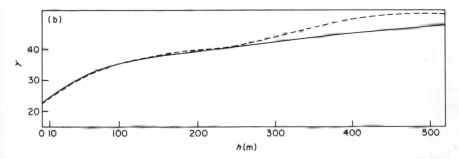

FIG. VII.9. Conditional simulation P_L, $t_c = 0\%$ Ni (a) ———, Experimental values, $m = 23\cdot7$, $D^2/m^2 = 0\cdot08$; $----$, conditional simulation values, $m = 23\cdot8$, $D^2/m^2 = 0\cdot08$. (b) ———, Fitted model; $----$, conditional simulation.

It is a little more difficult to draw any conclusions about the agreement of the histograms. The histograms of the 17 600 simulated values are, as expected, much more regular than the experimental histograms of only 176 data values. However, it seems reasonable to conclude that the irregularities of the experimental histograms are not significant and would have been smoother if several thousand drills had been available on the zone considered, rather than just 176.

(On a first reading, the two following addenda can be omitted.)

Addendum A. The Gaussian transform of a lognormal variable

Let Z be a lognormal variable with the following characteristics: mean $E\{Z(x)\} = m$, variance $\mathrm{Var}\{Z\} = \sigma^2 = C(0)$, covariance $C(h)$, semi-variogram $\gamma_Z(h) =$

$C(0) - C(h)$. $U = \log Z$ is then a normal variable with the following characteristics, given by formulae (VI.9):

mean, $E\{U\} = m' = \log m - \sigma'^2/2$;

variance, $\mathrm{Var}\,\{U\} = \sigma'^2 = \log(1 + \sigma^2/m^2) = K(0)$;

covariance, $K(h) = \log[1 + C(h)/m^2]$;

semi-variogram, $\gamma_U(h) = K(0) - K(h)$.

Suppose a spherical model with range a_1, nugget constant K_0 and sill K_1 is adopted for $\gamma_U(h)$, i.e.,

$$\gamma_U(h) = K_0 + K_1\left(\frac{3}{2}\frac{h}{a_1} - \frac{1}{2}\frac{h^3}{a_1^3}\right), \qquad \forall h \in \,]0, a_1]. \qquad (VII.26)$$

The total sill is, thus, $K(0) = K_0 + K_1$, and the corresponding covariance is given by

$$K(h) = K(0) - \gamma_U(h) = K_1\left(1 - \frac{3}{2}\frac{h}{a_1} + \frac{1}{2}\frac{h^3}{a_1^3}\right), \qquad \forall h \in \,]0, a_1].$$

The relative semi-variogram of the initial lognormal variable Z is then

$$\gamma_Z(h)/m^2 = [C(0) - C(h)]/m^2 = e^{K_0 + K_1} - e^{K(h)}, \qquad \forall h. \qquad (VII.27)$$

This relative semi-variogram has the following characteristics:

nugget effect, $C_0 = \gamma_Z(\varepsilon)/m^2 = e^{K_0 + K_1} - e^{K_1}$;

total sill, $C_0 + C_1 = \gamma_Z(\infty)/m^2 = e^{K_0 + K_1} - 1$;

range a_1 identical to that of $\gamma_U(h)$.

In practice, an experimental relative semi-variogram $\gamma_Z(h)/m^2$ will be available, to which the parameters C_0, C_1 and a_1 will be fitted. The parameters K_0, K_1 and a_1 of the semi-variogram $\gamma_U(h)$ can then be deduced by the relations

$$\left.\begin{array}{l} K_1 = \log(1 + C_1), \\ K_0 = \log(1 + C_0 + C_1) - K_1. \end{array}\right\} \qquad (VII.28)$$

Numerical applications to the lognormal variable P_0 *for* $t_c = 1\cdot0$ The following values are estimated from the experimental relative semi-variogram $\gamma_{P_0}^*/m^2$ on Fig. VII.6(a):

$$C_0 = 0\cdot09, \qquad C_0 + C_1 = 0\cdot33, \qquad a_1 = 260 \text{ m};$$

from which

$$K_1 = 0\cdot215, \qquad K_0 = 0\cdot070.$$

It can be verified on Fig. VII.6(a) that the theoretical model (VII.27) thus determined is a good approximation of the experimental curve.

Addendum B. The practical steps in a non-*conditional simulation*

The envelope surface of the zone to be simulated is a square with 2 km sides, and the simulation grid is $10 \text{ m} \times 10 \text{ m}$, making a total of $201 \times 201 = 40\,401$ values to be simulated for each service variable. This number will be reduced to $17\,600$ at the conditioning stage when the simulation is limited to the precise polygonal contour defined by the 176 conditioning drills on a $100 \text{ m} \times 100 \text{ m}$ grid.

The turning bands method (*cf. section VII.A.2*)   Each of the 15 generating lines D are considered as NGMAX $= 300$ points spaced at a constant interval of $u = 10$ m. Next, 300 bands (slices) are centred on each of these 300 points, see Fig. VII.10. The number NGMAX $= 300$ was chosen so that each simulated point of the square of dimensions $200u \times 200u$ would correspond to a band on each of the 15 generating lines, for which it is enough that NGMAX $u > 200\sqrt{(2u)}$ (length of the diagonal).

For each line D, the subroutine RANDU(SSP–IBM) is used to generate 300 independent realizations of a random variable T with a uniform distribution on the interval $[0, 1]$. By definition, this variable T is such that $E\{T\} = 1/2$ and Var $\{T\} = 1/12$.

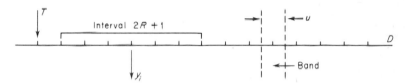

FIG. VII.10. One-dimensional simulations on the generating lines.

The values of T are then regularized by a discrete approximation of the weighting function $f(x) = x$ on the interval $[-R, +R]$ to provide a realization of the random function $\Omega = T * f$ with a spherical covariance, cf. formulae (VII.10) and (VII.14). The range $(2R + 1)$ is an approximation of the range of the spherical covariance to be simulated. The one-dimensional generating RF defined on each line D is then $Y = k\Omega + q$, where the constants k and q are determined by the expectation and variance imposed on Y.

For the coregionalization $P_L - A_{Ni}$, for $t_c = 0$, the three-dimensional variograms (or covariances) to be simulated are nested: $\gamma = C_0 + \gamma_1 + \gamma_2$, cf. model (VII.24). The procedure is thus to take three independent realizations T_0, T_1, T_2 and then regularize them by their respective weighting functions: $\Omega_0 = T_0$, $\Omega_1 = T_1 * f_1$, $\Omega_2 = T_2 * f_2$. The three one-dimensional partial generating RF's can then be calculated:

$$Y_0 = k_0\Omega_0 + q_0$$

for the nugget effect;

$$Y_1 = k_1\Omega_1 + q_1 \quad \text{and} \quad Y_2 = k_2\Omega_2 + q_2$$

for the two spherical structures γ_1 and γ_2.

The final one-dimensional generating RF, defined on each of the lines D, is then the sum of the three independent partial generating RF's: $Y = Y_0 + Y_1 + Y_2$.

The 15 lines D are then rotated in the three-dimensional space to obtain uniform partitions, cf. rotation matrix (VII.8).

Each of the 40 401 points of the square envelope falls within one of the bands on each of the 15 different lines. The sum of the 15 independent realizations of Y corresponding to these 15 bands is then assigned to each point, which results in the required three-dimensional non-conditional simulation.

Determination of the parameters   For $t_c = 0$, the two variables to be simulated are P_L and A_{Ni}, which are coregionalized according to the linear model (VII.24).

From section III.B.3, formula (III.23), the linear model can be generated by simulating two independent realizations z and z' of the same RF Z with zero expectation and a semi-variogram $\gamma(r)$. The required realizations p_L and a_{Ni} are then obtained by taking linear combinations of these two realizations z and z', plus a purely random variable that accounts for the nugget effect:

$$p_L = m_p[\varepsilon_p + A_1 z + A_2 z'], \atop a_{Ni} = m_a[\varepsilon_a + A_3 z + A_4 z'].}$$

(VII.29)

The identification of the expectation $E\{P_L\} = m_p$ and the relative semi-variogram $\gamma_{P_L}/\bar{m}_p^2 = C_0 + \gamma(r)$ leads to the following conditions:

$$A_1^2 + A_2^2 = 1, \quad \text{and we take, for example, } A_1 = 1, A_2 = 0,$$

$$E\{\varepsilon_p\} = 1 \quad \text{and} \quad \text{Var}\{\varepsilon_p\} = C_0 = 0\cdot04.$$

The identification of the other terms of the coregionalization model results in the conditions

$$E\{\varepsilon_a\} = 1 \quad \text{and} \quad \text{Var}\{\varepsilon_a\} = \alpha_a C_0 = 0\cdot06,$$

$$A_3^2 + A_4^2 = \beta_a = 2\cdot67,$$

$$A_1 A_3 + A_2 A_4 = \beta_{ap} = 1.34,$$

and we take, for example, $A_3 = 1\cdot34, A_4 = 0\cdot935$.

Simulation of the purely random variables ε   For each of the 40 401 points of the simulation zone, the subroutine RANDU provides two independent realizations t_0 and t_p of the RV T evenly distributed on the interval $[0, 1]$. The two purely random variables defining the two nugget effects are, thus, given by

$$\varepsilon_p = k_p T_p + 1 - k_p/2,$$

$$\varepsilon_a = \varepsilon_p + \alpha\varepsilon_0,$$

where

$$\varepsilon_0 = k_p T_0 - k_p/2.$$

The expectations of these two nugget effects are satisfied: $E\{\varepsilon_p\} = E\{\varepsilon_a\} = 1$ and, by identifying the variances, the parameters k_p and α can be determined:

$\text{Var}\{\varepsilon_p\} = k_p^2 \, \text{Var}\{T_p\} = k_p^2/12 = C_0, \quad \text{i.e., } k_p = 0\cdot69;$

$\text{Var}\{\varepsilon_0\} = k_p^2 \, \text{Var}\{T_0\} = C_0;$

$\text{Var}\{\varepsilon_a\} = \text{Var}\{\varepsilon_p\} + \alpha^2 \, \text{Var}\{\varepsilon_0\} = C_0 + \alpha^2 C_0 = \alpha_a C_0 = 0\cdot06, \quad \text{i.e., } \alpha = 0\cdot707.$

In addition, the nugget constant $\alpha_{ap} C_0$ of the relative cross-variogram $\gamma_{A-P_L}/m_a \cdot m_p$ can be verified:

$$\text{Var}\{\varepsilon_a \varepsilon_p\} = \text{Var}\{\varepsilon_p\} = C_0 = \alpha_{ap} \cdot C_0,$$

since $\alpha_{ap} = 1$.

Simulation of the regionalized variables z and z'   The semi-variogram $\gamma(r)$ of the RF Z consists of two nested spherical models: $\gamma(r) = C_1 \gamma_1(r) + C_2 \gamma_2(r)$, cf. model (VII.25). Thus, the required simulation can be achieved by simulating two independent RF's Y_1 and Y_2 on each of the 15 lines D and then taking the sum $Y = Y_1 + Y_2$:

$$Y_1 = k_1\Omega_1 = k_1(T_1 * \check{f}_1) \quad \text{and} \quad Y_2 = k_2\Omega_2 = k_2(T_2 * \check{f}_2).$$

The simulated range $(2R+1)u$ must approximate the real range and, thus, R is an integer equal to $a/2u$. For the first structure γ_1 with a range $a_1 = 80$ m, $R_1 = 80/20 = 4$ and the simulated range is, thus, $a_1 = (8+1)10 = 90$ m, which is a close enough approximation of the real 80-m range. Similarly, for the second structure γ_2 with a range $a_2 = 800$ m, $R_2 = 40$ and the simulated range is $a_2' = 810$ m. Note that the two ranges a_1 and a_2 can be reproduced exactly by altering the spacing of the simulated points along the lines D. In practice, the approximations (a_1, a_1') and (a_2, a_2') are largely justified, as can be seen by comparing the variogram on Fig. VII.11.

As the variables T_1 and T_2 are regularized by an odd weighting function $f(x) = x$ on the interval $[-R, +R]$, the expectations of the RF's Ω_1 and Ω_2 are zero:

$$E\{\Omega_1\} = E\{\Omega_2\} = 0, \quad \forall R,$$

which entails

$$E\{Y_1\} = E\{Y_2\} = 0.$$

The parameters k_1 and k_2 can be determined from the variances Var$\{Y_1\}$ and Var$\{Y_2\}$. For a range $a = 2R$, the regularized RF Ω can be expressed according to (VII.10) as

$$\Omega(t) = \int_{-R}^{+R} xT(t+x)\,dx$$

and, thus,

$$\text{Var}\{\Omega\} = E\{\Omega^2\} = \int_{-R}^{+R} x\,dx \int_{-R}^{+R} y \,.\, E\{T(t+x)T(t+y)\}\,dy.$$

Now,

$$E\{T(t+x)T(t+y)\} = m_T^2 + \sigma_T^2 \,.\, \delta_{xy},$$

where

$$m_T = E\{T\} = 0.5, \quad \text{Var}\{T\} = 1/12 \quad \text{and} \quad \delta_x(y) = \begin{cases} 1 & \text{if } x = y, \\ 0 & \text{if not.} \end{cases}$$

Thus, the variance becomes

$$\text{Var}\{\Omega\} = 0 + \sigma_T^2 \int_{-R}^{+R} x^2\,dx = \sigma_T^2 \,.\, \frac{2R^3}{3} = \sigma_T^2 \,.\, \frac{a^3}{12},$$

where $a = 2R$. But the simulated range is $a' = 2R+1$, and, thus,

$$\text{Var}\{\Omega\} = \frac{a'^3}{12}\sigma_T^2 = \frac{(2R+1)^3}{12}\sigma_T^2 = \frac{(2R+1)^3}{144}.$$

The variance of the RF $Y = k\Omega$ can then be deduced from this latter variance:

$$\text{Var}\{Y\} = k^2 \text{Var}\{\Omega\} = k^2\frac{(2R+1)^3}{144}.$$

The three-dimensional simulation of the RF Z is then obtained by summing the 15 independent realizations of Y on the 15 generating lines, yielding

$$\text{Var}\{Z\} = 15k^2 \frac{(2R+1)^3}{144}.$$

This last relation can be used to determine k. For $t_c = 0$ and for the structure γ_1 with a sill $C_1 = 0\cdot015$,

$$C_1 = 15k_1^2 \frac{(2R_1+1)^3}{144} \quad \text{and} \quad k_1 = 1\cdot405 \times 10^{-2}.$$

For the second structure γ_2 with a sill $C_2 = 0\cdot036$,

$$C_2 = 15k_2^2 \frac{(2R_2+1)^3}{144} \quad \text{and} \quad k_2 = 0\cdot806 \times 10^{-3}.$$

Results of the non-conditional simulations   As a verification, the semi-variograms of the non-conditionally simulated variables were calculated in the two main directions of the simulated zone. Figure VII.11(a) and (b) give the simulated semi-variograms and the imposed theoretical models for P_L and A_{Ni} for $t_c = 0$. For a step size $u = 10$ m, the first points of these simulated variograms are calculated from 40 200 pairs of simulated values.

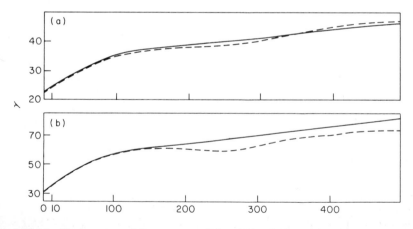

FIG. VII.11. Variograms of the non-conditional simulation. (a) P_L, $t_c = 0$; (b) A_{Ni}, $t_c = 0$. ———, theoretical model; – – – –, nonconditioned simulation.

2. *Simulation of a subhorizontal deposit with vertical drift, after M. Alfaro and Ch. Huijbregts (1974)*

The objective of this non-conditional simulation is to provide a three-dimensional data base which can be used to compare various estimation methods (cf. section VI.A.4) and various mining and blending methods (cf. section VII.B.3).

This simulated data base represents a zone D consisting of 500 blocks of dimensions $11\,\text{m} \times 11\,\text{m} \times 10\,\text{m}$, each of which consists of $11 \times 11 \times 20 = 2420$ elementary units of dimensions $1\,\text{m} \times 1\,\text{m} \times 0{\cdot}5\,\text{m}$, cf. Fig. VII.12. A vertical profile of 20 elementary data (grades, for instance) may represent a profile of data from 24 pits analysed in $0{\cdot}5$-m samples. The data base thus consists of $1\,210\,000$ elementary values, from which one can extract the vertical profile of grades of any of the $550 \times 110 = 60\,500$ pits, or the vertical profile of the true grades of the 20 samples of $0{\cdot}5$ m thickness of any of the 500 component blocks.

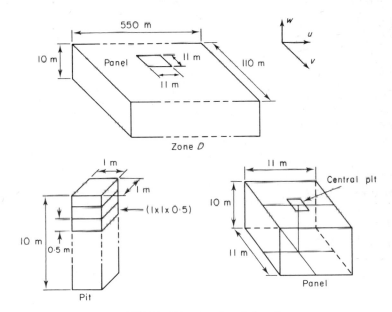

FIG. VII.12. The simulated data base.

Structural model The object in establishing the structural model is to represent a zone of a subhorizontal deposit with a vertical drift in grades, such as, for instance, a New Caledonian type of lateritic nickel deposit (cf. Fig. II.1) or a West African type of sedimentary phosphate deposit (cf. section V.B.4, case study of the Togo deposit).

Let $Z(u, v, w)$ be the grade of an elementary unit of $1\,\text{m} \times 1\,\text{m} \times 0{\cdot}5\,\text{m}$ centred on a point with three-dimensional coordinates (u, v, w):

$$Z(u, v, w) = D(u, v, w) + R(u, v, w). \qquad \text{(VII.30)}$$

The spatial variability of this grade Z is characterized by two terms. The first one – $D(u, v, w)$ – represents the vertical drift, i.e., the average shape

of a vertical profile of grades, cf. Figs II.1 and VII.13. The second term –
$R(u, v, w)$ – represents the residual fluctuations around this drift.

We impose the condition $E\{Z\} = E\{D\}$, which entails that $E\{R\} = 0$. The
realizations $d(u, v, w)$ and $r(u, v, w)$ of the two RF's, drift and residual, are
simulated independently. In this particular study, it was assumed that the
elementary grades followed a lognormal distribution, and so a Gaussian
transform was used to ensure that the final realization $z = d + r$ produced a
lognormal histogram.

Simulation of the drift The vertical drift along each profile of grades
represents a progressive enrichment followed by an impoverishment, and,
finally, a further enrichment toward the bottom of the deposit, as shown on
Fig. VII.13. The following RF was chosen to characterize this drift:

$$D(u, v, w) = K(u, v)[\lambda_3 w^{3B(u, v)} + \lambda_2 w^{2B(u, v)} + \lambda_1 w^{B(u, v)}], \qquad \text{(VII.31)}$$

where $K(u, v)$ and $B(u, v)$ are two stationary and independent RF's
defined in the horizontal plane (u, v), and λ_1, λ_2, λ_3 are three positive
constants.

The cubic function $f(w) = \lambda_3 w^3 + \lambda_2 w^2 + \lambda_1 w$ is, thus, modulated by the
two RF's $K(u, v)$ and $B(u, v)$. The multiplicative factor $K(u, v)$ defines the
maximum value $f(w_0)$ of the function $f(w)$, cf. Fig. VII.13(a) and (b). The
power $B(u, v)$ defines the vertical coordinate w_0 of this maximum, cf. Fig.
VII.13(b), (c) and (d).

The horizontal variabilities of the two independent RF's $K(u, v)$ and
$B(u, v)$ are characterized by isotropic spherical semi-variograms without
nugget effect, with identical ranges equal to $a = 100$ m and sills C_K and C_B.
Thus, two sampling pits less than 100 m apart on a horizontal plane will
have correlated drift curves $d(u, v, w)$ and $d'(u, v, w)$, while pits further
than 100 m apart will be independent. The expectations and variances (sills

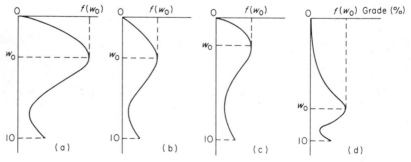

FIG. VII.13. Various realizations of the random drift. (a) $K = 2$, $B = 1$; (b) $K = 1$,
$B = 1$; (c) $K = 1$, $B = 0.8$; (d) $K = 1$, $B = 4$.

C_K and C_B) of the two RF's $K(u, v)$ and $B(u, v)$ were chosen so that the random drift $D(u, v, w)$ accounted for a fixed part of the total variability of the grade $Z = D + R$.

The positive constants λ_1, λ_2, λ_3 fix the form of the cubic $f(w) = \lambda_3 w^3 + \lambda_2 w^2 + \lambda_1 w$, i.e., the general form of the drift.

Simulation of the residuals The three-dimensional regionalization of the RF residual $R(u, v, w)$ is anisotropic with a more rapid variability in the vertical direction. The model chosen to represent this RF is the nested sum

$$R(u, v, w) = Q_0(u, v, w) + Q_1(u, v, w) + Q_2(u, v). \qquad \text{(VII.32)}$$

Q_0 is an isotropic three-dimensional nugget effect characterizing measurement errors and structures with ranges less than the smallest dimension considered, which in this case is 0.5 m.

Q_1 is a stationary RF characterizing an isotropic structure with a small range, in this case a spherical model with a range $a_1 = 3$ m.

$Q_2(u, v)$ is a stationary RF characterizing a horizontal structure with a larger range ($a_2 = 30$ m) corresponding, for example, to lenses in a sedimentary deposit. The model can be refined by imposing a geometric anisotropy on Q_2, cf. section III.B.4, Figs III.10 and III.11(b), to represent a pseudo-elliptical form of these lenses.

The three RF's Q_0, Q_1, Q_2 are independent, and their expectations are taken as zero, to ensure that the expectation of the grade is equal to that of the RF drift:

$$E\{R\} = 0 \Rightarrow E\{Z\} = E\{D\}.$$

Thus, on any given vertical grade profile, the residuals are correlated only if they are less than 3 m apart. On any given horizontal plane, the residuals are correlated only if they are less than 30 m apart. The horizontal correlation of the grades $Z = D + R$ for distances from 30 m to 100 m is due only to the correlation of the drifts.

Results The vertical profiles of the mean grades of the (11 m × 11 m × 0.5 m) slices of three blocks (full-line curves) and the grades of their respective central pits (broken-line curves) are shown on Fig. VII.14. The first two blocks ($u = 72$, $v = 6$) and ($u = 83$, $v = 6$) are contiguous and have very similar mean grade profiles, while the profiles of the grades of their central pits are notably different. The third block ($u = 424$, $v = 50$), which is more than 300 m away from the first two, has a completely different mean grade profile.

At the selection and mining stages, only the profiles of, for instance, the central pits are known in practice. If the central pit profile is used unwarily in place of the unknown block grades profile (polygonal influence estimation), then the resulting estimates of recoverable reserves will be biased, cf. the case study of section VI.A.4. Thus, a cut-off grade $z_0 = 0.03\%$ applied to the central pit profile of the second block on Fig. VII.14 would result in a recovered thickness of 2 m with a mean recovered grade of 0.036%, whereas no block slice has a mean greater than 0.03%. The aim of kriging is to provide an estimated profile as close as possible to the profile of the mean panel grades.

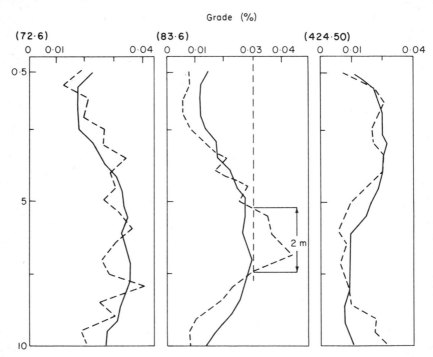

FIG. VII.14. Simulated profiles of block grades and central pit grades.

VII.B.2. Programming simulations

Basic steps in a conditional simulation

Before beginning any simulation, the model itself must be established through a structural data analysis. If the variable Z is to be simulated through its Gaussian transform $U = \varphi_Z^{-1}(Z)$, then the structural analysis is made on the variable U. The first step of the actual simulation is the non-conditional, normally distributed simulation using the turning bands method. Kriging is then used to condition the simulation to the real data values, and the Gaussian transform $Z = \varphi_Z(U)$ of this conditional simulation restores the required histogram. Finally, the simulation is verified by means of a number of statistics, such as calculations of the direct and cross-semi-variograms, histograms, scattergrams, etc.

Non-conditional Gaussian simulation

Only the *non*-conditional simulation by the turning bands method will be presented here, since the other steps in a conditional simulation all use programs that were given in preceding sections, such as GAMA (section III.C.4) for structural analysis and KRI3D (section V.B.3) for kriging. The problem of non-conditional simulation can always be reduced to one of simulating a number (NS) of independent realizations of Gaussian RF's with given covariances. This latter simulation can be broken down into the following three steps.

(i) The Gaussian variables $U_i(x)$ that are to be non-conditionally simulated are linear combinations of NS Gaussian components $V_k(x)$ which are independent of each other.

$$U_i(x) = \sum_{k=1}^{NS} a_{i,k} V_k(x), \qquad \forall i. \qquad (VII.33)$$

As all these Gaussian variables have a zero expectation and a unit variance, the coefficients $a_{i,k}$ satisfy the relations

$$\sum_{k=1}^{NS} (a_{i,k})^2 = 1, \qquad \forall i.$$

(ii) According to the dimension of the space in which $U_i(x)$ is defined and the type of zonal anisotropy that it exhibits, the components $V_k(x)$ are defined in one, two or three dimensions, cf. section III.B.5. The covariance of each of these components $V_k(x)$ is specified as an elementary covariance (only one structure); in the program SIMUL given below, this elementary covariance must be of the spherical or exponential type. Similarly, if the variables $U_i(x)$ come from a linear coregionalization model, the elementary covariances of the components $V_k(x)$ and the coefficients $a_{i,k}$ must be such that they reproduce the direct and cross covariances of the $U_i(x)$, cf. section III.B.3.

(iii) The simulation of the $U_i(x)$ is achieved by first simulating the NS components $V_k(x)$ in their respective spaces, using the turning bands method. The linear combinations (VII.33) are then used to form the required simulations of $U_i(x)$.

Program SIMUL

Program SIMUL below handles NS independent Gaussian components defined in spaces of the *same* dimension (one, two or three). Each time a space of different dimension is required, the program should be rerun. However, the program can be easily modified to accept components $V_k(x)$ that are defined in spaces of different dimension.

Each of the NS components $V_k(x)$ has an elementary covariance model that is either spherical or exponential. The nugget effect is represented by a zero range covariance, and is treated separately.

Correction of the discrete approximation

The first step is to generate the one-dimensional simulations along the 15 lines of the regular icosahedron. The one-dimensional elementary covariances on these lines are obtained by regularization of a stationary random measure, cf. formula (VII.9). This random measure is obtained in a discrete form T_i, by random drawings from a uniform distribution on the interval $[-0 \cdot 5, +0 \cdot 5]$, which has a zero expectation and a variance of one twelfth. This discrete approximation of the random measure on a grid b causes a slight bias in the one-dimensional covariance – cf. formula (VII.12) – which is corrected as follows.

First, for a spherical covariance with range a, the random variable T_i is regularized on the interval $[-a/2, +a/2]$ by the algorithm (VII.11):

$$Y_i = \sum_{-R}^{+R} k T_{i+k},$$

with $b = a/2R$. The RV Y_i has a covariance given by

$$C_d^{(1)}(s) = \tfrac{1}{144}[2R^3 - s(12R^2 + 12R + 2) + 8R^3 + 12R^2 + 4R]$$

and a variance given by

$$C_d^{(1)}(0) = R(R+1)(2R+1)/36.$$

Each of the values y_i can then be corrected by the multiplicative factor CK to obtain the imposed variance $C^{(1)}(0) = 1$, and, thus, $\overline{CK}^2 = 1/C_d^{(1)}(0)$.

Second, for an exponential covariance with parameter $\lambda = 1/a$ and, thus, a practical range of $a' = 3a$, the discrete random variable T_i is regularized on the interval $[0, 4a]$ by the algorithm (VII.11):

$$Y_i = \sum_{k=0}^{8R+1} (1 - k\lambda b) e^{-k\lambda b} . T_{i+k},$$

with $b = a/2R$.

If the bias due to limiting the regularization interval to $4a$ (instead of infinity since the true range is infinite) is disregarded, then the covariance of Y_i is

$$C_d^{(1)}(s) = \frac{1}{12A}\left(1 - \frac{\lambda b}{A}e^{-2\lambda b}\right)e^{-\lambda s}(1 - \lambda s)$$

$$+ \frac{1}{12A}e^{-2\lambda b}\frac{\lambda b\, e^{-2\lambda b}}{A}\left[-1 + \frac{\lambda b}{A}(1 - e^{-2\lambda b})\right]e^{-\lambda s},$$

where $A = 1 - e^{-2\lambda b}$. The second term, which represents the additive bias, is negligible and each value y_i can be corrected by the multiplicative factor CK to obtain the imposed variance $C^{(1)}(0) = 1$. Thus,

$$\overline{CK}^2 = 1/C_d^{(1)}(0) \simeq 12A\Big/\left(1 - \frac{\lambda b}{A}e^{-2\lambda b}\right).$$

Operations of program SIMUL

The three-dimensional network of points to be simulated (the same for all the realizations) is defined by its number of rows, columns and levels, and by the grid dimensions (parallelepipedic in rectangular coordinates).

The grid UN of the discrete approximation on lines (i.e., the thickness of the turning bands) is taken as the smallest dimension of the three-dimensional simulation grid. The number NGMAX of simulated points per line is such that the length of the simulated part of the line is greater than the main diagonal of the parallelepipedic network to be simulated.

A spherical covariance is indicated by a range entered with a positive sign; an exponential covariance is indicated by a parameter a entered with a negative sign (the practical range is, thus, $3a$). The values DX on the lines are obtained by regularization of a random measure. There are KD values of this random measure for each range. The regularization is carried out over KD = 2*NR + 1 values for the spherical model and KD = 8*NR + 1 values for the exponential model, with NR = AA/(2*UN), and NR > 20, AA being the range of the spherical model or the distance parameter a of the exponential model.

Each point of the parallelepipedic network to be simulated is defined by its three integer coordinates (row, column and level) with respect to the origin located at the centre of the network. These three coordinates correspond to the trihedron formed by the first three of the 15 lines. The four other trihedrons, corresponding to the 12 remaining lines, are deduced from the first by the rotation matrix SX.

The projections XI onto each of the 15 lines of each point of the parallelepipedic network are determined. The three-dimensional simulation is then obtained by summing the 15 one-dimensional simulated values DX at these 15 locations XI on the 15 lines.

The values simulated on the 15 lines are stored in the matrix DX (NGMAX, 15, NS).

The three-dimensional simulated values are stored in a direct access file in the following manner.

(i) Each of the NS independent simulations correspond to NP = NLI*NCO*NIV values in the file. The simulations are stored in sequence starting at address 0, so that simulation no. 1 corresponds to the first NP values, simulation no. 2 to the next NP values, etc.

(ii) Each of the simulations can be read as a matrix YS(i, j, k) with three indices, row, column and level: $i = 1$ to NLI, $j = 1$ to NCO and $k = 1$ to NIV.

```
        SUBROUTINE SIMUL(NS,A,NLI,NCO,NIV,CL,CC,CN
       1,NHAS,IUT,Y,DX,US,CS,G)
C
C ...........................................................
C          THREE-DIMENSIONAL SIMULATION
C SUBROUTINE OF NS INDEPENDENT GAUSSIAN VARIATES
C  (NORMAL 0,1) WITH EXPONENTIAL OR SPHERICAL
C SCHEMES (ISOTROPIC IN THREE DIMENSIONS)..RESULTS
C ARE STORED ON A DIRECT-ACCESS FILE..
C
C ....PARAMETERS
C   Y              WORKING VECTOR:
C          1-CONTAINS THE RANDOM VALUES TO BE
C               DILUTED ON KD TERMS.TAKE A DIMENSION
C               GREATER THAN 8*NR+1
C               NR=RANGE/SUBDIVISION ON BANDS
C               SUBDIVISION IS EQUAL TO MIN. OF CL,CC,CN
C          2-CONTAINS THE VALUES OF THE NS SIMULATIONS
C               FOR A COLUMN OF NLI TERMS:Y(NLI*NS)
C   NS  NUMBER OF INDEPENDENT SIMULATIONS
C   DX(NGMAX*15*NS)  WORKING VECTOR CONTAINING THE
C                    VALUES ON THE TURNING BANDS:
C               NGMAX IS EQUAL TO THE NUMBER OF POINTS
C               SIMULATED ON A TURNING BAND=DIAGONAL
C               OF TOTAL PARALLELEPIPED NLI,NCO,NIV
C               COUNTED IN NUMBER OF SUBDIVISIONS +6
C   A(NS)      RANGES: >0. SPHERICAL SCHEME
C                      =0. NUGGET EFFECT
C                      <0. EXPONENTIAL SCHEME
C   NLI,NCO,NIV  NUMBER OF LINES,COLUMNS,LEVELS OF
C               SIMULATED GRID
```

```
C   CL,DC,DN  SIZE OF SIMULATED GRID ALONG LINES,
C             CCLUMNS AND LEVELS
C   NFAS   INITIALISATION PARAMETEF FOR RANDCM
C          NUMBER GENERATION.TAKE ANY NUMBER
C            CIV=16**-8 IS A GIVEN CONSTANT
C   IUT  UNIT NUMBER OF OUTPUT FILE
C        THE SIMULATIONS ARE STCREC IN SUCCESSICN:
C        (((YS(I,J,L),I=1,NLI),J=1,NCC),L=1,NIV),
C        K=1,NS
C   US(NS),CS(NS*NS)  MEANS AND CCVARIANCES CF
C                     SIMULATED VALUES
C   G(NS*30)  SEMI-VARIOGRAM ALCNG COLUMNS
C
C   ....COMMONS
C   IOUT       LINE PRINTER UNIT NUMBER
C   ...............................................
C
      DIMENSION Y(1),DX(1),A(1),LS(1),CS(1),G(1)
      DIMENSION X(3),XI(3),S(3,3,4),SX(45),S1(9)
C
      CIMENSION F(401)
C
      COMMCN ICUT
      EQUIVALENCE(S(1,1,1),S1(1)),(SX(1C),S(1,1,1))
      DATA S1/0.5,-C.809017,0.309017,0.809C17,
     10.309017,-0.5,0.309017,0.5,0.809017/
C
C     SUBRCUTINE GENERATES ITS CWN RANCCM ALMBERS
C
      DATA CIV/Z39100000/,IAR/-1153374675/,
     1IBR/850000001/
C
C     COMPUTE ROTATION MATRIX SX OF THE 15 BANDS
C
      DC 1 K=1,8
    1 SX(K)=C.
      SX(1)=1.
      SX(5)=1.
      SX(9)=1.
      CO 10 K=2,4
      KO=K-1
      DO 11 I=1,3
      DO 11 J=1,3
      S(I,J,K)=0.
      DO 12 I1=1,3
   12 S(I,J,K)=S(I,J,K)+S(I,I1,KC)*S(I1,J,1)
   11 CONTINUE
   10 CONTINUE
C
C     INITIALIZE ANC COMPUTE SIMLLATION PARAMETERS
C     AND SUBDIVISION OF BANDS
C
```

```
      IH=IAR*NHAS+IBR
      LS=0
      DO 200 IS=1,NS
      US(IS)=0.
      DO 200 JS=1,NS
      LS=LS+1
  200 CS(LS)=0.
C
      UN=AMIN(DL,DC,DN)
      WRITE(IOUT,2000) UN,NLI,DL,NCO,DC,NIV,DN
      DL=DL/UN
      DC=DC/UN
      DN=DN/UN
      NGMAX=SQRT(NLI*NLI*DL*DL+NCO*NCO*DC*DC+
     1NIV*NIV*DN*DN)+6
      WRITE(IOUT,2001)NGMAX,NS
C
C     SIMULATICN ON 15 REFERENCE BANDS
C
      IP=0
      DO 3 IS=1,NS
      IF(A(IS).NE.0.)GO TO 30
C                      NUGGET EFFECT
      AA=0.
      IPAS=3
      NR=1
      EPS=0.
      PAS=UN/IPAS
      GO TO 302
C
   30 AA=ABS(A(IS))
      NR=AA/(2*UN)
      IPAS=1
      IF(NR.GE.20)GO TO 301
      NR=MAXO(1,NR)
      DO 300 IPAS=2,20
      IF(NR*IPAS.GE.20)GO TO 301
  300 CONTINUE
  301 PAS=UN/IPAS
      EPS=PAS/AA
      NR=NR*IPAS
      IF(A(IS).LT.0.)GO TO 31
C                    SPHERICAL SCHEME
  302 CK=SQRT(36./NR/(NR+1)/(2*NR+1))
      KD=2*NR+1
      DO 303 K=1,KD
  303 F(K)=K-NR-1
      IND=0
      GO TO 33
C                  EXPONENTIAL SCHEME
   31 KD=8*NR+1
      DO 32 K=1,KD
      XT=EPS*(K-1)
```

```
   32 F(K)=(1.-XT)*EXP(-XT)
      C1=1.-EXP(-2.*EPS)
      CK=SQRT(12.*C1*C1/(C1-EPS*EXP(-2.*EPS)))
      IND=1
C
   33 DO 34 ID=1,15
      CO 340 K=1,KC
      IH=IAR*IH+IBR
  340 Y(K)=IH*DIV
C
C
C         NEW POINT: DILUTICN ON KD RANDCM VALUES
C
      DO 35 J=1,NGMAX
      IP=IP+1
      AD=0.
      KK=MOD((J-1)*IPAS,KD)
      DO 350 K=1,KD
      IF(KK.EQ.KD)KK=0
      KK=KK+1
  350 AD=AD+F(K)*Y(KK)
C                      TRANSLATE RANDCM MEASLRE
      DO 351 K=1,IPAS
      IF(KK.EQ.KD)KK=0
      KK=KK+1
      IH=IAR*IH+IBR
  351 Y(KK)=IH*DIV
      AD=AD*CK
      US(IS)=US(IS)+AD
      DX(IP)=AD
      DO 352 JS=1,IS
      IJS=IS+NS*(JS-1)
      JP=IP+NGMAX*15*(JS-IS)
  352 CS(IJS)=CS(IJS)+AD*DX(JP)
   35 CONTINUE
   34 CONTINUE
      IF(INC.EQ.1)GO TO 36
      WRITE(IOUT,2002)IS,AA,PAS,EPS,IPAS,CK
      GO TO 3
   36 WRITE(IOUT,2003)IS,AA,PAS,EPS,IPAS,CK
    3 CONTINUE
      WRITE(IOUT,2004)(IS,IS=1,NS)
      DO 37 IS=1,NS
      IIS=IS+NS*(IS-1)
      CS(IIS)=(CS(IIS)-US(IS)*US(IS)/(15*NGMAX))/
     1(15*NGMAX)
      US(IS)=US(IS)/(15*NGMAX)
      DO 370 JS=1,IS
      IF(JS.EQ.IS)GO TO 370
      IJS=IS+NS*(JS-1)
      JJS=JS+NS*(JS-1)
      CS(IJS)=(CS(IJS)/(15*NGMAX)-US(IS)*US(JS))/
     1SQRT(CS(IIS)*CS(JJS))
  370 CONTINUE
```

```
      WRITE(IOUT,2005)IS,US(IS),(CS(IJS),
     1IJS=1,NS)
   37 CONTINUE
C
C     THREE-DIMENSIONAL SIMULATION ON NIV LEVELS
C             NCO CCLUMNS AND NLI LINES
C
      N1=(NLI+1)/2
      N2=(NCO+1)/2
      N3=(NIV+1)/2
      NG=NGMAX/2
      IK=0
      LS=0
      DO 38 IS=1,NS
      US(IS)=0.
      DO 380 JS=1,NS
      LS=LS+1
  380 CS(LS)=0.
      DO 38 I=1,28
      IK=IK+1
   38 G(IK)=0.
      SQ15=SCRT(15.)
C
      IAC=0
      NP=NLI*NCO+NIV
C
C     NEW POINT:PROJECT ON THE 15 BANDS AND SUM
C               UP CONTRIBUTIONS
C
      DO 4 M=1,NIV
      X(3)=-0.5+(M-N3)*DN
      DO 4 J=1,NCO
      X(2)=-0.5+(J-N2)*DC
      DO 40 I=1,NLI
      X(1)=-0.5+(I-N1)*DL
      DO 400 IS=1,NS
      I1=I+NLI*(IS-1)
  400 Y(I1)=0.
      INO=0
      DO 41 IR=1,5
      KO=(IR-1)*3
      DO 41 JO=1,3
      XI(JO)=0.
      DO 42 IL=1,3
      INO=INO+1
   42 XI(JO)=XI(JO)+X(IL)*SX(INO)
      ARG=XI(JO)+0.5
      LK=ARG
      IF(ARG.LT.0.)LK=LK-1
      LK=LK+NG
      KO=KO+1
      DO 43 IS=1,NS
```

```
      IZ=LK+NGMAX*(KO-1)+NGMAX*15*(IS-1)
      I1=I+NLI*(IS-1)
   43 Y(I1)=Y(I1)+DX(I2)
   41 CONTINUE
      DO 44 IS=1,NS
      I1=I+NLI*(IS-1)
      Y(I1)=Y(I1)/SC15
      US(IS)=US(IS)+Y(I1)
      DO 44 JS=1,IS
      I2=I+NLI*(JS-1)
      IJS=IS+NS*(JS-1)
   44 CS(IJS)=CS(IJS)+Y(I1)*Y(I2)
   40 CONTINUE
C
C     WRITE COLUMN CN DIRECT ACCESS FILE
C
      DO 45 IS=1,NS
      I1=1+NLI*(IS-1)
      I2=NLI*IS
      IAD1=IAD+NP*(IS-1)
   45 WRITE DISC IUT,IAD1,(Y(I),I=I1,I2)
      IAD=IAD+NLI
C
C     COMPUTE SEMI-VARIOGRAM
C
      KV=0
      CO 46 IN=1,2
      GO TO (461,462),IN
  461 I1=1
      I2=19
      IP=1
      GO TO 463
  462 I1=20
      I2=100
      IP=10
  463 DO 46 K=I1,I2,IP
      KL=NLI-K
      IF(KL.LE.0)GO TO 4
      KV=KV+1
      DO 464 I=1,KL
      CO 464 IS=1,NS
      J1=I+NLI*(IS-1)
      J2=J1+K
      T=Y(J1)-Y(J2)
      IK=IS+NS*(KV-1)
  464 G(IK)=G(IK)+T*T
   46 CONTINUE
    4 CONTINUE
C
C     EDIT STATISTICS AND VARIOGRAM ALCNG CCLUMNS
C
      WRITE(IOUT,2006)(IS,IS=1,NS)
```

```
      DO 5 IS=1,NS
      IIS=IS+NS*(IS-1)
      CS(IIS)=(CS(IIS)-US(IS)*US(IS)/NP)/NP
      US(IS)=US(IS)/NP
      DO 51 JS=1,IS
      IF(IS.EQ.JS)GC TO 51
      JJS=JS+NS*(JS-1)
      IJS=IS+NS*(JS-1)
      CS(IJS)=(CS(IJS)/NP-US(IS)*US(JS))/
     1SQRT(CS(IIS)*CS(JJS))
   51 CONTINUE
      WRITE(IOUT,2005)IS,US(IS),(CS(IJS),
     1IJS=IS,IIS,NS)
    5 CONTINUE
      IMP=(NS-1)/18+1
      ISM=FLOAT(NS)/FLOAT(IMP)*C.9999
      DO 52 IM=1,IMP
      IS1=1+ISM*(IM-1)
      IS2=MINO(NS,ISM*IM)
      WRITE(IOUT,2008)(IS,IS=IS1,IS2)
      DO 53 K=1,KV
      IK1=(K-1)*NS+IS1
      IK2=(K-1)*NS+IS2
      DO 54 IK=IK1,IK2
      NC=NIV*NCO*(NLI-K)
   54 G(IK)=0.5*G(IK)/NC
      D=K*UN*DL
      IF(K.GT.20)D=(K-20)*UN*DL*10
      WRITE(IOUT,2007)K,D,NC,(G(IK),IK=IK1,IK2)
   53 CONTINUE
   52 CONTINUE
C
 2000 FORMAT(1H1,'       CARACTERISTICS OF THE ',
     1'SIMULATION '/1H ,5X,14('*'),1X,'** *** ',
     210('*')//1H ,' SUBDIVISICN ON BANDS=',F7.3,
     310X,'SIZE OF GRID  : ',I4,' LINES    DL =',
     4F7.3/1H ,56X,I4,' COLUMNS  DC =',F7.3/1H ,
     556X,I4,' LEVELS   DN = ',F7.3)
 2001 FORMAT(1H0,' NUMBER OF POINTS SIMULATED/'
     1,'TURNING BAND = ',I5,' NUMBER OF INCEPENDENT'
     2,' SIMULATIONS = ',I2//)
 2002 FORMAT(1H ,'*SIMULATION*',I3,' SPHERICAL ',
     1'  SCHEME RANGE =',F8.4,' LAG =',F8.3,
     2' PRECISION =',F5.3,' LAGS/SUBD. =',I4,
     3' FACTOR CK =',F8.4)
 2003 FORMAT(1H ,'*SIMULATION*',I3,' EXPCNENTIAL ',
     1'SCHEME   RANGE =',F8.4,' LAG =',F8.3,
     2' PRECISION =',F5.3,' LAGS/SUBD. =',I4,
     3' FACTOR CK =',F8.4)
 2004 FORMAT(1H0,'  STATISTICS CF RESULTS CN BANDS'
     1,'(CORRELATION MATRIX)'/1H ,2X,10('-'),
     2' -- ------- -- -----'/1H ,'SIMULATICN'/1H ,
```

```
      3'NO AVERAGE    ',20(2X,I2,2X))
 2005 FORMAT(1H ,I2,2X,F6.3,2X,2C(1X,F5.2))
 2006 FORMAT(1H1,'     STATISTICS OF SIMULATION '/
    11H ,5X,1C('*'),1X,'** **********'//1H ,
    2'CORRELATION MATRIX'/1H ,11('-'),1X,6('-')/
    31H ,'SIMULATION'/1H ,'NO AVERAGE    ',20(2X,I2,
    42X))
 2007 FORMAT(1H ,I3,1X,F7.2,3X,I6,2X,18F6.3)
 2008 FORMAT(//1H ,'SEMI-VARIOGRAMS ALONG COLUMNS'
    1/1H ,15('-'),' ----- --------'/1H ,'LAG ',
    2'DISTANCE N.PAIRS  ',18(2X,I2,2X))
      RETURN
      END
 /*
```

VII.B.3. Typical uses of simulated deposits

A simulated deposit represents a known numerical model on a very dense grid. As the simulation can only reproduce known, modelled structures, the simulation grid is limited to the dimensions of the smallest modelled structure. Various methods of sampling, selection, mining, haulage, blending, etc., can be applied to this numerical model, to test their efficiency before applying them to the real deposit. The great advantage of such tests is that they do not require the large investment or the delays involved in setting up a pilot operation; they can be carried out for the cost of computer time, and delays are limited to the response time of the computing equipment used.

Some examples of typical uses of simulated deposits are as follows.

(i) Comparative studies of various estimation methods and various approaches to mine planning problems, including geostatistics, cf. P. Dowd and M. David (1975), A. Marechal (1975d).

(ii) Studies of the sampling level necessary for any given objective, cf. R. L. Sandefur and D. C. Grant (1976). In this particular work, the numerical model consisted of the true data from a very dense sampling grid over a given zone.

(iii) A study of the influence of the amount of information available at the time of selection on the recovery of the *in situ* resources, cf. O. Bernuy and A. Journel (1977) and section (VI.A.4).

(iv) A study of the size of the minimum selection unit on the recovery of the *in situ* resources. A study of the influence of spatial concentrations of rich and poor panels on this same recovery and on the dilution of the recovered ore. All these studies are involved in problems of choice of mining method, equipment size, type of haulage, etc., cf. J. L. Rebollo (1977), I. Clark and B. White (1976), P. Switzer, and H. Parker (1975).

(v) A study of blending from various stopes to stabilize the tonnage and quality of production (grades, hardness, mineralogy). A study of the requirement for blending stockpiles, cf. J. Deraisme (1977), whose work is partially presented in the following case study.

(vi) A study of the determination of production rates and cut-off grades, which vary in time, and their impact on recovered benefits, cf. P. Dowd (1976).

It must be understood that results obtained from simulated deposits will apply to reality only to the extent to which the simulated deposit reproduces the essential characteristics of the real deposit. Thus, the more the real deposit is known, the better its model will be, and the closer the *conditional* simulation will be to reality. As the quality of the conditional simulation improves, not only will the reproduced structures of the variability become closer to those of reality, but so also will the qualitative characteristics (geology, tectonics, morphology, etc.) that can be deterministically introduced into the numerical model. Once again, it must be stressed that simulation cannot replace a good sampling campaign of the real deposit.

Case study. Choice of a mining method from a numerical† model of a deposit, after J. Deraisme (1977)

The numerical model   The deposit model used in this study does not represent an actual deposit, but rather a type of deposit having characteristics similar to the homogeneous zone in a porphyry-copper deposit (e.g., the sulphide zone).

The zone G of homogeneous mineralization considered here has horizontal dimensions $450 \, \text{m} \times 450 \, \text{m}$ and consists of four levels of parallelepipedic blocks v with a square horizontal section $18 \, \text{m} \times 18 \, \text{m}$ and a $5 \, \text{m}$ height. The numerical model consists of the true grades of the blocks and the grades of vertical drill intersections centred on these blocks. The following characteristics were imposed on the model:

(i) mean grade $m = 1\%$ Cu;
(ii) dispersion variance of the true block grades $D^2(v/G) = 0 \cdot 235\%^2$, i.e., a relative standard deviation of 48%, $D(v/G) = 0 \cdot 48$;
(iii) a strongly asymmetrical histogram of the true block grades, as shown on Fig. VII.15;

----

† To avoid confusion, the term "simulation" will be used here only for the simulation of the various methods of mining and blending; the simulated deposit will be referred to as the numerical model.

(iv) an isotropic regionalization of the grades characterized by a spherical model with nugget effect, and a range of approximately 70 m.

Figure VII.15 also shows the histogram of the block values kriged from the central drill core intersections on a horizontal 18 m × 18 m grid. This histogram of kriged values has a mean $m^* = 1\%$ Cu and a variance $D_K^{2*}(v/G) = 0.185\%^2$; the kriging variance is $\sigma_K^2 = 0.032\%^2$.

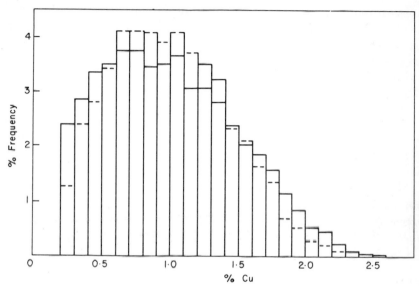

FIG. VII.15. Histograms of true and kriged block grades. ———, True grades, $m = 1\%$ Cu, $D^2(v/G) = 0.235\%^2$; – – – –, kriged grades, $m^* = 1\%$, $D_K^{2*}(v/G) = 0.185\%^2$.

Outline of problems   The minimum mining units are blocks 18 m × 18 m × 5 m, each containing 4730 tons of ore. The mining rate is fixed at 35 000 tons day^{-1} which is equivalent to eight blocks. A year's production consists of 280 days. The mill is planned to treat this daily production. The metal recovery is a function of the mill capacity to absorb fluctuations in the feed grade. Automatic process control of the milling operation is a possible solution, but it is expensive and not adapted to all scales of variability, since it is usually designed to respond to small scale (hourly), and not to large scale fluctuations (daily, fortnightly).

In this study, the period of overall feed grade regulation was considered to be two weeks. The miner predicts the quality of the ore to be sent to the mill during the next fortnight and during this period, it is the miner's responsibility to ensure that the fluctuations in the daily mean grade are not significantly different from his prediction. The short-term (e.g., hourly) fluctuations of the mill feed grade are the responsibility of the metallurgist and the process controller.

It is assumed here that grade is the primary variable affecting mill operation. Any other regionalized parameter (impurity grades, crushing indexes, etc.) can also be simulated and used in an analogous manner.

Projected mining methods   Three different mining methods were considered to meet the required degree of homogenization of daily mean grades over a fortnight's production. These methods are all *non-selective*: every block is eventually sent to the mill.

Method 1. Zone G is mined by two 10-m-high benches, cf. Fig. VII.16(a). From each bench, a big shovel of limited manoeuvrability can extract two contiguous $18 \text{ m} \times 18 \text{ m} \times 10 \text{ m}$ blocks per day. The two shovels advance parallel to the mining front and cannot reverse: they must extract all the ore along any given front ($18 \times 25 = 450 \text{ m}$ long) before beginning on the next front. All extracted ore is sent directly to the mill. Since the motion of the shovels is predetermined in this method, there is no choice for the mining engineer, who must relinquish his part in the homogenization process.

Method 1'. Mining is identical to the previous method, but some of the mined ore can be sent to a stockpile where homogenization can take place, cf. Fig. VII.16(b).

The stockpile consists of two subpiles, one rich, the other poor. The maximum capacity of the total stockpile is two day's production (70 000 tons), which is the equivalent of eight large blocks. A block will be sent to the stockpile when there is a risk that its mean grade affects the day's mean feed grade. When a block is sent to the rich subpile, the equivalent of a block is taken away from the poor subpile to meet the daily required tonnage (35 000 tons).

In addition, the existence of the stockpile provides some degree of flexibility to the mining operation, since on certain days, three blocks may be extracted from one bench and one on the other, provided that equilibrium is restored within eight days; alternatively, five blocks may be extracted within any given day, provided that this over-production is balanced by an equivalent under-production within the next eight days. But mining is expected to adhere to the norm as often as possible, i.e., daily production of two blocks per bench.

Method 2. The same zone G is mined by four 5-m-high benches, cf. Fig. VII.17. From each of these benches, a mobile shovel can extract two small

18 m × 18 m × 5 m blocks per day. The resulting daily production of 35 000 tons is sent directly to the mill.

The mobility of the shovels means that the eight blocks to be extracted can be chosen so as best to stabilize the daily mean grades over the fortnight. All the ore of any given mining front (450 m long) must be extracted before beginning on the next, and the total moving distance of each shovel is to be kept to a minimum.

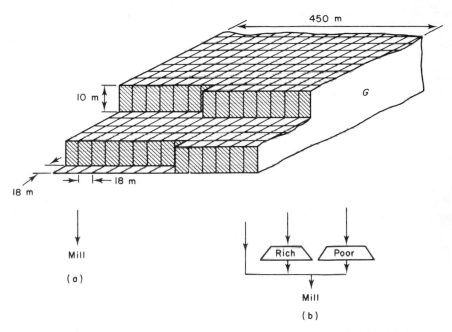

FIG. VII.16. View of the mining fronts (blocks 10 m high). (a) Method 1; (b) method 1'.

The homogenization obtained by blending the production coming from the four working levels will result in higher mining costs.

Simulation of the mining processes   Only method 1, for which the path of the big shovels is entirely determined, is completely simulated by computer. For methods 1' and 2, the daily decisions such as which block to mine, and whether to send it to the mill or to the stockpile, are made by an operating mining engineer, just as they would be in a *real situation*. The decisions were made through visualization of the state of the fronts and the stockpiles, taking the past and immediate future mill feed grades into

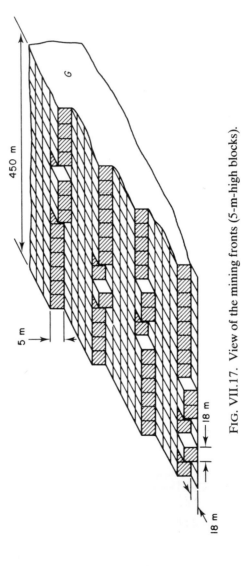

FIG. VII.17. View of the mining fronts (5-m-high blocks).

account. The decision of the mining engineer is certainly not always optimal, but the object of this study is precisely to reproduce reality (including, in particular, human errors) and not an absolute optimum which is often inaccessible in practice.

The engineer makes all his decisions on the basis of kriged estimates of the mining blocks, the blocks being kriged from the central drill core intersections. However, it is the true block grades that are delivered to the mill, cf. section VI.A.3.

As for the type of blending performed within the subpiles considered in method 1′, it is assumed that the true mean grade of the unit taken from a subpile is a normally distributed random variable, with the following characteristics.

(i) An expectation m_S equal to the true mean grade of the subpile at the time considered. This instant mean grade varies around 1·6% Cu for the rich subpile, and 0·6% Cu for the poor subpile.

(ii) A fixed relative standard deviation σ_S/m_S equal to $0·25\%^2$ for the rich subpile and $0·3\%^2$ for the poor. These values were chosen so that the confidence interval $[m_S \pm 2\sigma_S]$ for the rich subpile, for example, includes all individual true block grades sent to this particular subpile during the course of one year.

This model of the blending effect of the stockpile is very simplistic and pessimistic, but is enough to give an order of magnitude of the influence of the stockpile on the homogenization of the mill feed grades. In a real case, it would be a simple matter to replace this model either by a deterministic system or a more precise probabilistic model.

Results For the three simulated methods of mining zone G, Fig. VII.18 shows the variation in daily mean grades during the first two months (60 production days). J. Deraisme's study was carried out for a year's production (280 days) but there is not enough space here to show all the results.

The full-line curves show the variations of the true daily grades (i.e., the homogenization actually achieved), while the broken-line curves represent the variation of the estimated daily grades (i.e., the homogenization predicted on the basis of the kriged block grades).

The mill operator will obviously prefer methods 1′ and 2 to method 1, which the mining engineer may prefer because of its lower mining cost. Thus, there is a real problem involved, which can only be solved by assigning costs to the mill recovery losses caused by method 1, and to balance these costs with the ones caused by stockpiling (method 1′) or by the more flexible mining operation of method 2.

Table VII.5 gives the different dispersion variances observed at different scales, and, thus, gives an overall picture of the effect of each mining method on the fluctuations of the mill feed grades.

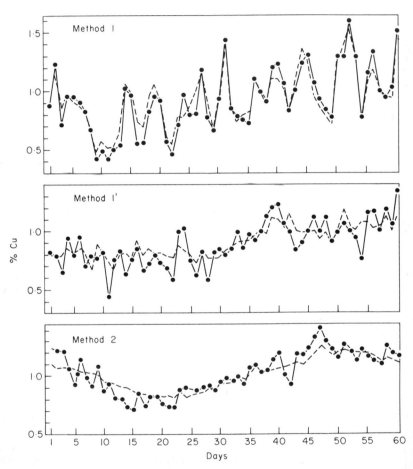

FIG. VII.18. Fluctuations of the daily mean grades at the mill-feed.•———•, True grades; – – – –. estimated grades.

The following remarks can be made about these results.

There is a considerable variation in the fluctuations of the daily grades sent to the mill, depending on the mining method used. The dispersion variance of the daily grades in 15 days for method 1 is six times greater than for method 2. These orders of magnitude of the mill feed grades variabilities at various scales can be predicted only by simulation. Standard estimation procedures (including kriging) are irrelevant for predicting these variabilities, for they take no account of the destructuring effect of mining, trucking, stockpiling, etc.

TABLE VII.5. *Dispersion variances for the three mining methods*

	Method 1	Method 1'	Method 2
Dispersion variance	0·0452	0·00115	0·0074
1 day within 7 days	0·0362	0·0040	0·0018
Dispersion variance	0·0501	0·0136	0·0082
1 day within 15 days	0·0422	0·0052	0·0016
Dispersion variance	0·0525	0·0156	0·0106
1 day within 1 month	0·0474	0·0070	0·0038
Dispersion variance	0·0816	0·0405	0·0310
1 day within 1 year			
Dispersion variance	0·0260	0·0249	0·0227
15 days within 1 year			

When there are two values, the upper one refers to the true grades while the lower refers to estimates.

The deviation between the dispersions resulting from the different mining methods increases as the time interval decreases. Thus, whatever method is used, the amount and location of the ore extracted during a given fortnight will be approximately constant, and the dispersion variances of a fortnight's mean grade over one year will also be approximately the same. On the other hand, methods 1' and 2 are particularly efficient in stabilizing the daily grades within any week or fortnight.

The dispersions of estimated grades are always less than those of true grades. This is a result of the classical smoothing property of the kriging estimator, cf. relation (VI.1) or (VI.2): the dispersion of estimators is not a good estimate of the real dispersion. For method 2, even though the homogenization of the estimated grades is nearly perfect, there is still an irreducible fluctuation of the true grades, due to the estimation variance of the daily grades, cf. Fig. VII.18.

The experimental estimation variances of the daily grades for each method are:

$$\text{method 1,} \quad \sigma_E^2 = 0.0089\%^2;$$
$$\text{method 1',} \quad \sigma_E^2 = 0.0131\%^2;$$
$$\text{method 2,} \quad \sigma_E^2 = 0.0061\%^2.$$

Note that for methods 1 and 2, the dispersion variance D^2 of the true grades can be deduced from that of the estimated grades D_K^{2*} by the approximate smoothing relation (VI.2):

$$D^2 \simeq D_K^{2*} + \sigma_E^2.$$

This relation is not applicable for method 1' because daily production does not come only from the mine but also from the stockpile.

For one year of simulated production, 17% of the production from method 1' passed through the stockpile. This represents a considerable proportion for a stockpile with a maximum capacity of two days' production. In fact, to obtain the homogenization shown on Fig. VII.18, almost 50% of production days involved departure from normal production (two blocks per bench). One solution would be to increase the capacity of the stockpile by, for example, adding a third subpile of average quality ore. In any case, it would appear that the capacity of the stockpile should not be greater than three or four days' production. Method 1' produces dispersions of the same order of magnitude as those of method 2 and since its mining cost is lower (10-m benches instead of 5 m) a feasibility study would probably result in method 1' being chosen.

Globally, there is no selection involved in any of the methods considered in this study. However, in his more complete study, J. Deraisme (1977) studied various selection criteria for method 2 and compared the results of applying these criteria on the simulations to the results that could be predicted on the basis of grade histograms alone. The differences were significant, and illustrate the well-known effects of concentration and dilution of reserves: a rich block with a grade above the cut-off will not be selected when it is located in an overall poor zone; conversely, a poor block with a grade below the cut-off will be selected when located in an overall rich zone. The prediction made on the basis of the grade histogram (cf. the various techniques outlined in section VI.A) are unable to take account of this spatial localization of rich and poor grades; more generally, they cannot take the technological conditions of the mining operation into account. Once again, a solution is afforded by the simulation of the mining operation on a numerical model of the deposit: the simulations can be used to evaluate the effect of these technological conditions on the actual recovery of the reserves.

VIII

Introduction to Non-linear Geostatistics

SUMMARY

The estimators that have been considered until now are linear combinations of the available data, and their prerequisites are limited to only the moments of order 2 of the RF $Z(x)$. If more is known about the RF $Z(x)$, for example, if its 1-, 2- or k-variate distributions can be inferred, then non-linear estimators can be built. These estimators are more accurate than the linear kriging estimators, and allow the estimation of unknown functions which are not simply linear combinations of unknown values.

The word "kriging" designates the procedures of selecting the estimator with a minimum estimation variance within a given class of possible estimators (linear or non-linear). This estimation variance can be viewed as a squared distance between the unknown value and its estimator; the process of minimization of this distance can then be seen as the projection of the unknown value onto the space within which the search for an estimator is carried out. In terms of these projections, section VIII.A introduces the various non-linear kriging estimators such as the conditional expectation or "disjunctive kriging".

Disjunctive kriging is developed in section VIII.B, together with its application to the estimation of non-linear functions, such as the so-called "transfer function" defined as the proportion of unknown values (e.g., grades of support v) greater than a given cut-off within a given domain V.

Warning At the time this book was designed (1976–77), non-linear geostatistics could not be considered as truly operational. Except for lognormal kriging, the practical applications of non-linear estimation procedures are too few to allow a fair appraisal of their efficiency and, above all, of their robustness with regard to the underlying hypotheses.

This chapter should be considered and read as an overview of recent research and, as such, it can be omitted by practitioners who wish to limit themselves to well tried and proven techniques.

WHY NON-LINEAR GEOSTATISTICS?

To evaluate an unknown value $z(x_0)$ from neighbouring data $\{z_\alpha = z(x_\alpha)$, $\alpha = 1$ to $n\}$, linear kriging (as well as the standard procedures of weighting by polygons of influence or inverse (square) distances) uses linear combinations of the available data: $z_K^* = \sum_\alpha \lambda_\alpha z_\alpha$. Now a wider choice of possible estimators can be made, for example, by considering a sum of n functions $f_\alpha(z_\alpha)$ of each available data, i.e.,

$$z_{DK}^* = \sum_\alpha f_\alpha(z_\alpha)$$

or, better still, considering a unique function of the n available data, i.e., $z_E^* = f(z_1, \ldots, z_n)$.

The linear estimator z_K^* is but a particular expression of the estimator z_{DK}^*, which, in turn, is a particular expression of the estimator z_E^*. Consequently, the best estimator of the z_{DK}^* type – best in the sense of minimum estimation variance $E\{[Z(x_0) - Z_{DK}^*]^2\}$ – is certainly better than (or equal to) the best linear estimator of the z_K^* type. Similarly, the best estimator of the z_E^* type is certainly better than (or equal to) the best estimators of the z_{DK}^* or z_K^* types. Thus, the larger the domain within which the search for a best estimator is carried out, the more accurate the corresponding estimator will be for a given amount of data. But there is no easy way; it is shown in section VIII.A that the gain in accuracy of non-linear estimators is balanced by a greater complexity, and, above all, by stronger prerequisites (which are not always met in practice).

But this gain in accuracy is not the only reason for developing researches in non-linear geostatistics. It has been shown, cf. the smoothing effect (formulae (VI.1) and (VI.2)), that one drawback of linear estimators is that they do not reproduce the spatial variability of the true values. Consequently, it is not possible to derive such features as the proportion of true values above a given cut-off from the distribution of these estimators, cf. Fig. VI.3. Two solutions have been proposed.

The first consists in building an estimator of the true grades histogram by modifying the variance of the experimental kriged grades, cf. section VI.A.2. This solution is efficient for the estimation of a global recovery over the whole deposit, or at least a wide field. But, if a study of local recoveries of units v within N blocks V is desired, with N around 100 or

1000, this solution would require the construction of N such histograms, and is, therefore, not practical.

The second solution consists of elaborating a numerical model of the true grades spatial variability, cf. Chapter VII, and evaluating the various required dispersions and recovery rates on this model. But a conditional simulation is always a costly and complex procedure.

Research has been focused on a direct procedure which would allow, for each block V (and *conditional* to the available neighbouring data), the estimation of the distribution of the true grades $\{z_v(x_i), x_i \in V\}$ of units v within V, i.e., the so-called "transfer function":

$$F_{v/V}(z_c) = \text{Prob}\,\{Z_v(x) < z_c / Z_1, \ldots, Z_n\}.$$

The problem now is to estimate, not a single value $z(x)$ or $z_v(x)$, but a whole function, and this cannot be done through a mere linear estimation procedure. It is shown in section VIII.B.3 that this function $F_{v/V}(z_c)$ can be estimated through "disjunctive kriging", i.e., by an estimator of the type

$$F_{DK}^* = \sum_{\alpha=1}^{n} f_\alpha(z_\alpha).$$

Similarly, disjunctive kriging allows estimation of the mean grade of the recovered ore for any cut-off z_c, and, thus, provides a direct solution to the problem of evaluating local recoveries.

It will be noted that the hypotheses underlying the construction of the estimator F_{DK}^* and the representativeness of the conditional simulation are of the same type, i.e., the Gaussian transform $Z = \varphi_Z(U)$ allows the reproduction not only of the univariate distribution, but also of the multivariate (at least of the bivariate) distribution of the RF $Z(x)$, cf. section VII.A.4. This point is important from a methodological point of view, since it shows the parallel nature of the two approaches of selection and recoverable reserves – through disjunctive kriging and transfer function, or through conditional simulation.

VIII.A. KRIGING IN TERMS OF PROJECTIONS

The initial kriging procedure, as introduced by D. G. Krige (1951) and formalized by G. Matheron (1962), is the process of building the best linear estimator under the hypothesis of second-order stationarity. Since then, various other estimation procedures have been introduced.

On one hand, the prerequisites were weakened by considering non-stationarity and the existence of a drift, $E\{Z(x)\} = m(x) = \sum_l a_l f^l(x)$, unknown but of known form. This gave way to the so-called "universal kriging", or "unbiased kriging of order k".

On the other hand, the prerequisites were strengthened by requiring the knowledge, not only of the covariance or variogram, but also of the 2- or k-variate distributions of the RF $Z(x)$. *Non*-linear estimators could then be considered, such as the conditional expectation and the disjunctive kriging.

The common name of "kriging" was kept, since all these estimators can be viewed as projections of the unknown value $Z(x_0)$ on sets of possible estimators. The wider the set onto which the projection is done, the nearer will be the corresponding kriging estimator to the unknown value, and, conversely, the more prerequisites will be needed.

Vectorial space and projection

Let $A = \{Z(x), x \in D\}$ be a set of random variables $Z(x)$ defined at each point x of, for example, a tridimensional deposit. Let $\mathscr{E}$ be the vector space defined by the set of all finite linear combinations of the elements of A plus the limits of such finite linear combinations (closure property):

$$\mathscr{E} = \left\{ \sum_\beta \lambda_\beta Z(x_\beta); Z(x_\beta) \in A, \lambda_\beta \text{ real} \right\}.$$

The neutral element $0 = (\lambda = 0) . Z(x)$ of $\mathscr{E}$ is the random variable almost certainly null.

The vector space $\mathscr{E}$ is provided with a scalar product equal to the *non-centred* covariance (which is not necessarily stationary):

$$\langle Z(x), Z(y) \rangle = E\{Z(x)Z(y)\} = \sigma_{xy}.$$

The norm $\|Z(x)\|$ of a vector $Z(x)$ is thus defined as the positive square root of $\langle Z(x), Z(x) \rangle$, i.e. $\|Z(x)\|^2 = \langle Z(x), Z(x) \rangle$. The distance between two elements $Z(x)$ and $Z(y)$ is defined as the norm $\|Z(x) - Z(y)\|$ of the vector $Z(x) - Z(y)$.

Let $Z(x_0) = Z_0 \in \mathscr{E}$ be any unknown variable, and let $\mathscr{E}' \subset \mathscr{E}$ be any vector subspace (or, less restrictively, any closed linear manifold). The Projection Theorem shows that there exists one and only one element $Z^* \in \mathscr{E}'$ which minimizes the distance $\|Z_0 - Z^*\|$; this unique element is then called the projection of Z_0 onto the subspace $\mathscr{E}'$, cf. Fig. VIII.1.

The kriging process is simply this projection of an unknown value onto a particular subspace $\mathscr{E}'$, within which the search for an estimator Z^* is carried out. The corresponding minimum squared distance, $\|Z_0 - Z^*\|^2 = E\{[Z_0 - Z^*]^2\}$, or minimum estimation variance, is then called "kriging variance".

There are as many kriging processes and corresponding kriging estimators Z^* as there are different sets $\mathscr{E}' \subset \mathscr{E}$ within which the projection of the unknown $Z(x_0)$ is to be made. Consider the two sets $\mathscr{E}''$ and $\mathscr{E}'$ on Fig.

VIII.2, with $\mathscr{E}'' \subset \mathscr{E}' \subset \mathscr{E}$, and the corresponding kriging estimators $Z^{*''}$ and $Z^{*'}$: as $\mathscr{E}'$ includes $\mathscr{E}''$, the projection $Z^{*'}$ is nearer to the unknown than $Z^{*''}$, i.e., in terms of estimation variance

$$\|Z(x_0) - Z^{*'}\|^2 \leqslant \|Z(x_0) - Z^{*''}\|^2.$$

Hence, as the set where the search for the estimator is carried out is wider, the estimation is better. This remark is a prelude to the classification of the various kriging estimators, cf. section VIII.A.4.

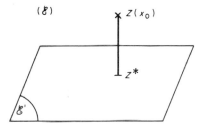

FIG. VIII.1. The kriging estimator defined as the projection of the unknown $Z(x_0)$ onto the subspace $\mathscr{E}' \subset \mathscr{E}$.

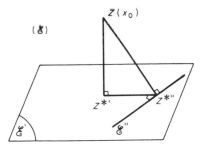

FIG. VIII.2. Projection of the unknown $Z(x_0)$ onto two subspaces $\mathscr{E}'' \subset \mathscr{E}'$.

VIII.A.1. Linear kriging processes

Let us first consider the class of linear estimators, i.e., the vector subspace $\mathscr{E}_{n+1} \subset \mathscr{E}$, of dimension $(n+1)$, generated by the linear combinations

$$\sum_{\alpha=1}^{n} \lambda_\alpha Z_\alpha + \lambda_0 . 1$$

of n particular variables $\{Z_\alpha = Z(x_\alpha), \alpha = 1 \text{ to } n\}$ called data plus the constant 1. The linear kriging processes are defined as the processes of projecting the unknown $Z(x_0)$, either onto $\mathscr{E}_{n+1}$ itself, or onto any linear manifold $C \subset \mathscr{E}_{n+1}$. The restrictions of $\mathscr{E}_{n+1}$ to various linear manifolds C guarantee the unbiasedness of the estimator Z^*, as will be shown.

Conditions of unbiasedness

Consider the expectation of any element $Z^* = \lambda_0 + \sum_\alpha \lambda_\alpha Z_\alpha \in \mathscr{E}_{n+1}$:

$$E\{Z^*\} = \lambda_0 + \sum_\alpha \lambda_\alpha E\{Z_\alpha\} = \lambda_0 + \sum_\alpha \lambda_\alpha m(x_\alpha).$$

The element Z^*, considered as an estimator of $Z(x_0)$ is unbiased if and only if

$$\lambda_0 + \sum_\alpha \lambda_\alpha m(x_\alpha) = m(x_0). \qquad \text{(VIII.1)}$$

Various cases are distinguished according to whether the expected values $m(x_0)$, $m(x_\alpha)$ are known or not, and, in the second case, whether $m(x)$ is stationary or not.

1. All the expectations are known (it does not matter then whether they are stationary or not). The unbiasedness is then characterized by the single condition (VIII.1), i.e.,

$$\lambda_0 = m(x_0) - \sum_\alpha \lambda_\alpha m(x_\alpha).$$

2. The expectation of the RF $Z(x)$ is stationary but unknown, i.e., $E\{Z(x)\} = m, \forall x$. The condition (VIII.1) is then satisfied if and only if

$$\lambda_0 = 0 \quad \text{and} \quad \sum_\alpha \lambda_\alpha = 1.$$

The first condition $\lambda_0 = 0$ amounts to restricting the set of the possible estimators to the vector subspace $\mathscr{E}_n \subset \mathscr{E}_{n+1}$ generated by the linear combinations $\sum_\alpha \lambda_\alpha Z_\alpha$ of the n data only. The second condition $\sum_\alpha \lambda_\alpha = 1$, amounts to restricting $\mathscr{E}_n$ to the linear manifold C_1 defined by the condition $\sum_\alpha \lambda_\alpha = 1$ on weights λ_α.

3. The expectation $m(x)$ is neither stationary nor known. We are then at a loss to express the unbiasedness relation (VIII.1). It is necessary to provide the form of the expectation $m(x)$, for example, $m(x)$ is an unknown linear combination of L known functions $f(x)$, cf. formula (V.9):

$$m(x) = \sum_{l=1}^{L} a_l f_l(x),$$

the L parameters a_l being unknown. The unbiasedness condition (VIII.1) can then be written as

$$\lambda_0 + \sum_l a_l \sum_\alpha \lambda_\alpha f_l(x_\alpha) = \sum_l a_l f_l(x_0).$$

This relation is satisfied, whatever the unknown parameters a_l, if and only if

$$\lambda_0 = 0,$$

which amounts to the restriction $\mathscr{E}_n \subset \mathscr{E}_{n+1}$, and

$$\sum_{\alpha=1}^n \lambda_\alpha f_l(x_\alpha) = f_l(x_0), \qquad \forall l = 1 \text{ to } L.$$

The last L conditions amount to a restriction of $\mathscr{E}_n$ to the linear manifold C_L, of dimension $(n-L)$.

Note the inclusions $C_L \subset C_{L-1} \ldots \subset C_1 \subset \mathscr{E}_n \subset \mathscr{E}_{n+1}$. The wider the set onto which the projection is made, the nearer the corresponding projected value Z^* will be to the unknown $Z(x_0)$, cf. Fig. VIII.2. Each of these projection sets gives rise to a particular kriging process.

1. Linear kriging with known expectations ("simple kriging")

The kriging estimator $Z^*_{K_0}$ is of the form

$$Z^*_{K_0} = \lambda_0 + \sum_\alpha \lambda_{K_0\alpha} Z_\alpha$$

and is the projection of the unknown $Z(x_0)$ onto the vector space $\mathscr{E}_{n+1}$, cf. Fig. VIII.3. This projection is unique and is characterized by the orthogonality of the vector $Z(x_0) - Z^*_{K_0}$ to each of the $(n+1)$ vectors generating $\mathscr{E}_{n+1}$:

$$\left.\begin{array}{l} \langle Z(x_0) - Z^*_{K_0}, 1 \rangle = 0, \\[2mm] \langle Z(x_0) - Z^*_{K_0}, Z_\alpha \rangle = 0, \qquad \forall \alpha = 1 \text{ to } n. \end{array}\right\} \qquad \text{(VIII.2)}$$

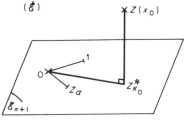

FIG. VIII.3. Simple kriging defined as the projection of the unknown $Z(x_0)$ onto $\mathscr{E}_{n+1}$.

Considering the centred covariance $\sigma'_{\alpha\beta} = \sigma_{\alpha\beta} - m(x_\alpha) \cdot m(x_\beta)$, the previous system (VIII.2) can then be expanded into

$$\begin{cases} \lambda_0 = m(x_0) - \sum_\beta \lambda_{K_0\beta} m(x_\beta), \\ \sum_{\beta=1}^{n} \lambda_{K_0\beta} \sigma'_{\alpha\beta} = \sigma'_{\alpha x_0}, \qquad \forall \alpha = 1 \text{ to } n. \end{cases}$$

Note that

$$m(x) = E\{Z(x)\} = \langle Z(x), 1 \rangle.$$

The first equation is simply the unbiasedness condition (VIII.1), the n last equations constitute a system of n linear equations with n unknowns, called the "system of simple kriging":

$$\sum_{\beta=1}^{n} \lambda_{K_0\beta} \sigma'_{\alpha\beta} = \sigma'_{\alpha x_0}, \qquad \forall \alpha = 1 \text{ to } n. \tag{VIII.3}$$

The kriging estimator is, thus, written as

$$Z^*_{K_0} = m(x_0) - \sum_\alpha \lambda_{K_0\alpha} m(x_\alpha) + \sum_\alpha \lambda_{K_0\alpha} Z_\alpha,$$

i.e.,

$$Z^*_{K_0} - m(x_0) = \sum_\alpha \lambda_{K_0\alpha} [Z_\alpha - m(x_\alpha)].$$

The centred estimator $[Z^*_{K_0} - m(x_0)]$ thus appears as a linear combination of the n centred data $[Z_\alpha - m(x_\alpha)]$; more precisely, $[Z^*_{K_0} - m(x_0)]$ is the projection of the unknown $[Z(x_0) - m(x_0)]$ onto the vector space $\mathscr{E}'_n$ generated by the linear combinations of the n centred data.

Kriging variance   The kriging variance $E\{[Z(x_0) - Z^*_{K_0}]^2\}$ is simply the minimum squared distance $\|Z(x_0) - Z^*_{K_0}\|^2$:

$$\sigma^2_{K_0} = \|Z(x_0) - Z^*_{K_0}\|^2 = \|Z(x_0) - m(x_0) - \sum_\alpha \lambda_{K_0\alpha} [Z_\alpha - m(x_\alpha)]\|^2,$$

i.e., by developing the calculations

$$\sigma^2_{K_0} = \sigma'_{x_0 x_0} - \sum_\alpha \lambda_{K_0\alpha} \sigma'_{\alpha x_0} \tag{VIII.4}$$

Prerequisites to obtain $Z^*_{K_0}$ The $(n+1)$ expected values $m(x_0)$, $m(x_\alpha)$, $\forall \alpha = 1$ to n, stationary or not, must be known, as well as the covariance function (centred or not).

2. *Linear kriging with unknown stationary expectation* ("*ordinary kriging*")

The unbiasedness condition (VIII.1) thus restricts the search for a linear estimator to the linear manifold $C_1 \subset \mathscr{E}_n \subset \mathscr{E}_{n+1}$, defined by the two conditions $\lambda_0 = 0$ and $\sum_\alpha \lambda_\alpha = 1$, on the weights. The kriging estimator is then defined as the projection of the unknown $Z(x_0)$ onto the linear manifold C_1, cf. Fig. VIII.4.

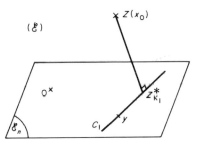

FIG. VIII.4. Ordinary kriging defined as the projection of the unknown $Z(x_0)$ onto the linear manifold C_1.

This projection, $Z_{K_1}^* = \sum_\alpha \lambda_{K_1\alpha} Z_\alpha$, is unique and is characterized by the following two conditions.

(a) $$Z_{K_1}^* \in C_1, \quad \text{i.e.} \sum_\alpha \lambda_{K_1\alpha} = 1.$$

(b) The vector $Z(x_0) - Z_{K_1}^*$ is orthogonal to each of the n vectors $Z_\alpha - Y$ generating C_1, with Z_α, $Y \in C_1$. As Y is an arbitrary element of C_1 (for example, the mean value $Y = (1/N)\sum_\alpha Z_\alpha$), the previous orthogonality is characterized by the n relations

$$\langle Z(x_0) - Z_{K_1}^*, \quad Z_\alpha - Y \rangle = 0, \quad \forall \alpha = 1 \text{ to } n.$$

By putting $\mu_1 = \langle Z(x_0) - Z_{K_1}^*, -Y \rangle$, these relations and, thus, condition (b) can be written as

$$\langle Z_{K_1}^*, Z_\alpha \rangle - \mu_1 = \langle Z(x_0), Z_\alpha \rangle, \quad \forall \alpha = 1 \text{ to } n.$$

Finally a system of $(n+1)$ linear equations with $(n+1)$ unknowns (the n weights $\lambda_{K_1\alpha}$ and the parameter μ_1) is obtained:

$$\left.\begin{aligned} &\sum_\beta \lambda_{K_1\beta} = 1, \\[2em] &\sum_\beta \lambda_{K_1\beta}\sigma_{\alpha\beta} - \mu_1 = \sigma_{\alpha x_0}, \quad \forall \alpha = 1 \text{ to } n. \end{aligned}\right\} \qquad \text{(VIII.5)}$$

This system is simply the "ordinary" kriging system (V.1) most commonly used, at least in mining practice.

Kriging variance The corresponding minimum squared distance $E\{[Z(x_0)-Z_{K_1}^*]^2\}=\|Z(x_0)-Z_{K_1}^*\|^2$, is found to be

$$\sigma_{K_1}^2 = \sigma_{x_0 x_0} - \sum_{\alpha} \lambda_{K_1\alpha} \sigma_{\alpha x_0} + \mu_1. \tag{VIII.6}$$

Note that the system (VIII.5), as well as the expression (VIII.6), can be restated by replacing the non-centred covariance by the centred covariance $\sigma'_{xy} = \sigma_{xy} - m^2$. The unknown value m^2 is eliminated from these equations, thanks to the unbiasedness condition $\sum_{\alpha} \lambda_{K_1\alpha} = 1$. Similarly, these equations can also be written by replacing the covariance function σ_{xy} by the semi-variogram function γ_{xy} defined by

$$2\gamma_{xy} = \sigma'_{xx} + \sigma'_{yy} - \sigma'_{xy} = E\{[Z(x)-Z(y)]^2\}.$$

Prerequisites to obtain $Z_{K_1}^*$ The expected value needs no longer to be known, but is assumed to be stationary. The statement of the kriging system (VIII.5) requires the knowledge of the covariance function (centred or not), or of the semi-variogram.

3. Linear kriging in the presence of a drift ("*universal kriging*")

The expectation of the RF $Z(x)$ is neither stationary nor known, but is of known form:

$$E\{Z(x)\} = m(x) = \sum_{l=1}^{L} a_l f_l(x).$$

The unbiasedness condition (VIII.1) thus restricts the search for a linear estimator to the linear manifold $C_L \subset \mathcal{E}_n$ defined by the L following conditions on the weights:

$$\sum_{\alpha} \lambda_\alpha f_l(x_\alpha) = f_l(x_0), \qquad \forall l = 1 \text{ to } L.$$

The kriging estimator $Z_{K_L}^* = \sum_{\alpha} \lambda_{K_L\alpha} Z_\alpha$, is then defined as the projection of the unknown $Z(x_0)$ onto the linear manifold C_L. This projection is characterized, cf. section V.A.2, by the system (V.12) of $(n+L)$ linear equations with $(n+L)$ unknowns, i.e.,

$$\begin{cases} \displaystyle\sum_{\beta} \lambda_{K_L\beta} f_l(x_\beta) = f_l(x_0), & \forall l = 1 \text{ to } L, \\[2mm] \displaystyle\sum_{\beta} \lambda_{K_L\beta} \sigma_{\alpha\beta} - \sum_{l} \mu_l f_l(x_\alpha) = \sigma_{\alpha x_0}, & \forall \alpha = 1 \text{ to } n. \end{cases} \tag{VIII.7}$$

This linear kriging process in the presence of a drift is called "unbiased kriging of order L" (for the unbiasedness is expressed by the L first equations of system (VIII.7)) or, for short, "universal kriging".

Kriging variance   The corresponding minimum squared distance is written, according to (V.13), as

$$\sigma_{K_L}^2 = \sigma_{x_0 x_0} - \sum_\alpha \lambda_{K_L \alpha} \sigma_{\alpha x_0} + \sum_l \mu_l f_l(x_0). \tag{VIII.8}$$

Once again, note that the relations (VIII.7) and (VIII.8) can be written in terms of the centred covariance $\sigma'_{\alpha\beta}$ or the semi-variogram $\gamma_{\alpha\beta}$.

Prerequisites to obtain $Z_{K_L}^*$.   The form of the drift $m(x)$, i.e., the L functions $f_l(x)$, is assumed to be known, and the statement of the kriging system (VIII.7) requires the knowledge of the covariance function (centred or not) or of the semi-variogram. In practice, since only one realization of the non-stationary RF $Z(x)$ is available, problems arise in the inference of the non-stationary covariance $E\{Z(x)Z(y)\}$, cf. section V.A.2, G. Matheron (1971, pp. 188), P. Delfiner, (1975).

VIII.A.2. Conditional expectation

If $\{Z_\alpha, \alpha = 1$ to $n\}$ are the n available data, the most general form for an estimator Z^* is a measurable function $f(Z_1, Z_2, \ldots, Z_n)$ of these data. The set H_n of all these measurable functions of n data is a vector space which contains, in particular, the vector subspace $\mathscr{E}_n$ generated by all the linear combinations $\sum_\alpha \lambda_\alpha Z_\alpha$ of the n data. As H_n is the widest space where the search for an estimator can be carried out, the projection of the unknown $Z(x_0)$ onto H_n is the *best possible* estimator of $Z(x_0)$ that can be deduced from the n data Z_α, cf. Fig. VIII.5. By definition, the best possible estimator is simply the conditional expectation of $Z(x_0)$, given the n data, cf. J. Neveu (1964, pp. 116), i.e.,

$$E_n Z_0 = E\{Z(x_0)/Z_1, Z_2, \ldots, Z_n\}.$$

This projection is characterized by the orthogonality of $Z(x_0) - E_n Z_0$ to any vector $Y = f(Z_1, \ldots, Z_n)$ belonging to H_n, i.e., by the relation

$$\langle Z(x_0) - E_n Z_0, Y \rangle = 0, \qquad \forall Y \in H_n. \tag{VIII.9}$$

Prerequisites for obtaining $E_n Z_0$ In order to build the projection $E_n Z_0$, i.e., to find the n-variable function $E_n Z_0 = f_K(Z_1, \ldots, Z_n) \in H_n$, it is necessary to be able to express scalar products of the type

$$\langle Z(x_0) - f_K, f \rangle = E\{[Z(x_0) - f_K(Z_1, \ldots, Z_n)] \cdot f(Z_1, \ldots, Z_n)\}, \qquad \forall f \in H_n.$$

To do this, the distribution of the $(n+1)$ variables $\{Z(x_0), Z(x_1), \ldots, Z(x_n)\}$ must be known. In the general case, when the information is limited to a single realization of the RF $Z(x)$, such an inference is not possible and the conditional expectation is inaccessible.

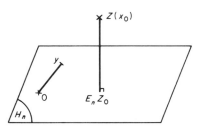

FIG. VIII.5. The conditional expectation $E_n Z_0$ defined as the projection of the unknown $Z(x_0)$ onto H_n.

Stationary and Gaussian case

In practice, it is only when the RF $Z(x)$ is stationary and multivariate Gaussian that the conditional expectation can easily be obtained. All the $(n+1)$-variate distributions are then Gaussian, and it is a classical result – cf. J. Neveu, (1964, pp. 123) – that the conditional expectation of a Gaussian stationary RF is identical to the best linear estimator, i.e., the unknown $Z(x_0)$ has the same projection onto H_n and $\mathscr{E}_{n+1}$, cf. Fig. VIII.3. This projection $Z_{K_0}^* = E_n Z_0 = \lambda_0 + \sum_\alpha \lambda_{K_0\alpha} Z_\alpha$ is then determined from the system (VIII.3) of simple kriging.

Simple kriging applied to the transformed variable

Because the stationary Gaussian case is so favourable (the best possible estimator of any unknown $Z(x_0)$ can then be obtained), it would be convenient to transform the stationary random variable $Z(x)$, drawn from any distribution, into a stationary centred Gaussian random variable $U(x)$: $Z(x) = \varphi_Z[U(x)]$ and $U = \varphi_Z^{-1}(Z)$. The transform function φ_Z can be obtained, either by a graphical comparison of the two cumulative distributions of $Z(x)$ and $U(x)$, or an expansion of Hermite polynomials, cf. section VI.B.2.

The random function $U(x)$ so obtained, has a univariate Gaussian distribution. It now remains to assert a stronger hypothesis: that all the multivariate distributions of the RF $U(x)$ are also Gaussian. Under this hypothesis, the conditional expectation $E_n U_0 = E\{U(x_0)/U(x_1), \ldots, U(x_n)\}$ is identical to the simple linear kriging estimator $U_{K_0}^* = \sum_\alpha \lambda_{K_0\alpha} U_\alpha$, provided by the system (VIII.3). Now, the conditional distribution of $U(x_0)$ given the n transformed data U_α is also Gaussian, with an expected value

$E_n U_0 \equiv U^*_{K_0}$ and a variance given by the kriging variance $\sigma^2_{K_0}$ provided by relation (VIII.4). This conditional distribution is, thus, completely determined, and it is easy to obtain the required conditional expectation of the initial variable $Z(x_0) = \varphi_Z[U(x_0)]$, i.e.,

$$E_n Z_0 = E\{\varphi_Z[U(x_0)/U(x_1), \ldots, U(x_n)]\}. \qquad \text{(VIII.10)}$$

Note that, because the transform function φ_Z is generally not linear, the conditional expectation $E_n Z_0$ is not linear with respect to the initial data $(Z_1, Z_2, \ldots, Z_n)$.

Prerequisites to obtain $E_n Z_0$ through the process of simple linear kriging applied to the transformed variable   The transform function φ_Z must be known, and this requires the stationarity of the RF $Z(x)$ and the knowledge of its univariate distribution function. Then, the linear kriging estimator $U^*_{K_0}$ must be built, which requires the knowledge of the stationary covariance function $E\{U(x)U(y)\}$; the inference of this covariance must be made from the available transformed data $U(x_\alpha)$. Ultimately, the linear kriging estimator $U^*_{K_0}$ must be assumed to be identical to the conditional expectation $E_n U_0$, which requires the assumption that all the multivariate distributions of the transformed RF $U(x)$ are Gaussian.

In practice, these prerequisites are seldom met all together, and the non-linear estimator $Z^*_0 = E\{\varphi_Z[U(x_0)/U_1, \ldots, U_n]\}$ provided by this non-linear estimation procedure differs from the true conditional expectation $E_n Z_0$. However, and under the condition that the univariate distribution of $Z(x)$ is well known, practice has shown that this non-linear estimator Z^*_0 is generally better than the estimators $Z^*_{K_0}$ or $Z^*_{K_1}$ provided by the direct linear kriging processes applied to the initial data, i.e., by projection of the unknown $Z(x_0)$ onto either $\mathscr{E}_{n+1}$ or the linear manifold $C_1 \subset \mathscr{E}_{n+1}$.

Example   If the initial RF $Z(x)$ has a lognormal multivariate distribution, the simple linear kriging process is to be applied to the RF $U(x) = \log Z(x)$, more precisely to the RF $U(x)$ given by formula (VI.8). The expression (VIII.10) thus provides, if not the true conditional expectation $E_n Z_0$, at least a non-linear estimator of the unknown value Z_0. This process, called "lognormal kriging" is developed in the addendum at the end of the present section, VIII.A.

VIII.A.3. Disjunctive kriging

Between the conditional expectation $E_n Z_0$ (best possible estimator of the unknown $Z(x_0)$, but most often inaccessible) and the various linear kriging estimators, one can look for an intermediate estimator, more accurate than the linear estimators but with less severe prerequisites than the conditional expectation. For this, a projection space included in H_n and including $\mathscr{E}_n$ must be considered. A good choice for this intermediate space is the vector space D_n generated by the sums of n *single-variable* measurable functions, i.e.,

$$D_n = \{g_1(Z_1) + g_2(Z_2) + \ldots + g_n(Z_n)\}.$$

This space satisfies the inclusions $\mathscr{E}_n \subset D_n \subset H_n$. The "disjunctive kriging process" is then, by definition, the process of projecting the unknown $Z(x_0)$ on to D_n, cf. Fig. VIII.6.

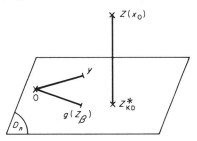

FIG. VIII.6. Disjunctive kriging defined as the projection of the unknown $Z(x_0)$ onto D_n.

This projection is characterized by the two standard conditions:

(a) $Z_{\mathrm{DK}}^* \in D_n$, i.e., $Z_{\mathrm{DK}}^* = \sum_{\alpha=1}^n f_\alpha(Z_\alpha)$;
(b) the vector $Z(x_0) - Z_{\mathrm{DK}}^*$ is orthogonal to any vector Y of D_n, i.e.,

$$\langle Z(x_0) - Z_{\mathrm{DK}}^*, Y \rangle = 0, \qquad \forall Y \in D_n.$$

Now, D_n is generated by the measurable functions $g(Z_\beta)$ depending on only *one* of the data Z_β, so that condition (b) can be written

$$\langle Z(x_0), g(Z_\beta) \rangle = \langle Z_{\mathrm{DK}}^*, g(Z_\beta) \rangle$$

for any $\beta = 1$ to n and any single-variable measurable function g. This condition is satisfied if and only if $Z(x_0)$ and Z_{DK}^* admit the same conditional expectation upon each of the Z_β, $\beta = 1$ to n, taken *separately*, i.e.,

$$\left.\begin{aligned}
E\{Z(x_0)/Z_1\} &= E\{Z_{\mathrm{DK}}^*/Z_1\}, \\
E\{Z(x_0)/Z_2\} &= E\{Z_{\mathrm{DK}}^*/Z_2\}, \\
\vdots \quad & \quad \vdots \\
E\{Z(x_0)/Z_n\} &= E\{Z_{\mathrm{DK}}^*/Z_n\}.
\end{aligned}\right\} \qquad \text{(VIII.11)}$$

(The proof, although classical, is given later on.)

Considering the expression $Z^*_{\mathrm{DK}} = \sum_\alpha f_\alpha(Z_\alpha)$ given by the first condition (a), the disjunctive kriged estimator Z^*_{KD} is finally characterized by the following system of n equations:

$$\sum_{\alpha=1}^{n} E\{f_\alpha(Z_\alpha)/Z_\beta\} = E\{Z(x_0)/Z_\beta\}, \qquad \forall \beta = 1 \text{ to } n. \quad \text{(VIII.12)}$$

This system provides the solution functions f_α, $\alpha = 1$ to n, in the form of integral equations. In section VIII.B.2, it will be shown how these solution functions can be approximated by limited expansions in Hermite polynomials.

Prerequisites to obtain Z^*_{DK} The statement of system (VIII.12) requires the knowledge of conditional expectations of the type $E\{f_\alpha(Z_\alpha)/Z_\beta\}$ and $E\{Z(x_0)/Z_\beta\}$, $\forall \beta = 1$ to n and $\forall$ the function f_α. For this, it is necessary to know all the bivariate distributions of the pairs (Z_α, Z_β) and (Z_0, Z_β).

Thus, Z^*_{DK} is not as good an estimator as the conditional expectation $E_n Z_0$. On the other hand, the disjunctive kriging process requires the knowledge of the bivariate distributions instead of the $(n+1)$-variate distribution.

Proof of system *(VIII.11)*

Consider the preceding condition (b);

$$\langle Z(x_0), g(Z_\beta) \rangle = \langle Z^*_{\mathrm{DK}}, g(Z_\beta) \rangle, \qquad \forall \beta = 1 \text{ to } n, \qquad \forall g \in D_n.$$

Let d_β be the vector space generated by all the single-variable measurable functions of the particular datum Z_β. The conditional expectation $E_\beta Z_0 = E\{Z(x_0)/Z_\beta\}$ upon the datum Z_β is, by definition, the projection of $Z(x_0)$ onto d_β, and, thus, the vector $Z(x_0) - E_\beta Z_0$ is orthogonal to any vector $g(Z_\beta) \in d_\beta$, i.e.,

$$\langle Z(x_0), g(Z_\beta) \rangle = \langle E_\beta Z_0, g(Z_\beta) \rangle, \qquad \forall g(Z_\beta) \in d_\beta.$$

Similarly, considering the projection of Z^*_{DK} onto d_β:

$$\langle Z^*_{\mathrm{DK}}, g(Z_\beta) \rangle = \langle E_\beta Z^*_{\mathrm{DK}}, g(Z_\beta) \rangle, \qquad \forall g(Z_\beta) \in d_\beta.$$

The preceding condition (b) can, thus, be written as

$$\langle E_\beta Z_0, g(Z_\beta) \rangle = \langle E_\beta Z^*_{\mathrm{DK}}, g(Z_\beta) \rangle, \qquad \forall g(Z_\beta) \in d_\beta.$$

This is satisfied if and only if

$$E_\beta Z_0 = E_\beta Z^*_{\mathrm{DK}}, \qquad \forall \beta = 1 \text{ to } n,$$

which is the system (VIII.11).

VIII.A.4. Hierarchy of the kriging estimators

When estimating an unknown value $Z(x_0)$ from n available data $\{Z(x_\alpha),$ $\alpha = 1$ to $n\}$, the search for an estimator can be carried out in various spaces of possible estimators. Once this space $\mathscr{E}'$ is defined, the best estimator, i.e., the element of $\mathscr{E}'$ nearest to the unknown $Z(x_0)$ is the projection of $Z(x_0)$ on $\mathscr{E}'$. The general term "kriging" is given to the various processes for obtaining the projections of the unknown $Z(x_0)$ on these spaces $\mathscr{E}'$ of possible estimators. The space $\mathscr{E}'$ on which the projection is to be carried out is determined by the amount of information that can be inferred from the RF $Z(x)$. The more that is known about $Z(x)$, i.e., the more severe the prerequisites that can be accepted, the wider the space $\mathscr{E}'$ of possible estimators, and, consequently, the better the corresponding kriging estimator. Thus, a hierarchy of the various kriging estimators can be established, cf. Table VIII.1.

Starting from the widest space of possible estimators, i.e., the vector space H_n of all measurable functions of the n data, one defines the conditional expectation which is the best possible estimator. Restricting the space of projection and, thus, reducing the prerequisites, disjunctive kriging, simple kriging, ordinary kriging and universal kriging are considered successively.

Addendum. Lognormal kriging

Let $Z(x)$ be a stationary RF with a multivariate lognormal distribution, an expectation m, a covariance function $C(h)$, and a variance $\sigma^2 = C(0)$. The RF $Y(x) = \log Z(x)$ thus has a multivariate normal distribution with expectation m', covariance $C'(h)$, variance $\sigma'^2 = C'(0)$. These arithmetic and logarithmic moments are related to each other by the classical formulae

$$m = e^{m' + \sigma'^2/2},$$
$$C(h) = m^2[e^{C'(h)} - 1] \Rightarrow \sigma^2 = m^2(e^{\sigma'^2} - 1). \left.\right\} \quad \text{(VIII.13)}$$

The problem consists in estimating the unknown $Z(x_0)$ from the n data $\{Z_\alpha = Z(x_\alpha), \alpha = 1$ to $n\}$.

Consider the n transformed data $\{Y_\alpha = \log Z_\alpha, \alpha = 1$ to $n\}$ and the conditional variable $Y(x_0)/Y_1, \ldots, Y_n$. The RF $Y(x)$ being multivariate normal, this conditional variable is also normal with expectation $E_n Y_0 = E\{Y(x_0)/Y_1, \ldots, Y_n\}$ and variance $E\{[Y(x_0) - E_n Y_0]^2\}$. Moreover, the conditional expectation $E_n Y_0$ is identical to the simple linear kriging estimator of $Y(x_0)$ from the n data Y_α, i.e.,

$$E_n Y_0 = Y_{K_0}^* = m' + \sum_\alpha \lambda_{K_0\alpha}[Y_\alpha - m'].$$

TABLE VIII.1. *Hierarchy of the various kriging estimators*

	Space of projection			
	Vector space H_n $\{f(Z_1, \ldots, Z_n)\}$	$\supset$ Vector space D_n $\{\sum_{\alpha=1}^{n} f_\alpha(Z_\alpha)\}$	$\supset$ Vector space $\mathscr{E}_{n+1}$ $\{\lambda_0 + \lambda^\alpha Z_\alpha\}$	$\supset$ Linear manifold C_L $\lambda^\alpha Z_\alpha + L$ conditions on λ_α
Kriging estimator	Conditional expectation $E_n Z_0$ (best possible estimator)	Disjunctive kriging estimator Z^*_{DK}	Simple kriging estimator $Z^*_{K_0}$	Universal kriging estimator $Z^*_{K_L}$
Estimation variance	$\|Z(x_0) - E_n Z_0\|^2$ $\leqslant$	σ^2_{DK} $\leqslant$	$\sigma^2_{K_0}$ $\leqslant$	$\sigma^2_{K_1} \leqslant \sigma^2_{K_2} \leqslant \cdots \leqslant \sigma^2_{K_L}$
Prerequisites	$(n+1)$-variate distributions, e.g., $Z(x)$ stationary and multivariate Gaussian	Bivariate distributions. In practice, stationarity of $Z(x)$.	The $(n+1)$ expectations. The covariance function. In practice, stationarity of $Z(x)$.	The form of the drift $m(x) = \sum_{l=1}^{L} a_l f_l(x)$. The covariance function. Stationarity of $Z(x)$ not required if $L > 1$.

Linear estimators

The n weights $\lambda_{K0\alpha}$ are determined from the simple linear kriging system (VIII.3), and the variance $E\{[Y(x_0)-E_nY_0]^2\}$ is simply the corresponding kriging variance $\sigma_{K_0}^{'2}$ given by expression (VIII.4).

Consider now the conditional variable $Z(x_0)/Z_1, \ldots, Z_n = \exp[Y(x_0)]/Z_1, \ldots, Z_n$. This variable is lognormally distributed, and its expectation E_nZ_0 is deduced from the parameters of the distribution of $Y(x_0)/Y_1, \ldots, Y_n$ through the formulae (VIII.13), i.e.,

$$E_nZ_0 = E\{Z(x_0)/Z_1, \ldots, Z_n\} = \exp[Y_{K_0}^* + \sigma_{K_0}^{'2}/2]. \quad \text{(VIII.14)}$$

The *non*-linear estimator $Z_0^* = \exp[Y_{K_0}^* + \sigma_{K_0}^{'2}/2]$ thus appears as the conditional expectation E_nZ_0 which is the best possible estimator of the unknown $Z(x_0)$ from the n available data Z_α. The corresponding estimation variance is expressed as

$$E\{[Z(x_0)-E_nZ_0]^2\} = m^2 \, e^{\sigma'^2}(1-e^{-\sigma_{K_0}^{'2}}). \quad \text{(VIII.15)}$$

(The proof is immediate by expanding the square $[Z(x_0)-E_nZ_0]^2$.)

Prerequisites to obtain E_nZ_0   The RF $Z(x)$ must be stationary and multivariate lognormal, and the expressions (VIII.14) and (VIII.15) require the knowledge of the expectation and covariance (of either $Z(x)$ or $Y(x) = \log Z(x)$).

In practice, it has been noticed that the estimator $Z_0^* = \exp[Y_{K_0}^* + \sigma_{K_0}^{'2}/2]$ does not always fulfill the non-bias condition, i.e., the arithmetic mean of the estimated values Z_0^* can differ noticeably from the expectation m estimated from the available data. A solution then consists in considering the estimator

$$Z_0^* = K_0 \exp[Y_{K_0}^* + \sigma_{K_0}^{'2}/2], \quad \text{(VIII.16)}$$

the corrective factor K_0 being determined by equating the arithmetic mean of the estimated values Z_0^* to the expectation m.

This divergence of the estimator (VIII.14) with regard to its theoretical non-bias is due to the lack of robustness of the exponential expression (VIII.14) with regard to the multivariate lognormal hypothesis. Even though the univariate distribution of $Z(x)$ can be fitted to a lognormal distribution, it does not necessarily follow that its multivariate distribution is also lognormal. And, in practice, it is not possible to test the multivariate lognormal character of $Z(x)$ from a single realization $z(x)$, particularly when one is limited to a few data points z_α.

However, experience has shown that, if the univariate distribution of $Z(x)$ is clearly lognormal, the preceding corrected estimator $Z_0^* = K_0 \exp[Y_{K_0}^* + \sigma_{K_0}^{'2}/2]$ is usually better (in the sense of a lesser experimental estimation variance) than the classical linear kriging estimators (simple or ordinary) that can be built directly from the initial data Z_α.

VIII.B. DISJUNCTIVE KRIGING AND ITS APPLICATIONS

Let us recall the results already obtained in section VIII.A.3. The disjunctive kriging estimator of an unknown $Z(x_0)$ from n available data $\{Z_\alpha, \alpha = 1 \text{ to } n\}$ is the sum of n measurable functions of each datum taken separately, i.e.,

$$Z_{DK}^* = \sum_{\alpha=1}^{n} f_\alpha(Z_\alpha).$$

The solution functions f_α are given by system (VIII.12):

$$\sum_\alpha E\{f_\alpha(Z_\alpha)/Z_\beta\} = E\{Z_0/Z_\beta\}, \qquad \forall \beta = 1 \text{ to } n.$$

Note that disjunctive kriging can be used to estimate any measurable function $\varphi(Z_0)$ of the unknown $Z_0 = Z(x_0)$. This unknown value $\varphi(Z_0)$ is projected onto the space D_n providing the estimator $\varphi_{DK}^* = \sum_\alpha f_\alpha(Z_\alpha)$ characterized by the system

$$\sum_\alpha E\{f_\alpha(Z_\alpha)/Z_\beta\} = E\{\varphi(Z_0)/Z_\beta\}, \qquad \forall \beta = 1 \text{ to } n. \qquad \text{(VIII.17)}$$

The statement of these disjunctive kriging systems requires calculation of conditional expectations of the type $E\{f(Z_\alpha)/Z_\beta\}$, i.e., the knowledge of the bivariate distributions of the RF $Z(x)$.

VIII.B.1. Bivariate Gaussian distribution

Let $Y(x)$ be a stationary RF with univariate and bivariate Gaussian distributions. This entails that:

1. all the random variables $Y(x)$, $\forall x$, have a Gaussian distribution (assumed to be centred with unit variance), i.e.,

$$G(du) = \text{Prob}\{u \leq Y(u) \leq u + du\} = \frac{1}{\sqrt{2\pi}} e^{-u^2/2} \, du;$$

2. all the bivariate distributions of $Y_i = Y(x_i)$ and $Y_j = Y(x_j)$ are Gaussian, i.e.,

$$G(du_i, du_j) = \text{Prob}\{u_i \leq Y_i \leq u_i + du_i; u_j \leq Y_j \leq u_j + du_j\}$$
$$= \emptyset(u_i, u_j) . G(du_i) . G(du_j). \qquad \text{(VIII.18)}$$

It can be shown that the density $\varnothing(u_i, u_j)$ can be expanded in a series of Hermite polynomials $\eta_k(u_i)$:

$$\varnothing(u_i, u_j) = \sum_{k=0}^{\infty} \rho_{ij}^k \eta_k(u_i)\eta_k(u_j). \qquad \text{(VIII.19)}$$

The standardized Hermite polynomials $\eta_k(u) = H_k(u)/\sqrt{k!}$ have already been introduced in section VI.B.3. The parameter ρ_{ij}^k is simply the correlation coefficient of Y_i, Y_j raised to the integer power k:$\rho_{ij}^k = [E\{Y_iY_j\}]^k$.

Properties of the Hermite polynomials

The following classical properties are recalled.
The polynomials $\eta_k(u)$ are orthogonal and standardized, i.e.,

$$\langle \eta_k(Y), \eta_{k'}(Y)\rangle = E\{\eta_k(Y)\eta_{k'}(Y)\} = \begin{cases} 1 & \text{if } k = k', \\ 0 & \text{if } k \neq k'. \end{cases}$$

Consequently, from the expressions (VIII.18) and (VIII.19) of the Gaussian bivariate distribution of Y_i, Y_j, it follows that

$$\langle \eta_k(Y_i), \eta_k(Y_j)\rangle = E\{\eta_k(Y_i)\eta_k(Y_j)\} = \rho_{ij}^k.$$

Any measurable function $f(u)$ with finite variance, i.e., such as $\int_{-\infty}^{+\infty} f^2(u)G(du) < \infty$, can be expanded in Hermite polynomials, cf. (VI.22):

$$f(u) = \sum_{k=0}^{\infty} f_k \eta_k(u),$$

the parameters f_k being given by the relations

$$f_k = \langle f(u), \eta_k(u)\rangle = \int_{-\infty}^{+\infty} f(u)\eta_k(u)G(du), \qquad \forall k.$$

The conditional expectation of any measurable function $f(u)$ has the following expansion:

$$E\{f(Y_i)/Y_j\} = \sum_{k=0}^{\infty} \rho_{ij}^k f_k \eta_k(Y_j). \qquad \text{(VIII.20)}$$

Thus under the univariate and bivariate Gaussian distribution hypothesis, the conditional expectations appearing in the disjunctive kriging systems can be calculated.

VIII.B.2. Equations of disjunctive kriging

Consider a stationary RF $Z(x)$ with any given univariate distribution and the Gaussian transform φ defining the stationary RF $Y(x)$ with a standardized univariate Gaussian distribution:

$$Z(x) = \varphi[Y(x)] \quad \text{and} \quad Y(x) = \varphi^{-1}[Z(x)].$$

It is *assumed* that the bivariate distribution of the RF $Y(x)$ is also Gaussian, i.e., of the type (VIII.18). Consider the disjunctive kriging estimation of the unknown function $Z_0 = \varphi(Y_0)$ from the n transformed data $Y_\alpha = \varphi^{-1}(Z_\alpha)$, $\alpha = 1$ to n.

First, the order K Hermite expansion of the function φ is determined according to the procedure indicated in section VI.B.3, i.e.,

$$\varphi(u) = \sum_{k=0}^{K} \varphi_k \cdot \eta_k(u).$$

Similarly, consider the order K Hermite expansion of the disjunctive kriging estimator:

$$Z_{DK}^* = \sum_{\alpha=1}^{n} f_\alpha(Y_\alpha) = \sum_{\alpha=1}^{n} \sum_{k=0}^{K} f_{\alpha,k} \cdot \eta_k(Y_\alpha).$$

The $n(K+1)$ unknown parameters $f_{\alpha,k}$ are determined by the disjunctive kriging system (VIII.17):

$$\sum_\alpha \sum_k \rho_{\alpha\beta}^k f_{\alpha,k} \eta_k(Y_\beta) = \sum_k \rho_{0\beta}^k \varphi_k \eta_k(Y_\beta), \qquad \forall \beta = 1 \text{ to } n.$$

This sytem is satisfied if and only if, for any $k = 0$ to K,

$$\sum_{\alpha=1}^{n} \rho_{\alpha\beta}^k f_{\alpha,k} = \rho_{0\beta}^k \varphi_k, \qquad \forall \beta = 1 \text{ to } n. \qquad \text{(VIII.21)}$$

Thus, for each k, the n unknown parameters $\{f_{\alpha,k}, \alpha = 1 \text{ to } n\}$ are determined by a system of n linear equations quite similar to the simple linear kriging system (VIII.3). The parameters φ_k of the transform function φ expansion are known as well as the various correlation coefficients $\rho_{\alpha\beta} = E\{Y_\alpha Y_\beta\}$ and $\rho_{0\beta} = E\{Y_0 Y_\beta\}$.

Note that, for $k = 0$, $\rho_{\alpha\beta}^0 = 1$ and the system (VIII.21) is reduced to the unique relation $\sum_\alpha f_{\alpha,0} = \varphi_0$, which is simply the non-bias condition. In fact, according to the first relation of (VI.24),

$$\sum_\alpha f_{\alpha,0} = E\{Z_{DK}^*\} = \varphi_0 = E\{\varphi(Y_0)\} = E\{Z(x_0)\}.$$

Thus, the disjunctive kriging process very simply reduces to the solving of K systems $\{k = 1 \text{ to } K\}$ of simple linear kriging type. The kriging matrix of each system is deduced from the correlation matrix $[\rho_{\alpha\beta}]$ by raising each element $\rho_{\alpha\beta}$ to the power k.

Kriging variance

Taking account of the orthogonality of the two vectors $Z_0 - Z^*_{DK}$ and Z^*_{DK}, i.e., $E\{[Z_0 - Z^*_{DK}]Z^*_{DK}\} = 0$, the corresponding estimation variance or kriging variance is written as

$$\sigma^2_{DK} = E\{[Z_0 - Z^*_{DK}]^2\} = E\{Z_0^2\} - E\{Z_0 Z^*_{DK}\}$$

with

$$E\{Z_0^2\} = \sum_k \varphi_k^2,$$

$$E\{Z_0 Z^*_{DK}\} = \sum_k \varphi_k \sum_{k'} \sum_\alpha f_{\alpha,k'} \langle \eta_k(Y_0), \eta_{k'}(Y_\alpha) \rangle$$

$$= \sum_k \varphi_k \sum_\alpha f_{\alpha,k} \rho_{0\alpha}^k.$$

It follows that

$$\sigma^2_{DK} = \sum_k \varphi_k \left[\varphi_k - \sum_\alpha f_{\alpha,k} \rho_{0\alpha}^k \right]. \tag{VIII.22}$$

Note that the expression in brackets, $[\varphi_k - \sum_\alpha f_{\alpha,k} \rho_{0\alpha}^k]$, is similar to the expression of a simple linear kriging variance, cf. expression (VIII.4).

Remark 1   If the univariate distribution of the initial RF $Z(x)$ is already Gaussian, the transform function φ is reduced to $\varphi(Z_0) = Z_0 = \eta_1(Z_0)$. Under the additional assumption that the bivariate distribution of $Z(x)$ is also Gaussian, the disjunctive kriging system (VIII.21) defining the estimator $Z^*_{DK} = \sum_\alpha \sum_k f_{\alpha,k} \eta_k(Z_\alpha)$ is written as

$$\sum_\alpha \rho_{\alpha\beta}^k f_{\alpha,k} = 0, \qquad \forall k \neq 1$$

(for this, it is enough to take $f_{\alpha,k} = 0$, $\forall \alpha$, $\forall k \neq 1$) and

$$\sum_\alpha \rho_{\alpha b} f_{\alpha,1} = -\rho_{0\beta}, \qquad \forall \beta = 1 \text{ to } n.$$

The last n equations (for $k = 1$) are the simple linear kriging system of the unknown Z_0, cf. (VIII.3).

Thus, the disjunctive kriging estimator is written as

$$Z_{DK}^* = \sum_\alpha f_{\alpha,1} \eta_1(Z_\alpha) = -\sum_\alpha f_{\alpha,1} Z_\alpha \equiv Z_{K_0}^*.$$

Hence, under the hypothesis of univariate and bivariate Gaussian distribution, the disjunctive kriging estimator is identical to the simple linear kriging estimator. If the multivariate distribution of $Z(x)$ is assumed to be Gaussian, the two estimators Z_{DK}^* and $Z_{K_0}^*$ are identical to the conditional expectation $E_n Z_0$, which is the best possible estimator.

Remark 2 The hypothesis underlying the disjunctive kriging process is that the Gaussian transformation $Y(x) = \varphi^{-1}\{Z(x)\}$ ensures the normality of not only the univariate distribution of $Y(x)$, but also its bivariate distribution. This hypothesis is of the same type as the hypothesis underlying the representativity of the conditional simulation obtained after a Gaussian transform, cf. section VII.A.4.

VIII.B.3. Disjunctive kriging of a transfer function

Let $Z(x)$ be a stationary RF and $Y(x)$ be the corresponding transformed RF with a standard normal univariate distribution

$$Z = \varphi(Y) \quad \text{and} \quad Y = \varphi^{-1}(Z).$$

It is required to estimate the proportion of values $Z(x)$ greater than any given cut-off z_c, among all the values located within a given panel V; there are n data available $\{Z(x_\alpha) = Z_\alpha, \alpha = 1 \text{ to } n\}$ both inside and outside panel V. In other words, the distribution of $Z(x)$, conditional to the n data Z_α when x varies within V, is required, i.e., the following probability:

$$P = \text{Prob}\{Z(\underline{x}) \geqslant z_c / Z_1, \ldots, Z_n\}$$

$$= \frac{1}{V} \int_V \text{Prob}\{Y(x) \geqslant y_c / Y_1, \ldots, Y_n\} \, dx \qquad \text{(VIII.23)}$$

with

$$y_c = \varphi^{-1}(z_c) \quad \text{and} \quad Y_\alpha = \varphi^{-1}(Z_\alpha).$$

The notation $Z(\underline{x})$ denotes that x is evenly distributed within V. If the variables $Z(x)$ are of point (or quasi-point) support, the previous probability P is also called the "point transfer function within V".

The direct calculation of this transfer function P would require the knowledge of the $(n + 1)$-variate distributions of $Y_1, \ldots, Y_n, Y(x), \forall x \in V$.

Disjunctive kriging provides an estimator P_{DK}^* which only requires the assumption that $Y(x)$ is bivariate normally distributed. Consider the function

$$f_{y_c}(y) = \begin{cases} 1 & \text{if } y \geq y_c, \\ 0 & \text{if } y < y_c. \end{cases}$$

The preceding transfer function (VIII.23) can be viewed as the conditional expectation $E\{f_{y_c}[Y(x)]/Y_1, \ldots, Y_n\}$, which is, by definition, the projection of the function $f_{y_c}[Y(x)]$ on the space H_n of measurable functions of the $(n+1)$ variables $Y_1, \ldots, Y_n, Y(x), \forall x \in V$, cf. Fig. VIII.7. The disjunctive kriging estimator P_{DK}^* of the function $f_{y_c}[Y(x)]$ is then the projection of this function on the vector subspace $D_n \subset H_n$ of the sum of measurable functions of each of the n data Y_α taken separately: $D_n = \{l_1(Y_1) + l_2(Y_2) + \ldots + l_n(Y_n)\}$. The classical theorem of the three perpendiculars thus shows that P_{DK}^* is the projection of $P = E\{f_{y_c}[Y(x)]/Y_1, \ldots, Y_n\}$ on D_n, cf. Fig. VIII.7.

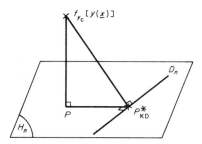

FIG. VIII.7. Disjunctive kriging of the transfer function
$P = E\{f_{y_c}[Y(x)]/Y_1, \ldots, Y_n\}$.

Consider the Kth-order Hermite expansion of function $f_{y_c}(y)$:

$$f_{y_c}(y) = \sum_{k=0}^{K} f_k \cdot \eta_k(y)$$

with

$$f_k = \langle f_{y_c}(y), \eta_k(y) \rangle = \frac{1}{\sqrt{(2\pi)}} \int_{-\infty}^{+\infty} f_{y_c}(u) \frac{H_k(u)}{\sqrt{k!}} e^{-u^2/2} \, du$$

$$= \frac{1}{\sqrt{(2\pi)}} \int_{y_c}^{+\infty} \frac{H_k(u)}{\sqrt{k!}} e^{-u^2/2} \, du.$$

Let $g(u) = (1/\sqrt{(2\pi)})e^{-u^2/2}$ be the Gaussian density and $G(y)$ the corresponding distribution function. Taking account of the definition (VI.21) of the Hermite polynomials $H_k(x)$, the parameters f_k can be expressed as

$$f_0 = \frac{1}{\sqrt{(2\pi)}} \int_{y_c}^{+\infty} e^{-u^2/2}\,du = 1 - G(y_c),$$

$$f_k = \frac{1}{\sqrt{k!}} \int_{y_c}^{+\infty} H_k(u)g(u)\,du = \frac{1}{\sqrt{k!}}[-g(y_c).H_{k-1}(y_c)], \qquad \forall k > 0.$$

Similarly, consider the Kth-order expansion of the estimator P_{DK}^*:

$$P_{DK}^* = \sum_{k=0}^{K} \sum_{\alpha=1}^{n} f_{\alpha,k}\eta_k(Y_\alpha).$$

The $n(K+1)$ unknown parameters $f_{\alpha,k}$ are determined from the disjunctive kriging systems (VIII.21), i.e., for each $k = 0$ to K,

$$\sum_{\alpha=1}^{n} \rho_{\alpha\beta}^k f_{\alpha,k} = f_k T_k(V, \beta), \qquad \forall \beta = 1 \text{ to } n,$$

where

$$T_k(V, \beta) = \frac{1}{V} \int_V [\rho_{x\beta}]^k\,dx.$$

By putting $\lambda_{\alpha,k} = f_{\alpha,k}/f_k$, these systems can be written as

$$\sum_{\alpha=1}^{n} \rho_{\alpha\beta}^k \lambda_{\alpha,k} = T_k(V, \beta), \qquad \forall \beta = 1 \text{ to } n, \qquad \text{(VIII.24)}$$

and the disjunctive kriging estimator is written as

$$P_{DK}^* = \sum_k f_k \sum_\alpha \lambda_{\alpha,k}\eta_k(Y_\alpha)$$

$$= 1 - G(y_c) + \sum_{k=1}^{K}\left[-g(y_c)\frac{H_{k-1}(y_c)}{\sqrt{k!}}\right]\sum_{\alpha=1}^{n}\lambda_{\alpha,k}\eta_k(Y_\alpha).$$

Note that this expression can be written as the integral

$$P_{DK}^* = \int_{y_c}^{-\infty} f_{DK}(u)\,du, \qquad \text{(VIII.25)}$$

where the "pseudo-density" function $f_{DK}(u)$ is

$$f_{DK}(u) = g(u)\left[1 + \sum_{k=1}^{K}\eta_k(u)\sum_{\alpha=1}^{n}\lambda_{\alpha,k}\eta_k(Y_\alpha)\right], \qquad \text{(VIII.26)}$$

the unknown parameters $\lambda_{\alpha,k}$ being determined from the systems (VIII.24).

More generally, the disjunctive kriging estimator of any measurable function $\Phi[Y(\underline{x})]$ is written as

$$\Phi^*_{DK} = \int_{-\infty}^{+\infty} \Phi(u) f_{DK}(u) \, du. \qquad (VIII.27)$$

Instead of the point transfer function, we could similarly estimate the "support v-transfer function" related to the distribution of the grades $Z_v(x)$ of unit blocks of support v within V, this distribution being conditional, either to the now available data Z_α, or to the future data Z_l, that will be available at the time of selection and mining. In the latter case, the transfer functions are said to be "indirect". The *exposé* of the calculation techniques of these indirect transfer functions would lead us beyond the scope of this introductory chapter on non-linear geostatistics; readers are referred to the basic papers of G. Matheron (1975d) and A. Maréchal (1975a) and to the forthcoming research notes of the Centre de Morphologie Mathématique, Fontainebleau.

Bibliography

This bibliography is divided into three sections.

Section A concerns the field of mathematical methods applicable to geology and, more generally, to analysis of data distributed in space. Most of the works cited here, and especially those of F. Agterberg, J. Davis and G. S. Koch, include remarkable bibliographical lists with adequate comments, to which the reader – geologist or miner – can refer.

Section B lists the theoretical and methodological basic references on geostatistics as viewed by G. Matheron.

Section C lists the main practical applications of geostatistics, covering various earth sciences fields, and essentially the mining field.

Abbreviations

CG Centre de Géostatistique, 35 rue St Honoré, 77305 Fontainebleau, France
ENSMP Ecole Nationale Supérieure des Mines de Paris, 60 Boulevard Saint-Michel, 75006 Paris, France.
* The asterisk indicates the works that, in our opinion, are a must in a somewhat complete geostatistical library.
"Geostat 75" NATO A.S.I. Congress "Geostat 75", Rome, Italy. The proceedings are edited by M. Guarascio, M. David, and Ch. Huijbregts, and published by Reidel Publishing Corporation, Dordrecht, Netherlands.

A. MATHEMATICAL METHODS

Agterberg, F. P. (1974). *Geomathematics – Mathematical Background and Geoscience Applications*, Elsevier, Amsterdam.

Angot, A. (1965). *Compléments de Mathématiques*, Revue d' Optique, Paris.

Bass, J. (1974) *Eléments de Calcul des Probabilités*, Masson, Paris.

Cailliez, F. and Pages, J. P. (1976). *Introduction à l'Analyse des Données*, Socièté Mathématiques Appliquées. Paris.

Cooley, W. (1971). *Multivariate Data Analysis*, Wiley and Sons, New York.

Davis, J. C. (1973). *Statistics and Data Analysis in Geology*, Wiley and Sons, New York.

Demidovitch, B. and Maron, I. (1973). *Eléments de Calcul Numérique*, Mir, Moscow.

Doob, J. L. (1953) *Stochastic Processes*, Wiley and Sons, New York.

Feller, W. (1966–68). *An Introduction to Probability Theory and its Applications*, Vols 1 and 2, Wiley and Sons, New York.

Formery, Ph. (1974). *Cours de Probabilités*, Vols 1 and 2, ENSMP, Paris.

Fraser, A. S. (1958) *Statistics: an Introduction*, Wiley and Sons, New York.

Gastinel, N. (1966). *Analyse Numérique Linéaire*, Hermann, Paris.

Hamming, R. W. (1971). *Introduction to Applied Numerical Analysis*, McGraw-Hill, New York.

Householder, (1953). *Principles of Numerical Analysis*, McGraw-Hill, New York.

Householder, (1964). *The Theory of Matrices in Numerical Analysis*, Blaisdell, New York.

Jenkins, G. M. and Watts, D. G. (1968). *Spectral Analysis and its Applications*, Holden-Day, San Francisco.

Koch, G. S. and Link, R. F. (1970–71). *Statistical Analysis of Geological Data*, Vols 1 and 2, Wiley and Sons, New York.

Neveu, J. (1964). *Bases Mathématiques du Calcul des Probabilités*, Masson, Paris.

Rohatgi, V. K. (1976). *An Introduction to Probability Theory and Mathematical Statistics*, Wiley and Sons, New York.

Ventsel, H. (1973). *Théorie des Probabilités*, Mir, Moscow.

Westlake, J. R. (1968). *A Handbook of Numerical Matrix Inversion and Solution of Linear Equations*, Wiley and Sons, New York.

B. GEOSTATISTICAL METHODOLOGY

Alfaro, M. (1974). *Introduccion practica a la Geoestadistica*, ETS Ingenieros de Minas, Madrid.

Alfaro, M. (1975). *Aplicaciones de la Teoria de la Variable Regionalizada*, ETS Ingenieros de Minas, Madrid.

Carlier, A. (1964). Contribution aux méthodes d'estimation des gisements d'uranium, Doctoral Thesis, Commissariat à l'Energie Atomique, Paris, Report no. 2352.

David, M. (1975). The practice of kriging, in "Geostat 75", pp. 31–48.

*David, M. (1977). *Geostatistical Ore Reserve Estimation*, Elsevier, Amsterdam.

Delfiner, P. (1973*a*). Optimum interpolation by kriging, in *Proceedings of NATO A.S.I. "Display and Analysis of Spatial Data"*, Wiley and Sons, London and New York, pp. 96–114.

*Delfiner, P. (1975). Linear estimation of non-stationary spatial phenomena, in "Geostat 75", pp. 49–68.

*Formery, Ph. and Matheron, G. (1963). Recherche d'optima dans la reconnaissance et la mise en exploitation des gisements miniers, *Annales des Mines*, **5**, 220–237; **6**, 260–277.

*Guarascio, M., David, M. and Huijbregts, Ch. (1975). *Advanced Geostatistics in the Mining Industry, Proceedings of NATO A.S.I. "Geostat 75", Rome, Oct. 75*, Reidel Publishing Corp., Dordrecht.

*Huijbregts, Ch. (1975*a*). Selection and grade–tonnage relationship, in "Geostat 75", pp. 113–135.

Huijbregts, Ch. (1975*b*). *Programmathèque de Géostatistique Minière*, CG, Fontainebleau.

Huijbregts, Ch. and Matheron, G. (1970). Universal kriging – an optimal approach to trend surface analysis, in *Decision-making in the Mineral Industry*, special volume no. 12, CIMM Montreal, pp. 159–169.

Journel, A. (1973*a*). Le formalisme des relations ressources-réserves – Simulations de gisements miniers. *Revue de l'Industrie Minérale*, Suppl. Mine no. 4, 214–226.

*Journel, A. (1974*a*). Geostatistics for conditional simulation of ore bodies, *Economic Geology*, **69**, 673–687.

Journel, A. (1974*b*) Simulations conditionnelles – théorie et pratique, Doctoral thesis, CG. Fontainebleau.

Journel, A. (1975*a*). From geological reconnaissance to exploitation – a decade of applied geostatistics, *CIMM Bulletin*, June.

Journel, A. (1975*b*). Convex analysis for mine scheduling, in "Geostat 75", pp. 185–194.

Journel, A. (1975*c*). *Guide Pratique de Géostatistique Minière*, CG, Fontainebleau.

*Journel, A. (1977). Kriging in terms of projections, *Mathematical Geology, Journal of the IAMG*, **9**, 563–586.

Jowett, G. H. (1955). Least-squares regression for trend-reduced time series, *Journal of the Royal Statistical Society B* **17**, 91–104.

Jowett, G. H. (1957). Statistical analysis using local properties of smoothly heteromorphic stochastic series, *Biometrika*, **44**, 454–463.

Krige, D. G. (1951). A statistical approach to some mine valuations and allied problems at the Witwatersrand, unpublished Master's Thesis, University of Witwatersrand.

Krige, D. G. (1970). The role of mathematical statistics in improved ore valuation techniques in South African gold mines, in *Topics in Mathematical Geology*, Consultants Bureau, New York and London.

Maréchal, A. (1970). Géostatistique et niveau de reconnaissance – applications aux gisements de bauxite métropolitains, Doctoral Thesis, CG, Fontainebleau.

*Maréchal, A. (1975*a*). The practice of transfer functions: numerical methods and their applications, in "Geostat 75", pp. 253–276.

Maréchal, A. (1976). Calcul numérique des variogrammes régularisés, Internal report N–467, CG, Fontainebleau.

*Matern, B. (1960). *Spatial Variation*. Almaenna Föerlaget, Stockholm.

*Matheron, G. (1962–1963). *Traité de Géostatistique Appliquée*, Vols 1 and 2, Technip, Paris.

Matheron, G. (1963*b*). Principles of geostatistics, *Economic Geology*, **58**, 1246–1266.

*Matheron, G. (1965). *Les Variables Régionalisées et leur Estimation*, Masson, Paris.

Matheron, G. (1967). Kriging, or polynomial interpolation procedures?, *CIMM Transactions*, **70**, 240–244.

*Matheron, G. (1968). *Osnovy Prikladnoï Geostatistiki*, Mir, Moscow.

*Matheron, G. (1969). *Le krigeage universel*, Les Cahiers du Centre de Morphologie Mathématique, Fasc. 1, CG, Fontainebleau.

*Matheron, G. (1971). *The Theory of Regionalized Variables and its Applications*, Les Cahiers du Centre de Morphologie Mathématique, Fasc. 5, CG Fontainebleau.

*Matheron, G. (1973). The intrinsic random functions and their applications, *Advances in Applied Probability*, **5**, 439–468.

Matheron, G. (1974). Effet proportional et lognormalité ou le retour du serpent de mer, Internal report N–374, CG, Fontainebleau.

*Matheron, G. (1975*a*). Le paramètrage technique des réserves, Internal report N–453, CG, Fontainebleau.

*Matheron, G. (1975*b*). Le Choix des modèles en géostatistique, in "Geostat 75", pp. 11–27.

*Matheron, G. (1975*c*). A simple substitute for conditional expectation: disjunctive kriging, in "Geostat 75", pp. 221–236.

*Matheron, G. (1975*d*). Forecasting block grade distributions: the transfer functions, in "Geostat 75", pp. 237–251.

*Matheron, G. (1977). Le choix et l'usage des modèles topoprobabilistes, Internal report, CG, Fontainebleau.

Royle, A. G. (1975). *A Practical Introduction to Geostatistics*, Mining Sciences Department, University of Leeds, Leeds.

*Serra, J. (1967*a*) Echantillonnage et estimation locale des phénomènes de transition miniers, Doctoral Thesis, IRSID and CG, Fontainebleau.

Serra, J. (1968). Les structures gigognes: morphologie mathématique et interprétation métallogénique, *Mineralium Deposita*, **3**, 135–154.

Sichel, H. S. (1951–1952). New methods in the statistical evaluation of mine sampling data, *Transactions of the Institution for Mining and Metallurgy*, March.

Sichel, H. S. (1966). The estimation of means and associate confidence limits for small samples from lognormal populations, in *Proceedings of the 1966, Symposium of the South African Institute of Mining and Metallurgy, Johannesburg*.

*Switzer, P. and Parker, H. (1975). The problem of ore versus waste discrimination for individual blocks: the lognormal model, in "Geostat 75", pp. 203–218.

C. PRACTICAL APPLICATIONS OF GEOSTATISTICS

Agterberg, F. (1970). Autocorrelation functions in geology, in *Geostatistics, a Colloquium* (D. Merriam, ed.), Plenum Press, New York, pp. 113–141.

Agterberg, F. (1975). New problems at the interface between geostatistics and geology, in "Geostat 75", pp. 403–421.

Albuisson, M. and Monget, J. M. (1975). The structure function as a criterion for data quality assessment of oceanographic time-series, *Mode Hot Line News*, no. 79.

Alfaro, M. (1976). Aplicacion de la Teoria de las Funciones Aleatorias al reconocimiento forestal, in *Trabajos de Estadistica y de Investigacion operativa*, Internal Report, ETS de Ingenieros de Minas, Madrid.

Alfaro, M. (1977). Reconnaissance hétérogène et géostatistique, in *Proceedings of the Pribam Mining Congress, Prague, Oct. 1977*, pp. 123–134.

Alfaro, M. and Huijbregts, Ch. (1974). Simulation of a subhorizontal sedimentary deposit, in *Proceedings of the 12th International APCOM Symposium, Golden, Colorado, April 1974*, pp. F65–77.

Bernuy, O. and Journel, A. (1977). Simulation d'une reconnaissance séquentielle, *Revue de l'Industrie Minérale*, Oct., 472–478.

Berry, P., Guarascio, M. and Sciotti, M. (1974). Analisi geostatistica del grado di fratturazione della rocca per le previsione del rendimiento in blocchi in cave di tufo litoïde, in *Proceedings of 1st Convegno Internationale sulla coltivazione di pietre e minerali litoïdi, Torino, Otto. 74*, pp. 1–35.

Blais, R. and Carlier, A. (1968). Applications of Geostatistics in ore valuation, ore reserve estimation and grade control, in Special Volume no. 9, CIMM, Montreal, pp. 48–61.

Brooker, P. I. (1975a). Avoiding unnecessary drilling, *Australasian IMM. Proceedings*, no. 253.

Brooker, P. I. (1975b). Optimal block estimation by kriging. *Australasian IMM. Proceedings*, no. 253.

Brooker, P. I. (1976). Block estimation at various stages of deposit development, in *Proceedings of the 14th International APCOM Symposium, Pennsylvania State University*, pp. 995–1003.

Bubenicek, L. and Haas, A. (1964). Method of calculation of the iron ore reserve in the Lorraine deposit, *A Decade of Digital Computing*, AIME special volume, AIME, New York, pp. 179–210.

Clark, I. and White, B. (1976). Geostatistical modelling of an orebody as an aid to mine planning, in *Proceedings of the 14th International APCOM Symposium, Pennsylvania State University*, pp. 1004–1012.

Chauvet, P. (1977). Example d'application de la géostatistique non stationnaire: le cokrigeage, in *Proceedings of the Pribram Mining Congress, Prague, Oct. 1977*, pp. 225–234.

Chauvet, P. and Chiles, J. P. (1975). Kriging, a method for cartography of the sea floor. *International Hydrographic Review*, **52**, 25–41.

Chiles, J. P. (1975). How to adapt kriging to non-classical problems: three case studies, in "Geostat 75", pp. 69–89.

Chiles, J. P. and Delfiner, P. (1975). Reconstitution par krigeage de la surface topographique à partir de divers schèmas d'échantillonnage photogrammétrique, *Société Française de Photogrammétrie. Bulletin*, 57, 42–50.

Chiles J. P. and Matheron, G. (1975). Interpolation optimale et cartographie, *Annales des Mines*, Nov., 19–26.

Dagbert, M. and David, M. (1977). Predicting vanishing tons before production starts, or small blocks are no good for planning in porphyry type deposits, in *Proceedings of the AIME Annual Meeting, Atlanta*.

Damay, J. (1974). L'ingènieur des mines et la recherche opérationnelle – Les applications du groupe SLN–Peñarroya dans le domaine de la décision, in *Proceedings of the 8th World Mining Congress, Lima, Peru, Oct. 74*; in *Revue Industrie Minérale*, **57** (3), 127–133.

Damay, J. (1975). Application de la Géostatistique au niveau d'un groupe minier, in "Geostat 75", pp. 313–325.

Damay, J. and Durocher, M. (1970). Essai de traitement par méthodes géostatistiques du problème de l'éstimation et de la répartition des teneurs dans les gisements en amas, Rapport interne Pennaroya, January 1970.

David, M. (1970*a*) Geostatistical ore estimation – A step by step case study, in *Decision-making in the Mineral Industry* Special volume no. 12, CIMM Montreal, 185–191.

David, M. (1970*b*). The geostatistical estimation of porphyry type deposits and scale factor problems, in *Proceedings of the Pribram Mining Congress, Prague, Oct. 1977*, pp. 91–109.

David, M. (1972*a*). Grade tonnage curve: use and misuse in ore reserve estimation, *Transactions of the Institution for Mining and Metallurgy*, **81**, 129–132.

David, M. (1973). Tools for planning variances and conditional simulations, in *Proceedings of the 11th APCOM Symposium, University of Arizona, Tucson, April 1973*, pp. D10–D23.

David, M., Dowd, P. and Korobov, S. (1974). Forecasting departure from planning in open pit design and grade control, in *Proceedings of 12th APCOM International Symposium, Golden, Colo.*, pp. F131–F153.

Delfiner, P. (1973*b*) Analyse objective du géopotentiel et du vent géostrophique par krigeage universel, Notes internes, Etablissement d'Etudes et de Recherches Météorologiques, Paris, no. 321.

Delhomme, J. P. (1976). Application de la théorie des variables régionalisées dans les sciences de l'eau, Doctoral Thesis, Centre d'Informatique Géologique, ENSMP, Paris.

Delhomme, J. P. and Delfiner, P. (1973). Application du krigeage à l'optimisation d'une campagne pluviométrique en zone aride, *Proceedings of DEWARPID Symposium, Madrid, June 1973*, pp. 191–210.

Deraisme, J. (1977). Modélisation de gisement et choix d'une méthode d'exploitation, *Revue de l'Industrie Minérale*, Oct. 483–489.

Deraisme, J. (1976). Étude géostatistique du gisement de Nickel de Mea, CMM confidential report.

Deraisme, J. and Guarascio, M. (1973). Estimation géostatistique d'un gisement de zinc, *Revue de l'Industrie Minérale*, Suppl. Mine no. 5, 310–316.

Dowd, P. (1971). Applications of Geostatistics, Internal Report, Zinc Corporation, N.B.H.C., Broken Hill, Australia.

Dowd, P. (1975). Mine planning and ore reserve estimation with the aid of a digigraphic console, *CIMM Bulletin*, **68**, 39–43.

Dowd, P. (1976). Application of dynamic and stochastic programming to optimize cut-off grades and production rates, *Transactions of the Institution for Mining and Metallurgy*, **85**, A21–A31.

Dowd, P. (1977). Geostatistical applications in the Athabasca for sands, in *Proceedings of the 15th International APCOM Symposium, Brisbane, Australia*, pp. 235–242.

Dowd, P. and David, M. (1975). Planning from estimates: sensitivity of mine production schedules to estimation methods, in "Geostat 75", pp. 163–183.

Formery, Ph. (1963). Étude geostatistique du gisement d'Ity Mont Flotuo, Internal CG report.

*François–Bongarçon, D. and Maréchal, A. (1976). A new method for open-pit design and parametrization of the final pit contour in *Proceedings of the 14th International APCOM Symposium, Pennsylvania State University, Oct. 1976*, pp. 573–583.

Guarascio, M. (1974). Valutazione dei Giacimenti Minerari: l'approccio geostatistico, *L'Industria Mineraria*, March, 1–20.

Guarascio, M. (1975). Improving the uranium deposit estimations (the Novazza case), in "Geostat 75", pp. 351–367.

Guarascio, M. and Raspa, G. (1974). Valuation and production optimization of a metal mine, in *Proceedings of the 12th International APCOM Symposium, Golden, Colorado, April 1974*, pp. F50–F64.

Guarascio, M. and Turchi, A. (1976). Exploration data management and evaluation techniques for uranium mining projects, in *Proceedings of the 14th International APCOM Symposium. Pennsylvania State University, Oct. 1976*, pp. 451–464.

Guibal, D. (1973a). Les fonctions auxiliaires à deux dimensions pour le schéma sphérique, Internal, Report N–347, CG, Fontainebleau.

Guibal, D. (1973b). L'estimation des okoumés du Gabon, Internal Report N–333, CG, Fontainebleau.

Guibal, D. (1976). Elementos de geostadistica aplicada, Apuntes del Seminario de Minero–Peru, June 1976, Internal report, Minero–Peru, Lima.

Haas, A. and Jousselin, C. (1975). Geostatistics in the petroleum industry, in "Geostat 75", pp. 333–347.

Haas, A. and Mollier, M. (1974). Un aspect du calcul d'erreur sur les réserves en place d'un gisement: l'influence du nombre et de la disposition spatiale des puits, *Revue de l'I.F.P.*, **29**, 507–527.

Haas, A. and Viallix, J. R. (1976). Krigeage applied to geophysics, *Geophysical Prospecting*, **24**, 49–69.

Huijbregts, Ch. (1971a). Courbes d'isovariance en cartographie automatique, *Sciences de la Terre*, **16**, 291–301.

Huijbregts, Ch. (1971b) Reconstitution du variogramme ponctuel à partir d'un variogramme expérimental régularisé, Internal Report N–244, CG, Fontainebleau.

Huijbregts, Ch. (1973). Regionalized variables and application to quantitative analysis of spatial data, in *Proceedings of NATO A.S.I. "Display and Analysis of Spatial Data"*, Wiley and Sons, London and New York, pp. 38–53.

Huijbregts, Ch. (1975c) Estimation of a mass proved by random diamond drillholes, in *Proceedings of the 13th APCOM Symposium, Clausthal, West Germany, Oct. 1975*, pp. A.I. 1–17.

Huijbregts, Ch. and Journel, A. (1972) Estimation of lateritic type ore bodies, in *Proceedings of the 10th International APCOM Symposium, Johannesburg, April 1972*, pp. 207–212.

Huijbregts, Ch. and Segovia, R. (1973). Geostatistics for the valuation of a copper deposit, in *Proceedings of the 11th International APCOM Symposium, University of Arizona, Tucson, April 1973*, pp. D24–D43.

Journel, A. (1969). *Etude sur l'estimation d'une variable régionalisée: application à la cartographie sous-marine*. Service Hydrographique de la Marine, Paris.

Journel, A. (1973b). Geostatistics and sequential exploration. *Mining Engineering*, Oct., 44–48.

Journel, A. (1974c). Grade fluctuations at various scales of a mine output, in *Proceedings of the 12th International APCOM Symposium, Golden, Colorado, April 1974*, pp. F78–F94.

Journel, A. (1975d). Ore grade distributions and conditional simulations – two geostatistical approaches, in "Geostat 75", pp. 195–202.

*Journel, A. and Sans, H. (1974). Ore grade control in subhorizontal deposits, *Transactions of the Institution for Mining and Metallurgy. Bulletin 812*, A74–A84.

Journel, A. and Segovia, R. (1974). Improving block estimations at El Teniente copper mine (Chile), Internal Report N–371, CG, Fontainebleau.

Krige, D. G. (1966). Two-dimensional weighted moving average trend surfaces for ore valuation, in *Proceedings of the Symposium on Mathematics, Statistics and Computer Applications in Ore Valuation, Johannesburg, 1966.* South African Institute of Mining and Metallurgy, Johannesburg.

Krige, D. G. (1973). Computer applications in investment analysis, ore valuation and planning for the Prieska copper mine. *Proceedings of the 11th International APCOM Symposium, University of Arizona, Tucson, April 1973*, pp. G31–G47.

Krige, D. G. (1975). A review of the development of geostatistics in South Africa, in "Geostat 75", pp. 279–293.

Krige, D. G. (1976). Some basic considerations in the application of geostatistics to gold ore valuation, *Journal of the South African Institute of Mining and Metallurgy*, May.

Krige, D. G. and Rendu, J. M. (1975). The fitting of contour surfaces to hanging and footwall data for an irregular orebody, in *Proceedings of the 13th International APCOM Symposium, Clausthal, West Germany, Oct. 1975*, pp. C.V. 1–12.

Lallement, B. (1975). Geostatistical valuation of a gold deposit, in *Proceedings of the 13th International APCOM Symposium, Clausthal, West Germany, Oct. 1975*, pp. I. IV. 1–15.

Le Du, R. (1976). Etude géologique et géostatistique de la minéralisation en molybdène du gisement de Mont Copper, Mémoire M.Sc.A., Ecole Polytechnique, Montreal.

*Marbeau, J. P. (1976). Geostatistique forestière, Doctoral Thesis, CG, Fontainebleau.

Maréchal, A. (1975b). Forecasting a grade-tonnage distribution for various panel sizes, in *Proceedings of the 13th International APCOM Symposium, Clausthal, West Germany, Oct. 1975*, pp. E.I. 1–18.

Maréchal, A. (1975c). Géostatistique et applications minières, *Annales des Mines*, Nov., 1–12.

Maréchal, A. (1975d). Selecting mineable blocks: experimental results observed on a simulated orebody, in "Geostat 75", pp. 137–161.

Maréchal, A. and Roullier, J. P. (1970). Etude géostatistique des gisements de bauxite Français, *Revue de l'Industrie Minérale*, **52**, 492–507.

Maréchal, A. and Shrivastava, P. (1977). Geostatistical study of a lower proterozoïc iron orebody in the Pilbara region of Western Australia. *Proceedings of the 15th International APCOM Symposium, Brisbane, Australia, June 1977*, pp. 221–230.

Matheron, G. (1955). Application des Méthodes statistiques à l'estimation des gisements, *Annales de Mines*, Dec., 50–75.

Matheron, G. (1959). Remarques sur la loi de Lasky, *Chronique des Mines d'Outre-Mer*, Dec.

Matheron, G. (1970). Random functions, and their applications in geology, *"Geostatistics – A Colloquium"* (D. Merriam, ed.), Plenum Press, New York, pp. 79–87.

Minguet, J. M. and Rebollo–Alcantara, J. L. (1973). *Aplicacion de los Ordenadores a la Mineria de Cielo Abierto*, ETS de Ingenieros de Minas, Madrid.

Monget, J. M. (1971). A new statistical treatment of gravity data, *Journal of the International Association of Geodesy*, 102, 451–466.

Monget, J. M. (1972). The variogram as an aid to study the structure of two-dimensional field data, in *Proceedings of the International Congress of Geology, Montreal., 1972.*

Newton, M. J. (1973). The application of geostatistics to mine sampling patterns, in *Proceedings of the 11th International APCOM Symposium, University of Arizona, Tucson, April 1973*, pp. D44–D58.

Olea, R. (1972). Application of regionalized variable theory to automatic contouring, Special report to the American Petroleum Institute, Research Project 131.

Olea, R. (1972). Application of regionalized variable theory to automatic contouring, Special report to the American Petroleum Institute, Research Project 131.

Orfeuil, J. P. (1975). Géostatistique et analyse des données, in "Geostat 75", pp. 423–434.

Parker, H. M. (1975). The geostatistical evaluation of ore reserves, using conditional probability distribution. A case study for the Area 5 project, Warren, Maine, – Ph.D. Thesis, Stanford University, California, USA.

Parker, H. M. and Sandefur, R. L. (1976). A review of recent developments in geostatistics, in *Proceedings of the AIME Annual Meeting, Las Vegas, Nevada.*

Parker, H. M. and Switzer, P. (1975). Use of conditional probability distributions in ore reserve estimation, a case study, in *Proceedings of the 13th International APCOM Symposium, Clausthal, West Germany, Oct. 1975*, pp. M. II. 1–16.

Rebollo-Alcantara, J. L. (1977). Etude de l'exploitation du gisement de Fuentes Rosas à l'aide de la simulation non-conditionelle de blocs, *Revue de l'Industrie Minèrale*, Oct., 479–482.

Regazzaci, J. D. and Royer, J. J. (1975). Analyse structurale et modèle statistique des vents à 200 mb en région tropicale. Note interne, Service Météorologique National, Paris, Sept.

Rendu, J. M. (1970a). Some applications of geostatistics to decision-making in exploration, in *Decision-making in the Mineral Industry*, special. volume no. 12, CIMM, Montreal, pp. 175–184.

Rendu, J. M. (1970b). Geostatistical approach to ore reserve calculation, *Engineering and Mining Journal.*

Rendu, J. M. (1975). Bayesian decision theory applied to mineral exploration and mine valuation, in "Geostat 75", pp. 435–445.

Rendu, J. M. (1976). An attempt to solve the problem of optimization of sample spacing in the South African gold mines, *Journal of the South African Institute of Mining and Metallurgy.*

Royle, A. G. (1977). How to use geostatistics for ore reserve classification, *World Mining*, Feb. 52–56.

Royle, A. G., Newton, M. J. and Sarin, H. K. (1972). Geostatistical factors in design of mine sampling programmes, *Transactions of the Institution for Mining and Metallurgy*, **81**, A81–A88.

Royle, A. G. and Hosgit, E. (1974). Local estimation of sand and gravel reserves by geostatistical methods, *Transactions of the Institution for Mining and Metallurgy*. **83**, A53–A62.

Rutledge, R. W. (1973). Geostatistics, an introduction, in *Proceedings of the ANZAAS Meeting, August 1973, Perth, Australia*.

Rutledge, R. W. (1975). The potential of geostatistics in the development of mining, in "Geostat 75", pp. 295–311.

Sabourin, R. (1975). Geostatistical evaluation of the sulphur contents in Lingan coal mine, Cape Breton, in *Proceedings of the 13th International APCOM Symposium, Clausthal, West Germany, Oct. 1975*, pp. I. II. 1–16.

Sandefur, R. L. and Grant, D. C. (1976). Preliminary evaluation of uranium deposits. A geostatistical study of drilling density in Wyoming solution fronts, in *Proceedings of the 1976 IAEA Meeting, Vienna, Austria*.

Sichel, H. S. (1972). Statistical valuation of diamondiferous deposits, in *Proceedings of the 10th APCOM Symposium, Johannesburg, 1972* pp. 17–25.

Sinclair, A. J. and Deraisme, J. (1974). A geostatistical study of the Eagle copper vein, N. British Columbia, *CIMM Bulletin*, June, 1–12.

Sinclair, A. J. and Deraisme, J. (1975). A two-dimensional statistical study of a skarn deposit, Yukon Territory, Canada, in "Geostat 75", pp. 369–379.

Serra, J. (1967*b*). Un critère nouveau de découvertes de structures: le variogramme, *Sciences de la Terre*, **12**, 275–299.

Serra, J. et Al. (1974). Laws of linear homogenization in ore stockyards. *Proceedings of the 11th International mineral Processing Congress, Cagliari, Italy, April 1975*; Report N–387, CG, Fontainebleau.

Tulcanaza, E. (1972). Personal communication.

Ugarte, I. (1972). Ejemplos de modelos de estimacion a corto y largo plazo, *Boletin de Geoestadistica*, **4**, 3–22.

Walton, J. (1972). Resumen de un estudio de estimacion de bloque grande de block-caving, Boletin de Geostadistica, **4**, 43–54. Universidad de Chile.

Zwicky, R. W. (1975). Preliminary geostatistical investigations of tar-bearing sands, in *Proceedings of the 13th APCOM Symposium, Clausthal, West Germany, Oct. 1975*, pp. I. III. 1–15.

Index A.
Geostatistical Concepts

Index B. Quoted Deposits

Index C.
FORTRAN Programs

Index D. Notations

The notations appear as they do chronologically in the text.

(Ou, Ov, Ow)	indicates the reference system of rectangular coordinates.
x, or x'	indicate a point of coordinates x or x_u, (x_u, x_v), (x_u, x_v, x_w) according to whether a one-, two- or three-dimensional space is considered.
h, or h'	indicate a vector of modulus $|h| = r$, and three-dimensional coordinates (h_u, h_v, h_w).
α	generally indicates a direction in space, but can also be a data index.
R^1, R^2, R^3	indicate one-, two- or three-dimensional space, respectively. Thus $x \in R^3$ indicates a point x taken in the three-dimensional space.
ReV	abbreviation for regionalized variable.
RV	abbreviation for random variable.
RF	abbreviation for random function.
$z(x), y(x), a(x), \ldots$	in lower-case letters, indicate regionalized variables or realizations of random functions.
$Z(x), Y(x), A(x), \ldots$	in upper-case letters, indicate the corresponding random variables or functions.
$E\{Z(x)\} = m(x)$	indicates the expectation of the RF $Z(x)$.
$\mathrm{Var}\{Z(x)\}$	indicates the *a priori* variance of the RF $Z(x)$.
$C(h)$	indicates a stationary covariance function.
$\gamma(h)$	indicates a stationary semi-variogram function.
$p(h)$	indicates a stationary correlogram function.
λ_i, λ_j, or $\lambda_\alpha, \lambda_\beta$	indicates the weights of linear combinations.
a	generally indicates the range of a variogram model.
C, or K	generally indicate the sill of a variogram model.
$C_{kk'}(h)$	indicates a stationary cross-covariance.
$\gamma_{kk'}(h)$	indicates a stationary cross-semi-variogram.
$\rho_{kk'}$	indicates the correlation coefficient between two RV Z_k and $Z_{k'}$.
V, or $V(x)$	indicate a domain centred at point x
v, or $v(x)$	indicate a domain (generally of smaller dimensions than V) centred at point x. v could also indicate the second coordinate of the reference axes system (Ou, Ov, Ow).
$F(u)$	indicates the cumulative distribution function $F(u) = \mathrm{Prob}\{Z < u\}$, and, more often,

597

$F(l)$ indicates an auxiliary function.

$z^*, y^*, t^*, \ldots$ indicate the estimates of the unknown values z, y, t.

$Z^*, Y^*, T^*, \ldots$ in upper-case letters, indicate the random variables or functions, realizations of which are z^*, y^*, t^*.

$\int dx$ without limit on the integral sign, indicates the integration over the whole space of definition of x, e.g., if x is a point with three-dimensional coordinates (x_u, x_v, x_w), then

$$\int dx = \int_{-\infty}^{+\infty} dx_u \int_{-\infty}^{+\infty} dx_v \int_{-\infty}^{+\infty} dx_w.$$

$z_V(x) = \dfrac{1}{V} \displaystyle\int_{V(x)} z(y)\, dy$ indicates the mean value of the ReV $z(y)$ within the domain $V(x)$. The simple integral sign $\int_V$ representing a simple, double or triple integral according to the dimension of the domain v.

$\bar{\gamma}(V, v)$, or $\bar{C}(V, v)$ indicate the mean value of the function $\gamma(h)$ or $C(h)$ when one extremity (x) of the vector $h = (x - x')$ describes the domain V and the other (x') describes independently the domain v:

$$\bar{\gamma}(V, v) = \frac{1}{Vv} \int_V dx \int_v \gamma(x - x')\, dx'.$$

$\sigma_E^2(v, V)$ indicates the estimation variance of v to V, i.e., the estimation variance of the mean value over V by the mean value over v.

σ_K^2 indicates the minimum estimation variance provided by kriging.

σ_{CK}^2 indicates the minimum estimation variance provided by cokriging.

s^2, or $s^2(v/V)$ indicate the experimental dispersion variance of values of support v within the domain V.

$D^2(v/V)$ indicates the theoretical dispersion variance of values of support v within the domain V.

$D^2(0/V)$ indicates the dispersion variance of point support values (or quasi-point) within V.

$D^2(v/\infty)$ indicates the dispersion variance of values of support v within a field $(V \to \infty)$ very large with respect to the various ranges of the underlying variogram.

$\gamma_v(h)$, or $C_v(h)$ indicate the semi-variogram (or the covariance) regularized over the support v.

$\gamma_l(h)$ indicates the semi-variogram regularized over aligned cores of length l.

$\gamma_G(h)$ indicates the semi-variogram graded over a constant thickness.

$g(x) = f_1 * f_2$
$\quad = \displaystyle\int f_1(x - y) f_2(y)\, dy$ indicates the convolution product of the two functions f_1 and f_2.

$\check{f}(x) = f(-x)$ indicates the transposed of the function $f(x)$.

$k(x) = \begin{cases} 1 \text{ if } x \in V \\ 0 \text{ if not} \end{cases}$ indicates the indicator function of domain v.

$z_k(x) = z * k$

$\qquad = \int z(x+y)k(y)\,dy$ indicates the regularization of function $z(x)$ by the weighting function $k(x)$. If $k(x)$ is the indicator function of a volume $v(x)$, then $z_k(x) = z_v(x) \cdot v$.

$\chi(L)$ and $\chi(L; l)$ indicate the auxiliary function χ.

$F(L)$ and $F(L; l)$ indicate the auxiliary function F.

$\alpha(L; l)$ and $H(L; l)$ indicate other auxiliary functions. (An auxiliary function is a precalculated mean value $\bar{\gamma}$ function of the parameters L and l.)

$\mathscr{E}(\gamma)$ indicates a geostatistical operator (e.g., the estimation variance or the dispersion variance) depending on the semi-variogram γ.

$v \cap V$ indicates the intersection of v and V (the common part of the two domains v and V).

$v \cup V$ indicates the union (sum) of the two domains v and V.

ϕ indicates the empty set.

$v \cap V = \phi$ indicates that v and V have no common part, they are disjoint.

$v \equiv V$ indicates that the two domains are identical and merged.

$\gamma_0(h) = [C_0(0) - C_0(h)]$ indicates a pure nugget effect model, the corresponding semi-variogram $\gamma_0(h)$ and covariance $C_0(h)$ being transition-type models with a range ($a_0 = \varepsilon$) very small with respect to all the distances $|h|$ of the available experimental observation.

$C_0 = C_0(0) = \gamma_0(h)$, $\forall |h| > \varepsilon$ indicates the sill of a pure nugget effect model, ε being a very small distance.

$C_{0v} = \gamma_{0v}(h), \forall |h| > \varepsilon$ indicates the nugget constant appearing on a semi-variogram regularized over the support v.

$\gamma(h) = \gamma_0(h) + \gamma_1(h) + \ldots$ in a nested sum, the subscript 0 generally denotes the pure nugget effect model

$[K]$ indicates the first member matrix of a kriging system.

$[M2]$ indicates the second member matrix of a kriging system.

$[\lambda]$ indicates the solution column matrix of a kriging system.

μ, μ_l when used in a kriging system, indicate the Lagrange parameters.

$m(x) = \sum_{l=1}^{k} a_l f_l(x)$ indicates a non-stationary expectation, unknown but of known form, the functions $f_l(x)$ being known. The stationary case corresponds to $k = 1$, with $m(x) = a_1 f_1(x) = a_1, \forall x$, i.e., $f_1(x) = 1, \forall x$.

$b_v^l = \dfrac{1}{v} \int_v f_l(x)\,dx$ indicates the mean value of the function $f_l(x)$ over the support or domain v.

Poly   abbreviation for the polygon of influence weighting method.

ID, ID2   abbreviation for the inverse (squared) distance weighting method.

MS	abbreviation for the interpolation method by fitting mean square polynomials.
$K(h)$	indicates a geometric or transitive covariogram, or simply a covariance function.
$\{Z/Z^* > z\}$	indicates the set of (true) values Z conditioned by the fact that their estimators Z^* are superior to a given value z (e.g., cut-off grade).
$E\{Z/Z^* = z\}$	indicates the conditional expectation of Z given $Z^* = z$.
$Z = \varphi_Z(U)$	indicates the Gaussian transform function by means of which a RV Z with any given distribution corresponds to a standardized Gaussian RV U.
$H_k(u)$	indicates the Hermite polynomials of order k.
$\eta_k(u)$	indicates the standardized Hermite polynomial of order k.
$z_s(x)$	indicates a non-conditional simulated realization of a RF $Z(x)$.
$z_{sK}^*(x)$	indicates the kriged value at point x deduced from non-conditionally simulated data values $z_s(x_\alpha)$
$z_{sc}(x)$	indicates a conditional simulation
$C_{(1)}(s)$	indicates the unidimensional covariance imposed on the lines defined by the turning band method of simulation.
$\langle Z(x), Z(y) \rangle$ $= E\{Z(x)Z(y)\} = \sigma_{xy}$	indicates the scalar product of the two elements $Z(x)$ and $Z(y)$, this scalar product being taken as the *non*-centred covariance.
$\|Z(x)\|^2 = <Z(x),Z(x)>$	indicates the squared norm of the vector (element) $Z(x)$.
$Z_{K_0}^*$	indicates a simple linear kriging estimator.
$Z_{K_1}^*$	indicates an ordinary linear kriging estimator.
$Z_{K_L}^*$	indicates an unbiased linear kriging estimator of order L, or, more shortly, a universal kriging estimator.
$E_n Z_0$ $= E\{Z(x_0)/Z_1, Z_2, \ldots, Z_n\}$	indicates the conditional expectation of the RV $Z(x_0)$ given the n data values $Z_1, Z_2, \ldots, Z_n$.
$E_\beta Z_0 = E\{Z(x_0)/Z_\beta\}$	indicates the conditional expectation of the RV $Z(x_0)$ given the particular data value $Z_\beta = Z(x_\beta)$ at point x_β.
Z_{DK}^*	indicates the disjunctive kriging estimator.